PRINCIPLES OF
FOUNDATION ENGINEERING

PWS-KENT Series in Engineering

Askeland, *The Science and Engineering of Materials*, Second Edition
Askeland, *The Science and Engineering of Materials*, Alternate Edition
Bauld, *Mechanics of Materials*, Second Edition
Borse, *FORTRAN 77 and Numerical Methods for Engineers*
Das, *Principles of Foundation Engineering*, Second Edition
Das, *Principles of Geotechnical Engineering*, Second Edition
Duff, *Concepts of Technical Graphics*
Gere and Timoshenko, *Mechanics of Materials*, Third Edition
Keedy, *An Introduction to CAD Using CADKey*
McGill and King, *Engineering Mechanics: Statics*, Second Edition
McGill and King, *Engineering Mechanics: An Introduction to Dynamics*, Second
    Edition
McGill and King, *Engineering Mechanics: Statics and an Introduction to Dynamics*,
    Second Edition
Reynolds, *Unit Operations and Processes in Environmental Engineering*
Sack, *Matrix Structural Analysis*
Schmidt and Wong, *Fundamentals of Surveying*, Third Edition
Segui, *Fundamentals of Structural Steel Design*
Weinman, *FORTRAN for Scientists and Engineers*

PRINCIPLES OF FOUNDATION ENGINEERING

Second Edition

Braja M. Das

Southern Illinois University at Carbondale

PWS-KENT Publishing Company
BOSTON

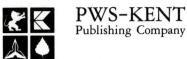

PWS-KENT
Publishing Company

20 Park Plaza
Boston, Massachusetts 02116

PWS-KENT Publishing Company is a division of Wadsworth, Inc.

Library of Congress Cataloging-in-Publication Data

Das, Braja M.
    Principles of foundation engineering / Braja M. Das. — 2nd ed.
      p. cm.
    Includes bibliographical references.
    ISBN 0-534-92171-X
    1. Foundations. I. Title.
    TA775.D227 1990
    624.1'5—dc20   89-26477
                                           CIP

Sponsoring Editor: Jonathan Plant
Assistant Editor: Mary Thomas
Production Coordinator: Robine Andrau
Manufacturing Coordinator: Margaret Sullivan Higgins
Production: Greg Hubit Bookworks
Interior Designer: Julia Gecha
Cover Designer: Steven Snider
Typesetter: Santype International Limited
Cover Printer: Henry N. Sawyer Co., Inc.
Printer and Binder: The Maple-Vail Book Manufacturing Group

Printed in the United States of America

92 93 94—10 9 8 7 6 5

In the memory of my father,
and to Janice and Valerie

Preface

The first edition of *Principles of Foundation Engineering* was published in 1984. The popularity of this text among students and practicing engineers has encouraged me to prepare this updated second edition. *Principles of Foundation Engineering* is primarily intended for undergraduate civil engineering students.

The first chapter, on geotechnical properties of soil and natural soil deposits, reviews the topics covered in the introductory soil mechanics course, which is a prerequisite for the foundation engineering course. The text comprises twelve chapters with examples and problems, two appendices, and an answer section for selected problems. The chapters are devoted mostly to the geotechnical aspects of foundation design. Appendix A covers reinforced concrete design of shallow foundations and retaining walls, and Appendix B gives the conversion factors from SI (Système International) units to English units and from English units to SI units. Both SI units and English units have been used in the text. All empirical equations have also been given in SI units, followed by the English units. This is a departure from the first edition of the text, in which equations using SI units were given in the main text and their English unit equivalents were given in the appendix.

In Chapter 1 the Unified Soil Classification System has been updated to conform to the most recent form suggested by the American Society for Testing and Materials. In Chapter 3 new graphs have been added to estimate the ultimate bearing capacity of foundations with two-way eccentricity. The section on the ultimate bearing capacity of shallow foundations on layered soil has been somewhat revised. A section on passive earth pressure on retaining walls with earthquake forces has been added to Chapter 5, along with a design procedure for gravity retaining walls for earthquake conditions.

In Chapter 6 some special cases for cantilever sheet pile walls penetrating sandy and clayey soils have been added. The computational pressure diagram method for estimating the penetration depth of sheet piles, the maximum moment, and the anchor force have also been added to this chapter. Some limited laboratory model test results for the ultimate uplift capacity of group piles embedded in sand and clay have been incorporated into Chapter 8.

Chapter 11 has been revised substantially to discuss the design procedure of retaining walls with metallic strips and geotextiles used as reinforcement in the backfill. Some recent findings on the construction of foundations on stone columns and granular trenches are discussed in Chapter 12. The reinforced concrete design of foundations and retaining walls given in Appendix A reflects the recommendations of ACI 318-83 (Revised 1986) code.

Because the text introduces civil engineering students to the application of the fundamental concepts of foundation analysis and design, the mathematical derivations of some equations are not always presented. Instead, just the final form of the equation is given. However, a list of references for further information and study is included in every chapter.

Each chapter provides ample example problems to help students understand the application of various equations and graphs. Practice problems are also given at the end of chapters. Answers to odd-numbered problems are at the back of the book.

Foundation analysis and design, as my colleagues in the geotechnical engineering area well know, is not just a matter of using theories, equations, and graphs from a textbook. Soil profiles found in nature are seldom homogeneous, elastic, and isotropic. The educated judgment needed to properly apply the theories, equations, and graphs to the evaluation of soils and foundation design cannot be overemphasized or completely taught by any textbook. Field experience must supplement classroom work.

Acknowledgments

The main driving force in the completion of the revised edition of this text is my wife, Janice, who appears to be the inexhaustible source of energy for me. She typed the revision of the manuscript and completed the original graphs and figures.

I wish to acknowledge Paul C. Hassler, formerly of University of Texas at El Paso; Gerald R. Seeley of Valparaiso University; Ronald B. McPherson of New Mexico State University; Said-Larbi Cherif of Navarro and Associates, El Paso, Texas; and Henry Ng, consulting engineer, El Paso, Texas, for their help, support, and encouragement during the preparation of the first edition of the manuscript.

I also wish to acknowledge the following reviewers of the second edition, whose helpful comments helped to shape this edition: M. Sherif Aggour, University of Maryland; Yong S. Chae, Rutgers—The State University; Samuel Clemence, Syracuse University; and George Gazetas, SUNY at Buffalo. In addition, I am thankful to one of my former students, Ronald P. Anderson, currently transportation marketing manager of Tensar

Corporation, Atlanta, Georgia, for his help in the development of the second edition.

Haskell Monroe, currently the chancellor of the University of Missouri at Columbia, has been my mentor for the past ten years. It is his continuous encouragement that has made my overall professional career satisfying. I am grateful for his generosity.

I extend thanks also to the staff of PWS-KENT for their interest and patience during the preparation and production of this text.

As a final note, I welcome suggestions from students and instructors for use in further editions.

Braja M. Das
Carbondale, Illinois

Contents

GEOTECHNICAL PROPERTIES OF SOIL AND NATURAL SOIL DEPOSITS

1.1 Introduction

The design of foundations of structures such as buildings, bridges, and dams generally requires a knowledge of such factors as (a) the load that will be transmitted by the superstructure to the foundation system, (b) the requirements of the local building code, (c) the behavior and stress-related deformability of soils that will support the foundation system, and (d) the geological conditions of the soil under consideration. To a foundation engineer, the last two factors listed are extremely important, because they concern soil mechanics.

The geotechnical properties of a soil—such as the grain-size distribution, plasticity, compressibility, and shear strength—can be determined by proper laboratory testing. And, recently, emphasis has been placed on *in situ* determination of strength and deformation properties of soil, because this process avoids the sample disturbances that occur during field exploration. However, under certain circumstances, all of the needed parameters cannot be determined, or are not determined because of economic or other reasons. In such cases, the engineer must make certain assumptions regarding the properties of the soil. In order to assess the accuracy of the soil parameters—whether they were made in the laboratory and the field or were assumed—the engineer must have a good grasp of the basic principles of soil mechanics. At the same time, he or she must realize that the natural soil deposits on which foundations are constructed are not homogeneous in most cases. Thus the engineer must have a thorough understanding of the geology of the area—that is, the origin and nature of soil stratification and also the ground water conditions. Foundation engineering is a clever combination of

1

soil mechanics, engineering geology, and proper judgment derived from past experience. To a certain extent, it may be called an "art."

When determining which foundation is the most economical, the engineer must consider the superstructure load, the subsoil conditions, and the desired tolerable settlement. In general, foundations of buildings and bridges may be divided into two major categories: (1) *shallow foundations* and (2) *deep foundations*. *Spread footings*, *wall footings*, and *mat foundations* are all shallow foundations. In most shallow foundations, the *depth of embedment can be equal to or less than three to four times the width of the foundation.* *Pile* and *caisson* foundations are deep foundations. These are used when top layers have poor load-bearing capacity and when use of shallow foundations will cause considerable damage and/or instability to the structure(s). The problems relating to shallow foundations and mat foundations are considered in Chapters 3 and 4, respectively. Chapter 8 discusses pile foundations, and Chapter 9 examines drilled shafts.

This chapter is divided into two major parts: (1) basic geotechnical properties of soils and (2) natural soil deposits. The first part reviews topics such as grain-size distribution, plasticity, soil classification, effective stress, consolidation, and shear strength parameters. It is assumed that the reader has already been exposed to these concepts in a basic soil-mechanics course. The second part deals with the formation of soil by water, wind, glaciers, and so on and discusses the characteristics of natural soil deposits. This subject is covered in detail in engineering geology courses.

GEOTECHNICAL PROPERTIES OF SOILS

1.2   Grain-Size Distribution

In any soil mass, the size of various soil grains varies greatly. To classify a soil properly, one must know its *grain-size distribution*. The grain-size distribution of *coarse-grained* soil is generally determined by means of *sieve analysis*. For a *fine-grained* soil, the grain-size distribution can be obtained by means of *hydrometer analysis*. The fundamental features of these analyses will be presented in this section. For detailed descriptions, readers are referred to any soil mechanics laboratory manual (for example, Das, 1989).

Sieve Analysis

A sieve analysis is conducted by taking a measured amount of dry, well-pulverized soil. The soil is passed through a stack of sieves with a pan at the bottom. The amount of soil retained on each sieve is measured, and the cumulative percentage of soil passing through each sieve is determined. This is generally referred to as *percent finer*. Following is a list of U.S. sieve numbers and the corresponding size of their hole openings. These sieves are commonly used for the analysis of soil for classification purposes.

U.S. sieve no.	Sieve opening, mm
4	4.75
10	2.00
20	0.85
40	0.425
60	0.25
100	0.150
200	0.075

The percent finer for each sieve determined by a sieve analysis is plotted on *semilogarithmic graph paper*, as shown in Figure 1.1. Note that the grain diameter, *D*, is plotted on the *logarithmic scale*, and the percent finer is plotted on the *arithmetic scale*.

Two parameters can be determined from the grain-size distribution curves of coarse-grained soils: (1) the *uniformity coefficient* (C_u) and (2) the *coefficient of gradation*, or *coefficient of curvature* (C_z). These coefficients can be defined as

$$C_u = \frac{D_{60}}{D_{10}}$$ (1.1)

$$C_z = \frac{D_{30}^2}{(D_{60})(D_{10})}$$ (1.2)

Figure 1.1 Grain-size distribution curve of a coarse-grained soil obtained from sieve analysis

where D_{10}, D_{30}, and D_{60} are the diameters corresponding to percents finer than 10, 30, and 60%, respectively.

For the grain-size distribution curve shown in Figure 1.1, $D_{10} = 0.08$ mm, $D_{30} = 0.17$ mm, and $D_{60} = 0.57$ mm. Thus, the values of C_u and C_z are as follows:

$$C_u = \frac{0.57}{0.08} = 7.13$$

$$C_z = \frac{0.17^2}{(0.57)(0.08)} = 0.63$$

Parameters (C_u and C_z) are used in the *Unified Soil Classification System* (Section 1.8).

Hydrometer Analysis

Hydrometer analysis is based on the principle of sedimentation of soil particles in water. In this test, one uses 50 grams of dry, pulverized soil. A *deflocculating* agent is always added to the soil. The most common deflocculating agent used for hydrometer analysis is 125 cc of 4% solution of sodium hexametaphosphate. The soil is allowed to soak for at least 16 hours in the deflocculating agent. After the soaking period, distilled water is added, and the soil-deflocculating agent mixture is thoroughly agitated. The sample is then transferred to a 1000-ml glass cylinder. More distilled water is added to the cylinder to fill it up to the 1000-ml mark, and then the mixture is again thoroughly agitated. A hydrometer is placed in the cylinder to measure— usually over a 24-hour period—the specific gravity of the soil-water suspension in the vicinity of its bulb (Figure 1.2). Hydrometers are calibrated to show the amount of soil that is still in suspension at any given time, t. The largest diameter of the soil particles still in suspension at time t can be determined by Stokes' law:

$$D = \sqrt{\frac{18\eta}{(G_s - 1)\gamma_w}} \sqrt{\frac{L}{t}} \qquad (1.3)$$

where D = diameter of the soil particle
G_s = specific gravity of soil solids
η = viscosity of water
γ_w = unit weight of water
L = effective length (that is, length measured from the water surface in the cylinder to the center of gravity of the hydrometer; see Figure 1.2)
t = time

Soil particles having diameters larger than those calculated by Eq. (1.3) would have settled beyond the zone of measurement. In this manner, with

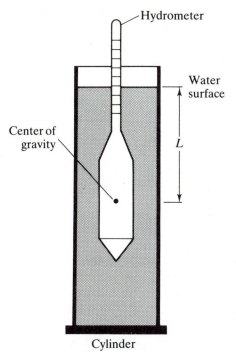

Figure 1.2   Hydrometer analysis

hydrometer readings taken at various times, the soil *percent finer* than a given diameter D can be calculated, and a grain-size distribution plot can be prepared.

1.3   Size Limits for Soils

Several organizations have attempted to develop the size limits for *gravel, sand, silt,* and *clay* based on the grain sizes present in soils. Table 1.1 presents the size limits as recommended by the Massachusetts Institute of Technology (MIT), the American Association of State Highway and Transportation Officials (AASHTO), and the Unified (Corps of Engineers, Department of the Army and Bureau of Reclamation) systems. From Table 1.1 it can be seen that soil particles smaller than 0.002 mm have been classified as *clay*. However, it should be kept in mind that clays by nature have a cohesive property and can be rolled into a thread when moist. This is caused by the presence of *clay minerals* such as *kaolinite, illite,* and *montmorillonite.* In contrast, some minerals such as *quartz* and *feldspar* may be present in a soil in particle sizes as small as clay minerals. But these particles will not have the cohesive property of clay minerals. Hence, they are called *clay-size particles,* not *clay particles.*

Table 1.1 Soil-Separate Size Limits

Classification system	Grain size (mm)
MIT	Gravel: 100 mm to 2 mm Sand: 2 mm to 0.06 mm Silt: 0.06 mm to 0.002 mm Clay: <0.002 mm
Unified	Gravel: 75 mm to 4.75 mm Sand: 4.75 mm to 0.075 mm Silt and clay (fines): <0.075 mm
AASHTO	Gravel: 75 mm to 2 mm Sand: 2 mm to 0.05 mm Silt: 0.05 mm to 0.002 mm Clay: <0.002 mm

1.4 Weight-Volume Relationships

In nature, soils are three-phase systems consisting of solid soil particles, water, and air (or gas). In order to develop the *weight-volume relationships* for a soil, the three phases can be separated as shown in Figure 1.3a. Based on this, the volume relationships can be defined as follows:

Void ratio, e, is the ratio of the volume of voids to the volume of soil solids in a given soil mass and can be written as

$$e = \frac{V_v}{V_s} \tag{1.4}$$

where V_v = volume of voids
V_s = volume of soil solids

Porosity, n, is the ratio of the volume of voids to the volume of the soil specimen, or

$$n = \frac{V_v}{V} \tag{1.5}$$

where V = total volume of soil

It can also be seen that

$$n = \frac{V_v}{V} = \frac{V_v}{V_s + V_v} = \frac{\dfrac{V_v}{V_s}}{\dfrac{V_s}{V_v} + \dfrac{V_v}{V_s}} = \frac{e}{1 + e} \tag{1.6}$$

Degree of saturation, S, is the ratio of the volume of water in the void spaces to the volume of voids, and it is generally expressed as a percentage.

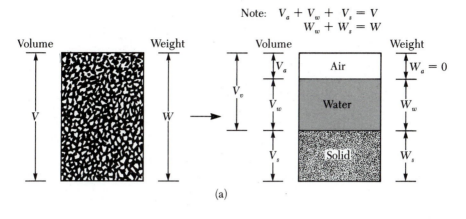

Note: $V_a + V_w + V_s = V$
       $W_w + W_s = W$

(a)

(b)   Unsaturated soil; $V_s = 1$

(c)   Saturated soil; $V_s = 1$

Figure 1.3   Weight-volume relationships

So

$$S = \frac{V_w}{V_v} \tag{1.7}$$

where V_w = volume of water

Note that, for saturated soils, the degree of saturation is 100%.

The weight relationships are *moisture content, moist unit weight, dry unit weight,* and *saturated unit weight.* They can be defined as follows:

$$\text{Moisture content} = w(\%) = \frac{W_w}{W_s} \times 100 \qquad (1.8)$$

where W_s = weight of the soil solids
W_w = weight of water

$$\text{Moist unit weight} = \gamma = \frac{W}{V} \qquad (1.9)$$

where W = total weight of the soil specimen = $W_s + W_w$

It is assumed that the weight of air, W_a, in the soil mass is negligible.

$$\text{Dry unit weight} = \gamma_d = \frac{W_s}{V} \qquad (1.10)$$

When a soil mass is completely saturated (that is, all the void volume is occupied by water), the moist unit weight of a soil [Eq. 1.9)] becomes equal to the *saturated unit weight* (γ_{sat}). So, $\gamma = \gamma_{\text{sat}}$ if $V_v = V_w$.

More useful relations can now be developed by considering a representative soil specimen in which the volume of soil solids is equal to *unity* as shown in Figure 1.3b. Note that if $V_s = 1$, from Eq. (1.4), $V_v = e$. The weight of the soil solids can be given as

$$W_w = G_s \gamma_w$$

where G_s = specific gravity of soil solids
γ_w = unit weight of water (9.81 kN/m^3)

Also, from Eq. (1.8), the weight of water $W_w = wW_s$. Thus, for the soil specimen under consideration, $W_w = wW_s = wG_s\gamma_w$. Now, using the general relation for moist unit weight given in Eq. (1.9)

$$\boxed{\gamma = \frac{W}{V} = \frac{W_s + W_w}{V_s + V_v} = \frac{G_s\gamma_w(1 + w)}{1 + e}} \qquad (1.11)$$

Similarly, the dry unit weight [Eq. (1.10)]

$$\boxed{\gamma_d = \frac{W_s}{V} = \frac{W_s}{V_s + V_v} = \frac{G_s\gamma_w}{1 + e}} \qquad (1.12)$$

From Eqs. (1.11) and (1.12), note that

$$\gamma_d = \frac{\gamma}{1 + w} \qquad (1.13)$$

If a soil specimen is completely saturated as shown in Figure 1.3c

$$V_v = e$$

Also, for this case

$$V_v = \frac{W_w}{\gamma_w} = \frac{wG_s\gamma_w}{\gamma_w} = wG_s$$

Thus

$$e = wG_s \qquad \text{(for saturated soil *only*)} \tag{1.14}$$

The saturated unit weight of soil can now be determined as

$$\boxed{\gamma_{\text{sat}} = \frac{W_s + W_w}{V_s + V_v} = \frac{G_s\gamma_w + e\gamma_w}{1 + e}} \qquad (1.15)$$

Relationships similar to Eqs. (1.11), (1.12), and (1.15) in terms of porosity can also be obtained by considering a representative soil specimen with a unit volume. These relationships are

$$\gamma = G_s\gamma_w(1 - n)(1 + w) \tag{1.16}$$

$$\gamma_d = (1 - n)G_s\gamma_w \tag{1.17}$$

$$\gamma_{\text{sat}} = [(1 - n)G_s + n]\gamma_w \tag{1.18}$$

1.5   Relative Density

In *granular soils*, the degree of compaction in the field can be measured according to *relative density*, D_r, which is defined as

$$\boxed{D_r = \frac{e_{\text{max}} - e}{e_{\text{max}} - e_{\text{min}}}} \qquad (1.19)$$

where e_{max} = void ratio of the soil in the loosest state
   e_{min} = void ratio in the densest state
      e = *in situ* void ratio

The values of e_{max} and e_{min} are determined in the laboratory in accordance with the test procedures outlined in the American Society for Testing and Materials, *ASTM Standards* (1986) (Test Designation D2040).

The relative density can also be expressed in terms of dry unit weight, or

$$D_r = \left\{ \frac{\gamma_d - \gamma_{d(\text{min})}}{\gamma_{d(\text{max})} - \gamma_{d(\text{min})}} \right\} \frac{\gamma_{d(\text{max})}}{\gamma_d} \tag{1.20}$$

where $\gamma_d = $ *in situ* dry unit weight

$\gamma_{d(max)} = $ dry unit weight in the *densest* state—that is, when the void ratio is e_{min}

$\gamma_{d(min)} = $ dry unit weight in the *loosest* state—that is, when the void ratio is e_{max}

Example 1.1

A representative soil specimen collected from the field weighs 1.8 kN and has a volume of 0.1 m³. The moisture content as determined in the laboratory is 12.6%. Given $G_s = 2.71$, determine the following:

a. Moist unit weight
b. Dry unit weight
c. Void ratio
d. Porosity
e. Degree of saturation

Solution
Part a: Moist Unit Weight

From Eq. (1.9)

$$\gamma = \frac{W}{V} = \frac{1.8 \text{ kN}}{0.1 \text{ m}^3} = \textbf{18 kN/m}^3$$

Part b: Dry Unit Weight

From Eq. (1.13)

$$\gamma_d = \frac{\gamma}{1+w} = \frac{18}{1 + \frac{12.6}{100}} = \textbf{15.99 kN/m}^3$$

Part c: Void Ratio

From Eq. (1.12)

$$\gamma_d = \frac{G_s \gamma_w}{1+e}$$

or

$$e = \frac{G_s \gamma_w}{\gamma_d} - 1 = \frac{(2.71)(9.81)}{15.99} - 1 = \textbf{0.66}$$

Part d: Porosity

From Eq. (1.6)

$$n = \frac{e}{1+e} = \frac{0.66}{1+0.66} = \textbf{0.398}$$

Part e: Degree of Saturation

Referring to Figure 1.3b

$$S = \frac{V_w}{V_v} = \frac{wG_s}{e} = \frac{(0.126)(2.71)}{0.66} \times 100 = \textbf{51.7}\%$$

Example 1.2

A moist soil has $e = 0.62$, $w = 16\%$, and $G_s = 2.68$. Determine the following:

   a. n
   b. S
   c. γ
   d. γ_d

Solution
Part a, Porosity

From Eq. (1.6)

$$n = \frac{e}{1+e} = \frac{0.62}{1+0.62} = \textbf{0.383}$$

Part b, Degree of Saturation

From Eq. (1.7)

$$S = \frac{V_w}{V_v} = \frac{wG_s}{e} = \frac{(0.16)(2.68)}{0.62} = \textbf{0.692}$$

Part c, Moist Unit Weight

From Eq. (1.11)

$$\gamma = \frac{G_s \gamma_w (1+w)}{1+e} = \frac{(2.68)(9.81)(1+0.16)}{1+0.62}$$
$$= \textbf{18.83 kN/m}^3$$

Part d, Dry Unit Weight

From Eq. (1.13)

$$\gamma_d = \frac{\gamma}{1+w} = \frac{18.83}{1+0.16} = \textbf{16.23 kN/m}^3$$

Example 1.3

A granular soil (sand) was tested in the laboratory and found to have maximum and minimum void ratios of 0.84 and 0.38, respectively. The value of G_s was determined to be 2.65. A natural soil deposit of the same sand has 9% moisture, and its moist unit weight is 18.64 kN/m³. Determine the relative density of the soil in the field.

Solution

From Eq. (1.13)

$$\gamma_d = \frac{\gamma}{1+w} = \frac{18.64}{1 + \dfrac{9}{100}} = 17.1 \text{ kN/m}^3$$

$$\gamma_d = \frac{G_s \gamma_w}{1+e}$$

or

$$e = \frac{G_s \gamma_w}{\gamma_d} - 1 = \frac{(2.65)(9.81)}{17.1} - 1 = 0.52$$

From Eq. (1.19)

$$D_r = \frac{e_{max} - e}{e_{max} - e_{min}} = \frac{0.84 - 0.52}{0.84 - 0.38} = 0.696 = \mathbf{69.6}\%$$

1.6 Representative Values of G_s, e, and γ_d for Natural Soils

Except for peat and highly organic soils, the general range of the values of specific gravity of soil solids (G_s) found in nature is not great. Table 1.2 gives some representative values. For practical purposes, a reasonable value can be assumed in lieu of running a test.

Table 1.2 Specific Gravities of Some Soils

Soil type	G_s
Quartz sand	2.64–2.66
Silt	2.67–2.73
Clay	2.70–2.9
Chalk	2.60–2.75
Loess	2.65–2.73
Peat	1.30–1.9

Table 1.3 presents some representative values for the void ratio, dry unit weight, and moisture content (in a saturated state) of some naturally occurring soils. From this table it can be seen that in most cohesionless soils the void ratio varies from about 0.4 to 0.8. The dry unit weights in these soils generally fall within a range of about 14–19 kN/m³ (90–120 lb/ft³).

Table 1.3 Typical Void Ratio, Moisture Content, and Dry Unit Weight for Some Soils

Type of soil	Void ratio e	Natural moisture content in saturated condition (%)	Dry unit weight, γ_d (kN/m³)	(lb/ft³)
Loose uniform sand	0.8	30	14.5	92
Dense uniform sand	0.45	16	18	115
Loose angular-grained silty sand	0.65	25	16	102
Dense angular-grained silty sand	0.4	15	19	120
Stiff clay	0.6 ✓	21	17	108 ·
Soft clay	0.9–1.4	30–50	11.5–14.5	73–92
Loess	0.9	25	13.5	86
Soft organic clay	2.5–3.2	90–120	6–8	38–51
Glacial till	0.3	10	21	134

1.7   **Atterberg's Limits**

When a clayey soil is mixed with an excessive amount of water, it may flow like a *semiliquid*. If the soil is gradually dried, it will lose moisture. Depending on its moisture content, it will behave like a *plastic, semisolid,* or *solid* material. The moisture content, in percent, at which the soil changes from a liquid to a plastic stage is defined as the *liquid limit (LL)*. Similarly, the moisture contents, in percent, at which the soil changes from a plastic to a semisolid state and from a semisolid to a solid state are defined as the *plastic limit (PL)* and the *shrinkage limit (SL)*, respectively. These limits are referred to as *Atterberg's limits* (Figure 1.4).

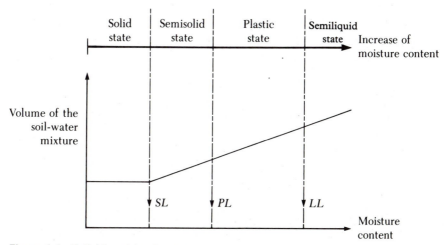

Figure 1.4 Definition of Atterberg's limits

The liquid limit of a soil is determined by Casagrande's liquid device (ASTM Test Designation D-4318) and is defined as the moisture content at which a groove closure of 12.7 mm ($\frac{1}{2}$ in.) occurs at 25 blows.

The *plastic limit* is defined as the moisture content at which the soil crumbles when rolled into a thread of 3.18 mm ($\frac{1}{8}$ in.) in diameter (ASTM Test Designation D-4318).

The *shrinkage limit* is defined as the moisture content at which the soil does not undergo further volume change with loss of moisture (ASTM Test Designation D-427). This limit is shown in Figure 1.4.

The difference between the liquid limit and the plastic limit of a soil is defined as the *plasticity index (PI)*, or

$$\boxed{PI = LL - PL} \qquad\qquad (1.21)$$

Table 1.4 gives some representative values of liquid limit and plastic limit for a number of clay minerals and soils. It should, however, be kept in mind that Atterberg's limits for various soils will vary considerably depending on the soil's origin and the nature and amount of clay minerals in it.

Table 1.4 Typical Liquid and Plastic Limits for Some Clay Minerals and Soils

Description	Liquid limit	Plastic limit
Kaolinite	35–100	25–35
Illite	50–100	30–60
Montmorillonite	100–800	50–100
Boston Blue clay	40	20
Chicago clay	60	20
Louisiana clay	75	25
London clay	66	27
Cambridge clay	39	21
Montana clay	52	18
Mississippi Gumbo	95	32
Loessial soils in north and northwest China	25–35	15–20

1.8   Soil Classification Systems

Soil classification divides soils into a number of groups and subgroups based on common engineering properties such as *grain-size distribution, liquid limit,* and *plastic limit*. The two major classification systems presently in use are (1) the *AASHTO (American Association of State Highway and Transportation Officials) System* and (2) the *Unified System*. The AASHTO classification system is mainly used for classification of highway subgrades. It is not used in foundation construction. This section will briefly describe each of these two systems.

AASHTO System

The AASHTO Soil Classification System was originally proposed by the Highway Research Board's Committee on Classification of Materials for Subgrades and Granular Type Roads (1945). According to the present form of this system, soils can be classified into eight major groups, A-1 through A-8, based on their grain-size distribution, liquid limit, and plasticity indices. Soils listed in the major groups—A-1, A-2, and A-3—are coarse-grained materials, and those in groups A-4, A-5, A-6, and A-7 are fine-grained materials. Peat, muck, and other highly organic soils are classified under A-8. They are identified by visual inspection.

The AASHTO classification system (for soils A-1 through A-7) is presented in Table 1.5. Note that under group A-7 there are two types of soil. For the A-7-5 type, the plasticity index of the soil should be less than or equal to the liquid limit minus 30. For the A-7-6 type, the plasticity index is greater than the liquid limit minus 30.

For qualitative evaluation of the desirability of a given soil as a highway subgrade material, a number referred to as the *group index* has also been developed. The higher the value of the group index for a given soil, the weaker will be the soil's performance as a subgrade. A group index of 20 or more indicates a very poor subgrade material. The formula for group index, GI, is

$$GI = (F_{200} - 35)[0.2 + 0.005(LL - 40)] \\ + 0.01(F_{200} - 15)(PI - 10) \qquad (1.22)$$

where F_{200} = percent passing no. 200 sieve, expressed as a whole number
LL = liquid limit
PI = plasticity index

When calculating the group index for a soil belonging to group A-2-6 or A-2-7, use only the partial group index equation relating to the plasticity index:

$$GI = 0.01(F_{200} - 15)(PI - 10)$$

The group index is rounded to the nearest whole number and written next to the soil group in parentheses; for example

A − 4 (5)
  Group index
Soil group

Table 1.5   AASHTO Soil Classification System

General classification	Granular materials (35% or less of total sample passing no. 200 sieve)						
	A-1		A-3	A-2			
Group classification	A-1-a	A-1-b		A-2-4	A-2-5	A-2-6	A-2-7
Sieve analysis (% passing)							
No. 10 sieve	50 max						
No. 40 sieve	30 max	50 max	51 min				
No. 200 sieve	15 max	25 max	10 max	35 max	35 max	35 max	35 max
For fraction passing No. 40 sieve							
Liquid limit (LL)				40 max	41 min	40 max	41 min
Plasticity index (PI)	6 max		Nonplastic	10 max	10 max	11 min	11 min
Usual type of material	Stone fragments, gravel, and sand		Fine sand	Silty or clayey gravel and sand			
Subgrade rating	Excellent to good						

General classification	Silt-clay materials (More than 35% of total sample passing no. 200 sieve)			
Group classification	A-4	A-5	A-6	A-7 A-7-5[a] A-7-6[b]
Sieve analysis (% passing)				
No. 10 sieve				
No. 40 sieve				
No. 200 sieve	36 min	36 min	36 min	36 min
For fraction passing No. 40 sieve				
Liquid limit (LL)	40 max	41 min	40 max	41 min
Plasticity index (PI)	10 max	10 max	11 min	11 min
Usual types of material	Mostly silty soils		Mostly clayey soils	
Subgrade rating	Fair to poor			

[a] If $PI \leq LL - 30$, it is A-7-5
[b] If $PI > LL - 30$, it is A-7-6

Unified System

This system of soil classification was originally proposed by A. Casagrande in 1942 and was later revised and adopted by the United States Bureau of Reclamation and the Corps of Engineers. This system is presently used in practically all geotechnical work.

In the Unified System, the following symbols are used for identification.

Symbol	Description
G	Gravel
S	Sand
M	Silt
C	Clay
O	Organic silts and clay
Pt	Peat and highly organic soils
H	High plasticity
L	Low plasticity
W	Well graded
P	Poorly graded

Table 1.6 and the plasticity chart (Figure 1.5) show the procedure of determining the group symbols for various types of soil. It is important that while classifying a soil, a group name that is a general description of the soil is provided along with the group symbol. Tables 1.7, 1.8, and 1.9, respectively, give the criteria for obtaining the group names for coarse-grained soil, inorganic fine-grained soil, and organic fine-grained soil. These tables are based on ASTM Designation D-2487.

Figure 1.5   Plasticity chart

Table 1.6 Group Symbols for Soil According to the Unified Classification System [Based on Material Passing 3-in. (75-mm) Sieve]

Major divisions	Criteria	Group symbol
Coarse-grained soil $R_{200} > 50$ Gravelly soil $R_4 > 0.5R_{200}$	$F_{200} < 5$, $C_u \geq 4$, $1 \leq C_z \leq 3$	GW
	$F_{200} < 5$, $C_u < 4$, and/or C_z not between 1 and 3	GP
	$F_{200} > 12$, $PI < 4$, or Atterberg's limits plot below A-line (Figure 1.5)	GM
	$F_{200} > 12$, $PI > 7$, and Atterberg's limits plot on or above A-line (Figure 1.5)	GC
	$F_{200} > 12$, $LL < 50$, $4 \leq PI \leq 7$, and Atterberg's limits plot on or above A-line	GC-GM[a]
	$5 \leq F_{200} \leq 12$; meets the gradation criteria of GW and the plasticity criteria of GM	GW-GM[a]
	$5 \leq F_{200} \leq 12$; meets the gradation criteria of GW and the plasticity criteria of GC	GW-GC[a]
	$5 \leq F_{200} \leq 12$; meets the gradation criteria of GP and the plasticity criteria of GM	GP-GM[a]
	$5 \leq F_{200} \leq 12$; meets the gradation criteria of GP and the plasticity criteria of GC	GP-GC[a]
Sandy soil $R_4 \leq 0.5R_{200}$	$F_{200} < 5$, $C_u \geq 6$, $1 \leq C_z \leq 3$	SW
	$F_{200} < 5$, $C_u < 6$, and/or C_z not between 1 and 3	SP
	$F_{200} > 12$, $PI < 4$, or Atterberg's limits plot below A-line (Figure 1.5)	SM
	$F_{200} > 12$, $PI > 7$, and Atterberg's limits plot on or above A-line (Figure 1.5)	SC
	$F_{200} > 12$, $LL > 50$, $4 \leq PI \leq 7$, and Atterberg's limits plot on or above A-line (Figure 1.5)	SC-SM[a]
	$5 \leq F_{200} \leq 12$; meets the gradation criteria of SW and the plasticity criteria of SM	SW-SM[a]
	$5 \leq F_{200} \leq 12$; meets the gradation criteria of SW and the plasticity criteria of SC	SW-SC[a]
	$5 \leq F_{200} \leq 12$; meets the gradation criteria of SP and the plasticity criteria of SM	SP-SM[a]
	$5 \leq F_{200} \leq 12$; meets the gradation criteria of SP and the plasticity criteria of SC	SP-SC[a]
Fine-grained soil (inorganic), $R_{200} \leq 50$ Silty and clayey soil $LL < 50$	$PI < 4$, or Atterberg's limits plot below A-line (Figure 1.5)	ML
	$PI > 7$, and Atterberg's limits plot on or above A-line (Figure 1.5)	CL
	$4 \leq PI \leq 7$, and Atterberg's limits plot above A-line (Figure 1.5)	CL-ML[a]
	Atterberg's limits plot below A-line (Figure 1.5)	MH
Silty and clayey soil $LL \geq 50$	Atterberg's limits plot on or above A-line (Figure 1.5)	CH
Fine-grained soil (organic) Organic silt and clay $LL < 50$	$\dfrac{LL_{\text{not oven dry}}}{LL_{\text{oven dry}}} < 0.75$	OL
Organic silt and clay $LL \geq 50$	$\dfrac{LL_{\text{not oven dry}}}{LL_{\text{oven dry}}} < 0.75$	OH

Note: F_{200} = percent finer than no. 200 sieve; R_{200} = percent retained on no. 200 sieve; R_4 = percent retained on no. 4 sieve; C_u = uniformity coefficient; C_z = coefficient of gradation; LL = liquid limit; PI = plasticity index; Atterberg's limits based on minus no. 40 fraction

[a] Borderline case; dual classification

Table 1.7 Group Names for Coarse-Grained Soils (Based on ASTM D-2487)

Group symbol	Gravel fraction (%)	Sand fraction (%)	Group name
GW		<15	Well-graded gravel
		≥15	Well-graded gravel with sand
GP		<15	Poorly graded gravel
		≥15	Poorly graded gravel with sand
GM		<15	Silty gravel
		≥15	Silty gravel with sand
GC		<15	Clayey gravel
		≥15	Clayey gravel with sand
GC-GM		<15	Silty clayey gravel
		≥15	Silty clayey gravel with sand
GW-GM		<15	Well-graded gravel with silt
		≥15	Well-graded gravel with silt and sand
GW-GC		<15	Well-graded gravel with clay
		≥15	Well-graded gravel with clay and sand
GP-GM		<15	Poorly graded gravel with silt
		≥15	Poorly graded gravel with silt and sand
GP-GC		<15	Poorly graded gravel with clay
		≥15	Poorly graded gravel with clay and sand
SW	<15		Well-graded sand
	≥15		Well-graded sand with gravel
SP	<15		Poorly graded sand
	≥15		Poorly graded sand with gravel
SM	<15		Silty sand
	≥15		Silty sand with gravel
SC	<15		Clayey sand
	≥15		Clayey sand with gravel
SM-SC	<15		Silty clayey sand
	≥15		Silty clayey sand with gravel
SW-SM	<15		Well-graded sand with silt
	≥15		Well-graded sand with silt and gravel
SW-SC	<15		Well-graded sand with clay
	≥15		Well-graded sand with clay and gravel
SP-SM	<15		Poorly graded sand with silt
	≥15		Poorly graded sand with silt and gravel
SP-SC	<15		Poorly graded sand with clay
	≥15		Poorly graded sand with clay and gravel

Note: Sand fraction = percent of soil passing no. 4 sieve but retained on no. 200 sieve = $R_{200} - R_4$; gravel fraction = percent of soil passing 3-in. sieve but retained on no. 4 sieve = R_4

Table 1.8 Group Names for Inorganic Fine-Grained Soils (Based on ASTM D-2487)

			Criteria		
		Sand fraction			
Group symbol	R_{200}	Gravel fraction	Gravel fraction	Sand fraction	Group name
CL	<15				Lean clay
	15 to 29	≥1			Lean clay with sand
		<1			Lean clay with gravel
	≥30	≥1	<15		Sandy lean clay
		≥1	≥15		Sandy lean clay with gravel
		<1		<15	Gravelly lean clay
		<1		≥15	Gravelly lean clay with sand
ML	<15				Silt
	15 to 29	≥1			Silt with sand
		<1			Silt with gravel
	≥30	≥1	<15		Sandy silt
		≥1	≥15		Sandy silt with gravel
		<1		<15	Gravelly silt
		<1		≥15	Gravelly silt with sand
CL-ML	<15				Silty clay
	15 to 29	≥1			Silty clay with sand
		<1			Silty clay with gravel
	≥30	≥1	<15		Sandy silty clay
		≥1	≥15		Sandy silty clay with gravel
		<1		<15	Gravelly silty clay
		<1		≥15	Gravelly silty clay with sand
CH	<15				Fat clay
	15 to 29	≥1			Fat clay with sand
		<1			Fat clay with gravel
	≥30	≥1	<15		Sandy fat clay
		≥1	≥15		Sandy fat clay with gravel
		<1		<15	Gravelly fat clay
		<1		≥15	Gravelly fat clay with sand
MH	<15				Elastic silt
	15 to 29	≥1			Elastic silt with sand
		<1			Elastic silt with gravel
	≥30	≥1	<15		Sandy elastic silt
		≥1	≥15		Sandy elastic silt with gravel
		<1		<15	Gravelly elastic silt
		<1		≥15	Gravelly elastic silt with sand

Note: R_{200} = percent of soil retained on no. 200 sieve; sand fraction = percent of soil passing no. 4 sieve but retained on no. 200 sieve = $R_{200} - R_4$; gravel fraction = percent of soil passing 3-in. sieve but retained on no. 4 sieve = R_4

Table 1.9   Group Names for Organic Fine-Grained Soils (Based on ASTM D-2487)

Group symbol	Plasticity	R_{200}	Criteria Sand fraction Gravel fraction	Gravel fraction	Sand fraction	Group name
OL	$PI \geq 4$, and	<15				Organic clay
	Atterberg's	15 to 29	≥1			Organic clay with sand
	limits on		<1			Organic clay with gravel
	or above	≥30	≥1	<15		Sandy organic clay
	A-line		≥1	≥15		Sandy organic clay with gravel
			<1		<15	Gravelly organic clay
			<1		≥15	Gravelly organic clay with sand
	$PI < 4$, and	<15				Organic silt
	Atterberg's	15 to 29	≥1			Organic silt with sand
	limits plot		<1			Organic silt with gravel
	below A-line	≥30	≥1	<15		Sandy organic silt
			≥1	≥15		Sandy organic silt with gravel
			<1		<15	Gravelly organic silt
			<1		≥15	Gravelly organic silt with sand
OH	Atterberg's	<15				Organic clay
	limits plot	15 to 29	≥1			Organic clay with sand
	on or above		<1			Organic clay with gravel
	A-line	≥30	≥1	<15		Sandy organic clay
			≥1	≥15		Sandy organic clay with gravel
			<1		<15	Gravelly organic clay
			<1		≥15	Gravelly organic clay with sand
	Atterberg's	<15				Organic silt
	limits plot	15 to 29	≥1			Organic silt with sand
	below A-line		<1			Organic silt with gravel
		≥30	≥1	<15		Sandy organic silt
			≥1	≥15		Sandy organic silt with gravel
			<1		<15	Gravelly organic silt
			<1		≥15	Gravelly organic silt with sand

Note: R_{200} = percent of soil retained on no. 200 sieve; sand fraction = percent of soil passing no. 4 sieve but retained on no. 200 sieve = $R_{200} - R_4$; gravel fraction = percent of soil passing 3-in. sieve but retained on no. 4 sieve = R_4

Example 1.4

Classify the following soil by the AASHTO classification system:

Percent passing no. 4 sieve = 82

Percent passing no. 10 sieve = 71

Percent passing no. 40 sieve = 64

Percent passing no. 200 sieve = 41

Liquid limit = 31

Plasticity index = 12

Solution

Referring to Table 1.5, since more than 35% passes through a no. 200 sieve, it is a silt-clay material. It could be A-4, A-5, A-6, or A-7. Since, for this soil, $LL = 31$ (that is, less than 40) and $PI = 12$ (that is, greater than 11), it falls in group A-6. From Eq. (1.22)

$$GI = (F_{200} - 35)[0.02 + 0.005(LL - 40)] + 0.01(F_{200} - 15)(PI - 10)$$

So

$$GI = (41 - 35)[0.02 + 0.005(31 - 40)] + 0.01(41 - 15)(12 - 10)$$
$$= 0.37 \approx 0$$

So, the soil is **A-6(0)**.

Example 1.5

Classify the soil described in Example 1.4 according to the Unified Soil Classification System.

Solution

Given $F_{200} = 41$. So

$$R_{200} = 100 - F_{200} = 100 - 41 = 59$$

Since $R_{200} > 50$, it is a coarse-grained soil.

$$R_4 = 100 - (\text{percent passing no. 4 sieve})$$
$$= 100 - 82 = 18$$

Since $R_4 = 18 < 0.5R_{200} = 29.5$, it is a sandy soil. Now, we refer to Table 1.6. Since F_{200} is greater than 12, the group symbol should be SM or SC. Since the PI is greater than 7 and the Atterberg's limits plot above the A-line in Figure 1.5, it is SC.

For group name, we refer to Table 1.7. The gravel fraction is more than 15%, so the group name is **clayey sand with gravel.**

1.9   Permeability of Soil

The void spaces or pores between soil grains allow water to flow through them. In soil mechanics and foundation engineering, one must know how much water is flowing through a given soil in unit time. This knowledge is required to design earth dams, to determine the quantity of seepage under hydraulic structures, and to dewater before and during the construction of foundations. In 1856, Darcy proposed the following equation (Figure 1.6) for calculating the velocity of flow of water through a soil.

$$v = ki$$

(1.23)

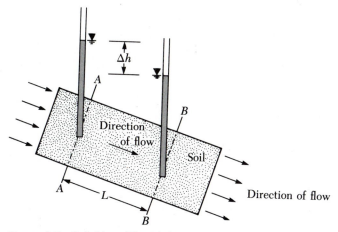

Figure 1.6   Definition of Darcy's law

where v = velocity (unit: cm/sec)

k = coefficient of permeability of soil (unit: cm/sec)

i = hydraulic gradient

The hydraulic gradient (i) may be defined as

$$i = \frac{\Delta h}{\Delta L} \qquad (1.24)$$

where Δh = piezometric head difference between the sections at AA and BB

ΔL = distance between the sections at AA and BB

(*Note:* The sections AA and BB are perpendicular to the direction of flow.)

Darcy's law [Eq. (1.23)] is valid for a wide range of soil types. However, with materials like clean gravel and open-graded rockfills, Darcy's law breaks down because of the turbulent nature of flow through them.

The value of the coefficient of permeability of various soils varies greatly. In the laboratory, it can be determined by means of *constant head* or *falling head* permeability tests. The constant head test is more suitable for granular soils. Table 1.10 provides the general range for the values of k for various soils. In granular soils, the value primarily depends on the void

Table 1.10   Range of the Coefficient of Permeability for Various Soils

Type of soil	Coefficient of permeability, k (cm/sec)
Medium to coarse gravel	Greater than 10^{-1}
Coarse to fine sand	10^{-1} to 10^{-3}
Fine sand, silty sand	10^{-3} to 10^{-5}
Silt, clayey silt, silty clay	10^{-4} to 10^{-6}
Clays	10^{-7} or less

ratio. In the past, several equations have been proposed to relate the value of k with the void ratio in the granular soil. They are as follows:

$$\frac{k_1}{k_2} = \frac{e_1^2}{e_2^2} \tag{1.25}$$

$$\frac{k_1}{k_2} = \frac{\left(\dfrac{e_1^2}{1 + e_1}\right)}{\left(\dfrac{e_2^2}{1 + e_2}\right)} \tag{1.26}$$

$$\frac{k_1}{k_2} = \frac{\left(\dfrac{e_1^3}{1 + e_1}\right)}{\left(\dfrac{e_2^3}{1 + e_2}\right)} \tag{1.27}$$

where k_1 and k_2 are the coefficients of permeability of a given soil at void ratios e_1 and e_2, respectively.

Hazen (1930) proposed an equation for the coefficient of permeability of fairly uniform sand as

$$k = AD_{10}^2 \tag{1.28}$$

where k is in mm/sec
A = a constant that varies between 10 and 15
D_{10} = effective soil size, in mm

For clayey soils in the field, a practical relationship for estimating the coefficient of permeability (Tavenas, et al., 1983) can be given as

$$\log k = \log k_0 - \frac{e_0 - e}{C_k} \tag{1.29}$$

where k = coefficient of permeability at a void ratio e
k_0 = *in situ* coefficient of permeability at a void ratio e_0
C_k = permeability change index $\approx 0.5e_0$

Example 1.6

The coefficient of permeability of a fine sand is 0.012 cm/sec at a void ratio of 0.57. Estimate the permeability coefficient of the sand at a void ratio of 0.72. Use Eq. (1.27).

Solution

From Eq. (1.27)

$$\frac{k_1}{k_2} = \frac{\dfrac{e_1^3}{1 + e_1}}{\dfrac{e_2^3}{1 + e_2}}$$

Given $k_1 = 0.012$ cm/sec, $e_1 = 0.57$, and $e_2 = 0.72$

$$\frac{0.012}{k_2} = \frac{\dfrac{(0.57)^3}{1+0.57}}{\dfrac{(0.76)^3}{1+0.72}} = 0.544$$

$$k_2 = \mathbf{0.022 \ cm/sec}$$

1.10   Steady State Seepage

For most cases of seepage under hydraulic structures, the flow path changes direction and is not uniform over the entire area. In such cases, one of the ways of determining the rate of seepage is by means of a graphical construction that is referred to as *flow net*. The flow net is based on Laplace's theory of continuity. According to this theory, for a steady flow condition, the flow at any point A (Figure 1.7) can be represented by the equation

$$k_x \frac{\partial^2 h}{\partial x^2} + k_y \frac{\partial^2 h}{\partial y^2} + k_z \frac{\partial^2 h}{\partial z^2} = 0 \tag{1.30}$$

where h = hydraulic head at point A (that is, the head of water that a piezometer placed at A would show with the *datum* as the *downstream water level* as shown in Figure 1.7)

k_x, k_y, k_z = coefficient of permeability of the soil in x, y, and z directions, respectively

Figure 1.7   Steady state seepage

For a two-dimensional flow condition as shown in Figure 1.7

$$\frac{\partial^2 h}{\partial^2 y} = 0$$

So, Eq. (1.30) takes the form

$$k_x \frac{\partial^2 h}{\partial x^2} + k_z \frac{\partial^2 h}{\partial z^2} = 0 \tag{1.31}$$

If the soil is isotropic with respect to permeability, $k_x = k_z = k$. So

$$\boxed{\frac{\partial^2 h}{\partial x^2} + \frac{\partial^2 h}{\partial z^2} = 0} \tag{1.32}$$

The preceding equation, which is referred to as Laplace's equation and is valid for confined flow, represents two orthogonal sets of curves that are known as *flow lines* and *equipotential lines*. A flow net is a combination of a number of equipotential lines and flow lines. A flow line is a path that a water particle would follow in traveling from the upstream side to the downstream side. An equipotential line is a line along which water in piezometers would rise to the same elevation (see Figure 1.7).

In drawing a flow net, one needs to establish the *boundary conditions*. For example, in Figure 1.7 the ground surfaces on the upstream (OO') and downstream (DD') sides are equipotential lines. The base of the dam below the ground surface, $O'BCD$, is a flow line. The top of the rock surface, EF, is also a flow line. Once the boundary conditions are established, a number of flow lines and equipotential lines are drawn by trial and error such that all of the flow elements in the net have the same length-to-width ratio (L/B). In most cases, the L/B ratio is kept as *one*—that is, the flow elements are drawn as curvilinear "squares." This is shown in a flow net drawn in Figure 1.8. Note that all flow lines must intersect all equipotential lines at *right angles*.

Once the flow net is drawn, the seepage in unit time per unit length of the structure can be calculated as

$$q = k h_{max} \frac{N_f}{N_d} n \tag{1.33}$$

where N_f = number of flow channels
N_d = number of drops
n = width-to-length ratio of the flow elements in the flow net (B/L)
h_{max} = difference in water level between the upstream and downstream sides

Note that the space between two consecutive flow lines is defined as a *flow channel*, and the space between two consecutive equipotential lines is called

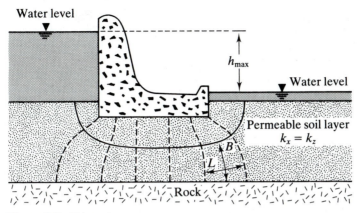

Figure 1.8 Flow net

a *drop*. In Figure 1.8, $N_f = 2$, $N_d = 7$, and $n = 1$. When square elements are drawn in a flow net

$$q = kh_{max} \frac{N_f}{N_d} \qquad (1.34)$$

1.11 Filter Design Criteria

In the design of earth structures one often encounters problems caused by the flow of water, such as soil erosion, which may result in structural instability. Erosion is generally prevented by building soil zones that are referred to as *filters* (see Figure 1.9). Two main factors influence the choice of filter

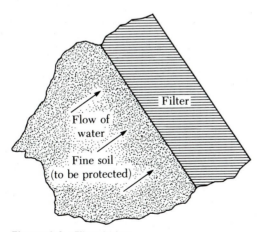

Figure 1.9 Filter design

material: the grain-size distribution of the filter materials should be such that (a) the soil to be protected is not washed into the filter and (b) excessive hydrostatic pressure head is not created in the soil that has a lower coefficient of permeability.

In order to satisfy the preceding conditions, the following requirements must be met (Terzaghi and Peck, 1967):

$$\frac{D_{15(F)}}{D_{85(B)}} < 5 \qquad \text{[to satisfy condition (a)]} \tag{1.35}$$

$$\frac{D_{15(F)}}{D_{15(B)}} > 4 \qquad \text{[to satisfy condition (b)]} \tag{1.36}$$

In these relations, the subscripts F and B refer to the *filter* and the *base* material (that is, the soil to be protected). D_{15} and D_{85} refer to the diameters through which 15% and 85% of the soil (filter/base, as the case may be) will pass.

The U.S. Department of the Navy (1971) provides some additional requirements for filter design to satisfy condition (a). They are as follows:

$$\frac{D_{50(F)}}{D_{50(B)}} < 25 \tag{1.37}$$

$$\frac{D_{15(F)}}{D_{15(B)}} < 20 \tag{1.38}$$

An example of filter design using these requirements [Eqs. (1.35)–(1.38)] is given in Chapter 5.

1.12   Effective Stress Concept

Consider the vertical stress at a point A located at a depth $h_1 + h_2$ below the ground surface, as shown in Figure 1.10a. The total vertical stress, σ, at A can be given as

$$\sigma = h_1 \gamma + h_2 \gamma_{\text{sat}} \tag{1.39}$$

where γ and γ_{sat} are unit weights of soil above and below the ground water table, respectively.

The total stress is carried partially by the *pore water* in the void spaces and partially by the *soil solids* at their points of contact. For example, consider a wavy plane AB drawn through point A (see Figure 1.10a) that passes through the points of contact of soil grains. The plan of this section is shown in Figure 1.10b. The small dots in Figure 1.10b represent the areas in which there is solid-to-solid contact. Let the sum of these areas be equal to A'. Hence, the area that is filled by water is equal to $XY - A$. The force carried by the pore water over the area shown in Figure 1.10b is

$$F_w = (XY - A')u \tag{1.40}$$

where u = pore water pressure = $\gamma_w h_2$ \qquad (1.41)

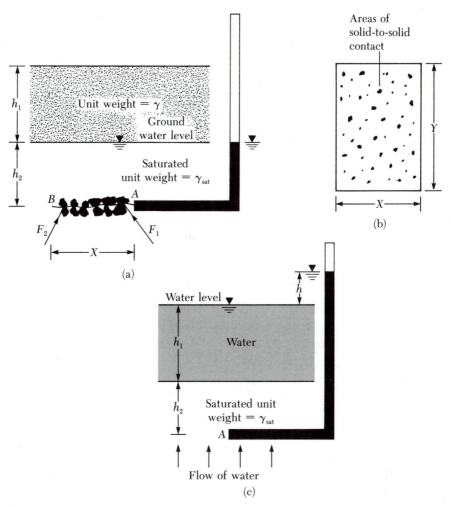

Figure 1.10 Effective stress calculation

Now let $F_1, F_2, \ldots$ be the forces of the contact points of the soil solids as shown in Figure 1.10a. The sum of the vertical components of these forces over a horizontal area XY can be expressed as

$$F_s = \Sigma F_{1(v)} + F_{2(v)} + \cdots \tag{1.42}$$

where $F_{1(v)}, F_{2(v)}, \ldots,$ are vertical components of forces $F_1, F_2, \ldots,$ respectively.

Based on the principles of statics

$$(\sigma)XY = F_w + F_s$$

or

$$(\sigma)XY = (XY - A')u + F_s$$

So

$$\sigma = (1 - a)u + \sigma' \tag{1.43}$$

where $a = A'/XY$ = fraction of the unit cross-sectional area occupied by solid-to-solid contact

$\sigma' = F_s/(XY)$ = vertical component of forces at solid-to-solid contact points over a unit cross-sectional area

The term σ' in Eq. (1.43) is generally referred to as the *vertical effective stress*. Also, the quantity a in Eq. (1.43) is very small. Thus, it can be written that

$$\boxed{\sigma = u + \sigma'} \tag{1.44}$$

Note that the effective stress is a *derived* quantity. Also, because the effective stress σ' is related to the contact between the soil solids, changes in effective stress will induce volume changes. It is also responsible for producing *frictional resistance* in soils and rocks. For dry soils, $u = 0$; hence, $\sigma = \sigma'$.

For the problem under consideration in Figure 1.10a, $u = h_2 \gamma_w$ (γ_w = unit weight of water). Thus, the effective stress at point A is

$$\begin{aligned}
\sigma' = \sigma - u &= (h_1 \gamma + h_2 \gamma_{sat}) - h_2 \gamma_w \\
&= h_1 \gamma + h_2(\gamma_{sat} - \gamma_w) = h_1 \gamma + h_2 \gamma'
\end{aligned} \tag{1.45}$$

where γ' = effective or the submerged unit weight of soil

$$= \gamma_{sat} - \gamma_w$$

From Eq. (1.15)

$$\gamma_{sat} = \frac{G_s \gamma_w + e\gamma_w}{1 + e}$$

So

$$\gamma' = \gamma_{sat} - \gamma_w = \frac{G_s \gamma_w + e\gamma_w}{1 + e} - \gamma_w = \frac{\gamma_w(G_s - 1)}{1 + e} \tag{1.46}$$

In the case of the problem in Figure 1.10a and 1.10b, there was *no seepage of water* in soil. Figure 1.10c shows a simple condition in a soil profile where there is an upward seepage. For this case, at point A

$$\sigma = h_1 \gamma_w + h_2 \gamma_{sat}$$
$$u = (h_1 + h_2 + h)\gamma_w$$

Thus, from Eq. (1.44)

$$\begin{aligned}
\sigma' = \sigma - u &= (h_1 \gamma_w + h_2 \gamma_{sat}) - (h_1 + h_2 + h)\gamma_w \\
&= h_2(\gamma_{sat} - \gamma_w) - h\gamma_w = h_2 \gamma' - h\gamma_w
\end{aligned}$$

or

$$\sigma' = h_2\left(\gamma' - \frac{h}{h_2}\gamma_w\right) = h_2(\gamma' - i\gamma_w) \qquad (1.47)$$

Note that in Eq. (1.47), h/h_2 is the hydraulic gradient, i. If the hydraulic gradient is very high, such that $\gamma' - i\gamma_w$ becomes equal to zero, *the effective stress will become equal to zero*. This means that there is no contact stress between the soil particles, and the soil structure will break up. This situation is referred to as the *quick condition*, or *failure by heave*. So, for heave

$$\boxed{\, i = i_{cr} = \frac{\gamma'}{\gamma_w} = \frac{G_s - 1}{1 + e} \,} \qquad (1.48)$$

where i_{cr} = critical hydraulic gradient

For most sandy soils, i_{cr} ranges from 0.9 to 1.1, with an average of about 1.

Example 1.7

Figure 1.11 shows a soil profile. Determine the total vertical stress, pore water pressure, and effective vertical stress at points A, B, and C. Draw the variation of the effective stress with depth.

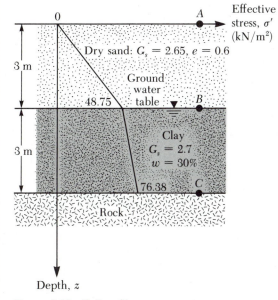

Figure 1.11　Soil profile

Solution

Determination of Unit Weights of Soil

$$\gamma_{d(sand)} = \frac{G_s \gamma_w}{1 + e} = \frac{(2.65)(9.81)}{1 + 0.6} = 16.25 \text{ kN/m}^3$$

$$\gamma_{sat(clay)} = \frac{G_s \gamma_w + w G_s \gamma_w}{1 + w G_s}$$

Note: For saturated soils, $e = w G_s$ [Eq. (1.14)]; so, for this case, $e = (0.3)(2.7) = 0.81$. So

$$\gamma_{sat(clay)} = \frac{(2.7)(9.81) + (0.81)(9.81)}{1 + 0.81} = 19.02 \text{ kN/m}^3$$

Total Stress Calculation

At $A: \sigma_A = \mathbf{0}$

At $B: \sigma_B = \gamma_{d(sand)} \times 3 = 16.25 \times 3 = \mathbf{48.75 \ kN/m^2}$

At $C: \sigma_C = \sigma_B + \gamma_{sat(clay)} \times 3 = 48.75 + (19.02)3 = \mathbf{105.81 \ kN/m^2}$

Pore Water Pressure Calculation

At $A: u_A = \mathbf{0}$

At $B: u_B = \mathbf{0}$

At $C: u_C = 3 \times \gamma_w = 3 \times 9.81 = \mathbf{29.43 \ kN/m^2}$

Effective Stress Calculation

At $A: \sigma'_A = \sigma_A - u_A = \mathbf{0}$

At $B: \sigma'_B = 48.75 - 0 = \mathbf{48.75 \ kN/m^2}$

At $C: \sigma'_C = 105.81 - 29.43 = \mathbf{76.38 \ kN/m^2}$

The variation of effective stress with depth is shown in Figure 1.11.

1.13   Capillary Rise in Soil

It is a well-known fact that when a capillary tube is placed in water, the water level in the tube rises (Figure 1.12a). This is caused by the *surface tension* effect. According to Figure 1.12a, the pressure at any point A in the capillary tube (with respect to the atmospheric pressure) can be given as

$$u = -\gamma_w z' \qquad \text{(for } z' = 0 \text{ to } h_c)$$

and

$$u = 0 \qquad \text{(for } z' \geq h_c)$$

In a given soil mass, the interconnected void spaces can behave like a number of capillary tubes with varying diameters. The surface-tension force may cause water in the soil to rise above the ground water table, as shown in

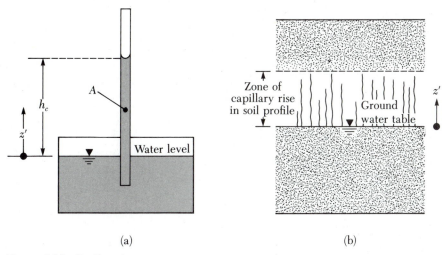

(a)   (b)

Figure 1.12 Capillary rise

Figure 1.12b. The height of the capillary rise will depend on the diameter of the capillary tubes. The capillary rise will *decrease* with the increase of the tube diameter. Because the capillary tubes in soil have variable diameters, the height of capillary rise will be nonuniform. The pore water pressure at any point in the zone of capillary rise in soil can be approximated as

$$u = -S\gamma_w z' \qquad (1.49)$$

where S = degree of saturation of soil (Eq. 1.7)
    z' = distance measured above the ground water table

Example 1.8

Refer to the soil profile shown in Figure 1.11 (Example 1.7). Assume that the capillary rise extends to 1.5 m above the ground water table, as shown in Figure 1.13a. Draw the variation of σ, u, and σ' with depth.

Solution

From Example 1.7, $\gamma_{d(\text{sand})} = 16.25$ kN/m³ and $\gamma_{\text{sat(clay)}} = 19.02$ kN/m³. In the zone of capillary rise

$$\gamma = \frac{\gamma_w(G_s + S_e)}{1 + e} = \frac{(9.81)[2.65 + (0.62)(0.6)]}{1 + 0.6}$$

$$= 18.53 \text{ kN/m}^3$$

Stress Calculation

At point A: $\sigma = 0$

$u = 0$

$\sigma' = 0$

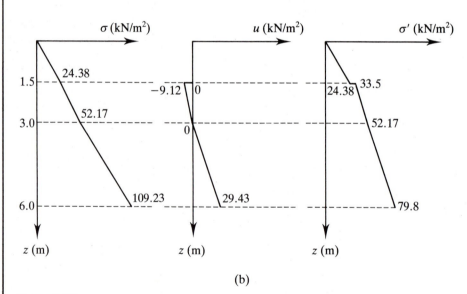

Figure 1.13

At point B: $\quad \sigma = (1.5)(16.25) = $ **24.38 kN/m²**

$\qquad$ Immediately $\Big\}\quad u = 0$
$\qquad$ above $B \qquad\quad \sigma' = 24.38 - 0 = $ **24.38 kN/m²**

$\qquad$ Immediately $\left.\begin{array}{l} \quad u = -(S)(\gamma_w)(1.5) \\[4pt] \qquad = -(0.62)(9.81)(1.5) \\[4pt] \qquad = \mathbf{-9.12\ kN/m^2} \\[4pt] \quad \sigma' = 24.38 - (-9.12) = \mathbf{33.5\ kN/m^2} \end{array}\right.$
$\qquad$ below B

At point C: $\quad \sigma = (1.5)(16.25) + (1.5)(18.53) = $ **52.17 kN/m²**

$\qquad\qquad u = 0$

$\qquad\qquad \sigma' = 52.17 - 0 = $ **52.17 kN/m²**

At point D: $\quad \sigma = (1.5)(16.25) + (1.5)(18.53) + 3(19.02) = $ **109.23 kN/m²**

$\qquad\qquad u = (9.81)(3) = $ **29.43 kN/m²**

$\qquad\qquad \sigma' = 109.23 - 29.43 = $ **79.80 kN/m²**

$\qquad$ The plot of the variation of σ, u, and σ' with depth is shown in Figure 1.13b.

1.14   Consolidation

When the stress on a saturated clay layer in the field is increased, for example, by the construction of a foundation, the pore water pressure in the clay will increase. Because the coefficients of permeability of clays are very small (see Table 1.10), it will take some time for the excess pore water pressure to dissipate and the stress increase to be transferred to the soil skeleton gradually. According to Figure 1.14, if Δp is a surcharge at the ground

Figure 1.14   Principles of consolidation

surface over a very large area, the increase of total stress ($\Delta\sigma$) at any depth of the clay layer will be equal to Δp, or

$$\Delta\sigma = \Delta p$$

However, at time $t = 0$ (that is, immediately after the stress application), the excess pore water pressure at any depth (Δu) will be equal to Δp, or

$$\Delta u = \Delta h_i \gamma_w = \Delta p \qquad \text{(at time } t = 0\text{)}$$

Hence, the increase of effective stress at time $t = 0$ will be

$$\Delta\sigma' = \Delta\sigma - \Delta u = 0$$

Theoretically, at time $t = \infty$, when all the excess pore water pressure in the clay layer is dissipated as a result of drainage into the sand layers

$$\Delta u = 0 \qquad \text{(at time } t = \infty\text{)}$$

This means that the increase of effective stress in the clay layer is

$$\Delta\sigma' = \Delta\sigma - \Delta u = \Delta p - 0 = \Delta p$$

This gradual increase in the effective stress in the clay layer will cause gradual settlement over a period of time and is referred to as *consolidation*.

Laboratory tests on undisturbed saturated clay specimens can be conducted (ASTM Test Designation D-2435) to determine the consolidation settlement caused by various incremental loadings. The test specimens are usually 63.5 mm (2.5 in.) in diameter and 25.4 mm (1 in.) in height. Specimens are placed inside a ring, with one porous stone at the top and one at the bottom of the specimen (Figure 1.15a). Load on the specimen is then applied such that the total stress is equal to p. Settlement readings for the specimen are taken for 24 hours. After that, the load on the specimen is doubled and settlement readings are taken. At all times during the test the specimen is kept under water. This procedure is continued until the desired limit of stress on the clay specimen is reached.

Based on the laboratory tests, a graph can be plotted showing the variation of the void ratio e at the *end* of consolidation against the corresponding stress p (semilogarithmic graph: e on the arithmetic scale and p on the log scale). The nature of variation of e against log p for a clay specimen is shown in Figure 1.15b. During the laboratory tests, after the desired consolidation pressure is reached, the specimen can be gradually unloaded. This will result in the swelling of the specimen. Figure 1.15b also shows the variation of the void ratio during the unloading period.

From the e-log p curve shown in Figure 1.15b, three parameters can be determined that will be necessary for calculation of the settlement in the field. They are as follows:

1. *Preconsolidation Pressure, p_c:* This is the *maximum past effective overburden pressure* to which the soil specimen has been subjected. It can be determined by using a simple graphical procedure as proposed by Casa-

(a)

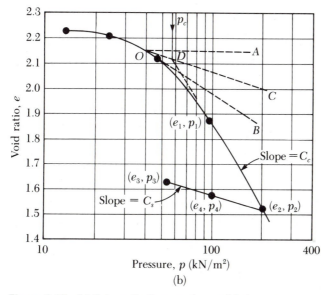

(b)

Figure 1.15 (a) Schematic diagram of consolidation test arrangement; (b) e-log p curve for a soft clay from East St. Louis, Illinois

grande (1936). This procedure for determination of the preconsolidation pressure, with reference to Figure 1.15b, has five steps:

a. Determine the point O on the e-log p curve that has the sharpest curvature (that is, the smallest radius of curvature).

b. Draw a horizontal line OA.

c. Draw a line OB that is tangent to the e-log p curve at O.

d. Draw a line OC that bisects the angle AOB.

e. Produce the straight line portion of the e-log p curve backward to intersect OC. This is point D. The pressure that corresponds to the point p is the preconsolidation pressure, p_c.

Natural soil deposits can be *normally consolidated* or *overconsolidated* (or *preconsolidated*). If the present effective overburden pressure $p = p_o$ is equal to the preconsolidation pressure p_c, the soil is *normally consolidated*. However, if $p_o < p_c$, the soil is *overconsolidated*.

Nagaraj and Murthy (1985) have suggested the following empirical relationship for p_c in overconsolidated soil:

$$\frac{e_o}{e_L} = 1.122 - 0.188 \log p_c - 0.0463 \log p_o \qquad (1.50)$$

$$\uparrow \qquad\qquad \uparrow$$
$$\text{kN/m}^2 \qquad\quad \text{kN/m}^2$$

where e_L = void ratio at liquid limit

e_o = void ratio at *in situ* effective overburden pressure p_o

However, for saturated soils [Eq. (1.14)], $e = wG_s$. So at liquid limit

$$e_L = \left(\frac{LL}{100}\right)G_s \qquad (1.51)$$

So

$$\frac{100e_o}{(LL)G_s} = 1.122 - 0.188 \log p_c - 0.0463 \log p_o$$

or

$$\log p_c = \frac{1.122 - \dfrac{100e_o}{(LL)G_s} - 0.0463 \log p_o}{0.188} \qquad (1.52)$$

In the preceding equation, the units of p_c and p_o are kN/m².

2. *Compression Index, C_c:* This is the slope of the straight-line portion (latter part of the loading curve), and it can be determined as

$$C_c = \frac{e_1 - e_2}{\log p_2 - \log p_1} = \frac{e_1 - e_2}{\log\left(\dfrac{p_2}{p_1}\right)} \qquad (1.53)$$

where e_1 and e_2 are the void ratios at the end of consolidation under stresses p_1 and p_2, respectively.

The *compression index*, as determined from the laboratory e-log p curve, will be somewhat different from that encountered in the field. This is primarily due to the remolding of soil to some degree during the field exploration. The nature of variation of the e-log p curve in the field for a normally consolidated clay is shown in Figure 1.16. This is generally referred to as the *virgin compression curve*. The virgin curve approximately intersects the laboratory curve at a void ratio of $0.42e_o$ (Terzaghi and Peck, 1967). Note that e_o is the void ratio of the clay in the field. Knowing the values of e_o and p_c, one can easily construct the virgin curve, as shown in Figure 1.16. The compression index of the virgin curve can now be calculated by using Eq. (1.53).

The value of C_c can vary widely depending on the soil. Skempton (1944) has given an empirical correlation for the compression index in which

$$C_c = 0.009(LL - 10) \qquad ,126$$

(1.54)

where LL = liquid limit

Several other correlations for the compression index have been summarized by Azzouz, Krizek, and Corotis (1976). Some of these are the following:

$$C_c = 0.01w_n \qquad \text{(Chicago clay)} \tag{1.55}$$

$$C_c = 0.208e_o + 0.0083 \qquad \text{(Chicago clay)} \tag{1.56}$$

$$C_c = 0.0115w_n \qquad \text{(organic soils, peats)} \tag{1.57}$$

$$C_c = 0.0046(LL - 9) \qquad \text{(Brazilian clay)} \tag{1.58}$$

where w_n = natural moisture content of soil in percent

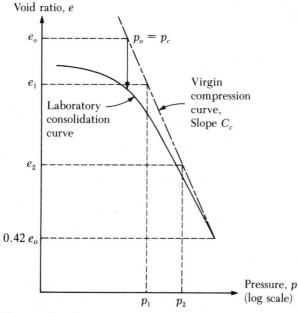

Figure 1.16 Construction of virgin compression curve for normally consolidated clay

Based on observation of several natural clays, Rendon-Herrero (1980) has given the relationship for compression index in the form

$G_s = 2.7$

$e_o =$

$$C_c = 0.141 G_s^{1.2}\left(\frac{1 + e_o}{G_s}\right)^{2.38}$$

(1.59)

Nagaraj and Murthy (1985) expressed compression index as

$$C_c = 0.2343 e_L = 0.2343\left(\frac{LL}{100}\right)G_s$$

(1.60)

In any case, if empirical correlations are used in calculation, a deviation of $\pm 15\%$ from the actual field values may be expected.

3. *Swelling Index, C_s:* This is the slope of the unloading portion of the *e*-log *p* curve. In Figure 1.15b, it can be defined as

$$C_s = \frac{e_3 - e_4}{\log\left(\dfrac{p_4}{p_3}\right)}$$

(1.61)

The value of the swelling index (C_s) is, in most cases, $\frac{1}{4}$ to $\frac{1}{5}$ of the compression index. Following are some representative values of C_s/C_c for natural soil deposits.

Description of soil	C_s/C_c
Boston Blue clay	0.24–0.33
Chicago clay	0.15–0.3
New Orleans clay	0.15–0.28
St. Lawrence clay	0.05–0.1

The swelling index determination is important in the estimation of consolidation settlement of *overconsolidated clays* (see Figure 1.17). In the field, depending on the pressure increase, an overconsolidated clay will follow an *e*-log *p* path *abc*, as shown in Figure 1.17. Note that, in this figure, the point *a* with coordinates of p_o and e_o corresponds to the field conditions before any pressure increase. The point *b* corresponds to the preconsolidation pressure (p_c) of the clay. The line *ab* is approximately parallel to the laboratory unloading curve *cd* (Schmertmann, 1953). Hence, with a knowledge of e_o, p_o, p_c, C_c, and C_s, one can easily construct the field consolidation curve.

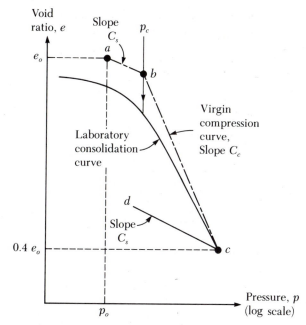

Figure 1.17 Construction of field consolidation curve for overconsolidated clay

The swelling index has been expressed by Nagaraj and Murthy (1985) as

$$C_s = 0.0463e_L = 0.0463\left(\frac{LL}{100}\right)G_s \tag{1.62}$$

Calculation of Settlement

The one-dimensional consolidation settlement, caused by an additional load, of a clay layer (Figure 1.18a) having a thickness H_c may be calculated as

$$\boxed{S = \frac{\Delta e}{1 + e_o} H_c} \tag{1.63}$$

where S = settlement

Δe = total change of void ratio caused by the additional load application

e_o = the void ratio of the clay before the application of load

Note that

$$\frac{\Delta e}{1 + e_o} = \epsilon_v = \text{vertical strain}$$

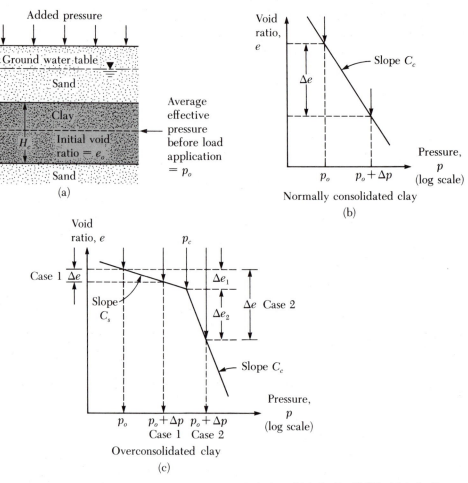

Figure 1.18 One-dimensional settlement calculation: (b) is for Eq. (1.64); (c) is for Eqs. (1.67) and (1.69).

For normally consolidated clay, the field e-log p curve will be like the one in Figure 1.18b. If p_o = initial average effective overburden pressure on the clay layer, and Δp = average pressure increase on the clay layer caused by the added load, the change of void ratio caused by the load increase is

$$\Delta e = C_c \log \frac{p_o + \Delta p}{p_o} \tag{1.64}$$

Now, combining Eqs. (1.63) and (1.64)

$$S = \frac{C_c H_c}{1 + e_o} \log \frac{p_o + \Delta p}{p_o} \tag{1.65}$$

For overconsolidated clay, the field e-log p curve will be like the one in Figure 1.18c. In this case, depending on the value of Δp, two conditions may arise. First, if $p_o + \Delta p < p_c$

$$\Delta e = C_s \log \frac{p_o + \Delta p}{p_o} \tag{1.66}$$

Combining Eqs. (1.63) and (1.66)

$$S = \frac{H_c C_s}{1 + e_o} \log \frac{p_o + \Delta p}{p_o} \tag{1.67}$$

Second, if $p_o < p_c < p_o + \Delta p$

$$\Delta e = \Delta e_1 + \Delta e_2 = C_s \log \frac{p_c}{p_o} + C_c \log \frac{p_o + \Delta p}{p_o} \tag{1.68}$$

Now, combining Eqs. (1.63) and (1.68)

$$S = \frac{C_s H_c}{1 + e_o} \log \frac{p_c}{p_o} + \frac{C_c H_c}{1 + e_o} \log \frac{p_o + \Delta p}{p_c} \tag{1.69}$$

Average Degree of Consolidation

Earlier in this section (see Figure 1.14) we saw that consolidation is the result of gradual dissipation of the excess pore water pressure from a clay layer. Pore water pressure dissipation, in turn, increases the effective stress, which induces settlement. Hence, to estimate the degree of consolidation of a clay layer at some time t after the load application, one should know the rate of dissipation of the excess pore water pressure.

Figure 1.19a shows a clay layer of thickness H_c that has highly permeable sand layers at its top and bottom. Let the excess pore water pressure at any point A at any time t after the load application be equal to $\Delta u = (\Delta h)\gamma_w$. For a vertical drainage condition (that is, in the direction of z only) from the clay layer, Terzaghi has derived the following differential equation:

$$\frac{\partial(\Delta u)}{\partial t} = C_v \frac{\partial^2(\Delta u)}{\partial z^2} \tag{1.70}$$

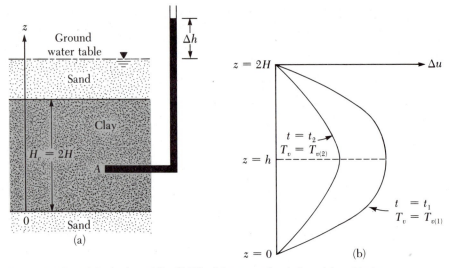

Figure 1.19   (a) Derivation of Eq. (1.72); (b) nature of variation of Δu with time

where C_v = coefficient of consolidation

$$C_v = \frac{k}{m_v \gamma_w} = \frac{k}{\dfrac{\Delta e}{\Delta p(1 + e_{\mathrm{av}})}\gamma_w}$$

(1.71)

where k = coefficient of permeability of the clay
   Δe = total change of void ratio caused by a stress increase of Δp
   e_{av} = average void ratio during consolidation
   m_v = volume coefficient of compressibility = $\Delta e/[\Delta p(1 + e_{\mathrm{av}})]$

Equation (1.70) can be solved to obtain Δu as a function of time t with the following boundary conditions:

1. Because highly permeable sand layers are located at $z = 0$ and $z = H_c$, the excess pore water pressure developed in the clay at those points will be immediately dissipated. So

$\Delta u = 0$ at $z = 0$

$\Delta u = 0$ at $z = H_c = 2H$

where H = length of maximum drainage path (due to two-way drainage condition—that is, at top and bottom of clay)

2. At time $t = 0$

   $\Delta u = \Delta u_o$ = initial excess pore water pressure after the load application

With the preceding boundary conditions, Eq. (1.70) yields

$$\Delta u = \sum_{m=0}^{m=\infty} \left[\frac{2(\Delta u_o)}{M} \sin\left(\frac{Mz}{H}\right) \right] e^{-M^2 T_v} \tag{1.72}$$

where $M = \dfrac{(2m + 1)\pi}{2}$

m = an integer = 1, 2, ...

T_v = nondimensional time factor = $\dfrac{C_v t}{H^2}$ (1.73)

The value of Δu for various depths (that is, $z = 0$ to $z = 2H$) at any given time t (thus, T_v) can be calculated from Eq. (1.72). The nature of this variation of Δu is shown in Figure 1.19b.

The *average degree of consolidation* of clay layer can be defined as

$$\boxed{U = \frac{S_t}{S_{\max}}} \tag{1.74}$$

where U = average degree of consolidation
S_t = settlement of a clay layer at time t after the load application
$S_{\max}$ = maximum consolidation settlement that the clay will undergo under a given loading

If the initial pore water pressure (Δu_o) distribution is constant with depth as shown in Figure 1.20a, the average degree of consolidation can also be expressed as

$$U = \frac{S_t}{S_{\max}} = \frac{\displaystyle\int_0^{2H} (\Delta u_o)dz - \int_0^{2H} (\Delta u)dz}{\displaystyle\int_0^{2H} (\Delta u_o)dz} \tag{1.75}$$

or

$$U = \frac{(\Delta u_o)2H - \displaystyle\int_0^{2H} (\Delta u)dz}{(\Delta u_o)2H} = 1 - \frac{\displaystyle\int_0^{2H} (\Delta u)dz}{2H(\Delta u_o)} \tag{1.76}$$

Now, combining Eqs. (1.72) and (1.76), one obtains

$$U = \frac{S_t}{S_{\max}} = 1 - \sum_{m=0}^{m=\infty} \left(\frac{2}{M^2}\right) e^{-M^2 T_v} \tag{1.77}$$

The variation of U with T_v can be calculated from Eq. (1.77) and is plotted in Figure 1.21. Note that Eq. (1.77) and, thus, Figure 1.21 are also

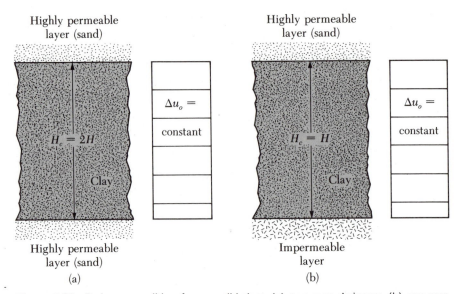

Figure 1.20 Drainage condition for consolidation: (a) two-way drainage; (b) one-way drainage

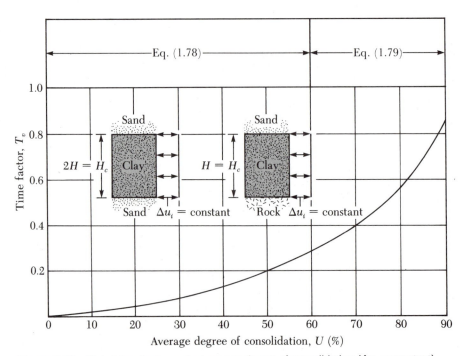

Figure 1.21 Plot of time factor against average degree of consolidation (Δu_o = constant)

valid when an impermeable layer is located at the bottom of the clay layer (Figure 1.20b). In such a case, excess pore water pressure dissipation can take place in one direction only. So, the length of the *maximum drainage path* is equal to $H = H_c$.

The variation of T_v with U shown in Figure 1.21 can also be approximated by the following relations:

$$T_v = \frac{\pi}{4}\left(\frac{U\%}{100}\right)^2 \quad \text{(for } U = 0\text{–}60\%) \tag{1.78}$$

and

$$T_v = 1.781 - 0.933 \log (100 - U\%) \quad \text{(for } U > 60\%) \tag{1.79}$$

Sivaram and Swamee (1977) have also developed an empirical relationship between T_v and U which is valid for U varying from 0 to 100%. It is of the form

$$T_v = \frac{\left(\frac{\pi}{4}\right)\left(\frac{U\%}{100}\right)^2}{\left[1 - \left(\frac{U\%}{100}\right)^{5.6}\right]^{0.357}} \tag{1.80}$$

Example 1.9

A laboratory consolidation test on a normally consolidated clay showed the following:

Load, p (kN/m²)	Void ratio at the end of consolidation, e
140	0.92
212	0.86

The specimen tested was 25.4 mm in thickness drained on both sides. The time required for the specimen to reach 50% consolidation was 4.5 min.

A similar clay layer in the field, 2.8 m thick and drained on both sides, is subjected to similar average pressure increase (that is, $p_o = 140$ kN/m² and $p_o + \Delta p = 212$ kN/m²). Determine the following:

a. The expected maximum consolidation settlement in the field

b. The length of time it will take for the total settlement in the field to reach 40 mm

Solution
Part a

For normally consolidated clay [Eq. (1.53)]

$$C_c = \frac{e_1 - e_2}{\log\left(\dfrac{p_2}{p_1}\right)} = \frac{0.92 - 0.86}{\log\left(\dfrac{212}{140}\right)} = 0.333$$

From Eq. (1.65)

$$S = \frac{C_c H_c}{1 + e_o} \log \frac{p_o + \Delta p}{p_o} = \frac{(0.333)(2.8)}{1 + 0.92} \log \frac{212}{140} = 0.0875 \text{ m}$$

$$= 87.5 \text{ mm}$$

Part b

From Eq. (1.74), the average degree of consolidation

$$U = \frac{S_t}{S_{\max}} = \frac{40}{87.5}(100) = 45.7\%$$

The coefficient of consolidation, C_v, can be calculated from the laboratory test. From Eq. (1.73)

$$T_v = \frac{C_v t}{H^2}$$

For 50% consolidation (Figure 1.21), $T_v = 0.197$, $t = 4.5$ min, $H = H_c/2 = 12.7$ mm. So

$$C_v = T_{50} \frac{H^2}{t} = \frac{(0.197)(12.7)^2}{4.5} = 7.061 \text{ mm}^2/\text{min}$$

Again, for field consolidation, $U = 45.7\%$. From Eq. (1.78)

$$T_v = \frac{\pi}{4}\left(\frac{U\%}{100}\right)^2 = \frac{\pi}{4}\left(\frac{47.5}{100}\right)^2 = 0.177$$

But

$$T_v = \frac{C_v t}{H^2}$$

or

$$t = \frac{T_v H^2}{C_v} = \frac{0.177\left(\dfrac{2.8 \times 1000}{2}\right)^2}{7.061} = 49132 \text{ min} = 34.1 \text{ days}$$

1.15   Shear Strength

The shear strength of a soil (s), in terms of effective stress, can be given by the equation

$$\boxed{s = c + \sigma' \tan \phi}$$

(1.81)

where σ' = effective normal stress on plane of shearing

c = cohesion, or apparent cohesion

ϕ = angle of friction

Equation (1.81) is referred to as the *Mohr-Coulomb failure criteria*. The value of c for sands and normally consolidated clays is equal to zero. For overconsolidated clays, $c > 0$.

For most day-to-day work, the shear strength parameters of a soil (that is, c and ϕ) are determined by two standard laboratory tests. They are (a) the *direct shear test* and (b) the *triaxial test*. Following is a brief description of each test.

Direct Shear Test

Dry sand can be conveniently tested by direct shear tests. These tests are conducted by placing the sand in a shear box that is split into two halves (Figure 1.22a). A normal load is first applied to the specimen. After that, a shear force is applied to the top half of the shear box to cause failure in soil. The normal and shear stresses at failure can be determined as

$$\sigma' = \frac{N}{A}$$

$$s = \frac{R}{A}$$

where A = area of the failure plane in soil—that is, the area of cross section of the shear box

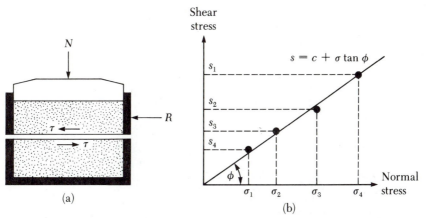

Figure 1.22   Direct shear test in sand: (a) schematic diagram of test equipment; (b) plot of test results to obtain the friction angle, ϕ

Several tests of this type can be conducted by varying the normal load. The angle of friction of the sand can be determined by plotting a graph of s against σ' as shown in Figure 1.22b, or

$$\phi = \tan^{-1}\left(\frac{s}{\sigma'}\right) \qquad\qquad (1.82)$$

For sands, the angle of friction usually ranges from 26° to 45°. It increases with the relative density of compaction. The approximate range of relative density of compaction and the corresponding range of the angle of friction for various sands are given in Table 2.4. Figure 1.23 shows a photograph of a direct shear machine.

Triaxial Tests

Triaxial compression tests can be conducted on sands and clays. Figure 1.24a shows a schematic diagram of the triaxial test arrangement. It essentially consists of placing a soil specimen confined by a rubber membrane

Figure 1.23   Direct shear test equipment (courtesy of John Hester, Carterville, Illinois)

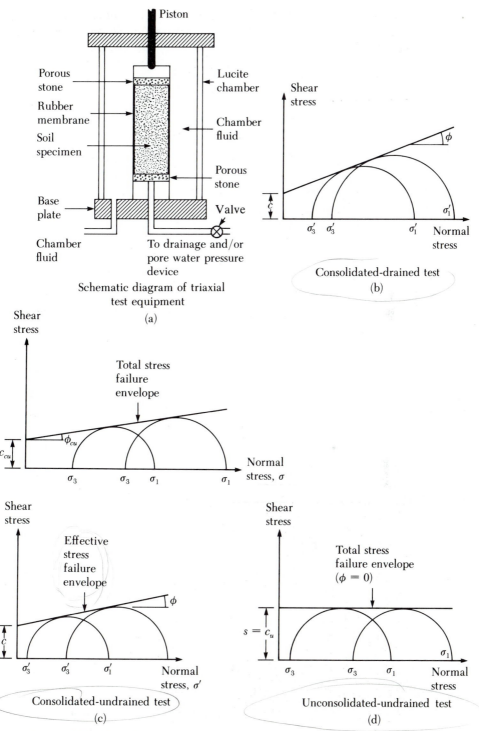

Figure 1.24   Triaxial test

inside a lucite chamber. An all-around confining pressure (σ_3) is applied to the specimen by means of the chamber fluid (generally water or glycerin). An added stress ($\Delta\sigma$) can also be applied to the specimen in the axial direction to cause failure ($\Delta\sigma = \Delta\sigma_f$ at failure). Drainage from the specimen can be allowed or stopped, depending on the test condition. For clays, three main types of tests can be conducted by means of triaxial equipment:

1. Consolidated-drained test (CD test)
2. Consolidated-undrained test (CU test)
3. Unconsolidated-undrained test (UU test)

Table 1.11 gives a summary of these three tests.
For *consolidated-drained tests*, at failure

$$\text{Major principal effective stress} = \sigma_3 + \Delta\sigma_f = \sigma_1 = \sigma'_1$$

$$\text{Minor principal effective stress} = \sigma_3 = \sigma'_3$$

Table 1.11   Summary of Triaxial Tests on Saturated Clays

Test type	Step 1	Step 2
Consolidated-drained	Apply chamber pressure, σ_3. Allow complete drainage, so pore water pressure ($u = u_a$) developed is zero.	Apply axial stress, $\Delta\sigma$, slowly. Allow drainage, so pore water pressure ($u = u_d$) developed through application of $\Delta\sigma$ is zero. At failure, $\Delta\sigma = \Delta\sigma_f$; total pore water pressure $u_f = u_a + u_d = 0$.
Consolidated-undrained	Apply chamber pressure, σ_3. Allow complete drainage, so pore water pressure ($u = u_a$) developed is zero.	Apply axial stress, $\Delta\sigma$. Do not allow drainage ($u = u_d \neq 0$). At failure, $\Delta\sigma = \Delta\sigma_f$; pore water pressure $u = u_f = u_a + u_d = 0 + u_{d(f)}$.
Unconsolidated-undrained	Apply chamber pressure, σ_3. Do not allow drainage, so pore water pressure ($u = u_a$) developed through application of σ_3 is not zero.	Apply axial stress, $\Delta\sigma$. Do not allow drainage ($u = u_d \neq 0$). At failure $\Delta\sigma = \Delta\sigma_f$; pore water pressure $u = u_f = u_a + u_{d(f)}$.

Several tests of this type can be conducted on a number of clay specimens by changing σ_3. The shear strength parameters (c and ϕ) can now be determined by plotting Mohr's circle at failure, as shown in Figure 1.24b, and drawing a common tangent to the Mohr's circles. This is the *Mohr-Coulomb failure envelope.* (*Note:* For normally consolidated clay, $c \approx 0$.) It can be

shown that, at failure

$$\sigma'_1 = \sigma'_3 \tan^2 \left(45 + \frac{\phi}{2} \right) + 2c \tan \left(45 + \frac{\phi}{2} \right) \qquad (1.83)$$

For *consolidated-undrained* tests, at failure

Major principal total stress $= \sigma_3 + \Delta\sigma_f = \sigma_1$
Minor principal total stress $= \sigma_3$
Major principal effective stress $= (\sigma_3 + \Delta\sigma_f) - u_f = \sigma'_1$
Minor principal effective stress $= \sigma_3 - u_f = \sigma'_3$

A number of tests of this type can be conducted on several soil specimens by changing σ_3. The total-stress Mohr's circles at failure can now be plotted as shown in Figure 1.24c, and then a common tangent can be drawn to define the *failure envelope*. This is the *total stress failure envelope* and can be defined by the equation

$$s = c_{cu} + \sigma \tan \phi_{cu} \qquad (1.84)$$

where c_{cu} and ϕ_{cu} are the *consolidated-undrained cohesion* and *angle of friction,* respectively. (*Note:* $c_{cu} \approx 0$ for normally consolidated clays.)

Similarly, effective-stress Mohr's circles at failure can be drawn to determine the *effective stress failure envelopes* (Figure 1.24c). This will follow the relation as given by Eq. (1.81).

Kenney (1959) has given a correlation between the friction angle, ϕ, and the plasticity index (*PI*) of normally consolidated clays based on the observations of over 60 soils. This correlation is shown in Figure 1.25. Based on the average plot, the value of ϕ generally decreases from about 37–38° with a plasticity index of about 10, to about 25° with a plasticity index of about 100. The consolidated undrained friction angle (ϕ_{cu}) of normally consolidated saturated clays generally ranges from 5° to 20°.

For *unconsolidated-undrained* triaxial tests

Major principal total stress $= \sigma_3 + \Delta\sigma_f = \sigma_1$
Minor principal total stress $= \sigma_3$

The total-stress Mohr's circle at failure can now be drawn as shown in Figure 1.24d. It can be shown that, for saturated clays, the value of $\sigma_1 - \sigma_3 = \Delta\sigma_f$ is a constant, irrespective of the chamber confining pressure, σ_3 (this is also shown in Figure 1.24d). The tangent to these Mohr's circles will be a horizontal line, which is called the $\phi = 0$ condition. The shear stress for this condition can be given as

$$s = c_u = \frac{\Delta\sigma_f}{2} \qquad (1.85)$$

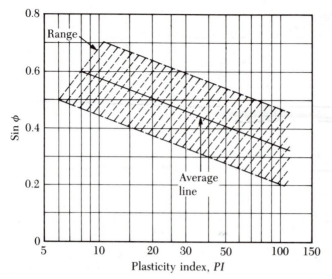

Figure 1.25   Variation of sin ϕ with the plasticity index of several clays (redrawn after Kenney, 1959)

The pore pressure developed in the soil specimen during the unconsolidated-undrained triaxial test is equal to

$$u = u_a + u_d \tag{1.86}$$

The pore pressure u_a is the contribution of the hydrostatic chamber pressure, σ_3. So, it can be written as

$$u_d = B\sigma_3 \tag{1.87}$$

where B = Skempton's pore pressure parameter

Similarly, the pore pressure u_d is the result of added axial stress, $\Delta\sigma$. The magnitude of u_d can be given by the equation

$$u_a = A\,\Delta\sigma \tag{1.88}$$

where A = Skempton's pore pressure parameter

However

$$\Delta\sigma = \sigma_1 - \sigma_3 \tag{1.89}$$

Combining Eqs. (1.86), (1.87), (1.88), and (1.89)

$$u = u_a + u_d = B\sigma_3 + A(\sigma_1 - \sigma_3) \tag{1.90}$$

The pore water pressure parameter B in saturated soils is equal to one. So,

$$\boxed{u = \sigma_3 + A(\sigma_1 - \sigma_3)} \tag{1.91}$$

Figure 1.26 Triaxial test equipment

The value of the pore water pressure parameter A at failure will vary with the type of soil. Following is a general range of the values of A at failure for various types of clayey soil encountered in nature:

Type of soil	A at failure
Sandy clays	0.5–0.7
Normally consolidated clays	0.5–1
Overconsolidated clays	−0.5–0

A photograph of laboratory triaxial equipment is shown in Figure 1.26.

1.16   Unconfined Compression Test

The *unconfined compression test* (Figure 1.27a) is a special type of *unconsolidated-undrained triaxial test* in which the confining pressure $\sigma_3 = 0$, as shown in Figure 1.27b. In this test, an axial stress, $\Delta\sigma$, is applied to the specimen to cause failure (that is, $\Delta\sigma = \Delta\sigma_f$). The corresponding Mohr's circle is shown in Figure 1.27b. Note that, for this case

Major principal total stress $= \Delta\sigma_f = q_u$

Minor principal total stress $= 0$

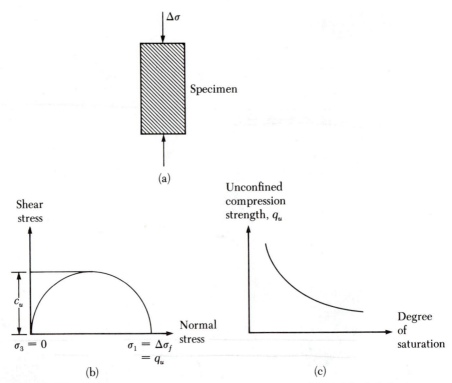

Figure 1.27 Unconfined compression test: (a) soil specimen; (b) Mohr's circle for the test; (c) variation of q_u with the degree of saturation

The axial stress at failure, $\Delta\sigma_f = q_u$, is generally referred to as the *unconfined compression strength*. The shear strength of saturated clays under this condition ($\phi = 0$) can be given by Eq. (1.81) as

$$s = c_u = \frac{q_u}{2} \tag{1.92}$$

The unconfined compression strength can be used as an indicator for the consistency of clays. This is shown in Table 2.3.

Unconfined compression tests are sometimes conducted on unsaturated soils. With the void ratio of a soil specimen remaining constant, the unconfined compression strength rapidly decreases with the degree of saturation (Figure 1.27c).

Figure 1.28 shows an unconfined compression test in progress.

The undrained cohesion, c_u, is a very important parameter in the design of foundations. For normally consolidated clay deposits (Figure 1.29), the

Figure 1.28 Unconfined compression test in progress (courtesy of Soiltest Inc., Lake Bluff, Illinois)

magnitude of c_u increases almost linearly with the increase of effective overburden pressure, p. Skempton (1957) correlated c_u with p in the form

$$\frac{c_u}{p} = 0.11 + 0.0037(PI) \qquad \text{(normally consolidated clay)} \qquad (1.93)$$

where PI = plasticity index

The variation of c_u/p with overconsolidation ratio, $OCR = p/p_c$ (p_c = preconsolidation pressure) for some natural clays is shown in Figure 1.30 (Ladd and Foot, 1974). Based on the observations shown in Figure 1.30, we can say that

$$\left(\frac{c_u}{p}\right)_{\text{overconsolidated}} = \alpha \left(\frac{c_u}{p}\right)_{\text{normally consolidated}} \qquad (1.94)$$

Figure 1.31 shows the range of variation of α with OCR based on the experimental results shown in Figure 1.30. For initial estimation purposes,

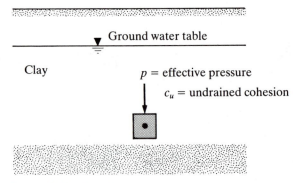

Figure 1.29 Clay deposit

Curve	Soil	PI
1	Marine organic clay	34
2	Bangkok clay	41
3	Atchafalaya clay	75
4	Boston Blue clay	21
5	Connecticut varve	
	Clay layer	39
	Silt layer	12

Figure 1.30 Variation of c_u/p with *OCR* for five clay soils (after Ladd and Foot, 1974)

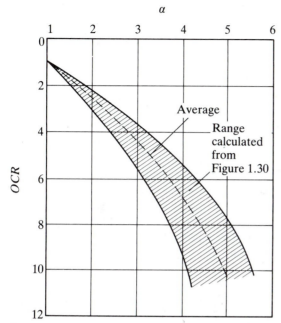

Figure 1.31 Variation of α with *OCR* [Eq. (1.94)] based on the results shown in Figure 1.30

the value of α from the average plot of Figure 1.30 can be taken and used in Eq. (1.94).

Example 1.10

At a point in a saturated overconsolidated clay deposit, $p = 2300$ lb/ft². The plasticity index of the clay is 24, and the overconsolidation ratio, *OCR*, is 2.5. Estimate the undrained cohesion, c_u.

Solution

From Eq. (1.93), for normally consolidated clay

$$\frac{c_u}{p} = 0.11 + 0.0037(PI) = 0.11 + 0.0037(24)$$

$$= 0.199$$

For *OCR* = 2.5, the value of α from Figure 1.31 is about 1.9. Now, using Eq. (1.94)

$$\left(\frac{c_u}{p}\right)_{\text{overconsolidated}} = \alpha \left(\frac{c_u}{p}\right)_{\text{normally consolidated}}$$

$$= (1.9)(0.199) = 0.378$$

So

$$c_u = (0.378)(p) = (0.378)(2300) = \textbf{869.4 lb/ft}^2$$

NATURAL SOIL DEPOSITS

1.17   Soil Origin

Most of the soils that cover the earth are formed by the weathering of various rocks. There are two general types of weathering: (1) *mechanical weathering* and (2) *chemical weathering*.

Mechanical weathering is the process by which rocks are broken down into smaller and smaller pieces by physical forces. These physical forces may be running water, wind, ocean waves, glacier ice, frost action, and expansion and contraction caused by gain and loss of heat.

Chemical weathering is the process of chemical decomposition of the original rock. In the case of mechanical weathering, the rock breaks down into smaller pieces without a change of chemical composition. However, in chemical weathering, the original material may be changed to something entirely different. For example, the chemical weathering of feldspar can produce clay minerals.

The soil that is produced by the weathering process of rocks can be transported by physical processes to other places. These soil deposits are called *transported soils*. In contrast, some soils stay in the place of their formation and cover the rock surface from which they derive. These soils are referred to as *residual soils*.

Based on the *transporting agent*, transported soils can be subdivided into three major categories:

1. *Alluvial*, or *fluvial:* deposited by running water
2. *Glacial:* deposited by glacier action
3. *Aeolian:* deposited by wind action

In addition to transported and residual soils, there are *peats* and *organic soils*, which derive from the decomposition of organic materials.

The following sections discuss the general physical characteristics of residual soils, the three types of transported soil, and organic soils.

1.18   Residual Soil

Residual soil deposits are common in humid tropics, Hawaii, and the southeastern United States. The nature of a residual soil deposit will generally depend on the parent rock. When hard rocks such as granite and gneiss undergo weathering, most of the materials are likely to remain in place. These soil deposits generally have a top layer of clayey or silty-clay material followed by silty and/or sandy soil layers. These are generally followed by a partially weathered rock and then sound bedrock. The depth of the sound bedrock may vary widely from place to place, even within a distance of a few meters. Figure 1.32 shows the log of a boring in a residual soil deposit derived from the weathering of granite.

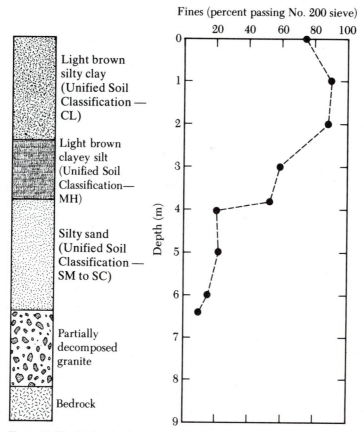

Figure 1.32 Boring log for a residual soil derived from granite

In contrast to hard rocks, there are some chemical rocks, such as limestone, that are chiefly made up of calcite ($CaCO_3$)mineral. Chalk and dolomite have large concentrations of dolomite minerals [$CaMg(CO_3)_2$]. These rocks have large amounts of soluble materials, some of which are removed by ground water, leaving behind the insoluble fraction of the rock. Residual soils that derive from chemical rocks do not possess a gradual transition zone to the bedrock as seen in Figure 1.32. The residual soils derived from the weathering of limestone-like rocks are mostly red in color. Although uniform in kind, the depth of weathering may vary greatly. The residual soils immediately above the bedrock may be normally consolidated. Large foundations with heavy loads may be susceptible to large consolidation settlements on these soils.

1.19   Alluvial Deposits

Alluvial soil deposits derive from the action of streams and rivers. They can be divided into two major categories: (1) *braided-stream deposits* and (2) deposits caused by the *meandering belt of streams.*

Deposits from Braided Streams

Braided streams are high-gradient, rapidly flowing streams. They are highly erosive and contain large amounts of sediment load. Because of the high bed load, a minor change in the velocity of flow will cause deposit of sediments. By this process, these streams may build up a complex tangle of converging and diverging channels separated by sandbars and islands.

The deposits formed from braided streams are very irregular in stratification with a wide range of grain sizes. Figure 1.33 shows a cross section of such a deposit. These deposits share several characteristics:

1. The grain sizes usually range from gravel to silt. Clay-size particles are generally *not* found in these deposits.

2. Although there is a wide variation in the grain size, the soil in a given pocket or lens is rather uniform.

3. At any given depth, the void ratio and unit weight may vary over a wide range within a lateral distance of a few meters. This variation can be seen during a soil-exploration program (Chapter 2) for construction of a foundation of a structure. The standard penetration resistance (N-value; see Chapter 2) at a given depth obtained from various bore holes will be highly irregular and varies over a wide range.

Alluvial deposits can be found in several parts of the western United States, such as Southern California, Utah, and basin and range provinces of Nevada. Also, a large amount of sediment originally derived from the Rocky Mountain range was carried eastward to form the alluvial deposits of the Great Plains. On a smaller scale, this type of natural soil deposit, left by braided streams, can be encountered locally.

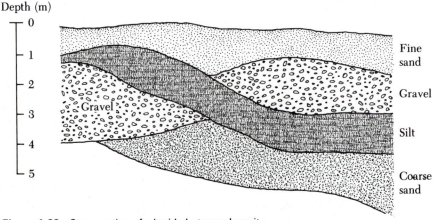

Figure 1.33   Cross section of a braided-stream deposit

Meander Belt Deposits

The term *meander* is derived from the Greek word *maiandros*, which means "bend." Mature streams in a valley curve back and forth. The valley floor in which a river meanders is referred to as the *meander belt*. In a meandering river, the soil from the bank is continually eroded from the points where it is concave in shape and deposited at points where the bank is convex in shape, as shown in Figure 1.34. These deposits are called *point bar deposits*, and they usually consist of sand and silt-size particles. Sometimes, during the process of erosion and deposition, the river abandons a meander and cuts a shorter path. The abandoned meander, when filled with water, is called an *oxbow lake* (see Figure 1.34).

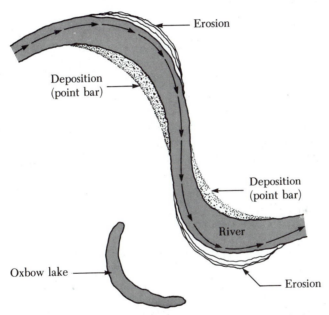

Figure 1.34 Formation of point bar deposits and oxbow lake in a meandering stream

During the flooding seasons, river water floods low-lying areas. The sand and silt-size particles carried by the river water are deposited along the banks to form ridges known as *natural levees* (Figure 1.35). Finer soil particles consisting of silts and clays are carried by the water farther into the flooded plains. These particles settle at different rates to form what is referred to as *backswamp deposits* (Figure 1.35). These clays may be highly plastic. Table 1.12 gives the properties of soil deposits found in natural levees, point bars, abandoned channels, backswamps, and swamps within the Mississippi alluvial valley.

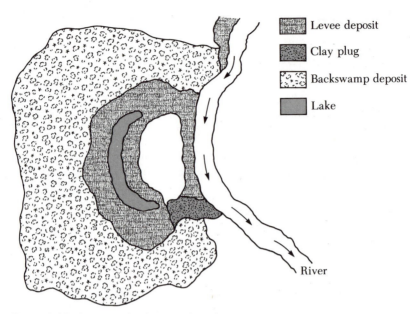

Figure 1.35 Levee and backswamp deposit

Table 1.12 Properties of Deposits within the Mississippi Alluvial Valley[a]

		Natural water content (%)	Liquid limit	Plasticity index	Shear strength	
					Cohesion[b] (kN/m^2)	Angle of friction (deg)
Environment	Soil texture					
Natural levees	Clay (CL)	25–35	35–45	15–25	17–57	0
	Silt (ML)	15–35	NP[c]–35	NP–5	9–33	10–35
Point bar	Silt (ML) and silty sand (SM)	25–45	30–55	10–25	0–41	25–35
Abandoned channel	Clay (CL, CH)	30–95	30–100	10–65	14–57	0
Backswamps	Clay (CH)	25–70	40–115	25–100	19–120	0
Swamp	Organic clay (OH)	100–265	135–300	100–165	—	—

[a] After Kolb and Shockley, 1959
[b] Rounded off
[c] NP = nonplastic

1.20   Glacial Deposits

During the great Pleistocene Ice Age, large areas of the earth were covered with glaciers. The glaciers advanced and retreated with time. During their advance, the glaciers carried large amounts of sand, silt, clay, gravel, and boulders. *Drift* is a general term usually applied to the deposits laid down by glaciers. Unstratified deposits laid down by glaciers when they melt are referred to as *till*. The physical characteristics of till may vary from glacier to glacier.

The land forms that developed from the deposits of till are called *moraines*. A *terminal moraine* (Figure 1.36) is a ridge of till that marks the maximum limit of advance of a glacier. *Recessional moraines* are ridges of till developed behind the terminal moraine at varying distances apart. They are the result of temporary stabilization of the glacier during the recessional period. The till deposited by the glacier between the moraines is referred to as *ground moraine* (Figure 1.36). Ground moraines constitute large areas in the central United States and are called *till plains*.

The sand, silt, and gravel that are carried by the melting water from the front of a glacier are called *outwash*. In a pattern similar to the braided-stream deposits, the melted water deposits the outwash, forming *outwash plains* (Figure 1.36). They are usually called *glaciofluvial deposits*.

The range of grain sizes present in a given till varies greatly. Figure 1.37 compares the grain-size distribution of *glacial-till* and *dune sand* (see Section 1.21). The amount of clay-size fractions present and the plasticity indices of tills also vary widely. During field-exploration programs (Chapter 2), erratic values of standard penetration resistances may also be expected.

Glacial water also carries with it silts and clays. The water finds its way to many basins and forms lakes. The silt particles initially tend to settle to the bottom of the lake when the water is still. During the winter, when the top of the lake freezes, the suspended clay particles gradually settle to the bottom of the lake. Again, during the summer, the snow on the lake melts.

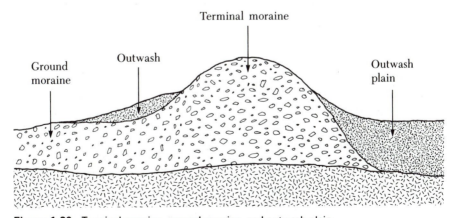

Figure 1.36 Terminal moraine, ground moraine, and outwash plain

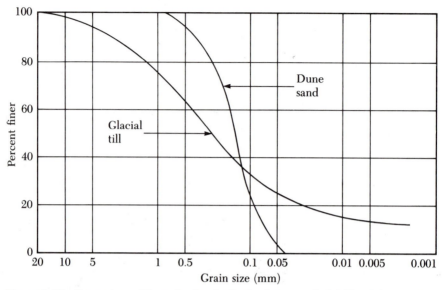

Figure 1.37 Comparison of the grain-size distribution between glacial till and dune sand

The supply of fresh water, loaded with sediments, repeats the process. As a result, the soil that is formed from such a deposit has alternate layers of silt and clay. This soil is called *varved clay*. The varves are usually a few millimeters thick; however, in some instances they can be 50–100 mm thick. Varved clays can be found in the Northeast and the Pacific Northwest of the United States. They are mostly normally consolidated and sensitive. The coefficient of permeability in the vertical direction is usually several times smaller than that in the horizontal direction. The load-bearing capacity of these deposits is quite low, and high settlement of structures with shallow foundations may be anticipated.

1.21   Aeolian Soil Deposits

Wind is also a major transporting agent in the formation of soil deposits. When large areas of sand lie exposed, wind can blow it away and redeposit it at some other place. Deposits of wind-blown sand generally take the shape of *dunes* (Figure 1.38). When dunes are formed, the sand is blown over the crest by the wind. Beyond the crest, the sand particles roll down the slope. This process tends to form a *compact sand deposit* on the *windward side* and a rather *loose deposit* on the *leeward side*. Dunes can be found along the southern and eastern shores of Lake Michigan, the Atlantic Coast, the southern coast of California, and at various places along the coasts of Oregon and Washington. Sand dunes can also be found in the alluvial and

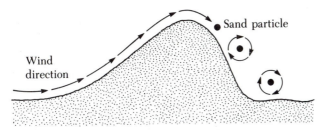

Figure 1.38   Sand dune

rocky plains of the western United States. Following are some of the typical properties of *dune sand*:

1. The grain-size distribution of the sand at any given place is surprisingly uniform. This can be attributed to the sorting action of the wind.

2. The general grain size decreases with the distance from the source because the wind carries the small particles farther than the large ones.

3. The relative density of sand deposited on the windward side of dunes may be as high as 50–65% and may decrease to about 0–15% on the leeward side.

Loess is an aeolian deposit consisting of silt and silt-size particles. The grain-size distribution of loess is rather uniform. The cohesion of loess is generally derived from a clay coating over the silt-size particles, which contributes to a stable soil structure in an unsaturated state. The cohesion may also be the result of the precipitation of chemicals leached by rain water. Loess is a *collapsing* soil, because when the soil becomes saturated, it loses its binding strength between the soil particles. Special precautions need to be taken for construction of foundations over loessial deposits (see Chapter 10). There are extensive deposits of loessial soils in the United States—mostly in the mid-western states, such as Iowa, Missouri, Illinois, and Nebraska.

1.22   Organic Soil

Organic soils are usually found in low-lying areas where the ground water table is near or above the ground surface. The presence of a high ground water table helps in the growth of aquatic plants that, when decomposed, form organic soil. This type of soil deposit is usually encountered in coastal areas and in glaciated regions. Organic soils show the following characteristics:

1. The natural moisture content may range from 200% to 300%.

2. They are highly compressible.

3. Laboratory tests have shown that, under loads, a large amount of settlement is derived from secondary consolidation.

Soils of the Filled Valleys and Great Plains Outwash Mantle

Predominantly sands and gravels with silts; sandy clays and clays

Residual Soils

All types

Recent Alluvium

Predominantly silts and clays

Nonsoil areas

Glacial Soils

Young and old drift, including associated sands and gravel

Lacustrine deposits, predominantly silts and clays

Loessial Soils

Silts and very fine sands

Soils of the Coastal Plain

Sand-clay; interbedded and mixed sands, gravels, clays, and silts; gravel and sand; or sand

Clay

Figure 1.39 Soil deposits of the United States (adapted from *Foundation Engineering*, Second Edition, by R. B. Peck, W. E. Hanson, and T. H. Thornburn. Copyright 1974 by John Wiley and Sons. Reprinted by permission.)

1.23   Some Local Terms for Soils

Soils are sometimes referred to by local terms. Following are a few of these terms with a brief description of each.

1. *Caliche:* a Spanish word derived from the Latin word *calix*, meaning *lime*. It is mostly found in the desert southwest of the United States. It is a mixture of sand, silt, and gravel bonded together by *calcareous deposits*. The calcareous deposits are brought to the surface by a net upward migration of water. The water evaporates in the high local temperature. Because of the sparse rainfall, the carbonates are not washed out of the top layer of soil.

2. *Gumbo:* a highly plastic, clayey soil.

3. *Adobe:* a highly plastic, clayey soil found in the southwestern United States.

4. *Terra Rossa:* residual soil deposits that are red in color and derive from limestone and dolomite.

5. *Muck:* organic soil with a very high moisture content.

6. *Muskeg:* organic soil deposit.

7. *Saprolite:* residual soil deposit derived from mostly insoluble rock.

8. *Loam:* a mixture of soil grains of various sizes, such as sand, silt, and clay.

9. *Laterite:* characterized by the accumulation of iron oxide (Fe_2O_3) and aluminum oxide (Al_2O_3) near the surface, and the leaching of silica. Lateritic soils in Central America contain about 80–90% of clay and silt-size particles. In the United States, lateritic soils can be found in the southeastern states, such as Alabama, Georgia, and the Carolinas.

Figure 1.39 shows the general nature of the various soil deposits encountered in the United States.

PROBLEMS

1.1 The results of a sieve analysis of a sand show that $D_{85} = 1.18$ mm, $D_{60} = 0.76$ mm, $D_{30} = 0.29$ mm, and $D_{10} = 0.14$ mm. Determine the uniformity coefficient and the coefficient of gradation of the soil.

1.2 For a soil, given: void ratio = 0.81, moisture content = 21%, and $G_s = 2.68$. Calculate the following:
a. Porosity
b. Degree of saturation
c. Moist unit weight in kN/m^3
d. Dry unit weight in kN/m^3

1.3 The moist unit weight of a soil is 119.5 lb/ft^3. If the moisture content is 12% and $G_s = 2.65$, calculate the following:
a. e
b. n

    c. S

    d. γ_d

1.4 For the soil described in Problem 1.3:

    a. What would be the saturated unit weight in lb/ft^3?

    b. How much water, in lb/ft^3, needs to be added to the soil for complete saturation?

    c. What would be the moist unit weight in lb/ft^3 when the degree of saturation is 80%?

1.5 The saturated unit weight of a soil is $19.2 \ kN/m^3$ at a moisture content of 23%. Determine (a) dry unit weight and (b) specific gravity of soil solids, G_s.

1.6 A soil specimen has a volume of $0.05 \ m^3$ and a mass of 87.5 kg. Given: $w = 15\%$, $G_s = 2.68$. Determine (a) void ratio, (b) porosity, (c) dry unit weight, (d) moist unit weight, and (e) degree of saturation.

1.7 For a granular soil, given: $\gamma = 108 \ lb/ft^3$, $D_r = 82\%$, $w = 8\%$, and $G_s = 2.65$. For this soil, if $e_{min} = 0.44$, what would be e_{max}? What would be the dry unit weight in the loosest state?

1.8 The laboratory test results of six soils are given in the following table. Classify the soils by the AASHTO Soil Classification System, and give the group indices.

Sieve Analysis—Percent Passing

			Soil			
Sieve no.	A	B	C	D	E	F
4	100	100	95	95	100	100
10	95	80	80	90	94	94
40	82	61	54	79	76	86
200	65	55	8	64	33	76
Liquid limit	42	38	NP[a]	35	38	52
Plastic limit	26	25	NP	26	25	28

[a] NP = nonplastic

1.9 Classify the soils given in Problem 1.8 by the Unified Soil Classification System and determine the group symbols and group names.

1.10 The permeability of a sand was tested in the laboratory at a void ratio of 0.63 and was determined to be 0.22 cm/sec. Estimate the coefficient of permeability of this sand at a void ratio of 0.43 by using Eqs. (1.25), (1.26), and (1.27).

1.11 A sand has a coefficient of permeability of 0.25 cm/sec at a void ratio of 0.7. Estimate the void ratio at which its coefficient of permeability would be 0.115 cm/sec. Use Eq. (1.27).

1.12 The *in situ* coefficient of permeability of a clay is 4.2×10^{-6} cm/sec at a void ratio of 0.92. What would be its coefficient of permeability at a void ratio of 0.72? Use Eq. (1.29).

1.13 A soil profile is shown in Figure P1.13. Given: $H_1 = 8$ ft, $H_2 = 4$ ft, $H_3 = 15$ ft. Determine the total stress, pore water pressure, and effective stress at A, B, C, and D.

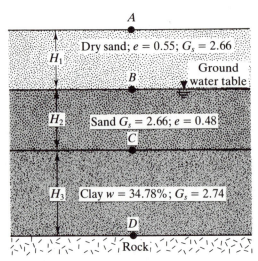

Figure P1.13

1.14 Repeat Problem 1.13 with the following: $H_1 = 4$ m, $H_2 = 2$ m, $H_3 = 6$ m.

1.15 A sandy soil ($G_s = 2.65$), in its densest and loosest states, has void ratios of 0.42 and 0.85, respectively. Estimate the range of the critical hydraulic gradient in this soil at which a quick condition might occur.

1.16 A saturated clay deposit in the field has the following: $G_s = 2.71$, moisture content = 32%, liquid limit = 55, average effective overburden pressure = 80 kN/m². Estimate the preconsolidation pressure, p_c. Use Eq. (1.52).

1.17 A normally consolidated clay layer 12 ft thick has a void ratio of 0.92. Given: $LL = 35$ and average effective stress on the clay layer = 2100 lb/ft². How much consolidation settlement would the clay undergo if the average effective stress on the clay layer is increased to 3200 lb/ft² as the result of the construction of a foundation? Use Eq. (1.54) to estimate C_c.

1.18 Refer to Problem 1.17. Assume the clay layer to be preconsolidated. Given: $p_c = 2900$ lb/ft². Assume that $C_s = \frac{1}{5}C_c$. Estimate the consolidation settlement.

1.19 Refer to the soil profile described in Problem 1.14. The clay is normally consolidated. A laboratory consolidation test on the clay gave the following results:

Pressure (kN/m²)	Void ratio
150	0.909
300	0.795

a. Calculate the average effective stress on the clay layer.
b. Determine the compression index, C_c.
c. If the average effective stress on the clay layer is increased $(p_o + \Delta p)$ to 220 kN/m², what would be the total consolidation settlement?

1.20 Refer to Problem 1.19c. For the clay soil, given $C_v = 5.6$ mm²/min, how long will it take to reach half the consolidation settlement?

1.21 For a normally consolidated soil, the following is given:

Pressure (kN/m^2)	Void ratio
120	0.82
360	0.64

Determine the following:

a. The compression index, C_c.

b. The void ratio corresponding to pressure of 200 kN/m^2.

1.22 A clay soil specimen, 1.5 in. thick (drained on top only) was tested in the laboratory. For a given load increment, the time for 60% consolidation was 8 min 10 sec. How long will it take for 50% consolidation of a similar clay layer in the field that is 10 ft thick and drained on both sides.

1.23 A direct shear test was conducted on dry sand. The results were as follows:

Area of the specimen = 2 in. × 2 in.

Normal force (lb)	Shear force at failure (lb)
50	43.5
110	95.5
150	132.0

Graph the shear stress at failure against normal stress and determine the soil friction angle.

1.24 A consolidated-drained triaxial test on a sand yielded the following results:

All-around confining pressure $= \sigma_3 = 30$ lb/in^2

Added axial stress at failure $= \Delta\sigma = 96$ lb/in.^{2}

Determine the shear stress parameters.

1.25 A consolidated-drained triaxial test on a normally consolidated clay yielded a friction angle, ϕ, of 28°. If the all-around confining pressure during the test was 140 kN/m^2, what was the major principal stress at failure?

1.26 Following are the results of two consolidated-drained triaxial tests on a clay:

Test I: $\sigma_3 = 82$ kN/m^2; $\sigma_{1(\text{failure})} = 329.2$ kN/m^2

Test II: $\sigma_3 = 165.6$ kN/m^2; $\sigma_{1(\text{failure})} = 558.6$ kN/m^2

Determine the shear strength parameters—that is, c and ϕ.

1.27 Repeat Problem 1.26 with the following:

Test I: $\sigma_3 = 140$ kN/m^2; $\sigma_{1(\text{failure})} = 368.16$ kN/m^2

Test II: $\sigma_3 = 280$ kN/m^2; $\sigma_{1(\text{failure})} = 700.86$ kN/m^2

1.28 A consolidated-undrained triaxial test on a saturated normally consolidated clay was performed with the following results:

$\sigma_{1(\text{failure})} = 28$ lb/in.^{2}

$\sigma_3 = 12$ lb/in.^{2}

Pore water pressure at failure $= u_f = 4.5$ lb/in.^{2}

Determine ϕ_u and ϕ.

1.29 A normally consolidated clay soil has $\phi = 28°$ and $\phi_u = 20°$. If a consolidated-undrained test is conducted on this clay with $\sigma_3 = 150$ kN/m^2, what would be the magnitude of the major principal stress, σ_1, and the pore water pressure, u, at failure?

1.30 A saturated clay layer has the following:

$$\text{Saturated unit weight, } \gamma_{\text{sat}} = 19.6 \text{ kN/m}^3$$
$$\text{Plasticity index} = 21$$

The water table coincides with the ground surface. If the clay is normally consolidated, estimate the magnitude of c_u (kN/m^2) at a depth of 8 m from the ground surface.

1.31 For the clay layer stated in Problem 1.30, assume that it is overconsolidated and that the $OCR = 2.5$. Estimate the magnitude of c_u (kN/m^2) at a depth of 6.5 m below the ground surface.

REFERENCES

American Society for Testing and Materials (1986). *Annual Book of ASTM Standards*, Vol. 04.08, Philadelphia, Pa.

Azzouz, A. S., Krizek, R. J., and Corotis, R. B. (1976). "Regression Analysis of Soil Compressibility," *Soils and Foundations*, Vol. 16, No. 2, pp. 19–29.

Casagrande, A. (1936). "Determination of the Preconsolidation Load and Its Practical Significance," *Proceedings*, First International Conference on Soil Mechanics and Foundation Engineering, Cambridge, Mass., Vol. 3, pp. 60–64.

Darcy, H. (1856). *Les Fontaines Publiques de la Ville de Dijon*, Paris.

Das, B. M. (1989). *Soil Mechanics Laboratory Manual*, 3rd ed., Engineering Press, San Jose, Calif.

Hazen, A. (1930). "Water Supply," *American Civil Engineers Handbook*, Wiley, New York.

Highway Research Board (1945). *Report of the Committee on Classification of Materials for Subgrades and Granular Type Roads*, Vol. 25, pp. 375–388.

Kenney, T. C. (1959). "Discussion," *Journal of the Soil Mechanics and Foundations Division*, American Society of Civil Engineers, Vol. 85, No. SM3, pp. 67–69.

Kolb, C. R., and Shockley, W. B. (1959). "Mississippi Valley Geology: Its Engineering Significance," *Proceedings*, American Society of Civil Engineers, Vol. 124, pp. 633–656.

Ladd, C. C., and Foot, R. (1974). "New Design Procedure for Stability of Soft Clays," *Journal of the Geotechnical Engineering Division*, American Society of Civil Engineers, Vol. 92, No. GT2, pp. 79–103.

Nagaraj, T. S., and Murthy, B. R. S. (1985). "Prediction of the Preconsolidation Pressure and Recompression Index of Soils," *Geotechnical Testing Journal*, American Society for Testing and Materials, Vol. 8, No. 4, pp. 199–202.

Peck, R. B., Hanson, W. E., and Thornburn, T. H. (1974). *Foundation Engineering*, 2nd ed., Wiley, New York.

Rendon-Herrero, O. (1980). "Universal Compression Index Equation," *Journal of the Geotechnical Engineering Division*, American Society of Civil Engineers, Vol. 106, No. GT11, pp. 1178–1200.

Schmertmann, J. H. (1953). "Undisturbed Consolidation Behavior of Clay," *Transactions*, American Society of Civil Engineers, Vol. 120, p. 1201.

Sivaram, B., and Swamee, P. (1977). "A Computational Method for Consolidation Coefficient," *Soils and Foundations*, Tokyo, Japan, Vol. 17, No. 2, pp. 48–52.

Skempton, A. W. (1944). "Notes on the Compressibility of Clays," *Quarterly Journal of Geological Society*, London, Vol. C, pp. 119–135.

Skempton, A. W. (1957). "The Planning and Design of New Hong Kong Airport," *Proceedings*, The Institute of Civil Engineers, London, Vol. 7, pp. 305–307.

Tavenas, F., Jean, P., Leblond, P., and Leroueil, S. (1983). "The Permeability of Natural Soft Clays. Part II: Permeability Characteristics," *Canadian Geotechnical Journal*, Vol. 20, No. 4, pp. 645–660.

Terzaghi, K., and Peck, R. B. (1967). *Soil Mechanics in Engineering Practice*, Wiley, New York.

U.S. Department of the Navy (1971). "Design Manual—Soil Mechanics, Foundations and Earth Structures," *NAVFAC DM-7*, U.S. Government Printing Office, Washington, D.C.

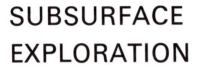

Chapter 2

SUBSURFACE EXPLORATION

2.1   Introduction

To design a foundation that will support adequate structural load, one must understand the nature of the soil(s) that will support the foundation. The process of determining the layers of deposits that will underlie a proposed structure and their physical properties is generally referred to as *subsurface exploration*. The purpose of a soil-exploration program is to obtain information that will aid the geotechnical engineer in the following:

1. Selection of the type and the depth of foundation suitable for a given structure

2. Evaluation of the load-bearing capacity of the foundation

3. Estimation of the probable settlement of a structure

4. Determination of potential foundation problems (for example, expansive soil, collapsible soil, sanitary landfill, and so on)

5. Determination of the location of ground water table

6. Prediction of lateral earth pressure for structures like retaining walls, sheet pile bulkheads, and braced cuts

7. Establishment of construction methods for changing subsoil conditions

Subsurface exploration may also be necessary when additions and alterations to existing structures are contemplated.

Foundation engineers should always remember that the soil at a given site is frequently nonhomogeneous; that is, the formation of the soil profile may be variable. Soil-mechanics theories concern idealized conditions. The application of these theories to foundation-engineering problems involves judicious evaluation of the information on site conditions and soil parameters obtained from field exploration programs. Good professional judgment constitutes a major part of geotechnical engineering, and it comes with practice.

76

2.2   **Subsurface Exploration Program**

Subsurface exploration comprises several steps, such as collection of preliminary information, reconnaissance, and site investigation. This section briefly covers each of these topics.

Collection of Preliminary Information

This step includes obtaining information regarding the type of structure to be built and its general use. For the construction of buildings, the approximate column loads and their spacings, and the local building-code and basement requirements should be known. In the construction of bridges, one should determine the span length as well as the loading on piers and abutments.

A general idea of the topography and the type of soil to be encountered near and around the proposed site can be obtained from the following sources:

1. United States Geological Survey maps.
2. State government geological survey maps.
3. United States Department of Agriculture's Soil Conservation Service county soil reports.
4. Agronomy maps published by the agriculture departments of various states.
5. The hydrological information published by the United States Corps of Engineers. These include the records of stream flow, high flood levels, tidal records, and so on.
6. Highway department soils manuals published by several states.

The information collected from these sources can be extremely helpful in the proper planning of site investigation. In some cases, substantial savings may be realized by anticipating problems that may be encountered later in the exploration program.

Reconnaissance

The engineer should always make a trip to the site for a visual inspection. The inspection is intended to obtain the following information:

1. The general topography of the site, possible existence of drainage ditches, abandoned dumps of debris, or other materials. Also, evidence of creep of slopes and deep, wide shrinkage cracks at regularly spaced intervals may be indicative of expansive soils.
2. Soil stratification from deep cuts, such as those made for construction of nearby highways and railroads.
3. Type of vegetation at the site. In several instances, the vegetation indicates the nature of the soil. For example, a mesquite cover in central

Texas can be an indication of the existence of expansive clays that can cause possible foundation problems.

4. High-water marks on nearby buildings and bridge abutments.

5. Ground water levels—these can be determined from checking nearby wells.

6. Types of construction nearby and existence of any cracks in walls or other problems.

The nature of stratification and physical properties of the soil nearby can also be obtained from any available soil-exploration reports for existing structures.

Site Investigation

This phase of the exploration program consists of planning, making some test boreholes, and collecting soil samples at desired intervals of the boreholes for subsequent observation and laboratory tests. The approximate required minimum depth of the borings should be predetermined. The estimated depths can be changed during the drilling operation, depending on the subsoil encountered. To determine the approximate minimum depth of boring, engineers may use the following rule (American Society of Civil Engineers, 1972):

1. Determine the net increase of stress, $\Delta\sigma$, under a foundation with depth as shown in Figure 2.1. (The general equations for estimation of stress increase are given in Chapter 3.)

2. Estimate the variation of the vertical effective stress, σ_v', with depth.

3. Determine the depth, $D = D_1$, at which the stress increase $\Delta\sigma$ is equal to $(1/10)q$ (q = estimated net stress on the foundation).

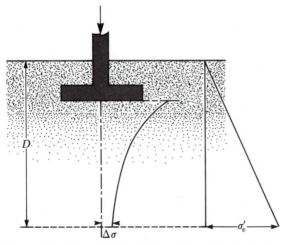

Figure 2.1   Determination of the minimum depth of boring

4. Determine the depth, $D = D_2$, at which $\Delta\sigma/\sigma_v' = 0.05$.

5. Unless bedrock is encountered, the smaller of the two depths, D_1 and D_2, just determined is the approximate minimum depth of boring required.

Table 2.1 shows the minimum depths of borings for buildings based on the preceding rule.

Table 2.1 Depth of Boring[a]

Number of stories	Boring depth		
	For building width = 30.5 m (100 ft)	For building width = 61 m (200 ft)	For building width = 122 m (400 ft)
1	3.5 m (11 ft)	3.7 m (12 ft)	3.7 m (12 ft)
2	6.1 m (20 ft)	6.7 m (22 ft)	7.0 m (23 ft)
3	10.1 m (33 ft)	12.5 m (41 ft)	13.7 m (45 ft)
4	16.2 m (53 ft)	20.7 m (68 ft)	24.7 m (81 ft)
5	24.1 m (79 ft)	32.9 m (108 ft)	41.5 m (136 ft)

[a] From *Introductory Soil Mechanics and Foundations*, Third Edition; by George B. Sowers and George F. Sowers, copyright © 1970 by Macmillan Pub. Co., Inc.

For hospitals and office buildings, Sowers and Sowers (1970) also use the following rule to determine boring depth:

$$D_b = 3S^{0.7} \quad \text{(for light steel or narrow concrete buildings)}$$

and

$$D_b = 6S^{0.7} \quad \text{(for heavy steel or wide concrete buildings)}$$

(2.1)

where D_b = depth of boring, in meters
S = number of stories

In English units, the preceding equations are of the form

$$D_b \text{ (ft)} = 10S^{0.7} \quad \text{(for light steel or narrow concrete buildings)}$$

and

$$D_b \text{ (ft)} = 20S^{0.7} \quad \text{(for heavy steel or wide concrete buildings)}$$

(2.2)

When deep excavations are anticipated, the depth of boring should be at least 1.5 times the depth of excavation.

Sometimes subsoil conditions are such that the foundation load may have to be transmitted to the bedrock. The minimum depth of core boring into the bedrock is about 3 m (10 ft). If the bedrock is irregular or weathered, the core borings may have to be extended to greater depths.

There are no hard and fast rules for the spacing of the boreholes. Table 2.2 gives some general guidelines for borehole spacing. These spacings can be increased or decreased, depending on the subsoil condition. If various soil strata are more or less uniform and predictable, the number of boreholes can be reduced.

Table 2.2 Approximate Spacing of Boreholes

	Spacing	
Type of project	(m)	(ft)
Multistory building	10–30	30–100
One story industrial plants	20–60	60–200
Highways	250–500	800–1600
Residential subdivision	250–500	800–1600
Dams and dikes	40–80	130–260

The ultimate cost of the structure should also be taken into account while making decisions regarding the extent of the field exploration. The exploration cost generally should be in the range of 0.1–0.5% of the cost of the structure.

Soil borings can be done by several methods, such as *auger boring, wash boring, percussion drilling,* and *rotary drilling.* The following section discusses these methods of advancing the test holes in the field and collecting soil samples.

2.3   Exploratory Borings in the Field

Auger boring is the simplest method of making exploratory boreholes. Figure 2.2 shows two types of hand auger—the *post hole auger* and the *helical auger.* Hand augers cannot be used for advancing holes to depths exceeding 3–5 m (10–16 ft). However, they can be used for soil exploration work for some highways and small structures. *Portable power-driven helical augers* (76.2 mm to 304.8 mm in diameter) are available for making boreholes to greater depths. The soil samples that are obtained from such boring operations are highly disturbed. In several noncohesive soils or soils having low cohesion, the walls of the boreholes will not stand unsupported. In such circumstances, a metal pipe is used as a *casing* to prevent the soil from caving in.

When power is available, *continuous-flight augers* are probably the most common method used for advancing a borehole. The power for drilling is delivered by truck- or tractor-mounted drilling rigs. Boreholes up to about 60–70 m (200–230 ft) can be easily made by this method. Continuous-flight

(a)   (b)

Figure 2.2   Hand tools: (a) post hole auger; (b) helical auger

augers are available in sections of about 1–2 m (3–6 ft) with the stem either
solid or hollow. Some of the commonly used solid stem augers have outside
diameters of 66.68 mm ($2\frac{5}{8}$ in.), 82.55 mm ($3\frac{1}{4}$ in.), 101.6 mm (4 in.), and
114.3 mm ($4\frac{1}{2}$ in.). Common hollow stem augers commercially available
have dimensions of 63.5 mm ID and 158.75 mm OD (2.5 in. × 6.25 in.),
69.85 mm ID and 177.8 OD (2.75 in. × 7 in.), 76.2 mm ID and 203.2 OD (3
in. × 8 in.), and 82.55 mm ID and 228.6 mm OD (3.25 in. × 9 in.).

     In making the boreholes, the tip of the auger is attached to a cutter head
(Figure 2.3). During the drilling operation (Figure 2.4), section after section
of auger can be added and the hole extended downwards. The flights of the
augers bring the loose soil from the bottom of the hole to the surface.
During the drilling, the driller can detect changes in soil type by noting
changes in the speed and sound of drilling. When solid stem augers are
used, it is necessary to withdraw the auger at regular intervals to obtain soil
samples and also to conduct other operations such as standard penetration
tests. Hollow stem augers have a distinct advantage over solid stem augers in
that they do not have to be removed at frequent intervals for sampling or
other tests. As shown schematically in Figure 2.5, the outside of the hollow
stem auger acts as a casing. A removable plug is attached to the bottom of
the auger by means of a center rod. During the drilling, the plug can be
pulled out with the auger in place, and soil sampling and standard penetration tests can be performed. When hollow stem augers are used in sandy
soils below the ground water table, the sand may be pushed several feet into

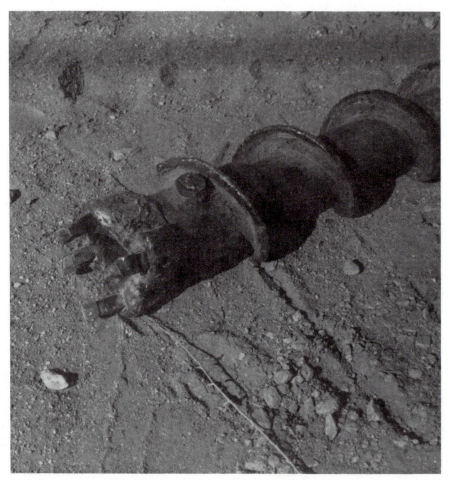

Figure 2.3 Carbide-tipped cutting head on auger flight attached with bolt (courtesy of William B. Ellis, El Paso Engineering and Testing, Inc., El Paso, Texas)

the stem of the auger by excess hydrostatic pressure immediately after the removal of the plug. In such conditions, the plug should not be used and, instead, water inside the hollow stem should be maintained at a higher level than the ground water table.

Wash boring is another method of advancing boreholes. In this method, a casing about 2–3 m (6–10 ft) long is driven into the ground. The soil inside the casing is then removed by means of a chopping bit that is attached to a drilling rod. Water is forced through the drilling rod, and it goes out at a very high velocity through the holes located at the bottom of the chopping bit (Figure 2.6). The water and the chopped soil particles rise upward in the drill hole and overflow at the top of the casing through a T-connection. The wash water is collected in a container. The casing can be extended with

Figure 2.4 Drilling with continuous-flight augers (courtesy of Danny R. Anderson, Danny R. Anderson Consultants, El Paso, Texas)

additional pieces as the borehole progresses; however, this is not required if the borehole will stay open and not cave in.

Rotary drilling is a procedure by which rapidly rotating drilling bits attached to the bottom of drilling rods cut and grind the soil and advance the borehole down. There are several types of drilling bit. Rotary drilling can be used in sand, clay, and rocks (unless badly fissured). Water, or *drilling mud*, is forced down the drilling rods to the bits, and the return flow forces the cuttings to the surface. Boreholes with diameters ranging from 50.8 mm to 203.2 mm (2 in. to 8 in.) can be easily made by this technique.

Center rod

Hollow-stem auger

Removable plug

Figure 2.5 Schematic diagram of the hollow stem auger with removable plug

The drilling mud is a slurry of water and bentonite. This is generally used when the soil encountered is such that boreholes are likely to cave in. When soil samples are needed, the drilling rod is raised and the drilling bit is replaced by a sampler.

Percussion drilling is an alternate method to advance a borehole, particularly through hard soil and rock. In this technique, a heavy drilling bit is raised and lowered to chop the hard soil. The chopped soil particles are brought up by circulation of water. Percussion drilling may require casing.

2.4   Procedures for Sampling Soil

Two types of soil sample can be obtained during subsurface exploration: *disturbed* and *undisturbed*. The disturbed but representative samples can generally be used for the following types of laboratory test:

1. Grain-size analysis
2. Determination of liquid and plastic limits
3. Specific gravity of soil solids
4. Organic content determination
5. Classification of soil

The disturbed soil samples cannot, however, be used for consolidation or shear strength tests. Undisturbed soil samples must be obtained for these

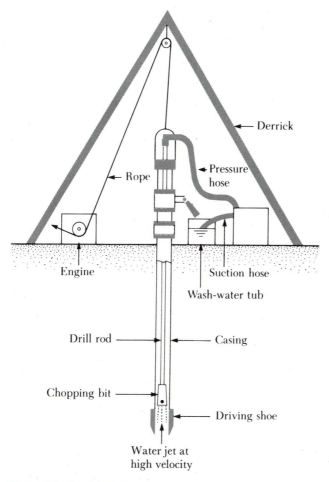

Figure 2.6 Wash boring

types of laboratory tests. The next sections discuss various types of soil sampling procedures.

Split-Spoon Sampling

Split-spoon samplers can be used in the field to obtain soil samples that are generally disturbed but still representative. A section of a *standard split-spoon sampler* is shown in Figure 2.7a. It consists of a tool-steel driving shoe, a steel tube that is split longitudinally in half, and a coupling at the top. The coupling connects the sampler to the drill rod. The standard split tube has an inside diameter of 34.93 mm ($1\frac{3}{8}$ in.) and an outside diameter of 50.8 mm (2 in.); however, samplers having inside and outside diameters up to 63.5 mm ($2\frac{1}{2}$ in.) and 76.2 mm (3 in.), respectively, are also available. When a borehole is extended to a given depth, the drill tools are removed and the

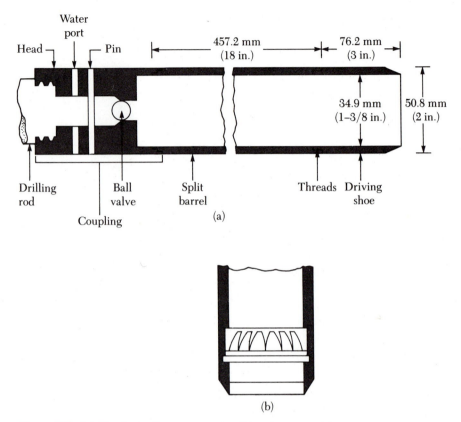

Figure 2.7   (a) Standard split-spoon sampler; (b) spring core catcher

sampler is lowered to the bottom of the borehole. The sampler is driven into the soil by hammer blows given at the top of the drill rod. The standard weight of the hammer is 622.72 N (140 lb), and, for each blow, the hammer drops a distance of 0.762 m (30 in.). The number of blows required for spoon penetration of three 152.4-mm (6-in.) intervals are recorded. The number of blows required for the last two intervals are added to give the *standard penetration number* at that depth. This is generally referred to as the *N-value* (American Society for Testing and Materials, 1986, Designation D-1586-67). The sampler is then withdrawn, and the shoe and the coupling are removed. The soil sample that is recovered from the tube is then placed in a glass bottle and transported to the laboratory.

The degree of disturbance for a soil sample is usually expressed by an equation

$$A_R(\%) = \frac{D_o^2 - D_i^2}{D_i^2} (100)$$

(2.3)

where A_R = area ratio

$\quad\quad D_o$ = outside diameter of the sampling tube

$\quad\quad D_i$ = inside diameter of the sampling tube

When the area ratio is 10% or less, the sample is generally considered to be undisturbed. For a standard split-spoon sampler

$$A_R(\%) = \frac{(50.8)^2 - (34.93)^2}{(34.93)^2}(100) = 111.5\%$$

Hence, these samples are highly disturbed. Split-spoon samples are generally taken at intervals of about 1.53 m (5 ft).

    When the material encountered on the field is sand (particularly fine sand below the water table), sample recovery by a split-spoon sampler may be difficult. In that case, a device such as a *spring core catcher* may have to be placed inside the split spoon (Figure 2.7b).

    Besides obtaining soil samples, standard penetration tests provide several useful correlations. For example, the consistency of clayey soils can often be estimated from the standard penetration number (N). This is shown in Table 2.3.

Table 2.3 Consistency of Clays and Approximate Correlation to the Standard Penetration Number, N

Standard penetration number, N	Consistency	Unconfined compression strength, q_u (kN/m^2)
0–2	Very soft	0–25
2–5	Soft	25–50
5–10	Medium stiff	50–100
10–20	Stiff	100–200
20–30	Very stiff	200–400
>30	Hard	>400

    In granular soils, the N-value is affected by the effective overburden pressure, σ_v'. For that reason, the N-value obtained from field exploration under different effective overburden pressures should be changed to correspond to a standard value of σ_v'. This means that

$$\boxed{N_{\text{cor}} = C_N N_F} \tag{2.4}$$

where N_{cor} = corrected N-value to a standard value of σ_v' [95.6 kN/m^2 (1 ton/ft^2)]

$\quad\quad C_N$ = correction factor

$\quad\quad N_F$ = N-value obtained from the field

Several authors, such as Bazaraa (1967); Peck and Bazaraa (1969); Peck, Hanson, and Thornburn (1974); Tang (1962); Seed (1976, 1979); and Tokimatsu and Yoshimi (1983), have proposed empirical relationships for C_N. However, the simplest relationship is that proposed by Liao and Whitman (1986), which has given as good a result as any other. This relationship can be given as

$$C_N = \sqrt{\frac{1}{\sigma_v'}} \qquad (2.5)$$

where σ_v' is in (U.S.) tons/ft^2. In SI units

$$C_N = 9.78\sqrt{\frac{1}{\sigma_v' \ (\text{kN/m}^2)}} \qquad (2.6)$$

In Eqs. (2.5) and (2.6), the standard value of σ_v' is 95.6 kN/m^2 (1 ton/ft^2). Figures 2.8 and 2.9 show the plot of C_N against σ_v'.

An approximate relation between the standard penetration number and the relative density of sand is given in Table 2.4.

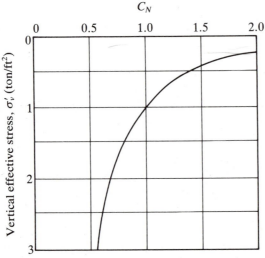

Figure 2.8   Variation of C_N with σ_v' [Eq. (2.5)]

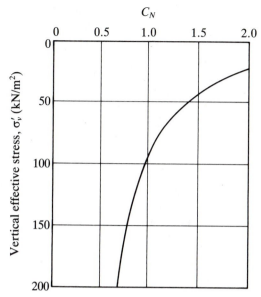

Figure 2.9   Variation of C_N with σ'_v [Eq. (2.6)]

Table 2.4   Relation between N-values, Relative Density, and Angle of Friction in Sands

Standard penetration number, N	Approximate relative density, D_r (%)	Approximate angle of friction of soil, ϕ (deg)
0–5	0–5	26–30
5–10	5–30	28–35
10–30	30–60	35–42
30–50	60–95	38–46

Note: See also Figure 9.9

When the standard penetration resistance values are used in the preceding correlations (Tables 2.3 and 2.4) for estimation of soil parameters, the following should be noted:

1. The equations are approximate.

2. Because the soil is not homogeneous, there may be a wide variation of the N-values obtained from a given borehole.

3. In soil deposits that contain large boulders and gravels, standard penetration numbers may be erratic and unreliable.

Although approximate, with correct interpretation the standard penetration test provides a good evaluation of soil properties.

Scraper Bucket

When soil deposits are sand mixed with pebbles, it may not be possible to obtain samples by a split spoon with a spring core catcher, because the pebbles may prevent the springs from closing. In such cases, a scraper bucket may be used to obtain disturbed representative samples from the field (Figure 2.10a). The scraper bucket has a driving point and can be attached to a drilling rod. The sampler is driven down into the soil and rotated, and the scrapings from the side fall into the bucket.

Thin Wall Tube

Thin wall tubes are sometimes referred to as *Shelby tubes*. They are made out of seamless steel and are commonly used to obtain undisturbed clayey soils. The commonly used thin wall tube samplers have outside diameters of 50.8 mm (2 in.) and 76.2 mm (3 in.). The bottom end of the tube is sharpened. The tubes can be attached to drilling rods (Figure 2.10b). The drilling rod with the sampler attached to it is lowered to the bottom of the borehole and the sampler is pushed into the soil. The soil sample inside the tube is then pulled out. The two ends of the sampler are sealed, and it is sent back to the laboratory for testing.

(a)

(b)

Figure 2.10 Sampling devices: (a) scraper bucket; (b) thin wall tube; (c) and (d) piston sampler

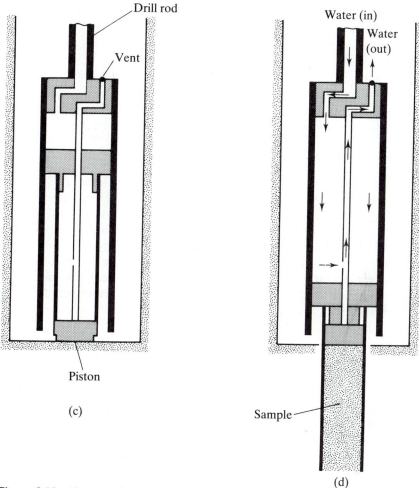

Figure 2.10   (Continued)

    Samples obtained in this manner may be used for consolidation or shear tests. A thin wall tube with 50.8 mm (2 in.) diameter has an inside diameter of about 47.63 mm ($1\frac{7}{8}$ in.). This gives an area ratio

$$A_R(\%) = \frac{D_o^2 - D_i^2}{D_o^2}\,(100) = \frac{(50.8)^2 - (47.63)^2}{(47.63)^2}\,(100) = 13.75\%$$

Note that by increasing the diameters of samples, one increases the cost of obtaining them.

Piston Sampler

When undisturbed soil samples become larger than 76.2 mm (3 in.) in diameter, they tend to fall out of the sampler. Piston samplers are particularly useful in such conditions. There are several types of piston sampler; however, the sampler proposed by Osterberg (1952) is the most advantageous (see Figure 2.10c and d). It consists of a thin wall tube with a piston. Initially, the piston closes the end of the thin wall tube. The sampler is lowered to the bottom of the borehole (Figure 2.10c), then the thin wall tube is pushed into the soil hydraulically, past the piston. After that, the pressure is released through a hole in the piston rod (Figure 2.10d). To a large extent, the presence of the piston prevents distortion in the sample by not letting the soil squeeze into the sampling tube very fast and by not admitting excess soil. Samples obtained in this manner are consequently less disturbed than those obtained by Shelby tubes.

2.5   **Observation of Ground Water Tables**

The presence of a ground water table near a foundation has a large effect on the foundation's load-bearing capacity and settlement, among other things. The ground water level will also change seasonally. In many cases, it may become necessary to establish the highest and lowest possible levels of ground water during the life of a project.

   If ground water is encountered in a borehole during a field exploration, it should be recorded. In soils with high coefficients of permeability, the level of ground water in a borehole will stabilize after about 24 hours of completion of the boring. The depth of the water table can then be recorded by lowering a chain or tape into the borehole.

   In highly impermeable layers, the ground water level in a borehole may not stabilize for several weeks. In such cases, if accurate ground water level measurements are required, a *piezometer* can be used. A piezometer basically consists of a porous stone or a perforated pipe with a plastic standpipe attached to it. The general nature of the placement of a piezometer in a borehole is shown in Figure 2.11.

   For silty soils, Hvorslev (1949) proposed a technique to determine the ground water level (see Figure 2.12). This technique uses the following steps:

   1. Bail out water in the borehole to a level below the estimated ground water table.
   2. Observe the water levels in the borehole at times

$$t = 0$$
$$t = t_1$$
$$t = t_2$$
$$t = t_3$$

Note that $t_1 - 0 = t_1 - t_2 = t_2 - t_3 = \Delta t$.
   3. Calculate Δh_1, Δh_2, and Δh_3 (see Figure 2.12).

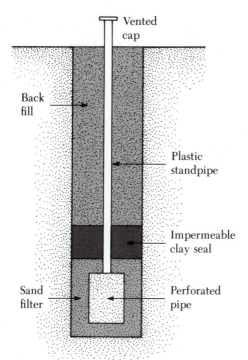

Figure 2.11   Casagrande-type porous stone piezometer

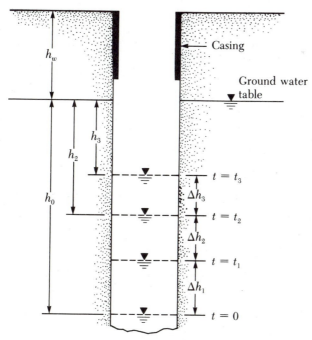

Figure 2.12   Determination of ground water table—Eq. (2.7)

4. Calculate

$$h_0 = \frac{\Delta h_1^2}{\Delta h_1 - \Delta h_2} \tag{2.7a}$$

$$h_2 = \frac{\Delta h_2^2}{\Delta h_1 - \Delta h_2} \tag{2.7b}$$

$$h_3 = \frac{\Delta h_3^2}{\Delta h_2 - \Delta h_3} \tag{2.7c}$$

5. Plot the values h_0, h_2, and h_3 above the water levels observed at times $t = 0$, t_2, and t_3, respectively, to determine the final water level in the borehole.

Example 2.1

Refer to Figure 2.12. For a borehole, given: $h_w + h_o = 9.5$ m

$$\Delta t = 24 \text{ hrs}$$
$$\Delta h_1 = 0.9 \text{ m}$$
$$\Delta h_2 = 0.70 \text{ m}$$
$$\Delta h_3 = 0.54 \text{ m}$$

Make the necessary calculations and locate the ground water level.

Solution

Using Eq. (2.7),

$$h_0 = \frac{\Delta h_1^2}{\Delta h_1 - \Delta h_2} = \frac{0.9^2}{0.9 - 0.70} = 4.05 \text{ m}$$

$$h_2 = \frac{\Delta h_2^2}{\Delta h_1 - \Delta h_2} = \frac{0.7^2}{0.9 - 0.7} = 2.45 \text{ m}$$

$$h_3 = \frac{\Delta h_3^2}{\Delta h_2 - \Delta h_3} = \frac{0.54^2}{0.7 - 0.54} = 1.82 \text{ m}$$

Figure 2.13 shows a plot of the preceding calculations as well as the estimated ground water table. From this figure it can be seen that $h_w = 5.5$ m.

2.6   Vane Shear Test

Vane shear is a type of test (ASTM D-2573) that may be used during the drilling operation to determine the *in situ* undrained shear strength (c_u) of clay soils—particularly soft clays. The vane shear apparatus consists of four blades on the end of a rod, as shown in Figure 2.14a. The vanes of the apparatus are pushed into the soil at the bottom of a borehole without disturbing the soil appreciably. Torque is applied at the top of the rod to rotate the vanes. This will induce failure in a soil of cylindrical shape surrounding

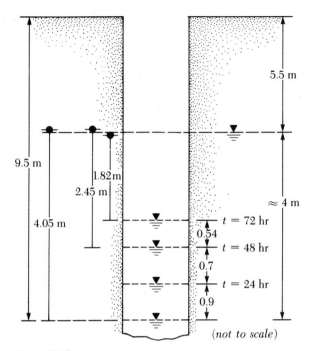

Figure 2.13

the vanes. The maximum torque applied can be related to the undrained strength of a clayey soil as

$$T = c_u \pi D^2 \left(\frac{H}{2} + \frac{D}{6} \right)$$

or

$$c_u = \frac{T}{\pi D^2 \left(\dfrac{H}{2} + \dfrac{D}{6} \right)} \tag{2.8}$$

where T = maximum torque applied
c_u = undrained shear strength ($\phi = 0$ concept)
D = diameter of the vanes
H = height of the vanes

The height-to-diameter ratio (H/D) of the vanes is usually kept at 2. The U.S. Bureau of Reclamation uses vanes with $D \times H$ dimensions of 50.8 mm × 101.6 mm (2 in. × 4 in.), 76.2 mm × 152.4 mm (3 in. × 6 in.), and 101.6 mm × 203.2 mm (4 in. × 8 in.). AASHTO Designation T-223 recommends the following dimensions of vanes for field tests.

Casing size	H (mm)	D (mm)	Blade thickness (mm)
AX	38.1	76.2	1.6
BX	50.8	101.6	1.6
NX	63.5	137.0	3.2

Bjerrum (1972) has recommended that, for actual design purposes, the field vane shear values should be corrected as

$$c_{u(\text{corrected})} = \lambda c_{u(\text{field})}$$ (2.9)

where λ = correction factor

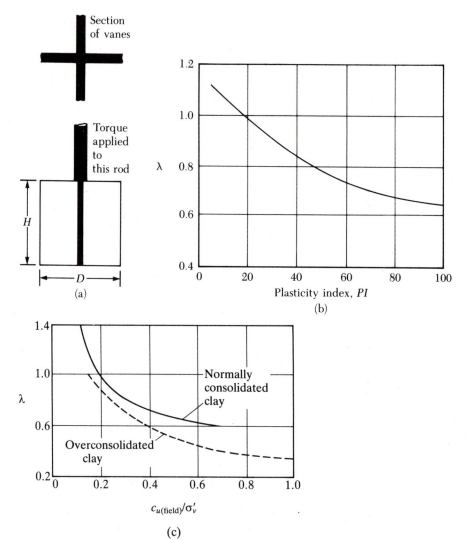

Figure 2.14 Vane shear test: (a) vane shear equipment; (b) and (c) correction factor, λ [Eq. (2.9)]. [*Note:* (b) from Bjerrum (1972); (c) from Aas, et al. (1986)]

The value of the correction factor, λ, as proposed by Bjerrum (1972) varies with plasticity index of soil and is shown in Figure 2.14b. More recently, Aas et al. (1986) have given the variation of λ as a function of $c_{u(\text{field})}$ and σ'_v (that is, present effective overburden pressure). This is shown in Figure 2.14c.

The field vane shear value, $c_{u(\text{field})}$, can also be used to estimate the overconsolidation ratio, OCR (preconsolidation pressure/present effective overburden pressure), of clay soils in the field. Mayne and Mitchell (1988) expressed an empirical relationship as

$$OCR = \beta \left[\frac{c_{u(\text{field})}}{\sigma'_v} \right] \tag{2.10}$$

where σ'_v = present effective overburden pressure

$$\beta = 22(PI)^{-0.48} \qquad (\text{where } PI = \text{plasticity index}) \tag{2.11}$$

Example 2.2

The magnitude of $c_{u(\text{field})}$ at a depth of 21 ft below the ground surface in a saturated clay soil is 820 lb/ft². The ground water table coincides with the ground surface. Given: saturated unit weight of clay, $\gamma_{\text{sat}} = 123$ lb/ft³ and $PI = 21$. Estimate the overconsolidation ratio of the clay.

Solution

Effective unit weight of clay

$$\gamma' = \gamma_{\text{sat}} - \gamma_w = 123 - 62.4 = 60.6 \text{ lb/ft}^3$$
$$\sigma'_v = \gamma'(21) = (60.6)(21) = 1272.6 \text{ lb/ft}^2$$

From Eq. (2.11)

$$\beta = 22(21)^{-0.48} = 5.1$$

From Eq. (2.10)

$$OCR = \beta \left[\frac{c_{u(\text{field})}}{\sigma'_v} \right] = 5.1 \left(\frac{820}{1272.6} \right) = 3.29$$

2.7   Dutch Cone Penetration Test

The *Dutch cone penetration test* (see Figure 2.15) is a technique by which a 60° cone with a base area of 10 cm² is pushed into the ground at a steady rate of about 20 mm/sec and the resistance to the penetration (the point resistance of the cone) and the frictional resistance of the casing to the soil around it are measured. This test is also called the *static penetration test,* and no boreholes are necessary to perform it. Refined Dutch cone penetrometers are also available to measure the point resistance of the cone at a certain

Figure 2.15 Dutch cone penetrometer

depth and the frictional resistance of the soil immediately above the point. The frictional resistance is obtained by means of a sleeve of limited length.

The point resistance, q_c, obtained from Dutch cone penetration tests has been related to the soil friction angle of granular soils and also to the consistency of cohesive soils (similar to that obtained for standard penetration tests; see Tables 2.3 and 2.4). The ratio of q_c/N (q_c in kN/m^2;

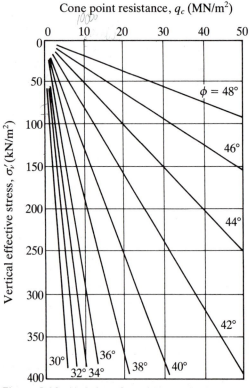

Figure 2.16 Variation of q_c with σ'_v and ϕ in quartz sand (after Robertson and Campanella, 1983)

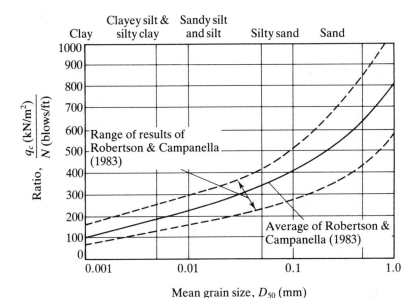

Figure 2.17   General range of variation of q_c/N for various types of soil

N = standard penetration resistance) for sands generally varies in a range of 400 to 600; for a gravelly deposit, it can be in a range of 800 to 1000.

The cone penetration test is more commonly used in Europe than in the United States. However, its application is rapidly growing in the United States. Schmertmann (1975) has provided an extensive bibliography of the published literature on cone penetration tests.

Figure 2.16 shows the nature of variation of the point resistance q_c with the vertical effective stress σ'_v and peak friction angle ϕ in quartz sand. The general range of variation of q_c/N for various types of soil is shown in Figure 2.17.

2.8   Coring of Rocks

When a rock layer is encountered during a drilling operation, rock coring may become necessary. For coring of rocks, a *core barrel* is attached to a drilling rod. A *coring bit* is attached to the bottom of the core barrel (Figure 2.18). The cutting elements may be diamond, tungsten, carbide, and so on. Table 2.5 gives a summary of the various types of core barrel and their sizes as well as the compatible drill rods commonly used for foundation exploration.

The coring is advanced by rotary drilling. Water is circulated through the drilling rod during coring, and the cutting is washed out.

Two types of core barrel are available: the *single-tube core barrel* (Figure 2.18a) and the *double-tube core barrel* (Figure 2.18b). Rock cores obtained by single-tube core barrels can be highly disturbed and fractured because of

Table 2.5 Standard Size and Designation of Casing, Core Barrel, and Compatible Drill Rod

Casing and core barrel designation	Outside diameter of core barrel bit		Drill rod designation
	(mm)	(in.)	
EX	36.51	$1\frac{7}{16}$	E
AX	47.63	$1\frac{7}{8}$	A
BX	58.74	$2\frac{5}{16}$	B
NX	74.61	$2\frac{15}{16}$	N

torsion. (Note also that rock cores smaller than the BX size tend to fracture during the coring process.)

When the core samples are recovered, the depth of recovery should be properly recorded for further evaluation in the laboratory. Based on the length of the rock core recovered from each run, the following quantities may be calculated for a general evaluation of the rock quality encountered.

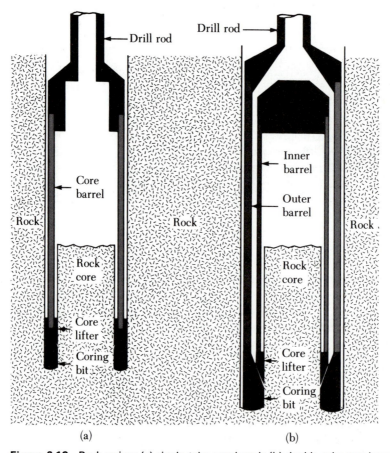

Figure 2.18 Rock coring: (a) single-tube core barrel; (b) double-tube core barrel

Table 2.5   (Continued)

Outside diameter of drill rod		Diameter of borehole		Diameter of core sample	
(mm)	(in.)	(mm)	(in.)	(mm)	(in.)
33.34	$1\frac{5}{16}$	38.1	$1\frac{1}{2}$	22.23	$\frac{7}{8}$
41.28	$1\frac{5}{8}$	50.8	2	28.58	$1\frac{1}{8}$
47.63	$1\frac{7}{8}$	63.5	$2\frac{1}{2}$	41.28	$1\frac{5}{8}$
60.33	$2\frac{3}{8}$	76.2	3	53.98	$2\frac{1}{8}$

They are

a. $\text{Recovery ratio} = \dfrac{\text{length of core recovered}}{\text{theoretical length of rock cored}}$ (2.12)

b. Rock quality designation $(RQD) =$

$$\frac{\Sigma \text{ length of recovered pieces equal to or larger than 101.6 mm (4 in.)}}{\text{theoretical length of rock cored}}$$ (2.13)

A recovery ratio of one will indicate the presence of intact rock; for highly fractured rocks, the recovery ratio may be 0.5 or smaller. Table 2.6 presents the general relationship (Deere, 1963) between the RQD and the *in situ* rock quality.

Table 2.6   Relation Between *in situ* Rock Quality and *RQD*

RQD	Rock quality
0–0.25	Very poor
0.25–0.5	Poor
0.5–0.75	Fair
0.75–0.9	Good
0.9–1	Excellent

2.9   Preparation of Boring Logs

The detailed information gathered from each borehole is presented in a graphical form that is referred to as the *boring log*. As the borehole is advanced downward in the field, the driller should generally record the following information in a standard log.

1. Name and address of the drilling company
2. Driller's name
3. Job description and number

4. Number and type of boring and boring location

5. Date of boring

6. Subsurface stratification. This can be obtained by visual observation of the soil brought out by auger, split-spoon sampler, and thin wall Shelby tube sampler.

7. Elevation of ground water table and date observed; use of casing and mud losses, and so on

8. Standard penetration resistance and the depth of *SPT*

Boring Log

Name of the Project   Two-story apartment building

Location Johnson & Olive St.   Date of Boring   March 2, 1982

Boring No.   3   Type of   Hollow stem auger   Ground Elevation   60.8 m
                    Boring

Soil description	Depth (m)	Soil sample type and number	N	w_n (%)	Comments
Light brown clay (fill)					
Silty sand (SM)	1 2	SS-1	9	8.2	
°G.W.T. ▽ 3.5 m	3 4	SS-2	12	17.6	$LL = 38$ $PI = 11$
Light gray silty clay (ML)	5	ST-1		20.4	$LL = 36$ $q_u = 112 \ kN/m^2$
	6	SS-3	11	20.6	
Sand with some gravel (SP)	7				
End of boring @ 8 m	8	SS-4	27	9	

N = standard penetration number (below/304.8 mm)
w_n = natural moisture content
LL = liquid limit; PI = plasticity index
q_u = unconfined compression strength
SS = split-spoon sample; ST = Shelby tube sample

°Ground water table observed after one week of drilling

Figure 2.19 A typical boring log

9. Number, type, and depth of soil sample collected

10. In case of rock coring, type of core barrel used should be recorded. For each run, the actual length of coring, length of core recovery, and the *RQD* should also be carefully noted.

None of this information should *ever* be left to memory, because this often results in erroneous boring logs.

After completion of the necessary laboratory tests, the geotechnical engineer prepares a new finished log that includes notes from the driller's field log and the results of tests conducted in the laboratory. Figure 2.19 shows a typical boring log. These logs have to be attached to the final soil exploration report submitted to the client. Note that Figure 2.19 also lists the classifications of the soils in the left-hand column, along with the description of each soil. This is based on the Unified Soil Classification System.

2.10   Determination of Coefficient of Permeability in the Field

There are several types of field test now available to determine the coefficient of permeability of soil. Two fairly easy test procedures described by the U.S. Bureau of Reclamation (1974) are the *open end test* and the *packer test*. Following are brief descriptions of these tests.

Open End Test

The first step in the open end test (Figure 2.20) is to advance a borehole to the desired depth. A casing is then driven to extend to the bottom of the borehole. Water is supplied at a constant rate from the top of the casing, and it escapes at the bottom of the borehole. The water level in the casing must remain constant. Once the steady state of water supply is established, the coefficient of permeability can be determined as

$$k = \frac{Q}{5.5rH} \qquad (2.14)$$

where k = coefficient of permeability
Q = constant rate of supply of water to the borehole
r = inside radius of the casing
H = differential head of water

Any type of consistent units may be used in Eq. (2.14).

The head, H, has been defined in Figure 2.20. Note that for pressure tests (Figure 2.20c and d) the value of H is given as

$$H = H_{(gravity)} + H_{(pressure)} \qquad (2.15)$$

The pressure head, $H_{(pressure)}$, given in Eq. (2.15) is expressed in meters (or feet) of water (1 kN/m^2 = 0.102 m; 1 $lb/in.^2$ = 2.308 ft).

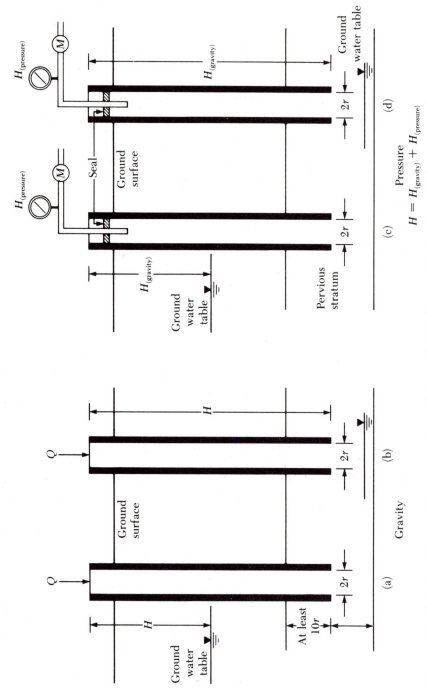

Figure 2.20 Coefficient of permeability—open end test (redrawn after U.S. Bureau of Reclamation, 1974)

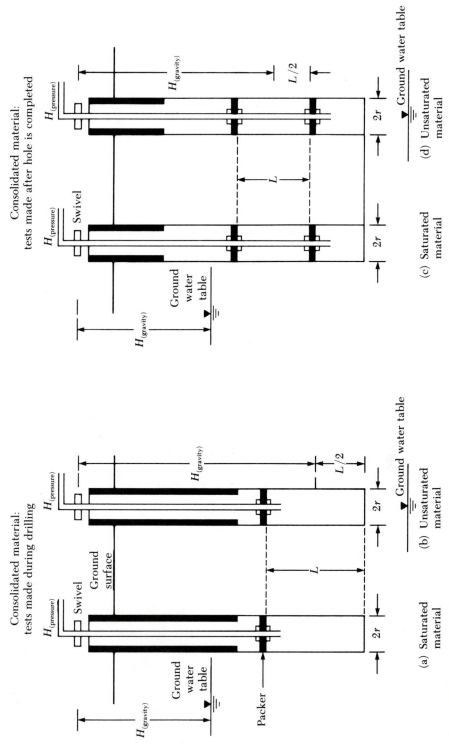

Figure 2.21 Coefficient of permeability determination—packer test (redrawn after U.S. Bureau of Reclamation, 1974)

Packer Test

The packer test (Figure 2.21) can be conducted in a portion of the borehole during the drilling or after the drilling has been completed. Water to the portion of the borehole under test is supplied under pressure at a constant rate. The coefficient of permeability can be determined from the following equations:

$$k = \frac{Q}{2\pi L H} \log_e \left(\frac{L}{r}\right) \qquad \text{(for } L \geq 10r\text{)} \tag{2.16}$$

$$k = \frac{Q}{2\pi L H} \sinh^{-1} \frac{L}{2r} \qquad \text{(for } 10r > L \geq r\text{)} \tag{2.17}$$

where k = coefficient of permeability
$\qquad Q$ = constant rate of flow into the hole
$\qquad L$ = length of portion of the hole under test
$\qquad r$ = radius of the hole
$\qquad H$ = differential pressure head

Note that the differential pressure head is the sum of the gravity head ($H_{(gravity)}$) and the pressure head ($H_{(pressure)}$).

The packer test is primarily used to determine the permeability of rock. However, as mentioned previously, it can also be used in soils.

2.11   Geophysical Exploration

Several types of geophysical exploration techniques are now available for a rapid evaluation of subsoil characteristics. They permit rapid coverage of large areas and are less expensive than conventional exploration by drilling. However, in many cases, definitive interpretation of the results is difficult. For that reason, these techniques should be used for preliminary works only. The following three sections cover three types of geophysical exploration technique: *seismic refraction survey*, *cross-hole seismic survey*, and *resistivity survey*.

Seismic Refraction Survey

Seismic refraction surveys are useful in obtaining preliminary information about the thickness of the layering of various soils at a given site and also the depth to rock or hard soil. Refraction surveys are conducted by impacting the surface, such as at point A in Figure 2.22a, and observing the first arrival of the disturbance (stress waves) at several points away from point A (such as B, C, D, . . .). The impact can be created by a hammer blow or by a small explosive charge. The observations of the first arrival of disturbances to arrive at various points can be recorded by geophones.

The impact on the ground surface creates two types of *stress wave*: *P-waves* (or *plane waves*) and *S-waves* (or *shear waves*). The *P*-waves travel faster than *S*-waves; hence, the first arrival of the disturbances will be

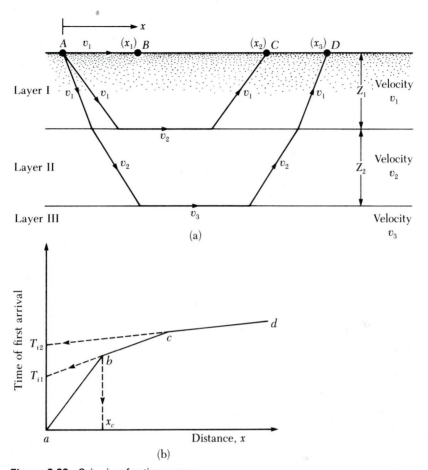

Figure 2.22 Seismic refraction survey

related to the velocities of the *P*-waves in various layers. The velocity of *P*-waves in a medium can be given by the equation

$$v = \sqrt{\frac{E}{\left(\dfrac{\gamma}{g}\right)}} \sqrt{\frac{(1-\mu)}{(1-2\mu)(1+\mu)}} \qquad (2.18)$$

where E = Young's modulus of the medium
   γ = unit weight of the medium
   g = acceleration due to gravity
   μ = Poisson's ratio

To determine the velocity (v) of *P*-waves in various layers and the thicknesses of those layers, use the following procedure:

1. Obtain the times of first arrival, $t_1, t_2, t_3, \ldots$, at various distances, $x_1, x_2, x_3, \ldots$, from the field.

2. Plot a graph of time (t) against distance (x). The graph will be like the one in Figure 2.22b.

3. Determine the slopes of the lines ab, bc, cd, ...

Slope of $ab = \dfrac{1}{v_1}$

Slope of $bc = \dfrac{1}{v_2}$

Slope of $cd = \dfrac{1}{v_3}$

where v_1, v_2, v_3, ..., are the P-wave velocities in layers I, II, III, ..., respectively (Figure 2.22a).

4. Determine the thickness of the top layer as

$$Z_1 = \frac{1}{2}\sqrt{\frac{v_2 - v_1}{v_2 + v_1}} \cdot x_c \tag{2.19}$$

The value of x_c can be obtained from the plot, as shown in Figure 2.22b.

5. Determine the thickness of the second layer, Z_2, shown in Figure 2.22a, as

$$Z_2 = \frac{1}{2}\left[T_{i2} - 2Z_1 \frac{\sqrt{v_3^2 - v_1^2}}{v_3 \cdot v_1} \right] \frac{v_3 \cdot v_2}{\sqrt{v_3^2 - v_2^2}} \tag{2.20}$$

where T_{i2} is the time intercept of the line cd in Figure 2.22b extended backward.

For detailed derivatives of these equations and other related information, refer to Dobrin (1960) and Das (1983).

Knowing the velocities of P-waves in various layers gives one an idea of the types of soil or rock that are present below the ground surface. The range of the P-wave velocity that is generally encountered in various types of soil and rock at shallow depths is given in Table 2.7.

In analyzing the results of a refraction survey, two limitations need to be kept in mind:

1. The basic equations for the refraction survey—that is, Eqs. (2.19) and (2.20)—are based on the assumption that the P-wave velocity $v_1 < v_2 < v_3 < \cdots$.

2. When a soil is saturated below the ground water table, the P-wave velocity may be deceptive. P-waves can travel through water with a velocity of about 1500 m/sec (5000 ft/sec). For dry, loose soils, the velocity may be well below 1500 m/sec. However, in a saturated condition, the waves will travel through water present in the void spaces with a velocity of about 1500 m/sec (5000 ft/sec). If the presence of ground water is not known, the P-wave velocity may be erroneously interpreted to be indicative of a stronger material (for example, sandstone) than what is present *in situ*.

Table 2.7   Range of P-Wave Velocity in Various Soils and Rocks

Type of soil or rock	P-wave velocity (m/sec)	ft/sec
Soil		
Sand, dry silt, and fine-grained top soil	200 to 1,000	650– 3,300
Alluvium	500 to 2,000	1,650– 6,600
Compacted clays, clayey gravel, and dense clayey sand	1,000 to 2,500	3,300– 8,200
Loess	250 to 750	800– 2,450
Rock		
Slate and shale	2,500 to 5,000	8,200–16,400
Sandstone	1,500 to 5,000	4,900–16,400
Granite	4,000 to 6,000	13,100–19,700
Sound limestone	5,000 to 10,000	16,400–32,800

Example 2.3

The results of a refraction survey at a site are given in the following table. Determine the velocity of P-waves and the thickness of the material encountered.

Distance from the source of disturbance (m)	Time of first arrival (sec × 10³)
2.5	11.2
5	23.3
7.5	33.5
10	42.4
15	50.9
20	57.2
25	64.4
30	68.6
35	71.1
40	72.1
50	75.5

Solution
Velocity

In Figure 2.23, the times of first arrival are plotted against the distance from the source of disturbance. The plot has three straight-line segments. The velocity of the top three layers can now be calculated in the following manner:

$$\text{Slope of segment } 0a = \frac{1}{v_1} = \frac{\text{time}}{\text{distance}} = \frac{23 \times 10^{-3}}{5.25}$$

or

$$v_1 = \frac{5.25 \times 10^3}{23} = 228 \text{ m/sec (top layer)}$$

$$\text{Slope of segment } ab = \frac{1}{v_2} = \frac{13.5 \times 10^{-3}}{11}$$

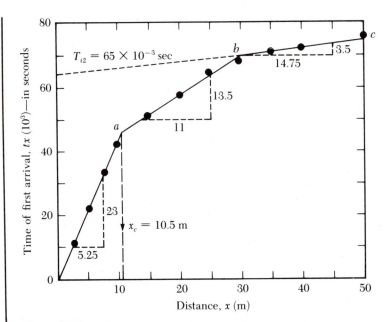

Figure 2.23

or

$$v_2 = \frac{11 \times 10^3}{13.5} = \textbf{814.8 m/sec (middle layer)}$$

$$\text{Slope of segment } bc = \frac{1}{v_3} = \frac{14.75 \times 10^{-3}}{3.5}$$

or

$$v_3 = \textbf{4214 m/sec (third layer)}$$

Comparing the velocities obtained here with those given in Table 2.7, it appears that the third layer is a *rock layer*.

Thickness of layers

From Figure 2.23, $x_c = 10.5$ m. So

$$Z_1 = \frac{1}{2}\sqrt{\frac{v_2 - v_1}{v_2 + v_1}}\, x_c \qquad \text{[Eq. (2.19)]}$$

So

$$Z_1 = \frac{1}{2}\sqrt{\frac{814.8 - 228}{814.8 + 228}} \times 10.5 = \textbf{3.94 m}$$

Again, from Eq. (2.20)

$$Z_2 = \frac{1}{2}\left[T_{i2} - \frac{2Z_1\sqrt{v_3^2 - v_1^2}}{(v_3 \cdot v_1)} \right] \frac{(v_3)(v_2)}{\sqrt{v_3^2 - v_2^2}}$$

The value of T_{i2} can be determined from Figure 2.23 as 65×10^{-3} sec. So

$$Z_2 = \frac{1}{2}\left[65 \times 10^{-3} - \frac{2(3.94)\sqrt{(4214)^2 - (228)^2}}{(4214)(228)}\right]\frac{(4214)(814.8)}{\sqrt{(4214)^2 - (814.8)^2}}$$

$$= \frac{1}{2}(0.065 - 0.0345)830.47 = \textbf{12.66 m}$$

Hence, the rock layer is located at a depth of $Z_1 + Z_2 = 3.94 + 12.66 = \textbf{16.60 m}$ **measured from the ground surface.**

Cross-Hole Seismic Survey

The velocity of shear waves created as the result of an impact to a given soil layer can be effectively determined by *cross-hole seismic survey* (Stokoe and Woods, 1972). The principle of this technique is illustrated in Figure 2.24, which shows two holes drilled into the ground at a certain distance, L, apart. A vertical impulse is created at the bottom of one borehole by means of an impulse rod. The shear waves thus generated are recorded by a vertically sensitive transducer. The velocity of shear waves, v_s, can be calculated as

$$v_s = \frac{L}{t} \tag{2.21}$$

where t = travel time of shear waves

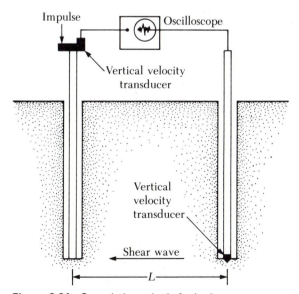

Figure 2.24 Cross-hole method of seismic survey

The shear modulus of the soil at the depth of the test can be determined from v_s as

$$v_s = \sqrt{\frac{G}{\left(\frac{\gamma}{g}\right)}}$$

$$G = \frac{v_s^2 \gamma}{g} \tag{2.22}$$

where G = shear modulus of soil
$\quad\quad \gamma$ = soil unit weight
$\quad\quad g$ = acceleration due to gravity

The values of shear modulus are useful in the design of foundations to support vibrating machinery and the like.

Resistivity Survey

Another geophysical method for subsoil exploration is the *electrical resistivity survey*. The electrical resistivity, ρ, of any conducting material having a length, L, and an area of cross section A can be defined as

$$\rho = \frac{RA}{L} \tag{2.23}$$

where R = electrical resistance

The unit of resistivity is generally expressed as *ohm-centimeter* or *ohm-meter*. The resistivity of various soils depends primarily on the moisture content and also on the concentration of dissolved ions. Saturated clays have a very low resistivity; in contrast, dry soils and rocks have a high resistivity. The general range of resistivity generally encountered in various soils and rocks is given in Table 2.8.

Table 2.8 Representative Values of Resistivity

Material	Resistivity (ohm-m)
Sand	500–1500
Clays, saturated silt	0– 100
Clayey sand	200– 500
Gravel	1500–4000
Weathered rock	1500–2500
Sound rock	>5000

The most common procedure for measuring electrical resistivity of a soil profile makes use of four electrodes that are driven into the ground, spaced equally along a straight line. This is generally referred to as the

Wenner method (Figure 2.25a). The two outside electrodes are used to send an electrical current, *I*, (usually a DC current with nonpolarizing potential electrodes) into the ground. The electrical current is typically in the range of 50–100 milliamperes. The voltage drop, *V*, is measured between the two inside electrodes. If the soil profile is homogeneous, its electrical resistivity can be expressed as

$$\rho = \frac{2\pi d V}{I} \tag{2.24}$$

In most cases, the soil profile may consist of various layers with different resistivities. In that case, Eq. (2.24) will yield the *apparent resistivity*. In order to obtain the *actual resistivity* of various layers and their thicknesses,

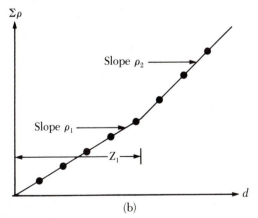

Figure 2.25 Electrical resistivity survey: (a) Wenner method; (b) empirical method for determination of resistivity and thickness of each layer

an empirical method may be used. This involves conducting a number of tests at various electrode spacings (that is, d is changed). The sum of the apparent resistivities (that is, $\Sigma\rho$) is plotted against the spacing d, as shown in Figure 2.25b. The plot thus obtained will have relatively straight segments. The slope of these straight segments will give the resistivity of individual layers. The thicknesses of various layers can be estimated as shown in Figure 2.25b.

The resistivity survey is particularly useful in locating gravel deposits within a fine-grained soil.

2.12   Soil Exploration Report

At the end of all subsoil exploration programs, the soil and/or rock specimens collected from the field are subject to visual observation and appropriate laboratory testing (the basic soil tests have been described in Chapter 1). After the compilation of all of the required information, a soil exploration report is prepared for the use of the design office and for reference during future construction work. Although the details and sequence of information in the report may vary to some degree depending on the structure under consideration and the person compiling the report, each report should include the following:

1. The scope of the investigation.
2. A description of the proposed structure for which the subsoil exploration has been conducted.
3. A description of the location of the site. This should include structure(s) nearby, drainage conditions of the site, nature of vegetation on the site and surrounding it, and any other feature(s) unique to the site.
4. Geological setting of the site.
5. Details of the field exploration—that is, number of borings, depths of borings, type of boring, and so on.
6. General description of the subsoil conditions as determined from soil specimens and from related laboratory tests, standard penetration resistance and cone penetration resistance, and so on.
7. Ground water table conditions.
8. Foundation recommendations. These should include the type of foundation recommended, allowable bearing pressure, and any special construction procedure that may be needed. Alternate foundation design procedures should also be discussed in this portion of the report.
9. Conclusions and limitations of the investigations.

Following are the graphical presentations that need to be attached to the report:

1. Site-location map
2. A plan showing the location of the borings with respect to the proposed structures and those existing nearby
3. Boring logs

4. Laboratory test results

5. Other special graphical presentations

The exploration reports should be well planned and documented. They will help in solving several questions and foundation problems that may arise at a later stage of design and construction.

PROBLEMS

2.1 The inside and outside diameters of a Shelby tube are 2 in. and $1\frac{7}{8}$ in., respectively. Determine the area ratio. What should be the inside diameter of the tube to give an area ratio of 10%?

2.2 A Shelby tube has an outside diameter of 3.5 in. and an inside diameter of 3.375 in. What is the area ratio of the Shelby tube?

2.3 According to Peck, Hanson, and Thornburn (1974), the correction factor, C_N [see Eq. (2.4)] can be given as

$$C_N = 0.77 \log \left(\frac{20}{0.0105\sigma_v'} \right) \qquad \text{for } \sigma_v' \geq 23.9 \text{ kN/m}^2$$

Make the necessary calculations and compare the above equation with Eq. (2.6) for $\sigma_v' = 25 \text{ kN/m}^2$ to $\sigma_v' = 150 \text{ kN/m}^2$.

2.4 A boring log in a sand soil is shown in Figure P2.4. The unit weight of soil above the ground water table is 18.08 kN/m³ and that below the ground water table is 19.34 kN/m³. Using Eq. (2.6), determine the corrected standard penetration numbers.

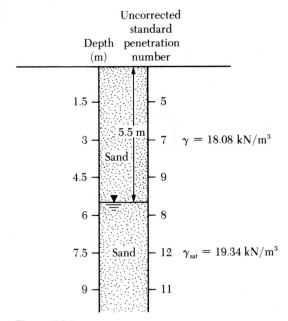

Figure P2.4

2.5 Following are the standard penetration numbers found from a sandy soil deposit in the field. Use Eq. (2.5) to determine the corrected standard penetration numbers.

Depth (ft)	Unit weight of soil (lb/ft^3)	N_F
10	106	7
15	106	9
20	106	11
25	118	16
30	118	18
35	118	20
40	118	22

Note: The ground water table was located at a depth of 43 ft.

2.6 For the sandy soil encountered in Problem 2.4, estimate the relative density and the angle of friction.

2.7 For the sandy soil encountered in Problem 2.5, estimate the relative density and the angle of friction.

2.8 Refer to Figure 2.12. The borehole is in a silty soil. Given: $h_w + h_0 = 12.24$ m, $t_1 = 24$ hours, $t_2 = 48$ hours, $\Delta h_1 = 1.2$ m, $\Delta h_2 = 0.86$ m, $\Delta h_3 = 0.6$ m. Determine h_w (the depth of the ground water table).

2.9 Repeat Problem 2.8 with the following:

$$h_w + h_0 = 42 \text{ ft}$$
$$t_1 = 24 \text{ hours}; \qquad \Delta h_1 = 6 \text{ ft}$$
$$t_2 = 48 \text{ hours}; \qquad \Delta h_2 = 4.8 \text{ ft}$$
$$t_3 = 72 \text{ hours}; \qquad \Delta h_3 = 3.8 \text{ ft}$$

2.10 The undrained shear strength value of a saturated clay deposit at a certain depth determined from a field vane shear test was 43 kN/m^2. Given:

Plasticity index of the clay, $PI = 60$

Effective overburden pressure, $\sigma'_v = 40$ kN/m^2

Estimate:
a. The corrected value of c_u for design purposes (refer to Figure 2.14b)
b. The overconsolidation ratio (OCR)

2.11 Redo Problem 2.10 with the following:

$$c_{u(\text{field})} = 31 \text{ kN/m}^2$$
$$PI = 25$$
$$\sigma'_v = 63 \text{ kN/m}^2$$

2.12 a. A vane shear test was conducted in a saturated clay. The height and diameter of the vane were 4 in. and 2 in., respectively. During the test the maximum torque applied was 12.4 lb-ft. Determine the undrained shear strength of the clay.
b. The clay soil described in (a) has a liquid limit of 64 and a plastic limit of 29. What would be the corrected undrained shear strength of this clay for design purposes? Use Figure 2.14b.

2.13 In a homogeneous deposit of dry sand, a cone penetration test was conducted. Following are the results:

Depth (m)	Point resistance of cone, q_c (kN/m²)
3	2.7
5	4.8
8	7.4
10	10.1
15	14.5

Assume the average unit weight of sand to be 16.8 kN/m³. Estimate the soil friction angle (refer to Figure 2.16).

2.14 During a field exploration, coring of rock was required. The core barrel was advanced 5 ft during the coring. The length of the core recovered was 2.8 ft. What was the recovery ratio?

2.15 An open end permeability test was conducted in a borehole (refer to Figure 2.20a). The inside diameter of the casing was 2 in. The differential head of water was 23.4 ft. In order to maintain a constant head $H = 23.4$ ft, a constant water supply rate of 4.8×10^{-2} ft³/min was required. Calculate the coefficient of permeability of soil.

2.16 The P-wave velocity in a given soil is 1900 m/sec. Assuming Poisson's ratio to be 0.32, calculate the Young's modulus of the soil. Assume the unit weight of soil to be equal to 18 kN/m³.

2.17 The results of a refraction survey (Figure 2.22a) at a site are given in the following table. Determine the thickness and the P-wave velocity of the materials encountered.

Distance from the source of disturbance (m)	Time of first arrival of P-waves (sec × 10³)
2.5	5.08
5.0	10.16
7.5	15.24
10.0	17.01
15.0	20.02
20.0	24.2
25.0	27.1
30.0	28.0
40.0	31.1
50.0	33.9

2.18 Repeat Problem 2.17 with the following:

Distance from the source of disturbance (ft)	Time of first arrival of P-waves (sec × 10³)
25	49.08
50	81.96
75	122.8
100	148.2
150	174.2
200	202.8
250	228.6
300	256.7

REFERENCES

Aas, G., Lacasse, S., Lunne, I., and Høeg, K. (1986). "Use of In Situ Tests for Foundation Design in Clay," *Proceedings, In Situ '86*, American Society of Civil Engineers, pp. 1–30.

American Association of State Highway and Transportation Officials (1978). *Standard Specifications for Transportation Materials and Methods of Sampling and Testing*, Part II, U.S. Government Printing Office, Washington, D.C.

American Society for Testing and Materials (1986). *Annual Book of ASTM Standards*, Vol. 04.08, Philadelphia, Pa.

American Society of Civil Engineers (1972). "Subsurface Investigation for Design and Construction of Foundations of Buildings," *Journal of the Soil Mechanics and Foundations Division*, American Society of Civil Engineers, Vol. 98, No. SM5, pp. 481–490.

Bazaraa, A. R. (1967). *Use of Standard Penetration Test for Estimating Settlements of Shallow Foundations on Sand*, Ph.D. Thesis, University of Illinois, Urbana.

Bjerrum, L. (1972). "Embankments on Soft Ground," *Proceedings of the Specialty Conference*, American Society of Civil Engineers, Vol. 2, pp. 1–54.

Das, B. M. (1983). *Fundamentals of Soil Dynamics*, Elsevier Science Publishing Co., New York.

Deere, D. U. (1963). "Technical Description of Rock Cores for Engineering Purposes," *Felsmechanik und Ingenieurgeologie*, Vol. 1, No. 1, pp. 16–22.

Dobrin, M. B. (1960). *Introduction to Geophysical Prospecting*, McGraw-Hill, New York.

Hvorslev, M. J. (1949). *Subsurface Exploration and Sampling of Soils for Civil Engineering Purposes*, Waterways Experiment Station, Vicksburg, Miss.

Liao, S. S. C., and Whitman, R. V. (1986). "Overburden Correction Factors for SPT in Sand," *Journal of Geotechnical Engineering*, American Society of Civil Engineers, Vol. 112, No. 3, pp. 373–377.

Mayne, P. W., and Mitchell, J. K. (1988). "Profiling of Overconsolidation Ratio in Clays by Field Vane," *Canadian Geotechnical Journal*, Vol. 25, No. 1, pp. 150–158.

Osterberg, J. O. (1952). "New Piston-Type Soil Sampler," *Engineering News-Record*, April 24.

Peck, R. B., and Bazaraa, A. S. (1969). "Discussion on Settlement of Spread Footings on Sand," *Journal of the Soil Mechanics and Foundations Division*, American Society of Civil Engineers, Vol. 95, No. SM3, pp. 905–909.

Peck, R. B., Hanson, W. E., and Thornburn, T. H. (1974). *Foundation Engineering*, 2nd ed., Wiley, New York.

Robertson, P. K., and Campanella, R. G. (1983). "Interpretation of Cone Penetration Tests. Part I: Sand," *Canadian Geotechnical Journal*, Vol. 20, No. 4, pp. 718–733.

Schmertmann, J. H. (1975). "The Measurement of *In-Situ* Shear Strength," *Proceedings*, American Society of Civil Engineers' Specialty Conference on *In-Situ* Measurement of Soil Properties, Raleigh, N.C., Vol. 2, pp. 57–138.

Seed, H. B. (1976). "Evaluation of Soil Liquefaction Effects on Level Ground During Earthquakes," *ASCE Specialty Session Preprint 2752*, American Society of Civil Engineers' National Convention, pp. 1–105.

Seed, H. B. (1979). "Soil Liquefaction and Cyclic Mobility Evaluation for Level Ground During Earthquakes," *Journal of Geotechnical Engineering*, American Society of Civil Engineers, Vol. 111, No. 12, pp. 1425–1445.

Sowers, G. B., and Sowers, G. F. (1970). *Introductory Soil Mechanics and Foundations*, 3rd ed., Macmillan, New York.

Stokoe, K. H., and Woods, R. D. (1972). "*In Situ* Shear Wave Velocity by Cross-Hole Method," *Journal of Soil Mechanics and Foundations Division*, American Society of Civil Engineers, Vol. 98, No. SM5, pp. 443–460.

Tang, W. C. (1962). *Foundation Design*, Prentice-Hall, Englewood Cliffs, N.J.

Tokimatsu, K., and Yoshimi, Y. (1983). "Empirical Correlation of Soil Liquefaction Based on SPT *N*-Value and Fines Content," *Soils and Foundations*, Vol. 23, No. 4, pp. 56–74.

U.S. Bureau of Reclamation (1974). *Design of Small Dams*, 2nd. ed., U.S. Government Printing Office, Washington, D.C.

Supplementary References for Further Study

Briaud, J. L., and Gambin, M. (1984). "Suggested Practice for Drilling of Boreholes for Pressuremeter Testing," *Geotechnical Testing Journal*, American Society for Testing and Materials, Vol. 7, No. 1, pp. 36–40.

DeMello, V. F. (1971). "The Standard Penetration Test," *Proceedings*, Fourth Panamerican Conference on Soil Mechanics and Foundation Engineering, San Juan, Puerto Rico, Vol. 1, pp. 1–86.

Schmertmann, J. H. (1975). "The Measurement of *In-Situ* Shear Strength," *Proceedings*, American Society of Civil Engineers' Specialty Conference on *In-Situ* Measurement of Soil Properties, Raleigh, N.C., Vol. 2, pp. 57–138.

Chapter 3

SHALLOW FOUNDATIONS

3.1   Introduction

To perform satisfactorily, shallow foundations must have two main characteristics:

1. The foundation should be safe against overall shear failure in the soil that supports it.

2. The foundation should not undergo excessive displacement, that is, settlement. (The term *excessive* is relative, because the degree of settlement allowable for a given structure is dependent on several considerations.) This chapter discusses in detail the evaluation of the safe load-bearing capacity and settlement of shallow foundations.

ULTIMATE BEARING CAPACITY
OF SHALLOW FOUNDATIONS

3.2   Ultimate Bearing Capacity

Consider a strip foundation resting on the surface of a dense sand or stiff cohesive soil, as shown in Figure 3.1a. Let the width of the foundation be equal to B. Now, if load is gradually applied on the foundation, the settlement of the foundation will also increase. The variation of the load per unit area on the foundation (q) with the foundation settlement is also shown in Figure 3.1a. At a certain point, when the load per unit area becomes equal to q_u, a sudden failure in the soil supporting the foundation will take place, and the failure surface in the soil will extend to the gound surface. This load per unit area, q_u, is usually referred to as the *ultimate bearing capacity of the foundation*. When this type of sudden failure in soil takes place, accompanied by the extension of the failure surface to the ground surface, it is called the *general shear failure*.

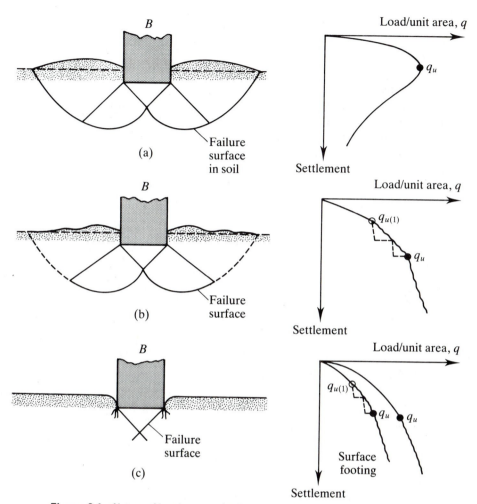

Figure 3.1   Nature of bearing capacity failure in soil: (a) general shear failure; (b) local shear failure; (c) punching shear failure (redrawn after Vesic, 1973)

In contrast, if the foundation under consideration is resting on sand or clayey soil of medium compaction (Figure 3.1b), an increase of load on the foundation will also be accompanied by an increase of settlement. However, in this case the failure surface in the soil will gradually extend outward from the foundation, as shown by the solid lines in Figure 3.1b. When the load per unit area on the foundation becomes equal to $q_{u(1)}$, the foundation movement will be accompanied by sudden jerks. A considerable movement of the foundation is then required for the failure surface in soil to extend to the ground surface (this part is shown by the broken lines in Figure 3.1b). The load per unit area at which this happens is the *ultimate bearing capacity*, q_u. Beyond this point, an increase of load will be accompanied by a large increase of foundation settlement. The load per unit area of the foundation,

$q_{u(1)}$, is referred to as the *first failure load* (Vesic, 1963). Note that a peak value of q is not realized in the case of this type of failure. This is called the *local shear failure* in soil.

If the foundation is supported by a fairly loose soil, the load-settlement plot will be like the one in Figure 3.1c. In this case, the failure surface in soil will not extend to the ground surface. Beyond the ultimate failure load, q_u, the load-settlement plot will be steep and practically linear. This type of failure in soil is called the *punching shear failure*.

Based on experimental results, Vesic (1973) has proposed a relationship for the mode of bearing capacity failure of foundations resting on sands. This is shown in Figure 3.2, which uses the following notations:

D_r = relative density of sand

D_f = depth of foundation measured from the ground surface

$$B^* = \frac{2BL}{B + L} \tag{3.1}$$

where B = width of foundation
L = length of foundation

(*Note:* L is always greater than B.)

For square foundations, $B = L$; for circular foundations, $B = L =$ diameter. So

$$B^* = B \tag{3.2}$$

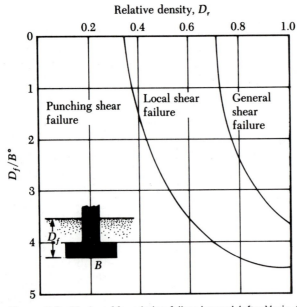

Figure 3.2 Modes of foundation failure in sand (after Vesic, 1973)

For foundations located at a shallow depth (that is, small D_f/B^*), the ultimate load may occur at a foundation settlement of 4–10% of B. This is true when general shear failure in soil occurs; however, in the case of local or punching shear failure, the ultimate load may occur at settlements of 15–25% of the width of foundation (B).

3.3 Terzaghi's Bearing Capacity Theory

Terzaghi (1943) was the first to present a comprehensive theory for the evaluation of the ultimate bearing capacity of rough shallow foundations. According to this theory, a foundation is *shallow* if the depth, D_f (Figure 3.3), of the foundation is less than or equal to the width of the foundation. Later investigators, however, have suggested that foundations with D_f equal to 3–4 times the width of the foundation may be defined as *shallow foundations*.

Terzaghi suggested that for a *continuous*, or *strip*, *foundation* (that is, width-to-length ratio of the foundation approaches zero), the failure surface in soil at ultimate load may be assumed to be similar to that shown in Figure 3.3. (Note that this is the case of general shear failure as defined in Figure 3.1a.) It may also be assumed that the effect of soil above the bottom of the foundation may be replaced by an equivalent surcharge equal to $q = \gamma D_f$ (where γ = unit weight of soil). The failure zone under the foundation can be separated into three parts (see Figure 3.3):

1. The *triangular zone ACD* immediately under the foundation
2. The *radial shear zones ADF* and *CDE*, with the curves *DE* and *DF* being arcs of a logarithmic spiral
3. Two triangular *Rankine passive zones AFH* and *CEG*

It is assumed that the angles *CAD* and *ACD* are equal to the soil friction angle, ϕ. Note that with the replacement of the soil above the bottom of the foundation by an equivalent surcharge q, the shear resistance of the soil along the failure surfaces *GI* and *HJ* was neglected.

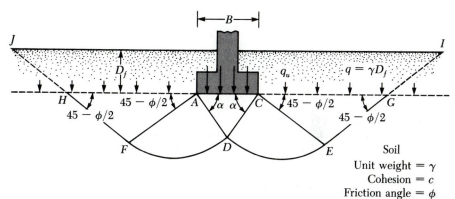

Figure 3.3 Bearing capacity failure in soil under a rough rigid continuous foundation

Using the equilibrium analysis, Terzaghi expressed the ultimate bearing capacity in the form

$$q_u = cN_c + qN_q + \frac{1}{2}\gamma BN_\gamma \quad \text{(strip foundation)} \tag{3.3}$$

where c = cohesion of soil
γ = unit weight of soil
$q = \gamma D_f$
N_c, N_q, N_γ = bearing capacity factors that are nondimensional and are only functions of the soil friction angle, ϕ

The bearing capacity factors, N_c, N_q, and N_γ are defined by the following equations:

$$N_c = \cot \phi \left[\frac{e^{2(3\pi/4 - \phi/2)\tan\phi}}{2\cos^2\left(\frac{\pi}{4} + \frac{\phi}{2}\right)} - 1 \right] \tag{3.4}$$

$$N_q = \frac{e^{2(3\pi/4 - \phi/2)\tan\phi}}{2\cos^2\left(45 + \frac{\phi}{2}\right)} \tag{3.5}$$

$$N_\gamma = \frac{1}{2}\left(\frac{K_{p\gamma}}{\cos^2\phi} - 1\right)\tan\phi \tag{3.6}$$

where $K_{p\gamma}$ = passive pressure coefficient

The variations of the bearing capacity factors defined by Eq. (3.3) are given in Figure 3.4.

For estimation of the ultimate bearing capacity of *square* or *circular foundations*, Eq. (3.1) may be modified to the following forms:

$$q_u = 1.3cN_c + qN_q + 0.4\gamma BN_\gamma \quad \text{(square foundation)} \tag{3.7}$$

and

$$q_u = 1.3cN_c + qN_q + 0.3\gamma BN_\gamma \quad \text{(circular foundation)} \tag{3.8}$$

In Eq. (3.7), B is equal to the dimension of each side of the foundation; in Eq. (3.8), B is equal to the diameter of the foundation.

Figure 3.4 Terzaghi's bearing capacity factors for general shear failure—Eq. (3.3)

For foundations that exhibit the local shear failure mode in soils, Terzaghi suggested the following modifications to Eqs. (3.3), (3.7), and (3.8):

$$q_u = \frac{2}{3}cN'_c + qN'_q + \frac{1}{2}\gamma BN'_\gamma \qquad \text{(strip foundation)} \qquad (3.9)$$

$$q_u = 0.867cN'_c + qN'_q + 0.4\gamma BN'_\gamma \qquad \text{(square foundation)} \qquad (3.10)$$

$$q_u = 0.867cN'_c + qN'_q + 0.3\gamma BN'_\gamma \qquad \text{(circular foundation)} \qquad (3.11)$$

N'_c, N'_q, and N'_γ are the *modified bearing capacity factors*. They can be calculated by using the bearing capacity factor equations (for N_c, N_q, and

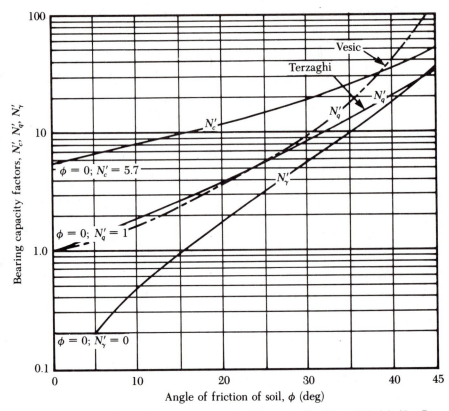

Figure 3.5   Terzaghi's modified bearing capacity factors: N'_c, N'_q, N'_γ; and Vesic's N'_q—Eq. (3.12)

N_γ) by replacing ϕ by $\phi' = \tan^{-1}(\tfrac{2}{3}\tan\phi)$. The variation of N'_c, N'_q, and N'_γ with the soil friction angle, ϕ, is given in Figure 3.5.

Based on the experience of several model tests, the author has observed (also see Ismael and Vesic, 1981) that the value of N'_q for shallow foundations in granular soil is underestimated by Terzaghi's modified procedure.

Table 3.1   Values of N'_q [Eq. (3.12)]

ϕ (deg)	N'_q
0	1.0
5	1.2
10	1.6
15	2.2
20	3.3
25	5.3
30	9.5
35	18.7
40	42.5
45	115.0

It is better represented by the equation given by Vesic (1963):

$$N_q' = (e^{3.8\phi \tan \phi}) \tan^2 \left(45 + \frac{\phi}{2} \right) \tag{3.12}$$

The values of N_q' as calculated from Eq. (3.12) are given in Table 3.1. For comparison purposes, these values are also plotted in Figure 3.5.

3.4   Modification of Bearing Capacity Equations for Water Table

Equations (3.3) and (3.7) to (3.11) have been developed for the determination of the ultimate bearing capacity based on the assumption that the water table is located well below the foundation. However, if the water table is close to the foundation, some modifications of the bearing capacity equations will be necessary, depending on the location of the water table (see Figure 3.6):

Case I: If the ground water table is located in such a way that $0 \leq D_1 < D_f$, then the factor q in the bearing capacity equations will take the form

$$q = \text{effective surcharge} = D_1\gamma + D_2(\gamma_{sat} - \gamma_w) \tag{3.13}$$

where γ_{sat} = saturated unit weight of soil
γ_w = unit weight of water

Also, the value of γ in the last term of the equations has to be replaced by $\gamma' = \gamma_{sat} - \gamma_w$.

Figure 3.6 Modification of bearing capacity equations for water table

Case II: For ground water table located in such a way that $0 \leq d \leq B$, the factor

$$q = \gamma D_f \tag{3.14}$$

The factor γ in the last term of the bearing capacity equations must be replaced by the factor

$$\boxed{\bar{\gamma} = \gamma' + \frac{d}{B}(\gamma - \gamma')} \tag{3.15}$$

The preceding modifications are based on the assumption that there is no seepage force in the soil.

Case III: When the ground water table is located such that $d \geq B$, the water will have no effect on the ultimate bearing capacity.

Example 3.1

A square foundation is 1.5 m × 1.5 m in plan. The soil supporting the foundation has a friction angle of $\phi = 20°$ and $c = 15.2$ kN/m². The unit weight of soil, γ, is 17.8 kN/m³. Determine the allowable gross load on the foundation with a factor of safety (*FS*) of 4. Assume that the depth of the foundation (D_f) is 1 m and that general shear failure occurs in the soil.

Solution

From Eq. (3.7)

$$q_u = 1.3cN_c + qN_q + 0.4\gamma BN_\gamma$$

From Figure 3.4, for $\phi = 20°$

$$N_c = 17.7$$
$$N_q = 7.4$$
$$N_\gamma = 5$$

Thus

$$q_u = (1.3)(15.2)(17.7) + (1 \times 17.8)(7.4) + (0.4)(17.8)(1.5)(5)$$
$$= 349.75 + 131.72 + 53.4 = 534.87 \approx 535 \text{ kN/m}^2$$

So, allowable load per unit area of the foundation =

$$q_{\text{all}} = \frac{q_u}{FS} = \frac{535}{4} = 133.75 \text{ kN/m}^2$$

Thus, the total allowable gross load

$$Q = (133.75)B^2 = (133.75)(1.5 \times 1.5) = 300.9 \approx \textbf{300 kN}$$

Example 3.2

Repeat Example 3.1, assuming that local shear failure occurs in the soil supporting the foundation.

Solution

From Eq. (3.10)

$$q_u = 0.867cN_c' + qN_q' + 0.4\gamma BN_\gamma'$$

From Figure 3.5, for $\phi = 20°$

$$N_c' = 12$$
$$N_q' = 4$$
$$N_\gamma' = 1.7$$

So

$$q_u = (0.867)(15.2)(12) + (1 \times 17.8)(4) + (0.4)(17.8)(1.5)(1.7)$$
$$= 158.1 + 71.2 + 18.2 = 247.5 \text{ kN/m}^2$$
$$q_{\text{all}} = \frac{247.5}{4} = 61.9 \text{ kN/m}^2$$

Allowable gross load $= Q = (q_{\text{all}})(B^2) = (61.9)(1.5^2) = \textbf{139 kN}$

3.5   General Bearing Capacity Equation

The ultimate bearing capacity equations presented in Eqs. (3.3), (3.7), and (3.8) are for continuous, square, and circular foundations only. They do not address the case of rectangular foundations ($0 < B/L < 1$). Also, the equations do not take into account the shearing resistance along the failure surface in soil located above the bottom of the foundation (portion of the failure surface marked as *GI* and *HĴ* in Figure 3.3. In addition, the load on the foundation may be inclined. In order to take all these shortcomings into account, the following form of general bearing capacity equation has been suggested by Meyerhof (1963).

$$q_u = cN_c \, F_{cs} \, F_{cd} \, F_{ci} + qN_q \, F_{qs} \, F_{qd} \, F_{qi} + \frac{1}{2} \, \gamma BN_\gamma \, F_{\gamma s} \, F_{\gamma d} \, F_{\gamma i} \qquad (3.16)$$

where c = cohesion
q = effective stress at the level of the bottom of foundation
γ = unit weight of soil
B = width of foundation (= diameter for a circular foundation)

F_{cs}, F_{qs}, $F_{\gamma s}$ = shape factors
F_{cd}, F_{qd}, $F_{\gamma d}$ = depth factors
F_{ci}, F_{qi}, $F_{\gamma i}$ = load inclination factors
N_c, N_q, N_γ = bearing capacity factors

The equations for determination of the various factors given in Eq. (3.16) will be briefly summarized in the following sections. Note that the original equation for ultimate bearing capacity is derived only for plane-strain case (that is, for continuous foundations). The shape, depth, and load inclination factors are empirical factors based on experimental data.

Bearing Capacity Factors

Based on laboratory and field studies done on bearing capacity, it appears now that the basic nature of the failure surface in soil as suggested by Terzaghi is correct (Vesic, 1973). However, the angle α as shown in Figure 3.3 is closer to $45 + \phi/2$ instead of ϕ. If this change is accepted, the values of N_c, N_q, and N_γ for a given soil friction angle will also change from those given in Figure 3.4. With $\alpha = 45 + \phi/2$, the relations for N_c and N_q can be derived as

$$N_q = \tan^2\left(45 + \frac{\phi}{2}\right)e^{\pi \tan \phi} \qquad (3.17)$$

$$N_c = (N_q - 1)\cot\phi \qquad (3.18)$$

The equation for N_c given by Eq. (3.18) was originally derived by Prandtl (1921), and the relation for N_q [Eq. (3.17)] was presented by Reissner (1924). Caquot and Kerisel (1953) and Vesic (1973) have given the relation for N_γ as

$$N_\gamma = 2(N_q + 1)\tan\phi \qquad (3.19)$$

The variation of the preceding bearing capacity factors with soil friction angles is given in Table 3.2.

In many texts and reference books, the relationship for N_γ may be recommended differently as compared with Eq. (3.19). This is because there is still some controversy about the variation of N_γ with the soil friction angle, ϕ. *In this text, Eq. (3.19) will be used.* Other relationships frequently found in various texts and references are shown in Table 3.3.

Table 3.2 Bearing Capacity Factors[a]

ϕ	N_c	N_q	N_γ	N_q/N_c	$\tan \phi$
0	5.14	1.00	0.00	0.20	0.00
1	5.38	1.09	0.07	0.20	0.02
2	5.63	1.20	0.15	0.21	0.03
3	5.90	1.31	0.24	0.22	0.05
4	6.19	1.43	0.34	0.23	0.07
5	6.49	1.57	0.45	0.24	0.09
6	6.81	1.72	0.57	0.25	0.11
7	7.16	1.88	0.71	0.26	0.12
8	7.53	2.06	0.86	0.27	0.14
9	7.92	2.25	1.03	0.28	0.16
10	8.35	2.47	1.22	0.30	0.18
11	8.80	2.71	1.44	0.31	0.19
12	9.28	2.97	1.69	0.32	0.21
13	9.81	3.26	1.97	0.33	0.23
14	10.37	3.59	2.29	0.35	0.25
15	10.98	3.94	2.65	0.36	0.27
16	11.63	4.34	3.06	0.37	0.29
17	12.34	4.77	3.53	0.39	0.31
18	13.10	5.26	4.07	0.40	0.32
19	13.93	5.80	4.68	0.42	0.34
20	14.83	6.40	5.39	0.43	0.36
21	15.82	7.07	6.20	0.45	0.38
22	16.88	7.82	7.13	0.46	0.40
23	18.05	8.66	8.20	0.48	0.42
24	19.32	9.60	9.44	0.50	0.45
25	20.72	10.66	10.88	0.51	0.47
26	22.25	11.85	12.54	0.53	0.49
27	23.94	13.20	14.47	0.55	0.51
28	25.80	14.72	16.72	0.57	0.53
29	27.86	16.44	19.34	0.59	0.55
30	30.14	18.40	22.40	0.61	0.58
31	32.67	20.63	25.99	0.63	0.60
32	35.49	23.18	30.22	0.65	0.62
33	38.64	26.09	35.19	0.68	0.65
34	42.16	29.44	41.06	0.70	0.67
35	46.12	33.30	48.03	0.72	0.70
36	50.59	37.75	56.31	0.75	0.73
37	55.63	42.92	66.19	0.77	0.75
38	61.35	48.93	78.03	0.80	0.78
39	67.87	55.96	92.25	0.82	0.81
40	75.31	64.20	109.41	0.85	0.84
41	83.86	73.90	130.22	0.88	0.87
42	93.71	85.38	155.55	0.91	0.90
43	105.11	99.02	186.54	0.94	0.93
44	118.37	115.31	224.64	0.97	0.97
45	133.88	134.88	271.76	1.01	1.00
46	152.10	158.51	330.35	1.04	1.04
47	173.64	187.21	403.67	1.08	1.07
48	199.26	222.31	496.01	1.12	1.11
49	229.93	265.51	613.16	1.15	1.15
50	266.89	319.07	762.89	1.20	1.19

[a] After Vesic (1973)

Table 3.3 Relationships for N_γ Frequently Referred to in Other Texts and References

Relationship	Source
$N_\gamma = (N_q - 1) \tan (1.4\phi)$ $\uparrow$ Eq. (3.17)	Meyerhof (1963)
$N_q = 1.5(N_q - 1) \tan \phi$ $\uparrow$ Eq. (3.17)	Hansen (1970)
By numerical methods from theory of plasticity	Lundgren and Mortensen (1953)

ϕ (deg)	N_γ
0	0
5	0.17
10	0.46
15	1.4
25	6.92
30	15.32
35	35.19
40	86.46
45	215.0

Shape, Depth, and Inclination Factors

The relationships for the shape factors, depth factors, and inclination factors *recommended for use in this text* are shown in Table 3.4. Other relationships generally found in many texts and references are shown in Table 3.5.

Net Ultimate Bearing Capacity

The net ultimate bearing capacity is defined as the ultimate pressure per unit area of the foundation that can be supported by the soil in excess of the pressure caused by the surrounding soil at the foundation level. Assuming that the difference between the unit weight of concrete used in the foundation and the unit weight of soil surrounding is negligible

$$q_{net(u)} = q_u - q \tag{3.20}$$

where $q_{net(u)}$ = net ultimate bearing capacity

General Comments

When the water table is present at or near the foundation, the factors q and γ given in the general bearing capacity equation will need modifications. The procedure for modifying them is the same as that described in Section 3.4.

Table 3.4 Shape, Depth, and Inclination Factors Recommended for Use

Factor	Relationship	Source
Shape[a]	$F_{cs} = 1 + \dfrac{B}{L} \cdot \dfrac{N_q}{N_c}$ $F_{qs} = 1 + \dfrac{B}{L} \tan \phi$ $F_{\gamma s} = 1 - 0.4 \dfrac{B}{L}$ where L = length of the foundation ($L > B$)	De Beer (1970)
Depth[b]	*Condition (a)*: $D_f/B \leq 1$ $F_{cd} = 1 + 0.4 \dfrac{D_f}{B}$ $F_{qd} = 1 + 2 \tan \phi \cdot (1 - \sin \phi)^2 \dfrac{D_f}{B}$ $F_{\gamma d} = 1$ *Condition (b)*: $D_f/B > 1$ $F_{cd} = 1 + (0.4) \tan^{-1}\left(\dfrac{D_f}{B}\right)$ $F_{qd} = 1 + 2 \tan \phi (1 - \sin \phi)^2 \cdot \tan^{-1}\left(\dfrac{D_f}{B}\right)$ $F_{\gamma d} = 1$	Hansen (1970)
Inclination	$F_{ci} = F_{qi} = \left(1 - \dfrac{\beta°}{90°}\right)^2$ $F_{\gamma i} = \left(1 - \dfrac{\beta}{\phi}\right)^2$ where β = inclination of the load on the foundation with respect to the vertical	Meyerhof (1963); Hanna and Meyerhof (1981)

[a] These shape factors are empirical relations based on extensive laboratory tests
[b] The factor $\tan^{-1}(D_f/B)$ is in radians

For undrained loading conditions ($\phi = 0$ concept) in clayey soils, the general load-bearing capacity equation [Eq. (3.16)] takes the form (vertical load)

$$q_u = cN_c \, F_{cs} \, F_{cd} + q \qquad\qquad (3.21)$$

Hence, the net ultimate bearing capacity is equal to (vertical load)

$$q_{net(u)} = q_u - q = cN_c \, F_{cs} \, F_{cd} \qquad\qquad (3.22)$$

Table 3.5 Shape, Depth, and Inclination Factors Recommended in Other Texts and References

Factor	Relationship	Source
✓ Shape[a]	*For* $\phi = 0$: $F_{cs} = 1 + 0.2\left(\dfrac{B}{L}\right)$ $F_{qs} = 1$ $F_{\gamma s} = 1$ *For* $\phi \geq 10°$: $F_{cs} = 1 + 0.2\left(\dfrac{B}{L}\right)\tan^2\left(45 + \dfrac{\phi}{2}\right)$ $F_{qs} = F_{\gamma s}$ $= 1 + 0.1\left(\dfrac{B}{L}\right)\tan^2\left(45 + \dfrac{\phi}{2}\right)$	Meyerhof (1953)
✓ Depth	*For* $\phi = 0$: $F_{cd} = 1 + 0.2\left(\dfrac{D_f}{B}\right)$ $F_{dq} = F_{\gamma d} = 1$ *For* $\phi \geq 10°$: $F_{cd} = 1 + 0.2\left(\dfrac{D_f}{B}\right)\tan\left(45 + \dfrac{\phi}{2}\right)$ $F_{qd} = F_{\gamma d}$ $= 1 + 0.1\left(\dfrac{D_f}{B}\right)\tan\left(45 + \dfrac{\phi}{2}\right)$	Meyerhof (1963)
Inclination	$F_{ci} = F_{qi} - \dfrac{(1 - F_{qi})}{(N_q - 1)}$ $F_{qi} = \left[1 - \dfrac{(0.5)(Q_u)\sin\beta}{Q_u\cos\beta + BLc\cot\phi}\right]^5$ $F_{\gamma i} = \left[1 - \dfrac{(0.7)(Q_u)\sin\beta}{Q_u\cos\beta + BLc\cot\phi}\right]^5$	Hansen (1970)

[a] L = length ($\geq B$)

Skempton (1951) has proposed the following equation for the net ultimate bearing capacity for clayey soils ($\phi = 0$ condition), which is similar to Eq. (3.22)

$$q_{net(u)} = 5c\left(1 + 0.2\,\frac{D_f}{B}\right)\left(1 + 0.2\,\frac{B}{L}\right) \tag{3.23}$$

Example 3.3

A square column foundation has to carry a gross allowable total load of 150 kN. The depth of the foundation is 0.7 m. The load is inclined at an angle of 20° to the

Figure 3.7

vertical (Figure 3.7). Determine the width of the foundation, B. Use Eq. (3.16) and a factor of safety of 3.

Solution

With $c = 0$, the ultimate bearing capacity [Eq. (3.16)] becomes

$$q_u = qN_q F_{qs} F_{qd} F_{qi} + \frac{1}{2} \gamma B N_\gamma F_{\gamma s} F_{\gamma d} F_{\gamma i}$$

$$q = (0.7)(18) = 12.6 \text{ kN/m}^2$$
$$\gamma = 18 \text{ kN/m}^3$$

From Table 3.2, for $\phi = 30°$

$$N_q = 18.4$$
$$N_\gamma = 22.4$$

From Table 3.4,

$$F_{qs} = 1 + \left(\frac{B}{L}\right) \tan \phi = 1 + 0.577 = 1.577$$

$$F_{\gamma s} = 1 - 0.4\left(\frac{B}{L}\right) = 0.6$$

$$F_{qd} = 1 + 2 \tan \phi (1 - \sin \phi)^2 \frac{D_f}{B} = 1 + \frac{(0.289)(0.7)}{B} = 1 + \frac{0.202}{B}$$

$$F_{\gamma d} = 1$$

$$F_{qi} = \left(1 - \frac{\beta°}{90°}\right)^2 = \left(1 - \frac{20}{90}\right)^2 = 0.605$$

$$F_{\gamma i} = \left(1 - \frac{\beta°}{\phi}\right)^2 = \left(1 - \frac{20}{30}\right)^2 = 0.11$$

Hence

$$q_u = (12.6)(18.4)(1.577)\left(1 + \frac{0.202}{B}\right)(0.605) + (0.5)(18)(B)(22.4)(0.6)(1)(0.11)$$

$$= 212.2 + \frac{44.68}{B} + 13.3B \tag{a}$$

Thus

$$q_{all} = \frac{q_u}{3} = 73.73 + \frac{14.89}{B} + 4.43B \qquad \text{(b)}$$

Given Q = total allowable load = $q_{all} \times B^2$ or

$$q_{all} = \frac{150}{B^2} \qquad \text{(c)}$$

Equating the right-hand sides of Eqs. (b) and (c)

$$\frac{150}{B^2} = 73.73 + \frac{14.89}{B} + 4.43B$$

By trial and error, $B \approx 1.3$ m

3.6   The Factor of Safety

In the preceding problems, in order to calculate the gross allowable load-bearing capacity of shallow foundations, a factor of safety (FS) has been applied to the gross ultimate bearing capacity, that is

$$q_{all} = \frac{q_u}{FS} \qquad (3.24)$$

However, some practicing engineers prefer to use a factor of safety as

$$\frac{\text{net stress increase}}{\text{on soil}} = \frac{\text{net ultimate bearing capacity}}{FS} \qquad (3.25)$$

Net ultimate bearing capacity has been defined in Eq. (3.20) as

$$q_{net(u)} = q_u - q$$

Substituting this equation in Eq. (3.25), one obtains

net stress increase on soil

= load from the superstructure per unit area of the foundation

$$= q_{all(net)} = \frac{q_u - q}{FS} \qquad (3.26)$$

The factor of safety as defined by Eq. (3.26) may be kept at least about 3 in all cases.

Another type of factor of safety for the bearing capacity of shallow foundations is often used. This is the factor of safety with respect to shear failure (FS_{shear}). In most cases, a value of $FS_{shear} = 1.4$–1.6 is desirable along with a *minimum* factor of safety of 3–4 against gross or net ultimate bearing capacity. In order to calculate the net allowable load on the basis of a given FS_{shear}, the following procedure should be adopted:

1. Let c and ϕ be the cohesion and the angle of friction of soil, and let FS_{shear} be the required factor of safety with respect to shear failure. So,

the developed cohesion and the angle of friction can be given as

$$c_d = \frac{c}{FS_{shear}} \tag{3.27}$$

$$\phi_d = \tan^{-1}\left(\frac{\tan\phi}{FS_{shear}}\right) \tag{3.28}$$

2. The gross allowable bearing capacity can now be calculated according to Eqs. (3.3), (3.7), (3.8) or the general bearing capacity equation [Eq. (3.16)] using c_d and ϕ_d as the shear strength parameters of the soil. For example, the gross allowable bearing capacity of a continuous foundation according to Terzaghi's equation can be written as

$$q_{all} = c_d N_c + q N_q + \frac{1}{2}\gamma B N_\gamma \tag{3.29}$$

where N_c, N_q, and N_γ = bearing capacity factors for friction angle, ϕ_d

3. The net allowable bearing capacity is thus

$$q_{net(all)} = q_{all} - q = c_d N_c + q(N_q - 1) + \frac{1}{2}\gamma B N_\gamma \tag{3.30}$$

Irrespective of the procedure by which the factor of safety is applied, the magnitude of FS should depend on the uncertainties and risks involved for a given condition.

Example 3.4

Refer to Example 3.1. Determine the net allowable load for the foundation using the definition of factor of safety given by Eq. (3.26). Use $FS = 4$.

Solution

From Example 3.1

$$q_u = 535 \text{ kN/m}^2$$

$$q = (1)(17.8) = 17.8 \text{ kN/m}^2$$

$$q_{net(all)} = \frac{q_u - q}{FS} = \frac{535 - 17.8}{4} = 129.3 \text{ kN/m}^2$$

Hence

$$Q_{net(all)} = (129.3)(1.5)(1.5) \approx 291 \text{ kN}$$

Example 3.5

Refer to Example 3.1. Determine the net allowable load for the foundation using a $FS_{shear} = 1.5$.

Solution

Given $c = 15.2 \text{ kN/m}^2$ and $\phi = 20°$,

$$c_d = \frac{c}{FS_{\text{shear}}} = \frac{15.2}{1.5} = 10.1 \text{ kN/m}^2$$

$$\phi_d = \tan^{-1}\left[\frac{\tan \phi}{FS_{\text{shear}}}\right] = \tan^{-1}\left[\frac{\tan 20}{1.5}\right] = 13.64°$$

$$q_{\text{all(net)}} = 1.3c_d N_c + q(N_q - 1) + 0.4\gamma B N_\gamma$$

For $\phi = 13.64°$, the values of the bearing capacity factors are as follows (Figure 3.4):

$$N_\gamma = 2$$
$$N_q = 3.8$$
$$N_c = 12$$

So

$$q_{\text{all(net)}} = 1.3(10.1)(12) + (17.8)(3.8 - 1) + (0.4)(17.8)(1.5)(2)$$
$$= 229.23 \text{ kN/m}^2$$

So

$$Q_{\text{all(net)}} = (229.23)(1.5)(1.5) = \textbf{515.8 kN}$$

Note: There appears to be a large discrepancy between the results of Examples 3.4 (or 3.1) and 3.5. By trial and error, it can be shown that when FS_{shear} is about 2, the results will be approximately equal.

3.7   Eccentrically Loaded Foundations

In several instances, such as with the base of a retaining wall, foundations are subjected to moments in addition to the vertical load, as shown in Figure 3.8a. In such cases, the distribution of pressure by the foundation on the soil is not uniform. The distribution of nominal pressure can be given by

$$q_{\text{max}} = \frac{Q}{BL} + \frac{6M}{B^2 L} \tag{3.31}$$

and

$$q_{\text{min}} = \frac{Q}{BL} - \frac{6M}{B^2 L} \tag{3.32}$$

where Q = total vertical load
  M = moment on the foundation

The exact distribution of pressure is difficult to estimate.

The factor of safety for such types of loading against bearing capacity failure can be evaluated by using the procedure suggested by Meyerhof (1953), which is generally referred to as the *effective area* method. The following is Meyerhof's step-by-step procedure for determination of the ulti-

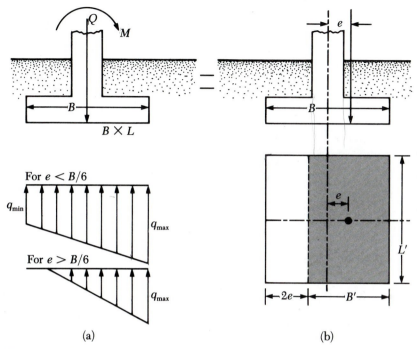

Figure 3.8 Eccentrically loaded foundations

mate load that the soil can support and the factor of safety against bearing capacity failure.

1. Figure 3.8b shows a force system equivalent to that shown in Figure 3.8a. The distance e is the eccentricity, or

$$e = \frac{M}{Q} \tag{3.33}$$

Substituting Eq. (3.33) in Eqs. (3.31) and (3.32) gives

$$q_{max} = \frac{Q}{BL}\left(1 + \frac{6e}{B}\right) \tag{3.34a}$$

and

$$q_{min} = \frac{Q}{BL}\left(1 - \frac{6e}{B}\right) \tag{3.34b}$$

Note that, in these equations, when the eccentricity, e, becomes equal to $B/6$, q_{min} is equal to zero. For $e > B/6$, q_{min} will be negative, which means that tension will develop. Because soil cannot take any tension, there will be a separation between the foundation and the soil underlying it. The nature of the pressure distribution on the soil will be as shown in Figure 3.8a. The value of q_{max} can be given by the expression

$$q_{max} = \frac{4Q}{3L(B - 2e)} \tag{3.35}$$

2. Determine the effective dimensions of the foundation as

$B' = $ effective width $= B - 2e$

$L' = $ effective length $= L$

Note that, if the eccentricity were in the direction of the length of the foundation, the value of L' would be equal to $L - 2e$. The value of B' would be equal to B. The smaller of the two dimensions (that is, L' and B') is the effective width of the foundation.

3. Use Eq. (3.16) for the ultimate bearing capacity as

$$q'_u = cN_c F_{cs} F_{cd} F_{ci} + qN_q F_{qs} F_{qd} F_{qi} + \frac{1}{2}\gamma B'N_\gamma F_{\gamma s} F_{\gamma d} F_{\gamma i} \tag{3.36}$$

For the evaluation of F_{cs}, F_{qs}, and $F_{\gamma s}$, Table 3.4 has to be used with *effective length* and *effective width* dimensions in place of L and B, respectively.

For determination of F_{cd}, F_{qd}, and $F_{\gamma d}$, use Table 3.4 (*do not* replace B with B').

4. The total ultimate load that the foundation can sustain is

$$Q_{ult} = q'_u \overbrace{(B')(L')}^{A'} \tag{3.37}$$

where $A' = $ effective area

5. The factor of safety against bearing capacity failure is given as

$$FS = \frac{Q_{ult}}{Q} \tag{3.38}$$

As we can see, eccentricity tends to decrease the load-bearing capacity of a foundation. In such cases, it is probably advantageous to place the foundation columns off center, as shown in Figure 3.9. This, in effect, produces a centrally loaded foundation with uniformly distributed pressure.

Foundations with Two-Way Eccentricity

Consider a situation in which a foundation is subjected to a vertical ultimate load Q_{ult} and a moment M as shown in Figure 3.10a and b. For this case, the

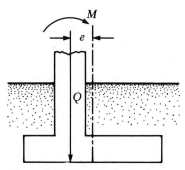

Figure 3.9 Foundation of columns with off-center loading

components of the moment, M, about the x and y axes can be determined as M_x and M_y, respectively (Figure 3.10c). This condition is equivalent to a load Q_{ult} placed eccentrically on the foundation with $x = e_B$ and $y = e_L$ (Figure 3.10d). Note that

$$e_B = \frac{M_y}{Q_{\text{ult}}} \tag{3.39}$$

and

$$e_L = \frac{M_x}{Q_{\text{ult}}} \tag{3.40}$$

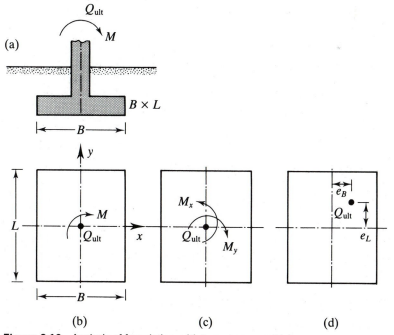

Figure 3.10 Analysis of foundation with two-way eccentricity

If it is necessary to find Q_{ult}, it can be obtained as follows [Eq. (3.37)]:

$$Q_{ult} = q'_u \cdot A'$$

where, from Eq. (3.36)

$$q'_u = cN_c F_{cs} F_{cd} F_{ci} + qN_q F_{qs} F_{qd} F_{qi} + \frac{1}{2} \gamma B' N_\gamma F_{\gamma s} F_{\gamma d} F_{\gamma i}$$

and

$$A' = \text{effective area} = B'L'$$

As before, for evaluation of F_{cs}, F_{qs}, and $F_{\gamma s}$ (Table 3.4), use the effective length (L') and effective width (B') dimensions in place of L and B, respectively. For calculating F_{cd}, F_{qd}, and $F_{\gamma d}$, use Table 3.4; however, do not replace B with B'. In determining the effective area (A'), effective width (B'), and effective length (L'), four possible cases may arise. They are discussed below (Highter and Anders, 1985).

Case I: $e_L/L \geq \frac{1}{6}$ and $e_B/B \geq \frac{1}{6}$. The effective area for this condition is shown in Figure 3.11, or

$$A' = \frac{1}{2} B_1 L_1 \tag{3.41}$$

where

$$B_1 = B\left(1.5 - \frac{3e_B}{B}\right) \tag{3.42a}$$

and

$$L_1 = L\left(1.5 - \frac{3e_L}{L}\right) \tag{3.42b}$$

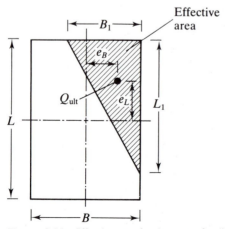

Figure 3.11 Effective area for the case of $e_L/L \geq \frac{1}{6}$ and $e_B/B \geq \frac{1}{6}$

Effective length, L', is equal to the larger of the two dimensions, that is, B_1 or L_1. So, effective width is equal to

$$B' = \frac{A'}{L'} \tag{3.43}$$

Case II: $e_L/L < 0.5$ and $0 < e_B/B < \frac{1}{6}$. The effective area for this case is shown in Figure 3.12a.

$$A' = \frac{1}{2}(L_1 + L_2)B \tag{3.44}$$

(a)

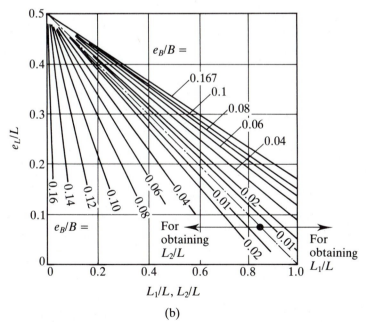

(b)

Figure 3.12 Effective area for the case of $e_L/L < 0.5$ and $0 < e_B/B < \frac{1}{6}$ (after Higher and Anders, 1985)

The magnitudes of L_1 and L_2 can be determined from Figure 3.12b. Effective width is equal to

$$B' = \frac{A'}{L_1 \text{ or } L_2} \quad \text{(whichever is larger)} \tag{3.45}$$

Effective length is equal to

$$L' = L_1 \text{ or } L_2 \quad \text{(whichever is larger)} \tag{3.46}$$

(a)

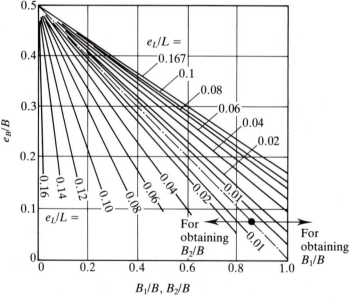

(b)

Figure 3.13 Effective area for the case of $e_L/L < \frac{1}{6}$ and $0 < e_B/B < 0.5$ (after Highter and Anders, 1985)

Case III: $e_L/L < \frac{1}{6}$ and $0 < e_B/B < 0.5$. The effective area is shown in Figure 3.13a.

$$A' = \frac{1}{2}(B_1 + B_2)L \tag{3.47}$$

Effective width is equal to

$$B' = \frac{A'}{L} \tag{3.48}$$

(a)

(b)

Figure 3.14 Effective area for the case of $e_L/L < \frac{1}{6}$ and $e_B/B < \frac{1}{6}$ (after Highter and Anders, 1985)

Effective length is equal to

$$L' = L \tag{3.49}$$

The magnitudes of B_1 and B_2 can be determined from Figure 3.13b.

Case IV: $e_L/L < \frac{1}{6}$ and $e_B/B < \frac{1}{6}$. Figure 3.14a shows the effective area for this case. The ratio B_2/B and thus B_2 can be determined by using the e_L/L curves sloping upward. Similarly, the ratio L_2/L and thus L_2 can be determined by using the e_L/L sloping downward. The effective area is then

$$A' = L_2 B + \frac{1}{2} (B + B_2)(L - L_2) \tag{3.50}$$

The effective width is

$$B' = \frac{A'}{L} \tag{3.51}$$

Effective length is equal to

$$L' = L \tag{3.52}$$

Example 3.6

A square foundation is shown in Figure 3.15. Assume that the load eccentricity $e = 0.15$ m. Determine the ultimate load, Q_{ult}.

Solution

With $c = 0$, Eq. (3.36) becomes

$$q'_u = q N_q F_{qs} F_{qd} F_{qi} + \frac{1}{2} \gamma B' N_\gamma F_{\gamma s} F_{\gamma d} F_{\gamma i}$$

$$q = (0.7)(18) = 12.6 \text{ kN/m}^2$$

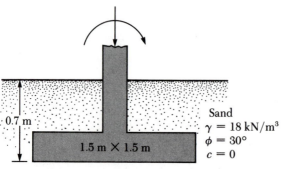

Sand
$\gamma = 18$ kN/m³
$\phi = 30°$
$c = 0$

0.7 m

1.5 m × 1.5 m

Figure 3.15

For $\phi = 30°$, from Table 3.2, $N_q = 18.4$ and $N_\gamma = 22.4$.

$$B' = 1.5 - 2(0.15) = 1.2 \text{ m}$$
$$L' = 1.5 \text{ m}$$

From Table 3.4

$$F_{qs} = 1 + \frac{B'}{L'} \tan \phi = 1 + \left(\frac{1.2}{1.5}\right) \tan 30° = 1.462$$

$$F_{qd} = 1 + 2 \tan \phi (1 - \sin \phi)^2 \frac{D_f}{B} = 1 - \frac{(0.289)(0.7)}{1.5} = 1.135$$

$$F_{\gamma s} = 1 - 0.4\left(\frac{B'}{L'}\right) = 1 - 0.4\left(\frac{1.2}{1.5}\right) = 0.68$$

$$F_{\gamma d} = 1$$

So

$$q'_u = (12.6)(18.4)(1.462)(1.135) + \frac{1}{2}(18)(1.2)(22.4)(0.68)(1)$$

$$= 384.3 + 164.50 = 548.8 \text{ kN/m}^2$$

Hence

$$Q_{\text{ult}} = B'L'(q'_u) = (1.2)(1.5)(548.8) = \textbf{988 kN}$$

Example 3.7

Refer to Example 3.6. Other quantities remaining the same, assume that the load has a two-way eccentricity. Given: $e_L = 0.3$ m and $e_B = 0.15$ m (Figure 3.16). Determine the ultimate load, Q_{ult}.

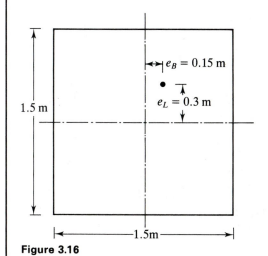

Figure 3.16

Solution

$$\frac{e_L}{L} = \frac{0.3}{1.5} = 0.2$$

$$\frac{e_B}{B} = \frac{0.15}{1.5} = 0.1$$

This is similar to the case shown in Figure 3.12a. From Figure 3.12b, for $e_L/L = 0.2$ and $e_B/B = 0.1$

$$\frac{L_1}{L} \approx 0.85; \qquad L_1 = (0.85)(1.5) = 1.275 \text{ m}$$

and

$$\frac{L_2}{L} \approx 0.21; \qquad L_2 = (0.21)(1.5) = 0.315 \text{ m}$$

From Eq. (3.44)

$$A' = \frac{1}{2}(L_1 + L_2)B = \frac{1}{2}(1.275 + 0.315)(1.5) = 1.193 \text{ m}^2$$

From Eq. (3.46)

$$L' = L_1 = 1.275 \text{ m}$$

From Eq. (3.45)

$$B' = \frac{A'}{L_1} = \frac{1.193}{1.275} = 0.936 \text{ m}$$

Now, from Eq. (3.36) $(c = 0)$

$$q'_u = qN_q F_{qs} F_{qd} F_{qi} + \frac{1}{2}\gamma B' N_\gamma F_{\gamma s} F_{\gamma d} F_{\gamma i}$$

From Example 3.6, $q = 12.6$ kN/m^2, $N_q = 18.4$, $N_\gamma = 22.4$, $F_{qd} = 1.135$, and $F_{\gamma d} = 1$.

$$F_{qs} = 1 + \left(\frac{B'}{L'}\right) \tan \phi = 1 + \left(\frac{0.936}{1.275}\right) \tan 30° = 1.424$$

$$F_{\gamma s} = 1 - 0.4\left(\frac{B'}{L'}\right) = 1 - 0.4\left(\frac{0.936}{1.275}\right) = 0.706$$

So

$$Q_{\text{ult}} = A'q'_u = A'\left(qN_q F_{qs} F_{qd} + \frac{1}{2}\gamma B' N_\gamma F_{\gamma s} F_{\gamma d}\right)$$

$$= (1.193)[(12.6)(18.4)(1.424)(1.135)$$
$$+ (0.5)(18)(0.936)(22.4)(0.706)(1)]$$
$$= 605.95 \text{ kN}$$

3.8   Some Special Cases of Ultimate Bearing Capacity

The bearing capacity equations presented in the preceding sections involve cases in which the soil supporting the foundation is homogeneous and extends to a considerable depth. It was assumed that cohesion, angle of friction, and unit weight of soil remained constant for the bearing capacity analysis. However, in practice, layered soil profiles are often encountered. In such instances, the failure surface at ultimate load may extend through two or more soil layers. Determination of ultimate bearing capacity in layered soils can be made in only a limited number of cases. In this section, the procedure for the estimation of bearing capacity for layered soils as proposed by Meyerhof and Hanna (1978) and Meyerhof (1974) will be elaborated upon.

Foundation on Layered Clay ($\phi = 0$ Condition)

Figure 3.17 shows a shallow foundation on a two-layered clay soil ($\phi = 0$). Soil layer I extends to a depth of H measured below the bottom of the foundation. Soil layer II extends to a great depth from that point downward. Since γ_1 and γ_2, the unit weights of the two soil layers, will not be too different, two possible cases may arise. They are as follows:

   1. If $c_{u(1)}/c_{u(2)} > 1$, the foundation is on a strong clay layer underlain by a weak clay layer.
   2. If $c_{u(1)}/c_{u(2)} < 1$, the foundation is on a weak clay layer underlain by a strong clay layer.

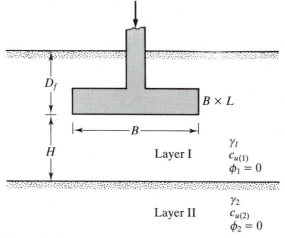

Figure 3.17   Foundation on layered soil

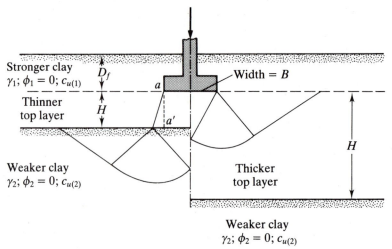

Stronger clay
γ_1; $\phi_1 = 0$; $c_{u(1)}$

Width = B

Thinner
top layer

Weaker clay
γ_2; $\phi_2 = 0$; $c_{u(2)}$

Thicker
top layer

Weaker clay
γ_2; $\phi_2 = 0$; $c_{u(2)}$

Figure 3.18 Bearing capacity of layered clay—stronger over weaker

The ultimate bearing capacity theories for each of the cases are described below.

Case I: $c_{u(1)}/c_{u(2)} > 1$. If the ratio of H/B is relatively small, failure in the soil under the foundation will take place by punching in the top soil layer followed by a general shear failure in the bottom soil layer, as shown in the left half of Figure 3.18. However, if H/B is relatively large, the failure surface in the soil at ultimate load will be fully contained in the top soil layer, as shown in the right half of Figure 3.18. For such a case (Meyerhof and Hanna, 1978), the ultimate bearing capacity can be expressed as

$$q_u = \underbrace{\left[1 + 0.2\left(\frac{B}{L}\right)\right]c_{u(2)}N_c}_{\substack{\text{From general shear} \\ \text{failure of bottom} \\ \text{soil layer}}} + \underbrace{\left(1 + \frac{B}{L}\right)\left(\frac{2c_a H}{B}\right) + \gamma_1 D_f}_{\substack{\text{From punching failure} \\ \text{of top soil layer}}}$$

$$\leqslant \underbrace{\left[1 + 0.2\left(\frac{B}{L}\right)\right]c_{u(1)}N_c + \gamma_1 D_f}_{\substack{\text{From general shear failure} \\ \text{of top soil layer}}} \tag{3.53}$$

where B = width of foundation
L = length of foundation
N_c = 5.4 (bearing capacity factor for $\phi = 0$; Table 3.2)
c_a = adhesion along the interface aa'

The variation of $c_a/c_{u(1)}$ with $c_{u(2)}/c_{u(1)}$ is shown in Figure 3.19.

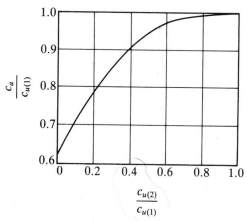

Figure 3.19 Variation of $c_a/c_{u(1)}$ with $c_{u(2)}/c_{u(1)}$ according to Meyerhof and Hanna's theory

Case II: $c_{u(1)}/c_{u(2)} < 1$. The ultimate bearing capacity for this can be given as follows (Meyerhof, 1974; Meyerhof and Hanna, 1978):

$$q_u = q_t + (q_b - q_t)\left(1 - \frac{H}{H_f}\right)^2 \geq q_t \tag{3.54}$$

where

$$q_t = \left[1 + 0.2\left(\frac{B}{L}\right)\right]c_{u(1)}N_c + \gamma_1 D_f \tag{3.55}$$

$$q_b = \left[1 + 0.2\left(\frac{B}{L}\right)\right]c_{u(2)}N_c + \gamma_2 D_f \tag{3.56}$$

$$H_f \approx B \tag{3.57}$$

Note that $N_c = 5.14$ since $\phi = 0$ (Table 3.2).

Foundations on Dense or Compacted Sand Overlying Soft Clay

In some cases, foundations are constructed on compacted sand layers that overlie soft clay soils. If the thickness of the sand layer under the foundation is relatively small, the failure surface may extend into the soft clay layer. This is shown in the left half of Figure 3.20. However, if the sand layer under the foundation is large, the failure surface will lie entirely in the sand layer, as shown in the right half of Figure 3.20. According to Meyerhof (1974), in this case the ultimate bearing capacity of a continuous foundation can be given by

$$q_u = cN_c + \gamma H^2\left(1 + \frac{2D_f}{H}\right)K_s\frac{\tan\phi}{B} + \gamma D_f \tag{3.58}$$

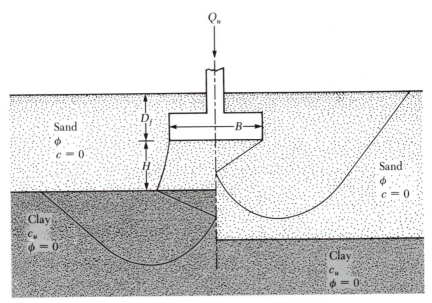

Figure 3.20 Foundation on compacted sand layer overlying soft clay

with a maximum of

$$q_u = \frac{1}{2}\gamma B N_\gamma + \gamma D_f N_q \tag{3.59}$$

where ϕ = angle of friction of top sand layer
γ = unit weight of sand
K_s = punching shear resistance coefficient

N_γ and N_q correspond to the angle of friction, ϕ, for sand (Table 3.2). *Note:* for $\phi = 0$ condition, $N_c = 5.14$ as determined from Table 3.2.

For rectangular foundations

$$q_u = \left(1 + 0.2\frac{B}{L}\right)c_u N_c + \left(1 + \frac{B}{L}\right)\gamma H^2\left(1 + \frac{2D_f}{H}\right)K_s\frac{\tan\phi}{B} + \gamma D_f \tag{3.60}$$

with a maximum of

$$q_u = \frac{1}{2}\left(1 - 0.4\frac{B}{L}\right)\gamma B N_\gamma + \gamma D_f N_q \tag{3.61}$$

The variation of the punching shear resistance factor, K_s, is given in Figure 3.21. At this point, it needs to be pointed out that Eqs. (3.59) and (3.61) are estimates of the values of q_u for continuous and rectangular foundations, respectively, in the upper sand layer. This corresponds to the condition shown in the right half of Figure 3.20.

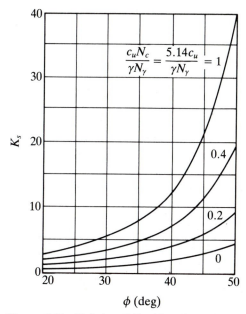

Figure 3.21   Variation of K_s with ϕ_1 based on Meyerhof and Hanna's theory

Example 3.8

Refer to Figure 3.17. A foundation 1.5 m × 1 m is located at a depth (D_f) of 1 m in a clay. A soft clay layer is located at a depth (H) of 1 m measured from the bottom of the foundation. Given:

For top clay layer:

Undrained shear strength = 120 kN/m²

Unit weight = 16.8 kN/m³

For bottom clay layer:

Undrained shear strength = 48 kN/m²

Unit weight = 16.2 kN/m³

Determine the gross allowable load for the foundation with a factor of safety of 4.

Solution

Given: $c_{u(1)} = 120$ kN/m²; $c_{u(2)} = 48$ kN/m². Since $c_{u(1)}/c_{u(2)} > 1$, Eq. (3.53) will apply, or

$$q_u = \left[1 + (0.2)\left(\frac{B}{L}\right)\right]c_{u(2)}N_c + \left(1 + \frac{B}{L}\right)\left(\frac{2c_a H}{B}\right) + \gamma_1 D_f$$

$$\leq \left[1 + (0.2)\left(\frac{B}{L}\right)\right]c_{u(1)}N_c + \gamma_1 D_f$$

where $B = 1$ m
$\quad L = 1.5$ m
$\quad D_f = 1$ m
$\quad H = 1$ m
$\quad N_c = 5.14$
$\quad \gamma_1 = 16.8$ kN/m^3

From Figure 3.19, for $c_{u(2)}/c_{u(1)} = 48/120 = 0.4$, the value of $c_a/c_{u(1)} \approx 0.9$. So

$$c_a = (0.9)(120) = 108 \text{ kN/m}^2$$

So

$$q_u = \left[1 + (0.2)\left(\frac{B}{L}\right)\right]c_{u(2)}N_c + \left(1 + \frac{B}{L}\right)\left(\frac{2c_a H}{B}\right) + \gamma_1 D_f$$

$$= \left[1 + (0.2)\left(\frac{1}{1.5}\right)\right](48)(5.14) + \left(1 + \frac{1}{1.5}\right)\left[\frac{(2)(108)(1)}{1}\right] + (16.8)(1)$$

$$= 279.6 + 360 + 16.8 = 656.4 \text{ kN/m}^2$$

Now check:

$$q_u = \left[1 + (0.2)\left(\frac{B}{L}\right)\right]c_{u(1)}N_c + \gamma_1 D_f$$

$$= \left[1 + (0.2)\left(\frac{1}{1.5}\right)\right](120)(5.14) + (16.8)(1)$$

$$= 699 + 16.8 = 715.8 \text{ kN/m}^2$$

So, $q_u = 656.4$ kN/m^2 (that is, the smaller of the two values calculated above)

$$q_{all} = \frac{q_u}{FS} = \frac{656.4}{4} = 164.1 \text{ kN/m}^2$$

Total allowable load is

$$(q_{all})(1 \times 1.5) = \textbf{246.15 kN}$$

$\equiv$

Example 3.9

Refer to Figure 3.20. Given:
  For sand:

$$\gamma = 117 \text{ lb/ft}^3$$
$$\phi = 40°$$

  For clay:

$$c_u = 400 \text{ lb/ft}^2$$

For foundation:

$$B = 3 \text{ ft}$$
$$L = 4.5 \text{ ft}$$
$$D_f = 3 \text{ ft}$$
$$H = 4 \text{ ft}$$

Determine the gross ultimate bearing capacity of the foundation.

Solution

Since the foundation is rectangular, Eqs. (3.60) and (3.61) will apply. For $\phi = 40°$, from Table 3.2, $N_\gamma = 109.41$.

$$\frac{c_u N_c}{\gamma N_\gamma} = \frac{(400)(5.14)}{(117)(109.41)} = 0.16$$

From Figure 3.21, for $c_u N_c/\gamma N_\gamma = 0.16$ and $\phi = 40°$, the value of $K_s \approx 3.1$. Using Eq. (3.60)

$$q_u = \left[1 + (0.2)\left(\frac{B}{L}\right)\right]c_u N_c + \left(1 + \frac{B}{L}\right)\gamma H^2 \left(1 + \frac{2D_f}{H}\right)K_s \frac{\tan \phi}{B} + \gamma D_f$$

$$= \left[1 + (0.2)\left(\frac{3}{4.5}\right)\right](400)(5.14) + \left(1 + \frac{3}{4.5}\right)(117)(4)^2$$

$$\times \left[1 + \frac{(2)(3)}{4}\right](3.1)\frac{\tan 40}{3} + (117)(3)$$

$$= 2330 + 6763 + 351 = 9444 \text{ lb/ft}^2$$

Again, from Eq. (3.61)

$$q_u = \frac{1}{2}\left[1 - (0.4)\left(\frac{B}{L}\right)\right]\gamma B N_\gamma + \gamma D_f N_q$$

For $\phi = 40°$, $N_q = 64.20$ (Table 3.2). So

$$q_u = (0.5)\left[1 - (0.4)\left(\frac{3}{4.5}\right)\right](117)(3)(109.41) + (117)(3)(64.20)$$

$$= 14,081 + 22,534 = 36,615 \text{ lb/ft}^2$$

Hence

$$q_u = 9444 \text{ lb/ft}^2$$

SETTLEMENT OF SHALLOW FOUNDATIONS

3.9   Types of Foundation Settlement

Foundation settlement under load can be classified according to two major types: *immediate* or *elastic, settlement*, S_e, and *consolidation settlement*, S_c. Immediate settlement of a foundation takes place during or immediately after the construction of the structure. Consolidation settlement is time dependent and takes place as the result of extrusion of the pore water from the void spaces of saturated clayey soils. The total settlement of a foundation is the sum of the elastic settlement and the consolidation settlement.

Consolidation settlement comprises two phases: *primary* consolidation settlement and *secondary* consolidation settlement. The fundamentals of primary consolidation settlement have been explained in detail in Section 1.14. Secondary consolidation settlement occurs after completion of the primary consolidation that is caused by slippage and reorientation of soil particles under sustained load. Primary consolidation settlement is more significant than secondary settlement in inorganic clays and silty clay soils. However, in organic soils, secondary consolidation settlement is more significant.

The settlement of foundations discussed in Section 3.2 for bearing capacity tests was primarily the immediate type. The procedure for calculating each type of foundation settlement is discussed in more detail in the following sections.

3.10   Immediate Settlement

Figure 3.22 shows a shallow foundation subjected to a net force per unit area equal to q_o. Let the Poisson's ratio and the Young's modulus of the soil supporting it be μ_s and E_s, respectively. Theoretically, if $D_f = 0$, $H = \infty$, and the foundation is perfectly flexible, the settlement can be expressed as follows (Harr, 1966):

$$S_e = \frac{Bq_o}{E_s}(1 - \mu_s^2)\frac{\alpha}{2} \qquad \text{(corner of the flexible foundation)} \qquad (3.62)$$

$$S_e = \frac{Bq_o}{E_s}(1 - \mu_s^2)\alpha \qquad \text{(center of the flexible foundation)} \qquad (3.63)$$

$$\text{where } \alpha = \frac{1}{\pi}\left[\ln\left(\frac{\sqrt{1 + m^2} + m}{\sqrt{1 + m^2} - m}\right) + m\ln\left(\frac{\sqrt{1 + m^2} + 1}{\sqrt{1 + m^2} - 1}\right)\right] \qquad (3.64)$$

$$m = B/L \qquad (3.65)$$

B = width of foundation
L = length of foundation

The values of α for various length-to-width (L/B) ratios are shown in Figure 3.23. The average immediate settlement for a flexible foundation can also be expressed as

$$\boxed{S_e = \frac{Bq_o}{E_s}(1 - \mu_s^2)\alpha_{av}} \qquad \text{(average for flexible foundation)} \qquad (3.66)$$

Figure 3.23 also shows the values of α_{av} for various types of foundation.

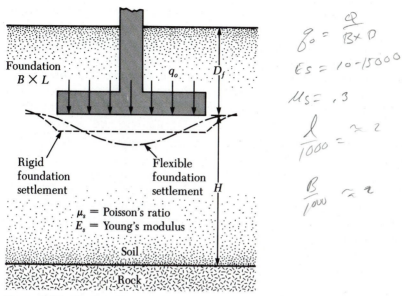

Figure 3.22 shows handwritten notes on the right side:

$$q_o = \frac{Q}{B \times D}$$

$$Es = 10\text{-}15000$$

$$\mu s = .3$$

$$\frac{\ell}{1000} \approx 2$$

$$\frac{B}{1000} \approx 2$$

Figure 3.22   Elastic settlement of flexible and rigid foundations

However, if the foundation shown in Figure 3.22 is rigid, the immediate settlement will be modified and can be expressed as

$$S_e = \frac{Bq_o}{E_s}(1 - \mu_s^2)\alpha_r \qquad \text{(rigid foundation)} \qquad (3.67)$$

The values of α_r for various types of foundation are given in Figure 3.23.

For circular foundation
$\alpha = 1$
$\alpha_{av} = 0.85$
$\alpha_r = 0.88$

Figure 3.23   Values of α, α_{av}, and α_r—Eqs. (3.62), (3.63), (3.66), (3.67)

The preceding equations for immediate settlement have been obtained by integrating the strain at any given depth below the foundations for limits of $z = 0$ to $z = \infty$. If an incompressible layer of rock is located at a limited depth, the actual settlement may be less than that calculated by the preceding equations. However, if the depth H in Figure 3.22 is greater than about $2B$ to $3B$, the actual settlement would not change considerably. Also note that the deeper the embedment, D_f, the less is the total elastic settlement.

3.11   Immediate Settlement of Foundations on Saturated Clay

Janbu, Bjerrum, and Kjaernsli (1956) proposed an equation for evaluation of the average settlement of flexible foundations on saturated clay soils (Poisson's ratio, $\mu_s = 0.5$). Referring to Figure 3.24 for notations, this equation can be written as

$$S_e = A_1 A_2 \frac{q_o B}{E_s}$$ (3.68)

where A_1 is a function of H/B and L/B, and A_2 is a function of D_f/B.
Christian and Carrier (1978) have modified the values of A_1 and A_2 to some extent, and these are presented in Figure 3.24.

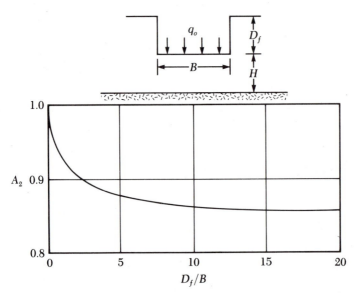

Figure 3.24 Values of A_1 and A_2 for elastic settlement calculation—Eq. (3.64) (after Christian and Carrier, 1978)

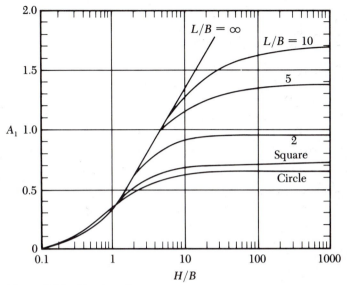

Figure 3.24   (Continued)

3.12   Immediate Settlement of Sandy Soil: Use of Strain Influence Factor

Immediate settlement of granular soils can also be evaluated by use of a semi-empirical *strain influence factor* (Figure 3.25) proposed by Schmertmann and Hartman (1978). According to this method, the settlement can be given by the equation

$$S_e = C_1 C_2 (\bar{q} - q) \sum_0^{z_2} \frac{I_z}{E_s} \Delta z \tag{3.69}$$

where I_z = strain influence factor
C_1 = a correction factor for the depth of foundation
embedment = $1 - 0.5[q/(\bar{q} - q)]$
C_2 = a correction factor to account for creep in soil
= $1 + 0.2 \log$ (time in years/0.1)
$\bar{q}$ = stress at the level of the foundation
$q = \gamma D_f$

The variation of the strain influence factor with depth below the foundation is shown in Figure 3.25a. Note that, for square or circular foundations,

$I_z = 0.1$ at $z = 0$
$I_z = 0.5$ at $z = z_1 = 0.5B$
$I_z = 0$ at $z = z_2 = 2B$

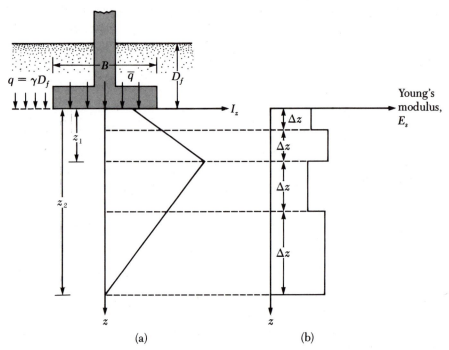

Figure 3.25   Elastic settlement calculation by using strain influence factor

Similarly, for foundations with $L/B \geq 10$,

$$I_z = 0.2 \text{ at } z = 0$$
$$I_z = 0.5 \text{ at } z = z_1 = B$$
$$I_z = 0 \text{ at } z = z_2 = 4B$$

where B = width of the foundation and L = length of the foundation. For values of L/B between 1 and 10, necessary interpolations can be made.

In order to use Eq. (3.69), one first needs to evaluate the approximate variation of Young's modulus with depth (Figure 3.25b). This can be done by using the standard penetration numbers or cone penetration resistances (Chapter 2). Then the soil layer can be divided into several layers up to a depth of $z = z_2$, and the immediate settlement of each layer can be estimated. The sum of the settlement of all layers is equal to S_e. The principle of settlement calculation using this procedure is demonstrated in Example 3.10.

3.13   Range of Material Parameters for Computing Immediate Settlement

Sections 3.10–3.12 presented the equations for the calculation of immediate settlement of foundations. These equations contain the elastic parameters, such as E_s and μ_s. If the laboratory test results for these parameters are not

available, certain realistic assumptions have to be made. Table 3.6 shows the
approximate range of the elastic parameters for various soils.

Table 3.6   Elastic Parameters of Various Soils

Type of soil	Young's modulus, E_s		Poisson's ratio, μ_s
	MN/m^2	$lb/in.^2$	
Loose sand	10.35– 24.15	1,500– 3,500	0.20–0.40
Medium dense sand	17.25– 27.60	2,500– 4,000	0.25–0.40
Dense sand	34.50– 55.20	5,000– 8,000	0.30–0.45
Silty sand	10.35– 17.25	1,500– 2,500	0.20–0.40
Sand and gravel	69.00–172.50	10,000–25,000	0.15–0.35
Soft clay	2.07– 5.18	300– 750	
Medium clay	5.18– 10.35	750– 1,500	0.20–0.50
Stiff clay	10.35– 24.15	1,500– 3,500	

Several investigators have correlated the values of Young's modulus
with the standard penetration number (N) and the cone penetration resistance (q_c). A list of these correlations has been compiled by Mitchell and
Gardner (1975). Schmertmann (1970) has indicated that Young's modulus
of sand can be given by the following correlations:

$$E_s \text{ (kN/m}^2) = 766N \tag{3.70}$$

where N = standard penetration number

In English units

$$E \text{ (U.S. ton/ft}^2) = 8N \tag{3.71}$$

Similarly

$$E_s = 2q_c \tag{3.72}$$

where q_c = static cone penetration resistance

Note: Any consistent set of units can be used in Eq. (3.72).
    The Young's modulus of normally consolidated clays can be estimated
as

$$E_s = 250c \text{ to } 500c \tag{3.73}$$

For overconsolidated clays

$$E_s = 750c \text{ to } 1000c \tag{3.74}$$

where c = undrained cohesion of clayey soil

Example 3.10

Figure 3.26a shows a shallow foundation on a deposit of sandy soil that is 3 m × 3 m in plan. The actual variation of the values of Young's modulus with depth determined by using the standard penetration numbers of Eq. (3.70) are also shown in Figure 3.26a. Using the strain influence factor method, estimate the settlement of the foundation five years after construction.

Solution

By observing the actual variation of Young's modulus with depth, one can plot an estimated idealized form of the variation of E_s, as shown in Figure 3.26a. Figure 3.26b shows the plot of the strain influence factor. The following table can now be prepared.

Depth (m)	Δz (m)	E_s (kN/m²)	Average I_z	$\dfrac{I_z}{E_s} \cdot \Delta z$ (m³/kN)
0–1	1	8,000	0.233	0.291×10^{-4}
1.0–1.5	0.5	10,000	0.433	0.217×10^{-4}
1.5–4	2.5	10,000	0.361	0.903×10^{-4}
4.0–6	2	16,000	0.111	0.139×10^{-4}
				$\Sigma = 1.55 \times 10^{-4}$

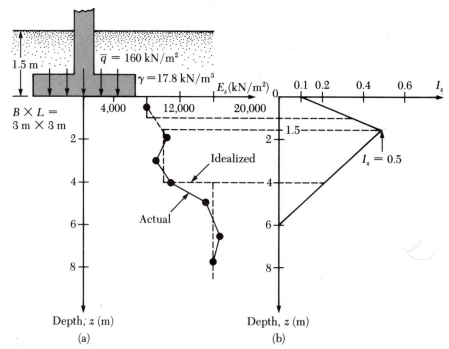

$\bar{q} = 160 \text{ kN/m}^2$

$\gamma = 17.8 \text{ kN/m}^3$

$B \times L = 3 \text{ m} \times 3 \text{ m}$

Idealized

Actual

$I_z = 0.5$

$E_s (\text{kN/m}^2)$

Depth, z (m)

(a)

Depth, z (m)

(b)

Figure 3.26

$$C_1 = 1 - 0.5\left(\frac{q}{\bar{q} - q}\right) = 1 - 0.5\left[\frac{17.8 \times 1.5}{160 - (17.8 \times 1.5)}\right] = 0.9$$

$$C_2 = 1 + 0.2 \log\left(\frac{5}{0.1}\right) = 1.34$$

Hence

$$S_e = C_1 \cdot C_2(\bar{q} - q) \sum_0^{2B} \frac{I_z}{E_s} \cdot \Delta z$$

$$= (0.9)(1.34)[160 - (17.8 \times 1.5)](1.55 \times 10^{-4})$$

$$= 249.2 \times 10^{-4} \text{ m} \approx \textbf{24.9 mm}$$

3.14 Immediate Settlement of Eccentrically Loaded Foundations

The settlement calculation procedure described in Sections 3.10, 3.11, and 3.12 relate to the settlement of centrally loaded foundations. An eccentrically loaded foundation will undergo vertical settlement and rotation as shown in Figure 3.27. Georgiadis and Butterfield (1988) have suggested a procedure for determining the settlement and rotation of a foundation under such loading conditions. This procedure is as follows:

1. Let the applied total load on the foundation (Q) and the load eccentricity (e) be known, and let it be required to determine the settlement, S_e, and the rotation angle, t (see Figure 3.27 for notations).

2. The ultimate load, $Q_{\text{ult}(e)}$, that the foundation can sustain can be evaluated by using Eq. (3.37) [Section 3.7; note the change of notation from Q_{ult} to $Q_{\text{ult}(e)}$].

3. Determine the factor for safety for the eccentrically loaded foundation as

$$FS = \frac{Q_{\text{ult}(e)}}{Q} = F_1 \tag{3.75}$$

4. Determine the ultimate load $Q_{\text{ult}(e=0)}$ for the same foundation with eccentricity $e = 0$ [centrally loaded foundation; Eq. (3.16)].

5. Determine:

$$\frac{Q_{\text{ult}(e=0)}}{F_1} = Q_{(e=0)} \tag{3.76}$$

Note that $Q_{(e=0)}$ is the allowable load for the foundation with a factor of safety $FS = F_1$ for central loading condition.

6. For the load $Q_{(e=0)}$ on the foundation, estimate the settlement by using the techniques presented in Sections 3.10–3.12. Let the settlement determined by any one of the methods be equal to $S_{e(e=0)}$.

Point of load application

L

B

Eccentricity = e

S_e

Rotation = t

Figure 3.27   Elastic settlement of eccentrically loaded foundation

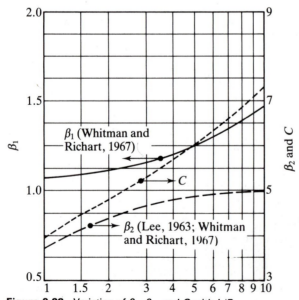

Figure 3.28   Variation of β_1, β_2, and C with L/B

7. Now, use the following equations to determine S_e and t:

$$S_e = S_{e(e=0)}\left[1 - 2\left(\frac{e}{B}\right)\right]^2 \qquad (3.77)$$

$\uparrow$

Step 6

$$t = \tan^{-1}\left[CS_e\left(\frac{e/B}{\sqrt{BL}}\right)\right] \qquad (3.78)$$

where $C = \beta_1\beta_2$ (3.79)

$\beta_1, \beta_2 =$ factors dependent on the L/B ratio   (3.80)
                    (Lee, 1963; Whitman and Richart, 1967)

Figure 3.28 shows the variation of β_1, β_2, and C. Note that the factor β_1 is related to vertical displacement and the factor β_2 is related to the rotation of the foundation.

Example 3.11

A square foundation is shown in Figure 3.29. It is subjected to a load of 180 kN and a moment of 27 kN-m. Determine the settlement of the foundation (S_e and t) according to the method presented in Section 3.14.

Solution
Step 1

Given: $Q = 180$ kN and moment $= M = 27$ kN-m, load eccentricity $= e = M/Q = 27/180 = 0.15$ m.

180 kN

27 kN

0.7 m

1.5 m × 1.5 m

Sand
$\gamma = 18$ kN/m³
$\phi = 30°$
$c = 0$
$\mu_s = 0.3$
$E_s = 15,000$ kN/m²

Figure 3.29

Step 2

The magnitude of $Q_{ult(e)}$ has been calculated in Example 3.6, or

$$Q_{ult(e)} = 988 \text{ kN}$$

Step 3

Determination of factor of safety, F_1: From Eq. (3.75)

$$\text{Factor of safety} = F_1 = \frac{Q_{ult(e)}}{Q} = \frac{988 \text{ kN}}{180} = 5.49$$

Step 4

Determination of $Q_{ult(e=0)}$: From Eq. (3.16), because $c = 0$,

$$q_u = qN_q F_{qs} F_{qd} + \frac{1}{2} \gamma B N_\gamma F_{\gamma s} F_{\gamma d}$$

$$q = 12.6 \text{ kN/m}^2$$

From $\phi = 30°$ (from Table 3.2), $N_q = 18.4$ and $N_\gamma = 22.4$.
   From Table 3.4,

$$F_{qs} = 1 + \frac{B}{L} \tan \phi = 1 + \left(\frac{1.5}{1.5}\right) \tan 30 = 1.577$$

$$F_{dq} = 1 + 2 \tan \phi (1 - \sin \phi)^2 \frac{D_f}{B} = 1 + \frac{(0.289)(0.7)}{1.5} = 1.135$$

$$F_{\gamma s} = 1 - 0.4\left(\frac{B}{L}\right) = 1 - 0.4\left(\frac{1.5}{1.5}\right) = 0.6$$

$$F_{\gamma d} = 1$$

$$q_u = (12.6)(18.4)(1.577)(1.135) + \left(\frac{1}{2}\right)(18)(1.5)(22.4)(0.6)(1)$$

$$= 414.97 + 181.44 = 596.41 \text{ kN/m}^2$$

So

$$Q_{ult(e=0)} = (596.41)(1.5 \times 1.5) = 1342 \text{ kN}$$

Step 5

Determination of $Q_{(e=0)}$: From Eq. (3.76)

$$Q_{(e=0)} = \frac{Q_{ult(e=0)}}{F_1} = \frac{1342}{5.49} = 244.4 \text{ kN/m}^2$$

Step 6

Determination of $S_{e(e=0)}$: From Eq. (3.67)

$$S_{e(e=0)} = \frac{B(Q_{e=0})}{E_s(B \times L)} (1 - \mu_s^2)\alpha_r$$

For $L/B = 1$, $\alpha_r \approx 0.82$ (Figure 3.23). Given $\mu_s = 0.3$ and $E_s = 15,000 \text{ kN/m}^2$,

$$S_{e(e=0)} = \frac{(1.5)(244.4)}{(15,000)(1.5 \times 1.5)} (1 - 0.3^2)0.82 = 0.0081 \text{ m} = 8.1 \text{ mm}$$

Step 7

From Eq. (3.77)

$$S_e = S_{e(e=0)}\left[1 - 2\left(\frac{e}{B}\right)\right]^2$$
$$= (8.1)[1 - (2)(0.1)]^2 = \textbf{5.18 mm}$$

From Eq. (3.78)

$$t = \tan^{-1}\left[CS_e\left(\frac{e/B}{\sqrt{BL}}\right)\right]$$

For $B/L = 1$, $C \approx 3.95$ (from Figure 3.28). So

$$t = \tan^{-1}\left[(3.95)\left(\frac{5.18 \text{ mm}}{1000}\right)\left(\frac{0.1}{\sqrt{(1.5)(1.5)}}\right)\right]$$
$$= \textbf{0.078 deg}$$

3.15 Consolidation Settlement

As mentioned before, consolidation settlement is time dependent, and it occurs in saturated clayey soils when they are subjected to increased load

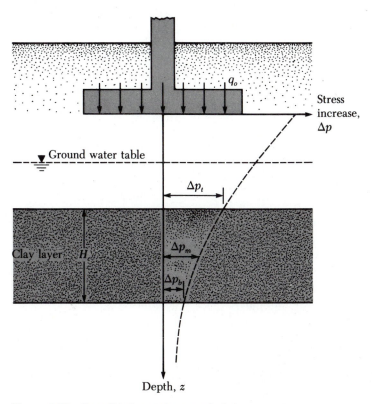

Figure 3.30 Consolidation settlement calculation

caused by foundation construction (Figure 3.30). The one-dimensional consolidation settlement equations are given in Chapter 1 as

$$S_c = \int \varepsilon_v \cdot dz$$

where ε_v = vertical strain

$$= \frac{\Delta e}{1 + e_o}$$

Δe = change of void ratio

$$= f(p_o, p_c, \text{ and } \Delta p)$$

So

$$S_c = \frac{C_c H_c}{1 + e_o} \log \frac{p_o + \Delta p_{av}}{p_o} \quad \text{(for normally consolidated clays)} \quad (1.65)$$

$$S_c = \frac{C_s H_c}{1 + e_o} \log \frac{p_o + \Delta p_{av}}{p_o} \quad \begin{array}{l}\text{(for overconsolidated clays} \\ \text{with } p_o + \Delta p_{av} < p_c)\end{array} \quad (1.67)$$

$$S_c = \frac{C_s H_c}{1 + e_o} \log \frac{p_c}{p_o} + \frac{C_c H_c}{1 + e_o} \log \frac{p_o + \Delta p_{av}}{p_c}$$

$$\begin{array}{l}\text{(for overconsolidated clays} \\ \text{with } p_o < p_c < p_o + \Delta p_{av})\end{array} \quad (1.69)$$

where p_o = average effective pressure on the clay layer before the construction of the foundation

Δp_{av} = average increase of pressure on the clay layer caused by the foundation construction

p_c = preconsolidation pressure

e_o = initial void ratio of the clay layer

C_c = compression index

C_s = swelling index

H_c = thickness of the clay layer

The procedures for the determination of the compression and swelling indexes have been discussed in Chapter 1.

Note that the increase of pressure, Δp, on the clay layer is not constant with depth. The magnitude of Δp will decrease with the increase of depth measured from the bottom of the foundation. However, the average increase of pressure can be approximated by the equation

$$\boxed{\Delta p_{av} = \frac{1}{6}(\Delta p_t + 4\Delta p_m + \Delta p_b)} \qquad (3.81)$$

where Δp_t, Δp_m, and Δp_b are the pressure increases at the *top*, *middle*, and *bottom* of the clay layer that are caused by the foundation construction.

The method of determination of the pressure increase caused by various types of foundation load is discussed in Section 3.17.

3.16   Skempton-Bjerrum Modification for Consolidation Settlement

The consolidation settlement calculation presented in the preceding section is based on Eqs. (1.65), (1.67), and (1.69). These equations, as shown in Chapter 1, are based on one-dimensional laboratory consolidation tests. The underlying assumption for these equations is that the increase of pore water pressure (Δu) immediately after the load application is equal to the increase of stress (Δp) at any depth. For this case

$$S_{c(oed)} = \int \frac{\Delta e}{1 + e_o} \, dz = \int m_v \, \Delta p_{(1)} \, dz \tag{3.82}$$

where $S_{c(oed)}$ = consolidation settlement calculated by using Eqs. (1.65), (1.67), and (1.69)

$\Delta p_{(1)}$ = vertical stress increase (note the change of notation from Δp as given in Section 3.15)

m_v = volume coefficient of compressibility (see Chapter 1)

In the field, however, when load is applied over a limited area on the ground surface, this assumption will not be correct. Consider the case of a circular foundation on a clay layer as shown in Figure 3.31. The vertical and the horizontal stress increase at a point in the clay layer immediately below the center of the foundation are $\Delta p_{(1)}$ and $\Delta p_{(3)}$, respectively. For a saturated clay, the pore water pressure increase at that depth can be given as follows (Chapter 1):

$$\Delta u = \Delta p_{(3)} + A[\Delta p_{(1)} - \Delta p_{(3)}] \tag{3.83}$$

where A = pore water pressure parameter (see Chapter 1). For this case, one can write that

$$S_c = \int m_v \, \Delta u \, dz = \int (m_v)\{\Delta p_{(3)} + A[\Delta p_{(1)} - \Delta p_{(3)}]\} \, dz \tag{3.84}$$

Combining Eqs. (3.82) and (3.84)

$$K_{cir} = \frac{S_c}{S_{c(oed)}} = \frac{\displaystyle\int_0^{H_c} m_v \, \Delta u \, dz}{\displaystyle\int_0^{H_c} m_v \, \Delta p_{(1)} \, dz} = A + (1 - A)\left[\frac{\displaystyle\int_0^{H_c} \Delta p_{(3)} \, dz}{\displaystyle\int_0^{H_c} \Delta p_{(1)} \, dz}\right] \tag{3.85}$$

where K_{cir} = settlement ratio for circular foundations

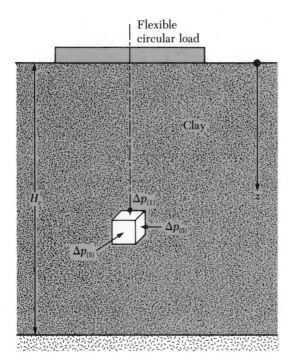

Figure 3.31 Circular foundation on a clay layer

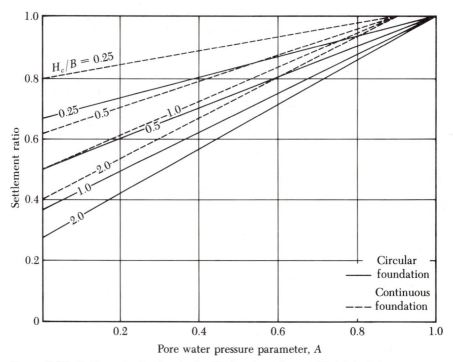

Figure 3.32 Settlement ratios for circular (K_{cir}) and continuous (K_{str}) foundations

The settlement ratio for a continuous foundation (K_{str}) can be determined in a manner similar to that for a circular foundation. The variation of K_{cir} and K_{str} with A and H_c/B is given in Figure 3.32. (*Note:* B = diameter of a circular foundation, and B = width of a continuous foundation.)

Following is the procedure for determination of consolidation settlement according to Skempton and Bjerrum (1957).

1. Determine the consolidation settlement using the procedure outlined in Section 3.15. This is $S_{c(oed)}$. (Note the change of notation from S_c.)
2. Determine the pore water pressure parameter, A.
3. Determine H_c/B.
4. Obtain the settlement ratio—in this case, from Figure 3.32.
5. Calculate the actual consolidation settlement as

$$S_c = S_{c(oed)} \times \text{settlement ratio} \qquad (3.86)$$
$$\uparrow$$
$$\text{Step 1}$$

This technique is generally referred to as the *Skempton-Bjerrum modification* for consolidation settlement calculation.

3.17   Vertical Stress Increase in a Soil Mass Caused by Foundation Load (For Consolidation Settlement Calculation)

Stress Due to a Concentrated Load

In 1885, Boussinesq developed the mathematical relationships for the determination of the normal and shear stresses at any point inside *homogeneous, elastic,* and *isotropic* mediums due to a *concentrated point load* located at the surface, as shown in Figure 3.33. According to his analysis, the *vertical stress increase* (Δp) at point A (Figure 3.33) caused by the point load of magnitude P can be given as

$$\Delta p = \frac{3P}{2\pi z^2 \left[1 + \left(\dfrac{r}{z} \right)^2 \right]^{5/2}} \qquad (3.87)$$

where $r = \sqrt{x^2 + y^2}$
x, y, z = coordinates of the point A

Note that Eq. (3.87) is not a function of the Poisson's ratio of the soil.

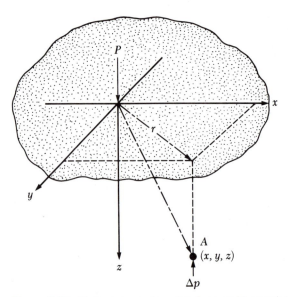

Figure 3.33 Vertical stress at a point, A, caused by a point load on the surface

(a)

Figure 3.34 Increase of pressure under a uniformly loaded flexible circular area

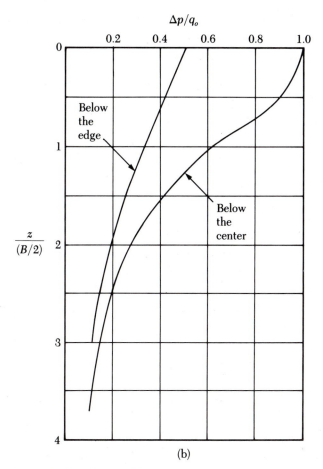

Figure 3.34 (Continued)

Stress Due to a Circularly Loaded Area

The Boussinesq equation [Eq. (3.87)] can also be used to determine the vertical stress below the center of a flexible circularly loaded area, as shown in Figure 3.34a. Let the radius of the loaded area be $B/2$, and let q_o be the uniformly distributed load per unit area. In order to determine the stress increase at a point A that is located at a depth z below the center of the circular area, consider an elementary area on the circle, as shown in Figure 3.34a. The load on this elementary area can be given to be equal to $q_o r \, d\theta \, dr$. This load can be considered as a point load. The stress increase at point A caused by this load can be determined by using Eq. (3.87):

$$dp = \frac{3(q_o r \, d\theta \, dr)}{2\pi z^2 \left[1 + \left(\dfrac{r}{z} \right)^2 \right]^{5/2}} \tag{3.88}$$

Thus, the total increase of stress caused by the entire loaded area can be obtained by integration of the above equation, or

$$\Delta p = \int dp = \int_{\theta=0}^{\theta=2\pi} \left.\int_{r=0}^{r=B/2} \frac{3(q_o r \, d\theta \, dr)}{2\pi z^2 \left[1 + \left(\dfrac{r}{z}\right)^2\right]^{5/2}}\right.$$

$$= q_o \left\{ 1 - \frac{1}{\left[1 + \left(\dfrac{B}{2z}\right)^2\right]^{3/2}} \right\} \qquad (3.89)$$

Based on the preceding equation, Figure 3.34b shows a plot of $\Delta p/q_o$ against $z/(B/2)$. A similar integration could be performed to obtain the stress increase at a point A' (Figure 3.34a) located at a depth z below the loaded area. The variation of $\Delta p/q_o$ for the point A' is also shown in Figure 3.34b.

Stress Below a Rectangular Area

Using the integration technique of Boussinesq's equation, the vertical stress at a given point A below a flexible rectangular loaded area (Figure 3.35a) can also be evaluated. In order to do that, consider an elementary area $dA = dx \cdot dy$ on the flexible loaded area. If the load per unit area is q_o, then the total load on the elementary area can be given as

$$dP = q_o \, dx \, dy \qquad (3.90)$$

This elementary load, dP, can be treated as a point load. The increase of vertical stress at point A caused by dP can be evaluated by using Eq. (3.87). Note that one needs to substitute $dP = q_o \, dx \, dy$ for P, and $x^2 + y^2$ for r^2, in Eq. (3.87). Thus

$$\text{The stress increase at } A \text{ caused by } dP = \frac{3q_o(dx \cdot dy)z^3}{2\pi(x^2 + y^2 + z^2)^{5/2}}$$

The total stress increase caused by the entire loaded area at point A can now be obtained by integrating the preceding equation:

$$\Delta p = \int_{y=0}^{L} \int_{x=0}^{B} \frac{3q_o(dx \, dz)z^3}{2\pi(x^2 + y^2 + z^2)^{5/2}} = q_o I \qquad (3.91)$$

where Δp = stress increase at A

$$I = \text{influence factor} = \frac{1}{4\pi}\left(\frac{2mn\sqrt{m^2 + n^2 + 1}}{m^2 + n^2 + m^2n^2 + 1} \cdot \frac{m^2 + n^2 + 2}{m^2 + n^2 + 1}\right.$$

$$\left. + \tan^{-1} \frac{2mn\sqrt{m^2 + n^2 + 1}}{m^2 + n^2 + 1 - m^2n^2}\right) \qquad (3.92)$$

(a)

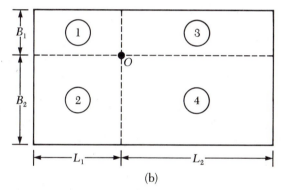

(b)

Figure 3.35   Determination of stress below a flexible rectangular loaded area

where

$$m = \frac{B}{z} \qquad (3.93)$$

and

$$n = \frac{L}{z} \qquad (3.94)$$

The variations of the influence values with m and n are given in Table 3.7. For convenience, they are also plotted in Figure 3.36.

The stress increase at any point below a rectangular loaded area can also be found by using Eq. (3.91). This can be demonstrated using Figure 3.35b.

Table 3.7 Variation of Influence Value, I [Eq. (3.92)]ᵃ

m	\multicolumn{12}{c}{n}
	0.1	0.2	0.3	0.4	0.5	0.6	0.7	0.8	0.9	1.0	1.2	1.4
0.1	0.00470	0.00917	0.01323	0.01678	0.01978	0.02223	0.02420	0.02576	0.02698	0.02794	0.02926	0.03007
0.2	0.00917	0.01790	0.02585	0.03280	0.03866	0.04348	0.04735	0.05042	0.05283	0.05471	0.05733	0.05894
0.3	0.01323	0.02585	0.03735	0.04742	0.05593	0.06294	0.06858	0.07308	0.07661	0.07938	0.08323	0.08561
0.4	0.01678	0.03280	0.04742	0.06024	0.07111	0.08009	0.08734	0.09314	0.09770	0.10129	0.10631	0.10941
0.5	0.01978	0.03866	0.05593	0.07111	0.08403	0.09473	0.10340	0.11035	0.11584	0.12018	0.12626	0.13003
0.6	0.02223	0.04348	0.06294	0.08009	0.09473	0.10688	0.11679	0.12474	0.13105	0.13605	0.14309	0.14749
0.7	0.02420	0.04735	0.06858	0.08734	0.10340	0.11679	0.12772	0.13653	0.14356	0.14914	0.15703	0.16199
0.8	0.02576	0.05042	0.07308	0.09314	0.11035	0.12474	0.13653	0.14607	0.15371	0.15978	0.16843	0.17389
0.9	0.02698	0.05283	0.07661	0.09770	0.11584	0.13105	0.14356	0.15371	0.16185	0.16835	0.17766	0.18357
1.0	0.02794	0.05471	0.07938	0.10129	0.12018	0.13605	0.14914	0.15978	0.16835	0.17522	0.18508	0.19139
1.2	0.02926	0.05733	0.08323	0.10631	0.12626	0.14309	0.15703	0.16843	0.17766	0.18508	0.19584	0.20278
1.4	0.03007	0.05894	0.08561	0.10941	0.13003	0.14749	0.16199	0.17389	0.18357	0.19139	0.20278	0.21020
1.6	0.03058	0.05994	0.08709	0.11135	0.13241	0.15028	0.16515	0.17739	0.18737	0.19546	0.20731	0.21510
1.8	0.03090	0.06058	0.08804	0.11260	0.13395	0.15207	0.16720	0.17967	0.18986	0.19814	0.21032	0.21836
2.0	0.03111	0.06100	0.08867	0.11342	0.13496	0.15326	0.16856	0.18119	0.19152	0.19994	0.21235	0.22058
2.5	0.03138	0.06155	0.08948	0.11450	0.13628	0.15483	0.17036	0.18321	0.19375	0.20236	0.21512	0.22364
3.0	0.03150	0.06178	0.08982	0.11495	0.13684	0.15550	0.17113	0.18407	0.19470	0.20341	0.21633	0.22499
4.0	0.03158	0.06194	0.09007	0.11527	0.13724	0.15598	0.17168	0.18469	0.19540	0.20417	0.21722	0.22600
5.0	0.03160	0.06199	0.09014	0.11537	0.13737	0.15612	0.17185	0.18488	0.19561	0.20440	0.21749	0.22632
6.0	0.03161	0.06201	0.09017	0.11541	0.13741	0.15617	0.17191	0.18496	0.19569	0.20449	0.21760	0.22644
8.0	0.03162	0.06202	0.09018	0.11543	0.13744	0.15621	0.17195	0.18500	0.19574	0.20455	0.21767	0.22652
10.0	0.03162	0.06202	0.09019	0.11544	0.13745	0.15622	0.17196	0.18502	0.19576	0.20457	0.21769	0.22654
∞	0.03162	0.06202	0.09019	0.11544	0.13745	0.15623	0.17197	0.18502	0.19577	0.20458	0.21770	0.22656

ᵃ After Newmark, 1935

Table 3.7   (Continued)

m	1.6	1.8	2.0	2.5	3.0	4.0	5.0	6.0	8.0	10.0	∞
0.1	0.03058	0.03090	0.03111	0.03138	0.03150	0.03158	0.03160	0.03161	0.03162	0.03162	0.03162
0.2	0.05994	0.06058	0.06100	0.06155	0.06178	0.06194	0.06199	0.06201	0.06202	0.06202	0.06202
0.3	0.08709	0.08804	0.08867	0.08948	0.08982	0.09007	0.09014	0.09017	0.09018	0.09019	0.09019
0.4	0.11135	0.11260	0.11342	0.11450	0.11495	0.11527	0.11537	0.11541	0.11543	0.11544	0.11544
0.5	0.13241	0.13395	0.13496	0.13628	0.13684	0.13724	0.13737	0.13741	0.13744	0.13745	0.13745
0.6	0.15028	0.15207	0.15326	0.15483	0.15550	0.15598	0.15612	0.15617	0.15621	0.15622	0.15623
0.7	0.16515	0.16720	0.16856	0.17036	0.17113	0.17168	0.17185	0.17191	0.17195	0.17196	0.17197
0.8	0.17739	0.17967	0.18119	0.18321	0.18407	0.18469	0.18488	0.18496	0.18500	0.18502	0.18502
0.9	0.18737	0.18986	0.19152	0.19375	0.19470	0.19540	0.19561	0.19569	0.19574	0.19576	0.19577
1.0	0.19546	0.19814	0.19994	0.20236	0.20341	0.20417	0.20440	0.20449	0.20455	0.20457	0.20458
1.2	0.20731	0.21032	0.21235	0.21512	0.21633	0.21722	0.21749	0.21760	0.21767	0.21769	0.21770
1.4	0.21510	0.21836	0.22058	0.22364	0.22499	0.22600	0.22632	0.22644	0.22652	0.22654	0.22656
1.6	0.22025	0.22372	0.22610	0.22940	0.23088	0.23200	0.23236	0.23249	0.23258	0.23261	0.23263
1.8	0.22372	0.22736	0.22986	0.23334	0.23495	0.23617	0.23656	0.23671	0.23681	0.23684	0.23686
2.0	0.22610	0.22986	0.23247	0.23614	0.23782	0.23912	0.23954	0.23970	0.23981	0.23985	0.23987
2.5	0.22940	0.23334	0.23614	0.24010	0.24196	0.24344	0.24392	0.24412	0.24425	0.24429	0.24432
3.0	0.23088	0.23495	0.23782	0.24196	0.24394	0.24554	0.24608	0.24630	0.24646	0.24650	0.24654
4.0	0.23200	0.23617	0.23912	0.24344	0.24554	0.24729	0.24791	0.24817	0.24836	0.24842	0.24846
5.0	0.23236	0.23656	0.23954	0.24392	0.24608	0.24791	0.24857	0.24885	0.24907	0.24914	0.24919
6.0	0.23249	0.23671	0.23970	0.24412	0.24630	0.24817	0.24885	0.24916	0.24939	0.24946	0.24952
8.0	0.23258	0.23681	0.23981	0.24425	0.24646	0.24836	0.24907	0.24939	0.24964	0.24973	0.24980
10.0	0.23261	0.23684	0.23985	0.24429	0.24650	0.24842	0.24914	0.24946	0.24973	0.24981	0.24989
∞	0.23263	0.23686	0.23987	0.24432	0.24654	0.24846	0.24919	0.24952	0.24980	0.24989	0.25000

n

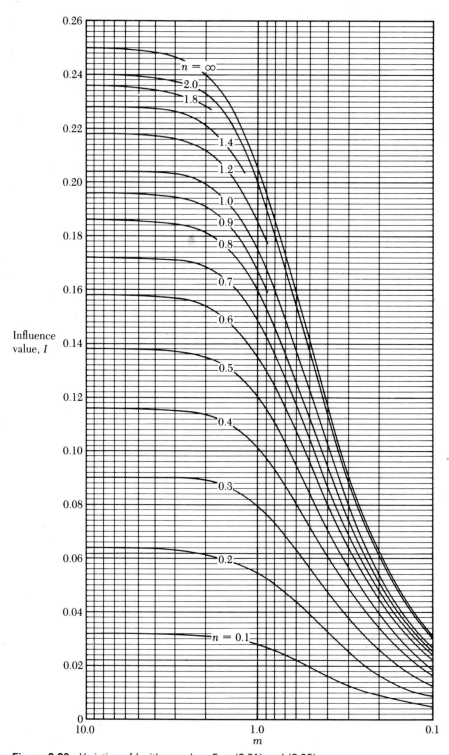

Figure 3.36 Variation of *I* with *m* and *n*—Eqs. (3.91) and (3.92)

To determine the stress at a depth z below point O, divide the loaded area into four rectangles. Point O is the common corner to each rectangle. The increase of stress at a depth z below point O caused by each rectangular area can now be calculated by using Eq. (3.91). The total stress increase caused by the entire loaded area can now be given by the expression

$$\Delta p = q_o(I_1 + I_2 + I_3 + I_4) \tag{3.95}$$

where I_1, I_2, I_3, and I_4 = the influence values of rectangles 1, 2, 3, and 4, respectively. The use of this technique is illustrated in Example 3.12. Also, Figure 3.37 shows the variation of $\Delta p/q_o$ below the center of rectangular areas with $L/B = 1, 1.5, 2$, and ∞. This has been calculated from Table 3.7.

Stress Increase Under a Rectangular Foundation—2 : 1 Method

Often, foundation engineers use an approximate method to determine the increase of stress with depth caused by the construction of a foundation. This is referred to as the *2 : 1 method* (Figure 3.38). According to this method, the increase of stress at a depth z can be obtained as follows:

$$\boxed{\Delta p = \frac{q_o \times B \times L}{(B + z)(L + z)}} \tag{3.96}$$

Note that Eq. (3.96) assumes that the stress from the foundation spreads out along lines with a *2 vertical to 1 horizontal slope*.

Stress Increase Under an Embankment

Figure 3.39 shows the cross section of an embankment of height H. This is a two-dimensional loading condition. The vertical stress increase caused by the embankment loading condition can be expressed as

$$\boxed{\Delta p = \frac{q_o}{\pi}\left[\left(\frac{B_1 + B_2}{B_2}\right)(\alpha_1 + \alpha_2) - \frac{B_1}{B_2}(\alpha_2)\right]} \tag{3.97}$$

where $q = \gamma H$
$\quad \gamma$ = unit weight of the embankment soil
$\quad H$ = height of the embankment

$$\alpha_1 \text{ (radians)} = \tan^{-1}\left(\frac{B_1 + B_2}{z}\right) - \tan^{-1}\left(\frac{B_1}{z}\right) \tag{3.98}$$

$$\alpha_2 = \tan^{-1}\left(\frac{B_1}{z}\right) \tag{3.99}$$

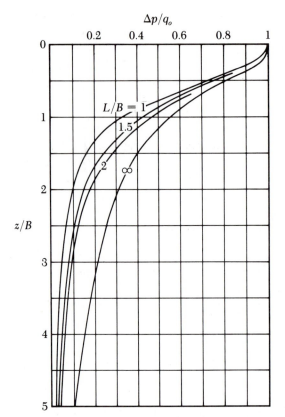

Figure 3.37   Increase of stress under the center of a flexible loaded rectangular area

For detailed derivation of the equation, see Das (1983). In a simplified form, Eq. (3.97) can be expressed as

$$\Delta p = q_o I' \tag{3.100}$$

where $I' = $ a function of B_1/z and B_2/z

The variation of I' with B_1/z and B_2/z is shown in Figure 3.40. Application of this diagram is shown in Example 3.13.

Stress Increase Due to Any Type of Loading

The increase of vertical stress under any type of flexible loaded area can be easily determined by the use of Newmark's *influence chart* (1942). The chart, in principle, is based on Eq. (3.89), which is for the estimation of vertical stress increase under the center of a circularly loaded area. According to Eq. (3.89)

$$\Delta p = q_o \left\{ 1 - \frac{1}{\left[1 + \left(\dfrac{B}{2z} \right)^2 \right]^{3/2}} \right\}$$

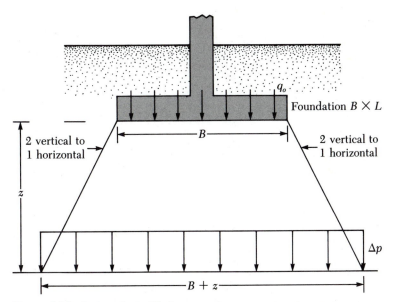

Figure 3.38   2 : 1 method of finding stress increase under a foundation

where $B/2$ = radius of the loaded area = R

The preceding equation can be rewritten as

$$\frac{R}{z} = \left[\left(1 - \frac{\Delta p}{q_o}\right)^{-2/3} - 1\right]^{1/2} \tag{3.101}$$

One can now substitute various values of $\Delta p/q_o$ into Eq. (3.101) to obtain corresponding values of R/z. Table 3.8 shows the calculated values of R/z for $\Delta p/q_o = 0, 0.1, 0.2, \ldots, 1$.

Figure 3.39   Embankment loading

Figure 3.40 Influence value of I' for embankment loading (after Osterberg, 1957)

Using the nondimensional values of R/z shown in Table 3.8, one can draw the concentric circles having radii equal to R/z. This is shown in Figure 3.41. Note that the distance AB in Figure 3.41 is equal to unity. The first circle is a point having a radius equal to zero. Similarly, the second circle has a radius of 0.2698 times AB. The last circle has a radius of infinity. These circles have been divided by a number of equally spaced radial lines. This is what is referred to as *Newmark's chart*. The influence value (IV) of this chart is equal to

$$IV = \frac{1}{\text{number of elements on the chart}} \qquad (3.102)$$

Table 3.8   Values of R/z for Various
Values of $\Delta p/q_o$ [Eq. (3.101)]

$\Delta p/q_o$	R/z
0	0
0.1	0.2698
0.2	0.4005
0.3	0.5181
0.4	0.6370
0.5	0.7664
0.6	0.9174
0.7	1.1097
0.8	1.3871
0.9	1.9084
1.0	∞

For the chart shown in Figure 3.41, $IV = 1/200 = 0.005$.

Following is a step-by-step procedure using Newmark's chart for determination of vertical stress under a loaded area of any shape.

1. Determine the depth z below the loaded area at which the stress is to be determined.

2. Adopt a scale $z = \overline{AB}$ (that is, unit length according to Newmark's chart).

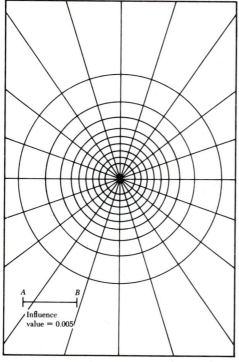

Influence
value = 0.005

Figure 3.41   Influence chart for vertical pressure calculation (after Newmark, 1942)

3. Draw the plan of the loaded area based on the scale adopted in Step 2.

4. Place the plan drawn in Step 3 on the Newmark's chart such that the point under which the stress is to be determined is directly above the center of the chart.

5. Count the number of elements of the chart that fall inside the plan. Let it be equal to N.

6. Calculate the stress increase as

$$\Delta p = (IV)(N)(q_o) \qquad (3.103)$$

where q_o = load per unit area on the loaded area.

Example 3.12

A flexible rectangular area, 2.5 m × 5 m, is located on the ground surface and loaded with $q_o = 145$ kN/m². Determine the stress increase caused by this loading at a depth of 6.25 m below the center of the rectangular area. Use Eq. (3.91).

Solution

Refer to Figure 3.35b.

$$B_1 = \frac{2.5 \text{ m}}{2} = 1.25 \text{ m}$$

$$L_1 = \frac{5}{2} = 2.5 \text{ m}$$

From Eqs. (3.93) and (3.94)

$$m_1 = \frac{B_1}{z} = \frac{1.25}{6.25} = 0.2$$

$$n_1 = \frac{L_1}{z} = \frac{2.5}{6.25} = 0.4$$

From Table 3.7, for $m_1 = 0.20$ and $n_1 = 0.4$, the value of $I_1 = 0.0328$. Also note that $I_1 = I_2 = I_3 = I_4$. Thus

$$\Delta p = q_o(4I_1) = (145)(4)(0.0328) = \textbf{19.024 kN/m}^2$$

Example 3.13

An embankment is shown in Figure 3.42a. Determine the stress increase under the embankment at points A_1 and A_2.

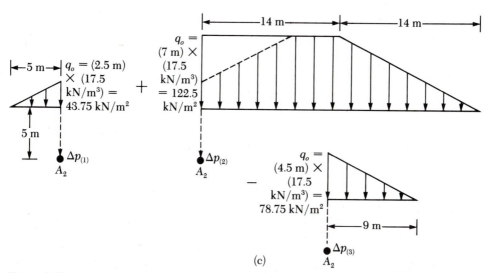

Figure 3.42

Solution

$$\gamma H = (17.5)(7) = 122.5 \text{ kN/m}^2$$

Stress Increase at A_1

Consider the left side of Figure 3.42b. For this, $B_1 = 2.5$ m and $B_2 = 14$ m. So

$$\frac{B_1}{z} = \frac{2.5}{5} = 0.5$$

$$\frac{B_2}{z} = \frac{14}{5} = 2.8$$

According to Figure 3.40, in this case, $I' = 0.445$. Because the two sides in Figure 3.42b are symmetrical, the value of I' for the right side will also be 0.445. So

$$\Delta p = \Delta p_1 + \Delta p_2 = q_o[I'_{(\text{left side})} + I'_{(\text{right side})}]$$
$$= 122.5[0.445 + 0.445] = \mathbf{109.03 \text{ kN/m}^2}$$

Stress Increase at A_2

Refer to Figure 3.42c. For the left side, $B_2 = 5$ and $B_1 = 0$. So

$$\frac{B_2}{z} = \frac{5}{5} = 1$$

$$\frac{B_1}{z} = \frac{0}{5} = 0$$

According to Figure 3.40, for these values of B_2/z and B_1/z, $I' = 0.25$. So

$$\Delta p_1 = 43.75(0.25) = 10.94 \text{ kN/m}^2$$

For the middle section

$$\frac{B_2}{z} = \frac{14}{5} = 2.8$$

$$\frac{B_1}{z} = \frac{14}{5} = 2.8$$

Thus, $I' = 0.495$. So

$$\Delta p_2 = 0.495(122.5) = 60.64 \text{ kN/m}^2$$

For the right side

$$\frac{B_2}{z} = \frac{9}{5} = 1.8$$

$$\frac{B_1}{z} = \frac{0}{5} = 0$$

$I' = 0.335$. So

$$\Delta p_3 = (78.75)(0.335) = 26.38 \text{ kN/m}^2$$

Total stress increase at point A_2 is

$$\Delta p = \Delta p_1 + \Delta p_2 - \Delta p_3 = 10.94 + 60.64 - 26.38 = \mathbf{45.2 \text{ kN/m}^2}$$

Example 3.14

Redo Example 3.12 using Newmark's chart (Figure 3.41).

Solution

For this problem, $z = 6.25$ m. So, the length $\overline{AB}$ in Figure 3.41 is equal to 6.25 m. With this scale, the plan of the loaded rectangular area can be drawn. Figure 3.43 shows this plan placed over the Newmark's chart such that the center of the loaded area falls above the center of the chart. This is because the stress increase is required at a point that is located immediately below the center of the rectangular area. The number of elements from the influence chart that are inside the plan is about 26. So

$$\Delta p = (IV)(N)(q_o) = (0.005)(26)(145) = \mathbf{18.85\ kN/m^2}$$

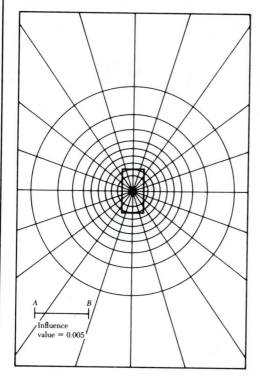

A ─ B

Influence
value = 0.005

Figure 3.43

This value of Δp is practically the same as that determined in Example 3.12. ≡

Example 3.15

A foundation 1 m × 2 m in plan is shown in Figure 3.44. Estimate the total settlement of the foundation.

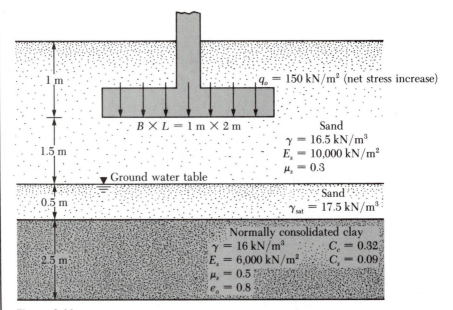

Figure 3.44

Solution

Elastic Settlement

The clay layer is located at a depth of 2 m—that is, $2B$ below the foundation. From Figure 3.25, it can be seen that the soil located at a depth $z > 2B$ has very little influence on the elastic settlement. Hence, if Eq. (3.67) is used for the elastic settlement calculation, it is reasonable to use the Young's modulus and Poisson's ratio values of the sand layer. Thus

$$S_e = \frac{Bq_o}{E_s}(1 - \mu_s^2)\alpha_r$$

Given: $q_o = 150 \text{ kN/m}^2$, $E_s = 10,000 \text{ kN/m}^2$, $\mu_s = 0.3$, and $\alpha_r \approx 1.2$ (Figure 3.23). So

$$S_e = \frac{(1)(150)}{10,000}(1 - 0.3^2)(1.2) = 0.0163 \text{ m} = \mathbf{16.38\ mm}$$

Consolidation Settlement

The clay is normally consolidated.

$$S_c = \frac{C_c H}{1 + e_o} \log \frac{p_o + \Delta p_{av}}{p_o}$$

$$p_o = (2.5)(16.5) + (0.5)(17.5 - 9.81) + 1.25(16 - 9.81)$$
$$= 41.25 + 3.85 + 7.74 = 52.84 \text{ kN/m}^2$$

From Eq. (3.81)

$$\Delta p_{av} = \frac{1}{6}(\Delta p_t + 4\Delta p_m + \Delta p_b)$$

Using the 2:1 method

$$\Delta p = \frac{q_o \times B \times L}{(B + z)(L + z)}$$

For the top of the clay layer, $z = 2$ m, so

$$\Delta p_t = \frac{(150)(1)(2)}{(1 + 2)(2 + 2)} = 25 \text{ kN/m}^2$$

Similarly

$$\Delta p_m = \frac{(150)(1)(2)}{(1 + 3.25)(2 + 3.25)} = 13.45 \text{ kN/m}^2$$

and

$$\Delta p_b = \frac{(150)(1)(2)}{(1 + 4.5)(2 + 4.5)} = 8.39 \text{ kN/m}^2$$

Thus

$$\Delta p_{av} = \frac{1}{6}[25 + 4(13.45) + 8.39] = 14.53 \text{ kN/m}^2$$

So

$$S_c = \frac{(0.32)(2.5)}{1 + 0.8} \log\left(\frac{52.84 + 14.53}{52.84}\right) = 0.0469 \text{ m} = \textbf{46.90 mm}$$

Hence, total settlement $= S = S_e + S_c = 16.38 + 46.90 = \textbf{63.28 mm}$

Note: The total settlement just calculated may exceed the tolerable settlement of the foundation. In order to reduce the settlement, the foundation size may be changed so that it will carry the same total load of 300 kN but cause less settlement. ≡

3.18   Allowable Bearing Pressure in Sand Based on Settlement Consideration

Meyerhof (1956) proposed a correlation for the *net allowable bearing pressure* for foundations with the standard penetration resistance (N). The net pressure has been defined in Eq. (3.20) as

$$q_{\text{net(all)}} = q_{\text{all}} - \gamma D_f \tag{3.104}$$

According to Meyerhof's theory, for one inch (25.4 mm) of estimated maximum settlement

$$q_{\text{net(all)}} \ (\text{kN/m}^2) = 11.98 \ N$$

(for $B \leq 1.22$ m and 25.4 mm of settlement) $\tag{3.105a}$

$$q_{\text{net(all)}} \ (\text{kN/m}^2) = 7.99 \ N\left(\frac{3.28B + 1}{3.28B}\right)^2$$

(for $B > 1.22$ m and 25.4 mm of settlement) $\tag{3.105b}$

where N = corrected standard penetration number

Note that in the preceding equations B is in meters.

In English units

$$q_{net(all)} \; (kips/ft^2) = \frac{N}{4} \qquad (for \; B \leq 4 \; ft) \qquad (3.106a)$$

and

$$q_{net(all)} \; (kips/ft^2) = \frac{N}{6}\left(\frac{B+1}{B}\right)^2 \qquad (for \; B > 4 \; ft) \qquad (3.106b)$$

Since Meyerhof proposed his correlation, researchers have observed that its results are rather conservative. Bowles (1977) suggested that the net allowable bearing pressure should be increased by about 50%. The modified form of the bearing pressure equations can then be expressed as follows:

$$q_{net(all)} \; (kN/m^2) = 19.16 N F_d\left(\frac{S}{25.4}\right)$$

$(for \; B \leq 1.22 \; m)$ \hfill (3.107a)

$$q_{net(all)} \; (kN/m^2) = 11.98 N\left(\frac{3.28B+1}{3.28B}\right)^2 F_d\left(\frac{S}{25.4}\right)$$

$(for \; B > 1.22 \; m)$ \hfill (3.107b)

where F_d = depth factor = $1 + 0.33(D_f/B) \leq 1.33$
S = tolerable settlement, in mm

The unit of B is meters.
   In English units

$$q_{net(all)} \; (kips/ft^2) = \frac{N}{2.5} F_d S \qquad (for \; B \leq 4 \; ft) \qquad (3.109a)$$

$$q_{net(all)} \; (kips/ft^2) = \frac{N}{4}\left(\frac{B+1}{B}\right)^2 F_d S \qquad (for \; B > 4 \; ft) \qquad (3.109b)$$

where F_d is given by Eq. (3.108) and S = tolerable settlement, in inches.
   The empirical relations just presented may raise some questions: for example, which value of the standard penetration number should be adopted, and what is the effect of the water table on the net allowable bearing capacity? The design value of N should be determined by taking into account the N values for a depth of $2B$ to $3B$, measured from the bottom of the foundation. Many engineers are also of the opinion that the N value should be reduced somewhat if the water table is close to the foundation. However, the author feels that this is not required, because the penetration resistance reflects the location of the water table.
   Meyerhof (1956) also prepared empirical relations for the net allowable bearing capacity of foundations based on the cone penetration resistance (q_c):

$$q_{net(all)} = \frac{q_c}{15} \qquad (for \; B \leq 1.22 \; m \; and \; settlement \; of \; 25.4 \; mm) \quad (3.110)$$

and

$$q_{net(all)} = \frac{q_c}{25}\left(\frac{3.28B + 1}{3.28B}\right)^2$$

(for $B > 1.22$ m and settlement of 25.4 mm)   (3.111)

In the preceding two equations, the unit of B is meters and the units of $q_{net(all)}$ and q_c are kN/m^2.

In English units

$$q_{net(all)}\ (lb/ft^2) = \frac{q_c\ (lb/ft^2)}{15}$$

(for $B \le 4$ ft and settlement of 1 in.)   (3.112a)

and

$$q_{net(all)}\ (lb/ft^2) = \frac{q_c\ (lb/ft^2)}{25}\left(\frac{B + 1}{B}\right)^2$$

(for $B > 4$ ft and settlement of 1 in.)   (3.112b)

In Eq. (3.112), the unit of B is feet.

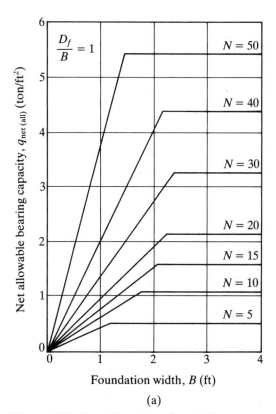

(a)

Figure 3.45 Correlation of net allowable bearing capacity in sand with standard penetration number for foundation settlements not exceeding 1 in. (25.4 mm) (after Peck, Hanson, and Thornburn, 1974) (Continued on page 192)

Figure 3.45 (Continued)

The basic philosophy behind the development of these correlations is that, if the maximum settlement is no more than 25.4 mm (1 in.) for any foundation, the differential settlement would be no more than 19.05 mm ($\frac{3}{4}$ in.). These are probably the allowable limits for most building foundation designs.

Peck, Hanson, and Thornburn (1974) have also provided charts of $q_{\text{net(all)}}$ for foundations on sand of varying widths (B), standard penetration numbers (N), and D_f/B. These are given in Figure 3.45a, b, and c. These values of $q_{\text{net(all)}}$ correspond to a maximum settlement of 25.4 mm (1 in.).

Example 3.16

A shallow square foundation for a column is to be constructed. It must carry a net vertical load of 1000 kN. The foundation soil is sand. The standard penetration numbers obtained from field exploration are given in Figure 3.46. Assume that the depth of the foundation will be 1.5 m and the tolerable settlement is 25.4 mm. Determine the size of the foundation.

Figure 3.46

Solution

The field standard penetration numbers need to be corrected by using Eq. (2.6). This is done in the following table.

Depth (m)	Field value of N	σ'_v (kN/m²)	Corrected N[a]
2	4	31.4	7
4	7	62.8	9
6	12	94.2	12
8	12	125.6	11
10	16	157.0	13
12	13	188.4	9
14	12	206.4	8
16	14	224.36	9
18	18	242.34	11

[a] Rounded off

From the table, it appears that a corrected average N value of about 10 would be appropriate. Using Eq. (3.107b)

$$q_{net(all)} = 11.98N\left(\frac{3.28B + 1}{3.28B}\right)^2 F_d\left(\frac{S}{25.4}\right)$$

Allowable $S = 25.4$ mm and $N = 10$. So

$$q_{net(all)} = 119.8\left(\frac{3.28B + 1}{3.28B}\right)^2 F_d$$

The following table can now be prepared for trial calculations.

B (m)	F_d[a]	$q_{net(all)}$ (kN/m²)	$Q = q_{net(all)} \times B^2$ (kN)
2	1.248	197.24	788.96
2.25	1.22	187.19	947.65
2.3	1.215	185.46	981.1
2.4	1.206	182.29	1050.0
2.5	1.198	179.45	1121.56

[a] $D_f = 1.5$ m.

Because Q_{net} required is 1000 kN, B will be approximately equal to 2.4 m.

Let us go on to see what the net allowable load is by using the ultimate bearing capacity equation [Eq. (3.16)] with $B = \mathbf{2.4}$ **m.**

$$q_{net(ult)} = q_{ult} - q = qN_q F_{qs} F_{qd} + \frac{1}{2}\gamma BN_\gamma F_{\gamma s} F_{\gamma d} - q$$

For an N value of about 10, the friction angle of soil is about 34° (Table 2.4). For $\phi = 34°$, from Table 3.2

$$N_q = 29.44$$
$$N_\gamma = 41.06$$

From Table 3.4

$$F_{qs} = 1 + \left(\frac{2.4}{2.4}\right) \tan 34° = 1.675$$

$$F_{\gamma s} = 1 - 0.4\left(\frac{2.4}{2.4}\right) = 0.6$$

$$F_{qd} = 1 + 0.262\left(\frac{1.5}{2.4}\right) = 1.164$$

$$F_{\gamma d} = 1$$

So, $q_{net(ult)} = (15.7 \times 1.5)(29.44)(1.675)(1.164)$

$$+ \frac{1}{2}(15.7)(2.4)(41.06)(0.6)(1) - (15.7 \times 1.5)$$

$$= 1351.8 + 464.14 - 23.55 \approx 1792 \text{ kN/m}^2$$

$$Q_{net(ult)} = 1792 \times B^2 = 1792 \times 2.4^2 = 10{,}322 \text{ kN}$$

So, the factor of safety required for a tolerable settlement of 25.4 mm is equal to $10{,}322/1000 = 10.3$. This demonstrates that, in most cases, the design is controlled by the tolerable settlement criterion.

3.19   Field Load Test

The ultimate load-bearing capacity of a foundation, as well as the allowable bearing capacity based on tolerable settlement considerations, can be effectively determined from the field load test. This is generally referred to as the *plate load test* (ASTM Test Designation D-1194-72). The plates that are used for tests in the field are usually made out of steel and are 25 mm thick and 150 mm to 722 mm in diameter. Occasionally, square plates that are 305 mm × 305 mm are also used.

To conduct a plate load test, a hole is excavated with a minimum diameter $4B$ (B = diameter of the test plate) up to a depth of D_f (D_f = depth of the proposed foundation). The plate is placed at the center of the hole. Load to the plate is applied in steps—about one-fourth to one-fifth of the estimated ultimate load—by means of a jack. A schematic diagram of the test arrangement is shown in Figure 3.47a. During each step load application, the settlement of the plate is observed using dial gauges. At least one hour elapses after the application of each step load before the next load is applied. The test should be conducted until failure, or at least until the plate has gone through 25 mm of settlement. Figure 3.47b shows the nature of the load-settlement curve obtained from such tests, from which the ultimate load per unit area can be determined.

For tests in clay

$$q_{u(F)} = q_{u(P)} \tag{3.113}$$

where $q_{u(F)}$ = ultimate bearing capacity of the proposed foundation
$q_{u(P)}$ = ultimate bearing capacity of the test plate

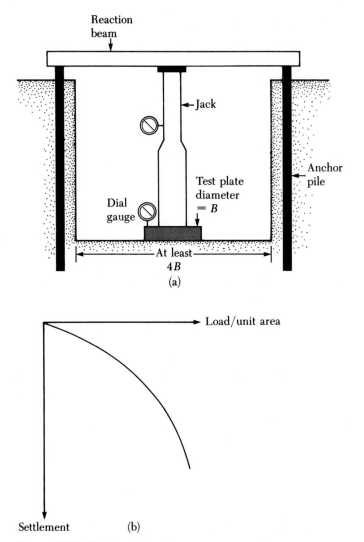

Figure 3.47   Plate load test: (a) test arrangement; (b) nature of load-settlement curve

Equation (3.113) implies that the ultimate bearing capacity in clay is practically independent of the size of the plate.

For tests in sandy soils

$$q_{u(F)} = q_{u(P)} \frac{B_F}{B_P}$$ (3.114)

where B_F = width of the foundation
B_P = width of the test plate

In order to obtain the allowable bearing capacity of a foundation based on settlement considerations, the following equations apply.

For a given intensity of load, q_o

$$S_F = S_P \frac{B_F}{B_P} \quad \text{(for clayey soil)} \tag{3.115}$$

and

$$S_F = S_P \left(\frac{B_F}{B_P}\right)^2 \left(\frac{3.28B_P + 1}{3.28B_F + 1}\right)^2 \quad \text{(for sandy soil)} \tag{3.116}$$

In Eq. (3.116), the units of B_P and B_F are meters.
In English units, Eq. (3.116) can be stated as

$$S = S_F \left(\frac{B_F}{B_P}\right)^2 \left(\frac{B_P + 1}{B_F + 1}\right)^2 \tag{3.117}$$

In the preceding equation, B_F and B_P are in feet.

Equations (3.116) and (3.117) are based on the works of Terzaghi and Peck (1967). Example 3.17 illustrates their application.

Housel (1929) proposed a different technique for determining the load-bearing capacity of shallow foundations based on settlement consideration. Following are the steps of this procedure:

1. Let it be required to find the dimensions of a foundation that will carry a load of Q_o with a tolerable settlement of S_{tol}.

2. Conduct two plate load tests with plates of diameters B_1 and B_2.

3. From the load-settlement curves obtained in Step 2, determine the total loads on the plates (Q_1 and Q_2) that correspond to the settlement of S_{tol}.

For plate No. 1, the total load can be expressed as

$$Q_1 = A_1 m + P_1 n \tag{3.118}$$

Similarly, for plate No. 2

$$Q_2 = A_2 m + P_2 n \tag{3.119}$$

where A_1, A_2 = areas of the plates No. 1 and No. 2, respectively
P_1, P_2 = perimeters of the plates No. 1 and No. 2, respectively
m, n = two constants that correspond to the bearing pressure and perimeter shear, respectively

The values of m and n can be determined by solving Eqs. (3.118) and (3.119).

4. For the foundation to be designed

$$Q_o = Am + Pn \tag{3.120}$$

where A = area of the foundation
P = perimeter of the foundation

Because Q_o, m, and n are known, Eq. (3.120) can be solved for determination of foundation width. The application of this procedure is given in Example 3.18.

Example 3.17

The results of a plate load test in a sandy soil are shown in Figure 3.48. The size of the plate is 0.305 m × 0.305 m. Determine the size of a square column foundation that should carry a load of 2500 kN with a maximum settlement of 25 mm.

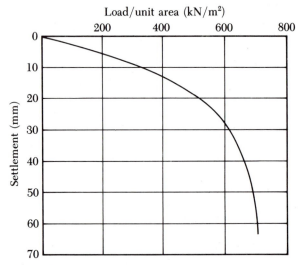

Figure 3.48

Solution

The problem has to be solved by a trial-and-error procedure. Use the following table and Eq. (3.116).

Q_o (kN) (1)	Assume width B_F (m) (2)	$q_o = \dfrac{Q_o}{B_F^2}$ (kN/m²) (3)	S_P corresponding to q_o in Col. 3 (mm) (4)	S_F from Eq. (3.116) (mm) (5)
2500	4.0	156.25	4.0	13.80
2500	3.0	277.80	8.0	26.35
2500	3.2	244.10	6.8	22.70
2500	3.1*	260.10	7.2	23.86

* So, a column footing with dimensions of **3.1 m × 3.1 m** will be appropriate.

Example 3.18

The results of two plate load tests are given in the following table.

Plate diameter, B (m)	Total load, Q (kN)	Settlement (mm)
0.305	32.2	20
0.610	71.8	20

A square column foundation has to be constructed to carry a total load of 715 kN. The tolerable settlement is 20 mm. Determine the size of the foundation.

Solution

Referring to Eqs. (3.118) and (3.119)

$$32.2 = \frac{\pi}{4}(0.305)^2 m + \pi(0.305)n \qquad (a)$$

$$71.8 = \frac{\pi}{4}(0.610)^2 m + \pi(0.610)n \qquad (b)$$

From the preceding two equations,

$$m = 50.68 \text{ kN/m}^2$$
$$n = 29.75 \text{ kN/m}$$

For the foundation to be designed [Eq. (3.120)]

$$Q_o = Am + Pn$$

or

$$Q_o = B_F^2 m + 4B_F n$$

Given: $Q_o = 715$ kN. So,

$$715 = B_F^2(50.68) + 4B_f(29.75)$$

or

$$50.68B_F^2 + 119B_F - 715 = 0$$

From the above equation, $B_F \approx 2.8$ **m.**

3.20 Presumptive Bearing Capacity

Several building codes (for example, Uniform Building Code, Chicago Building Code, New York City Building Code) specify the allowable bearing capacity of foundations on various types of soil. For minor construction works, they often provide fairly acceptable guidelines. However, these bearing capacity values are primarily based on the *visual* classification of near-surface soils. They generally do not take into consideration such factors as the stress history of the soil, ground water table location, the depth of the foundation, and the tolerable settlement. So, for large construction projects, the codes' presumptive values should be taken as a guide only.

3.21 Tolerable Settlement of Buildings

As has been emphasized in this chapter, settlement analysis plays an important part in the design and construction of foundations. Large settlements of various components of a structure may lead to considerable damage and/or may interfere with the proper functioning of the structure. Limited studies have been made to evaluate the conditions for tolerable settlement of various types of structure (for example, Bjerrum, 1963; Burland and Worth, 1974;

(a) Settlement without tilt

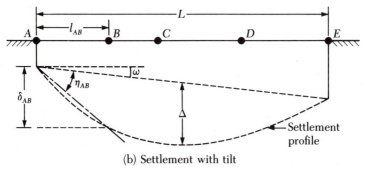

(b) Settlement with tilt

Figure 3.49   Parameters for definition of tolerable settlement (redrawn after Wahls, 1981)

Grant, Christian, and Vanmarcke, 1974; Polshin and Tokar, 1957; Wahls, 1981). Wahls (1981) has provided an excellent review of these studies.

Figure 3.49 provides the parameters for definition of tolerable settlement. Figure 3.49a is for a structure that has undergone settlement without tilt; Figure 3.49b is for a structure that has undergone settlement with tilt.

Table 3.9   Limiting Angular Distortion As Recommended by Bjerrum[a]

Category of potential damage	η
Danger to machinery sensitive to settlement	1/750
Danger to frames with diagonals	1/600
Safe limit for no cracking of buildings[b]	1/500
First cracking of panel walls	1/300
Difficulties with overhead cranes	1/300
Tilting of high rigid buildings becomes visible	1/250
Considerable cracking of panel and brick walls	1/150
Danger of structural damage to general buildings	1/150
Safe limit for flexible brick walls, $L/H > 4$[b]	1/150

[a] After Wahls, 1981
[b] Safe limits include a factor of safety

The parameters are as follows:

ρ_i = total vertical displacement at point i

δ_{ij} = differential settlement between points i and j

Δ = relative deflection

ω = tilt

$$\eta_{ij} = \frac{\delta_{ij}}{l_{ij}} - \omega = \text{angular distortion}$$

$$\frac{\Delta}{L} = \text{deflection ratio}$$

L = lateral dimension of the structure

Bjerrum (1963) has provided the conditions of *limiting* angular distortion, η, of various structures (see Table 3.9).

Polshin and Tokar (1957) have presented the settlement criteria of the 1955 U.S.S.R. Building Code. These were based on the experience gained from the observation of foundation settlement over 25 years. Tables 3.10 and 3.11 present their findings.

Table 3.10   Allowable Settlement Criteria: 1955 U.S.S.R. Building Code[a]

Type of structure	Sand and hard clay	Plastic clay
(a) η		
Civil- and industrial-building column foundations:		
For steel and reinforced concrete structures	0.002	0.002
For end rows of columns with brick cladding	0.007	0.001
For structures where auxiliary strain does not arise during nonuniform settlement of foundations	0.005	0.005
Tilt of smokestacks, towers, silos, and so on	0.004	0.004
Craneways	0.003	0.003
(b) Δ/L		
Plain brick walls:		
For multistory dwellings and civil buildings		
at $L/H \leq 3$	0.0003	0.0004
at $L/H \geq 5$	0.0005	0.0007
For one-story mills	0.0010	0.0010

[a] After Wahls, 1981

Table 3.11   Allowable Average Settlement for Different Building Types[a]

Kind of building	Allowable average settlement, in inches (millimeters)
Building with plain brick walls	
$L/H \geq 2.5$	3 (80)
$L/H \leq 1.5$	4 (100)
Building with brick walls, reinforced with reinforced concrete or reinforced brick	6 (150)
Framed building	4 (100)
Solid reinforced concrete foundations of smokestacks, silos, towers, and so on	12 (300)

[a] After Wahls, 1981

PROBLEMS

3.1 A continuous foundation is 3 ft wide. Given: $D_f = 3$ ft, $\gamma = 110$ lb/ft³, $\phi = 28°$, and $c = 400$ lb/ft². Using Terzaghi's equation, determine the allowable gross vertical load-bearing capacity (factor of safety = 4). Assume that general shear failure occurs in the soil.

3.2 Redo Problem 3.1 with the following: $B = 1.5$ m, $D_f = 1.2$ m, $\gamma = 17.8$ kN/m³, $\phi = 35°$, $c = 0$, factor of safety = 3.

3.3 A square column is 3 m × 3 m in plan. Given: $D_f = 2$ m, $\gamma = 16.5$ kN/m³, $\phi = 30°$, and $c = 0$. Using Terzaghi's equation, determine the gross allowable vertical load that the column could carry (factor of safety = 4). Assume that general shear failure occurs in the soil.

3.4 A gross allowable load of 1882 kN (factor of safety = 3) has to be carried by a square foundation. Given: $D_f = 1.4$ m, $\gamma = 15.9$ kN/m³, $\phi = 30°$, and $c = 0$. Use Terzaghi's equation and determine the width of the foundation.

3.5 Solve Problem 3.1 using Eq. (3.16).

3.6 Solve Problem 3.2 using Eq. (3.16).

3.7 Solve Problem 3.3 using Eq. (3.16).

3.8 For the foundation given in Problem 3.2, what will be the gross allowable load-bearing capacity if the load is inclined at an angle 10° to the vertical? Use Eq. (3.16).

3.9 For the foundation given in Problem 3.3, what will be the gross allowable load-bearing capacity if the load is inclined at an angle 10° to the vertical? Use Eq. (3.16).

3.10 For a square footing ($B \times B$) as shown in Figure P3.10, the following are given: $\gamma = 105$ lb/ft³, $\gamma_{sat} = 118$ lb/ft³, $c = 0$, $\phi = 35°$, $D_f = 4$ ft, $D_1 = 2$ ft, gross allowable load with a factor of safety of 3, $Q_{all} = 144{,}000$ lb. Determine the size of the footing. Use Eq. (3.16).

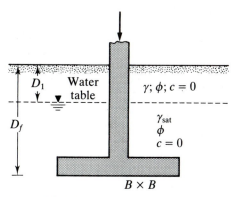

Figure P3.10

3.11 A column foundation is 8 ft × 4 ft in plan. Given: $D_f = 3$ ft, $c = 2500$ lb/ft^2, $\phi = 0$, and $\gamma = 121$ lb/ft^3. What is the net ultimate load that the column could carry? [Use Eq. (3.22).]

3.12 Repeat Problem 3.11 using Eq. (3.23).

3.13 Repeat Problem 3.11 with the following: foundation size = 3 m × 2 m, $D_f = 1.5$ m, $c = 110$ kN/m^2, $\gamma = 18.7$ kN/m^3, $\phi = 0$.

3.14 An eccentrically loaded foundation is shown in Figure P3.14. Use a factor of safety of 4 and determine the maximum allowable load that the foundation could carry.

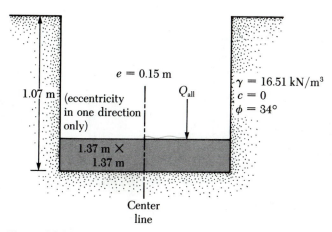

Figure P3.14

3.15 An eccentrically loaded foundation is shown in Figure P3.15. Determine the ultimate load, Q_u, that the load could carry.

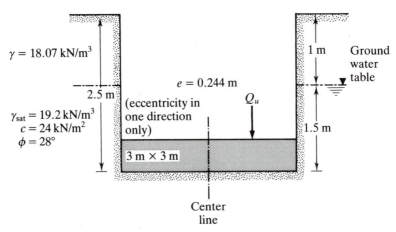

$\gamma = 18.07$ kN/m^3

1 m   Ground water table

2.5 m

$e = 0.244$ m
(eccentricity in one direction only)

Q_u

$\gamma_{sat} = 19.2$ kN/m^3
$c = 24$ kN/m^2
$\phi = 28°$

1.5 m

3 m × 3 m

Center line

Figure P3.15

3.16 For a square foundation that is $B \times B$ in plan, the following are given: $D_f = 3$ ft; vertical gross allowable load, $Q_{all} = 150,000$ lb, $\gamma = 115$ lb/ft^3, $\phi = 40°$, $c = 0$, and factor of safety $= 3$. Determine the size of the foundation.

3.17 A square footing is shown in Figure P3.17. Using a factor of safety of 6, determine the size of the footing.

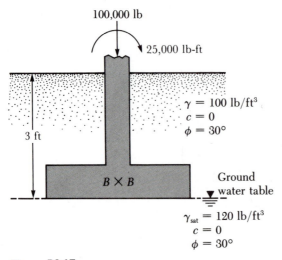

100,000 lb

25,000 lb-ft

$\gamma = 100$ lb/ft^3
$c = 0$
$\phi = 30°$

3 ft

$B \times B$

Ground water table

$\gamma_{sat} = 120$ lb/ft^3
$c = 0$
$\phi = 30°$

Figure P3.17

3.18 Refer to Figure 3.10d for a foundation with a two-way eccentricity. Given, for the soil: $\gamma = 18$ kN/m^3, $\phi = 35°$, $c = 0$; also $B = 1.5$ m, $L = 2$ m, $e_B = 0.3$ m, $e_L = 0.364$ m. Determine the gross ultimate load that the foundation could carry. Use $D_f = 1$ m.

3.19 Repeat Problem 3.18 with the following changes: $e_L = 0.4$ m, $e_B = 0.19$ m.

3.20 Repeat Problem 3.18 with the following changes: $e_B = 0.6$ m, $e_L = 0.214$ m.

3.21 Repeat Problem 3.18 with the following changes: $e_L = 0.286$ m, $e_B = 0.214$ m.

3.22 A continuous footing in a two-layered clay is shown in Figure P3.22. Find the gross allowable bearing capacity. Factor of safety = 3.

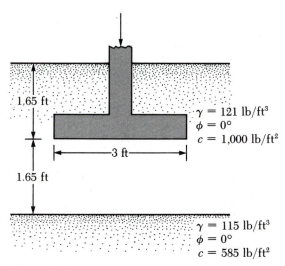

Figure P3.22

3.23 Find the gross ultimate load that the footing shown in Figure P3.23 can carry.

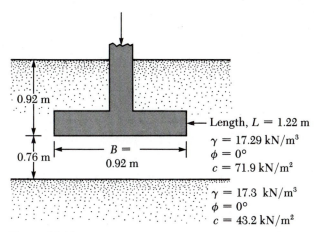

Figure P3.23

3.24 Refer to Figure 3.20. The foundation is 1 m × 2 m in plan. $D_f = 1$ m and $H = 1.5$ m. For the sand layer, $\phi = 35°$, $c = 0$, $\gamma = 17.8$ kN/m³; and for the clay layer, $\phi = 0$,

$c = 60 \text{ kN/m}^2$, $\gamma = 18.2 \text{ kN/m}^3$. Determine the gross allowable load that the foundation could carry. Use a factor of safety of 4 against bearing capacity failure. For bearing capacity factors shown in Eqs. (3.60) and (3.61), use Table 3.2.

3.25 Redo Problem 3.24 with the following:

Foundation: $B \times L = 3 \text{ ft} \times 6 \text{ ft}$

$D_f = 2.5 \text{ ft}$

$H = 3 \text{ ft}$

Sand: $\phi = 40°$

$c = 0$

$\gamma = 115 \text{ lb/ft}^3$

Clay: $\phi = 0$

$c = 750 \text{ lb/ft}^2$

$\gamma = 118 \text{ lb/ft}^3$

3.26 Refer to Figure 3.22. A foundation that is 10 ft × 6.5 ft in plan is resting on a sand deposit. The net load per unit area at the level of the foundation (q_o) is 3200 lb/ft². For the sand, given $\mu_s = 0.3$, $E_s = 3200$ lb/in.², $D_f = 2.95$ ft, and $H = 32$ ft. Assuming the foundation to be rigid, determine the elastic settlement that the foundation would undergo. Use Eq. (3.67).

3.27 Repeat Problem 3.26 with the following:

Foundation: 1.8 m × 1.8 m

$q_o = 190 \text{ kN/m}^2$

$D_f = 1 \text{ m}$

$H = 5 \text{ m}$

Soil: $\mu_s = 0.35$

$E_s = 16,500 \text{ kN/m}^2$

$\gamma = 16.5 \text{ kN/m}^2$

3.28 Solve Problem 3.26 using Eq. (3.69). For the correction factor, C_2, use a time of 5 years for creep.

3.29 Solve Problem 3.27 using Eq. (3.69). For the correction factor, C_2, use a time of 4 years for creep. Also use unit weight of soil, $\gamma = 110$ lb/ft³.

3.30 Refer to Figure P3.30, in which a square foundation is subjected to a load and a moment. Determine the settlement and the rotation of the foundation.

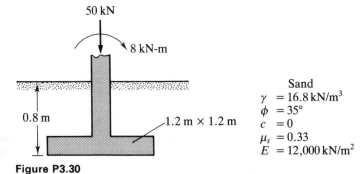

50 kN

8 kN-m

0.8 m

1.2 m × 1.2 m

Sand
$\gamma = 16.8 \text{ kN/m}^3$
$\phi = 35°$
$c = 0$
$\mu_s = 0.33$
$E = 12,000 \text{ kN/m}^2$

Figure P3.30

3.31 A flexible circular area is subjected to a uniformly distributed load of 3000 lb/ft^2. The diameter of the loaded area is 9.5 ft. Determine the stress increase in a soil mass at a point located 7.5 ft below the center of the loaded area.

3.32 Refer to Figure 3.35b, which shows a flexible rectangular area. Given: $B_1 = 1.2$ m, $B_2 = 3$ m, $L_1 = 3$ m, and $L_2 = 6$ m. If the area is subjected to a uniform load of 110 kN/m^2, determine the stress increase at a depth of 8 m located immediately below point O.

3.33 Solve Problem 3.32 using Newmark's chart.

3.34 A square column foundation is shown in Figure P3.34. Determine the average increase of pressure in the clay layer below the center of the foundation
a. By using Figure 3.37
b. By using the 2 : 1 method (Figure 3.38)

Figure P3.34

3.35 Solve Problem 3.34 using Newmark's chart.

3.36 Refer to Figure P3.34. Determine the average increase of stress in the clay layer below the corner of the foundation. Use Newmark's chart.

3.37 Redo Problem 3.36 using Table 3.7.

3.38 Estimate the consolidation settlement of the clay layer shown in Figure P3.34 using the results of part (a) of Problem 3.34.

3.39 Estimate the consolidation settlement of the clay layer shown in Figure P3.34 using the results of part (b) of Problem 3.34.

3.40 Refer to Problem 2.4. What will be the net allowable bearing capacity of a foundation 5 ft × 5 ft in plan? Given: $D_f = 3$ ft; allowable settlement = 1 in. Use Eq. (3.109).

3.41 Redo Problem 3.40 using Figure 3.45.

3.42 Two plate load tests with circular plates were conducted in the field. At 13 mm of settlement the results were as follows:

Diameter of plate (mm)	Load (kN)
304.8	45.5
762	169.7

What size square footing is required to carry a net load of 400 kN at a settlement of 13 mm?

REFERENCES

American Society for Testing and Materials (1982). *Annual Book of ASTM Standards*, Part 19, Philadelphia, Pa.

Bjerrum, L. (1963). "Allowable Settlement of Structures," *Proceedings*, European Conference on Soil Mechanics and Foundation Engineering, Wiesbaden, Germany, Vol. III, pp. 135–137.

Bowles, J. E. (1977). *Foundation Analysis and Design*, 2nd ed., McGraw-Hill, New York.

Burland, J. B., and Worth, C. P. (1974). "Allowable and Differential Settlement of Structures Including Damage and Soil-Structure Interaction," *Proceedings*, Conference on Settlement of Structures, Cambridge University, England, pp. 611–654.

Caquot, A., and Kerisel, J. (1953). "Sur le terme de surface dans le calcul des fondations en milieu pulverulent," *Proceedings*, Third International Conference on Soil Mechanics and Foundation Engineering, Vol. I, pp. 336–337.

Christian, J. T., and Carrier, W. D. (1978). "Janbu, Bjerrum, and Kjaernsli's Chart Reinterpreted," *Canadian Geotechnical Journal*, Vol. 15, pp. 124–128.

Das, B. M. (1983). *Advanced Soil Mechanics*, McGraw-Hill, New York.

De Beer, E. E. (1970). "Experimental Determination of the Shape Factors and Bearing Capacity Factors of Sand," *Geotechnique*, Vol. 20, No. 4, pp. 387–411.

Georgiadis, M., and Butterfield, R. (1988). "Displacement of Footings on Sand Under Eccentric and Inclined Loads," *Canadian Geotechnical Journal*, Vol. 25, No. 2, pp. 199–212.

Grant, R. J., Christian, J. T., and Vanmarcke, E. H. (1974). "Differential Settlement of Buildings," *Journal of the Geotechnical Engineering Division*, American Society of Civil Engineers, Vol. 100, No. GT9, pp. 973–991.

Hanna, A. M., and Meyerhof, G. G. (1981). "Experimental Evaluation of Bearing Capacity of Footings Subjected to Inclined Loads," *Canadian Geotechnical Journal*, Vol. 18, No. 4, pp. 599–603.

Hansen, J. B. (1970). "A Revised and Extended Formula for Bearing Capacity," Danish Geotechnical Institute, *Bulletin 28*, Copenhagen.

Harr, M. E. (1966). *Fundamentals of Theoretical Soil Mechanics*, McGraw-Hill, New York.

Higher, W. H., and Anders, J. C. (1985). "Dimensioning Footings Subjected to Eccentric Loads," *Journal of Geotechnical Engineering*, American Society of Civil Engineers, Vol. 111, No. GT5, pp. 659–665.

Housel, W. S. (1929). "A Practical Method for the Selection of Foundations Based on Fundamental Research in Soil Mechanics," *Research Bulletin No. 13*, University of Michigan, Ann Arbor.

Ismael, N. F., and Vesic, A. S. (1981). "Compressibility and Bearing Capacity," *Journal of the Geotechnical Engineering Division*, American Society of Civil Engineers, Vol. 107, No. GT12, pp. 1657–1676.

Janbu, N., Bjerrum, L., and Kjaernsli, B. (1956). "Veiledning ved losning av fundamentering—soppgaver," *Publication No. 16*, Norwegian Geotechnical Institute, pp. 30–32.

Lee, I. K. (1963). "Elastic Settlements of Footing with a Rough Interface," *Proceedings*, Fourth Australia–New Zealand Conference on Soil Mechanics and Foundation Engineering, pp. 225–230.

Lundgren, H., and Mortensen, K. (1953). "Determination by the Theory of Plasticity on the Bearing Capacity of Continuous Footings on Sand," *Proceedings*, Third International Conference on Soil Mechanics and Foundation Engineering, Vol. 1, pp. 409–412.

Meyerhof, G. G. (1953). "The Bearing Capacity of Foundations Under Eccentric and Inclined Loads," *Proceedings*, Third International Conference on Soil Mechanics and Foundation Engineering, Zürich, Vol. 1, pp. 440–445.

Meyerhof, G. G. (1956). "Penetration Tests and Bearing Capacity of Cohesionless Soils, *Journal of the Soil Mechanics and Foundations Division*, American Society of Civil Engineers, Vol. 82, No. SM1, pp. 1–19.

Meyerhof, G. G. (1963). "Some Recent Research on the Bearing Capacity of Foundations," *Canadian Geotechnical Journal*, Vol. 1, No. 1, pp. 16–26.

Meyerhof, G. G. (1974). "Ultimate Bearing Capacity of Footings on Sand Layer Overlying Clay," *Canadian Geotechnical Journal*, Vol. 11, No. 2, pp. 224–229.

Meyerhof, G. G., and Hanna, A. M. (1978). "Ultimate Bearing Capacity of Foundations on Layered Soil Under Inclined Load," *Canadian Geotechnical Journal*, Vol. 15, No. 4, pp. 565–572.

Mitchell, J. K., and Gardner, W. S. (1975). "*In Situ* Measurement of Volume Change Characteristics," *Proceedings*, Specialty Conference, American Society of Civil Engineers, Vol. 2, pp. 279–345.

Newmark, N. M. (1935). "Simplified Computation of Vertical Pressure in Elastic Foundation," *Circular 24*, University of Illinois Engineering Experiment Station, Urbana.

Newmark, N. M. (1942). "Influence Charts for Computation of Stresses in Elastic Foundations," *Bulletin No. 338*, University of Illinois Engineering Experiment Station, Urbana.

Osterberg, J. O. (1957). "Influence Values for Vertical Stresses in Semi-Infinite Mass Due to Embankment Loading," *Proceedings*, Fourth International Conference on Soil Mechanics and Foundation Engineering, Vol. 1, pp. 393–396.

Peck, R. B., Hanson, W. E., and Thornburn, T. H. (1974). *Foundation Engineering*, 2nd ed., Wiley, New York.

Polshin, D. E. and Tokar, R. A. (1957). "Maximum Allowable Nonuniform Settlement of Structures," *Proceedings*, Fourth International Conference on Soil Mechanics and Foundation Engineering, London, Vol. 1, pp. 402–405.

Prandtl, L. (1921). "Über die Eindringungsfestigkeit (Härte) plastischer Baustoffe und die Festigkeit von Schneiden," *Zeitschrift für angewandte Mathematik und Mechanik*, Vol. 1, No. 1, pp. 15–20.

Reedy, A. S., and Srinivasan, R. J. (1967). "Bearing Capacity of Footings on

Layered Clay," *Journal of the Soil Mechanics and Foundations Division*, American Society of Civil Engineers, Vol. 93, No. SM2, pp. 83–99.

Reissner, H. (1924). "Zum Erddruckproblem," *Proceedings*, First International Congress of Applied Mechanics, Delft, pp. 295–311.

Schmertmann, J. H. (1970). "Static Cone to Compute Settlement Over Sand," *Journal of the Soil Mechanics and Foundations Division*, American Society of Civil Engineers, Vol. 96, No. SM3, pp. 1011–1043.

Schmertmann, J. H., and Hartman, J. P. (1978). "Improved Strain Influence Factor Diagrams," *Journal of the Geotechnical Engineering Division*, American Society of Civil Engineers, Vol. 104, No. GT8, pp. 1131–1135.

Skempton, A. W. (1951). "The Bearing Capacity of Clays," *Proceedings*, Building Research Congress, London, pp. 180–189.

Skempton, A. W., and Bjerrum, L. (1957). "A Contribution to Settlement Analysis of Foundations in Clay," *Geotechnique*, London, Vol. 7, p. 178.

Terzaghi, K. (1943). *Theoretical Soil Mechanics*, Wiley, New York.

Terzaghi, K., and Peck, R. B. (1967). *Soil Mechanics in Engineering Practice*, 2nd ed., Wiley, New York.

Vesic, A. S. (1963). "Bearing Capacity of Deep Foundations in Sand," *Highway Research Record No. 39*, National Academy of Sciences, pp. 112–153.

Vesic, A. S. (1973). "Analysis of Ultimate Loads of Shallow Foundations," *Journal of the Soil Mechanics and Foundations Division*, American Society of Civil Engineers, Vol. 99, No. SM1, pp. 45–73.

Wahls, H. E. (1981). "Tolerable Settlement of Buildings," *Journal of the Geotechnical Engineering Division*, American Society of Civil Engineers, Vol. 107, No. GT11, pp. 1489–1504.

Whitman, R. V., and Richart, F. E. (1967). "Design Procedures for Dynamically Loaded Foundations," *Journal of the Soil Mechanics and Foundations Division*, American Society of Civil Engineers, Vol. 93, No. SM6, pp. 169–193.

Chapter 4

MAT FOUNDATIONS

4.1   Introduction

Mat foundations are primarily shallow foundations. They are a type of *combined footing*. There are four major types of combined footing (see Figure 4.1a):

1. *Rectangular Combined Footing:* In several instances, the load to be carried by a column and the soil bearing capacity are such that the standard spread footing design will require the extension of the column foundation beyond the property line. In such a case, two or more columns can be supported on a single rectangular foundation, as shown in Figure 4.1b. If the net allowable soil pressure is known, the size of the foundation $(B \times L)$ can be determined in the following manner:

a. Determine the area of the foundation, A, as

$$A = \frac{Q_1 + Q_2}{q_{\text{all(net)}}} \tag{4.1}$$

where Q_1, Q_2 = column loads

$q_{\text{all(net)}}$ = net allowable soil bearing capacity

b. Determine the location of the resultant of the column loads. Referring to Figure 4.1b,

$$X = \frac{Q_2 L_3}{Q_1 + Q_2} \tag{4.2}$$

c. For uniform distribution of soil pressure under the foundation, the resultant of the column loads should pass through the centroid of the foundation. Thus

$$L = 2(L_2 + X) \tag{4.3}$$

where L = length of the foundation

(a)

(b)

Figure 4.1 (a) Combined footing; (b) rectangular combined footing; (c) trapezoidal combined footing; (d) cantilever footing (continued on pages 213–214)

d. Once the length L is determined, the value of L_1 can be obtained as

$$L_1 = L - L_2 - L_3 \tag{4.4}$$

Note that the magnitude of L_2 will be known, depending on the extent of the property line.

e. The width of the foundation can now be obtained as

$$B = \frac{A}{L} \tag{4.5}$$

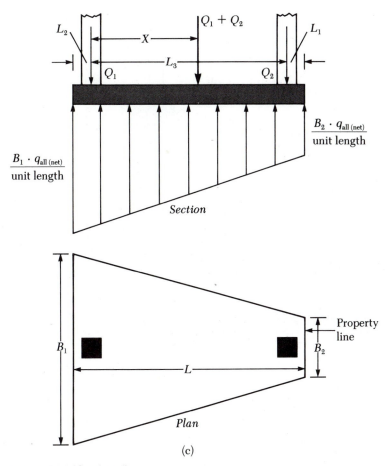

Section

Plan

(c)

Figure 4.1   (Continued)

2. *Trapezoidal Combined Footing:* This type of combined footing (Figure 4.1c) is sometimes used where there is not enough space for isolated spread foundation for a column carrying a large load. The size of the foundation that will uniformly distribute pressure on the soil can be obtained in the following manner:

a. If the net allowable soil pressure is known, determine the area of the foundation as

$$A = \frac{Q_1 + Q_2}{q_{all(net)}}$$

Referring to Figure 4.1c,

$$A = \frac{B_1 + B_2}{2} L \tag{4.6}$$

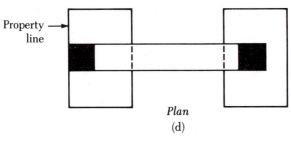

Figure 4.1 (Continued)

b. Determine the location of the resultant for the column loads as

$$X = \frac{Q_2 L_3}{Q_1 + Q_2}$$

c. From the property of a trapezoid

$$X + L_2 = \left(\frac{B_1 + 2B_2}{B_1 + B_2}\right)\frac{L}{3} \tag{4.7}$$

With known values of A, L, X, and L_2, solve Eqs. (4.6) and (4.7) to obtain B_1 and B_2. Note that, for a trapezoid

$$\frac{L}{3} < X + L_2 < \frac{L}{2}$$

3. *Cantilever Footing:* This type of combined footing construction uses a *strap beam* to connect an eccentrically loaded column foundation to the foundation of an interior column (Figure 4.1d). Cantilever footings may be used in place of trapezoidal or rectangular combined footings when the allowable soil bearing capacity is high and the distances between the columns are large.

4. *Mat Foundation:* A mat foundation, which is sometimes referred to as a *raft foundation,* is a combined footing that may cover the whole area under a structure supporting several columns and walls (Figure 4.1a). Mat foundations are sometimes preferred for soils that have low load-bearing capacity but that will have to support high column and/or wall loads. In some conditions, spread footings may cover more than half the building area; mat foundations may prove to be more economical. This chapter will discuss mat foundations in detail.

4.2   Common Types of Mat Foundation

There are several types of mat foundation in use nowadays. Some of the common types of construction are schematically shown in Figure 4.2. They are

1. Flat plate (Figure 4.2a): The mat is of uniform thickness.
2. Flat plate thickened under columns (Figure 4.2b).
3. Beams and slab (Figure 4.2c): The beams run both ways, and the columns are located at the intersection of the beams.
4. Slab with basement walls as a part of the mat (Figure 4.2d): The walls act as stiffeners for the mat.

(a)

Figure 4.2 Types of mat foundation: (a) flat plate; (b) flat plate thickened under column; (c) beams and slab; (d) slab with basement wall (continued on pages 216–218)

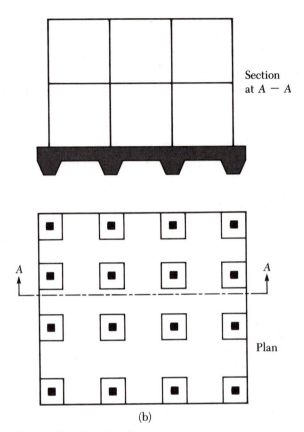

Figure 4.2 (Continued)

Mats are sometimes supported by piles. The piles help in reducing the settlement of a structure located over highly compressible soil. Where the ground water table is high, mats are often placed over piles to control buoyancy.

4.3   Bearing Capacity of Mat Foundations

The *gross ultimate bearing capacity* of a mat foundation can be determined by the same equation used for shallow foundations (see Section 3.5), or

$$q_u = cN_c F_{cs} F_{cd} F_{ci} + qN_q F_{qs} F_{qd} F_{qi} + \tfrac{1}{2}\gamma B N_\gamma F_{\gamma s} F_{\gamma d} F_{\gamma i} \qquad (3.16)$$

(Chapter 3 gives the proper values of the bearing capacity factors, as well as the shape, depth, and load inclination factors.) The term B in the preceding equation is the smallest dimension of the mat.

The *net ultimate capacity* can be given as

$$q_{\text{all(net)}} = q_u - q \qquad\qquad\qquad\qquad\qquad (3.20)$$

(c)

Figure 4.2   (Continued)

A suitable factor of safety should be used to calculate the net *allowable* bearing capacity. For rafts on clay, the factor of safety should not be less than 3 under dead load and maximum live load. However, under the most extreme conditions, the factor of safety should be at least 1.75 to 2. For rafts constructed over sand, a factor of safety of 3 should normally be used. Under most working conditions, the factor of safety against bearing capacity failure of rafts on sand is very large.

For saturated clays with $\phi = 0$ and vertical loading condition, Eq. (3.16) gives (*note:* $N_c = 5.14$, $N_q = 1$, and $N_\gamma = 0$)

$$q_u = c_u N_c F_{cs} F_{cd} + q \qquad (4.8)$$

where c_u = undrained cohesion

From Table 3.4, for $\phi = 0$ condition

$$F_{cs} = 1 + \frac{B}{L}\left(\frac{N_q}{N_c}\right) = 1 + \left(\frac{B}{L}\right)\left(\frac{1}{5.14}\right) = 1 + \frac{0.195B}{L}$$

Section
at $A - A$

Plan

(d)

Figure 4.2   (Continued)

and

$$F_{cd} = 1 + 0.4\left(\frac{D_f}{B}\right)$$

Substitution of the preceding shape and depth factors into Eq. (4.8) yields

$$q_u = 5.14c_u\left(\frac{1 + 0.195B}{L}\right)\left(1 + 0.4\,\frac{D_f}{B}\right) + q \qquad (4.9)$$

Hence, the net ultimate bearing capacity

$$q_{u(net)} = q_u - q = 5.14c_u\left(1 + \frac{0.195B}{L}\right)\left(1 + 0.4\,\frac{D_f}{B}\right) \qquad (4.10)$$

If a factor of safety of 3 is used, the net allowable soil bearing capacity can be given as

$$q_{all(net)} = \frac{q_{u(net)}}{FS} = 1.713c_u\left(1 + \frac{0.195B}{L}\right)\left(1 + 0.4\frac{D_f}{B}\right) \tag{4.11}$$

Figure 4.3 shows a plot of $q_{all(net)}/c_u$ for various values of L/B and D_f/B, based on Eq. (4.11).

The net allowable bearing capacity for mats constructed over granular soil deposits can be adequately determined from the standard penetration resistance numbers. Referring to Eq. (3.107b), for shallow foundations

$$q_{all(net)} \text{ (kN/m}^2) = 11.98N\left(\frac{3.28B + 1}{3.28B}\right)^2 F_d\left(\frac{s}{25.4}\right)$$

where N = corrected standard penetration resistance
        B = width (m)
        $F_d = 1 + 0.33(D_f/B) \leq 1.33$
        s = settlement, in mm

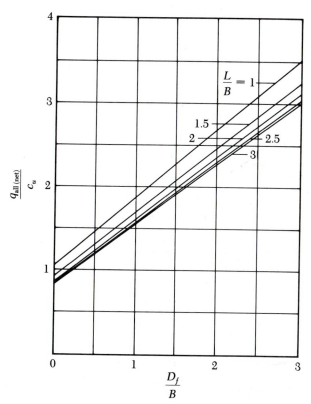

Figure 4.3   Plot of $q_{all(net)}/c_u$ against D_f/B [Eq. (4.11)] (*note:* factor of safety = 3)

When the width B is large, the preceding equation can be approximated (assuming $3.28B + 1 \approx 3.28B$) as

$$
q_{\text{all(net)}} \text{ (kN/m}^2) \approx 11.98 N F_d \left(\frac{s}{2.54}\right)
$$
$$
= 11.98N \left[1 + 0.33\left(\frac{D_f}{B}\right) \right] \left[\frac{s \text{ (mm)}}{25.4} \right]
$$
$$
\leq 15.93N \left[\frac{s \text{ (mm)}}{25.4} \right] \tag{4.12}
$$

In English units, Eq. (4.12) may be expressed as

$$
q_{\text{all(net)}} \text{ (kips/ft}^2) = 0.25N \left[1 + 0.33\left(\frac{D_f}{B}\right) \right] [s \text{ (in.)}]
$$
$$
\leq 0.33N[s \text{ (in.)}] \tag{4.13}
$$

Note that Eq. (4.13) could have been derived from Eqs. (3.108) and (3.109b).

Note that the original Eqs. (3.107) and (3.109) were for a settlement of 25.4 mm (1 in.) with a differential settlement of about 19 mm (0.75 in.). However, the widths of the raft foundations are larger than the isolated spread footings. As can be seen from Figure 3.37, the depth of significant stress increase in the soil below a foundation is dependent on the foundation width. Hence, for a raft foundation, the depth of the zone of influence is likely to be much larger than that of a spread footing. This means that the loose soil pockets under a raft may be more evenly distributed, resulting in a smaller differential settlement. Thus it is customary to assume that for a maximum raft settlement of 50.8 mm (2 in.), the differential settlement would be 19 mm (0.75 in.). Using this logic and conservatively assuming F_d to be equal to one, Eqs. (4.12) and (4.13) can be approximated as

$$
q_{\text{all(net)}} \text{ (kN/m}^2) \approx 23.96N \tag{4.14}
$$

and

$$
q_{\text{all(net)}} \text{ (kips/ft}^2) = 0.5N \tag{4.15}
$$

The net pressure applied on a foundation can be expressed as (Figure 4.4)

$$
q = \frac{Q}{A} - \gamma D_f \tag{4.16}
$$

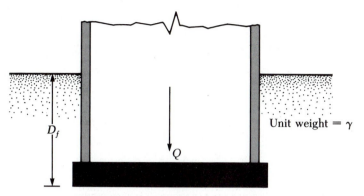

Figure 4.4   Definition of net pressure on soil caused by a mat foundation

where Q = dead weight of the structure and the live load
   A = area of the raft

Hence, in all cases, q should be less than or equal to $q_{\text{all(net)}}$.

Example 4.1

Determine the net ultimate bearing capacity of a mat foundation measuring 45 ft × 30 ft on a saturated clay with $c_u = 1950 \text{ lb/ft}^2$, $\phi = 0$, and $D_f = 6.5$ ft.

Solution

From Eq. (4.10)

$$q_{u(\text{net})} = 5.14c_u\left[1 + \left(\frac{0.195B}{L}\right)\right]\left[1 + 0.4\left(\frac{D_f}{B}\right)\right]$$

$$= (5.14)(1950)\left[1 + \left(\frac{0.195 \times 30}{45}\right)\right]\left[1 + \left(\frac{0.4 \times 6.5}{30}\right)\right]$$

$$= \textbf{12,307 lb/ft}^2$$

Example 4.2

What will be the net allowable bearing capacity of a mat foundation with dimensions of 15 m × 10 m constructed over a sand deposit? Given: $D_f = 2$ m, allowable settlement = 30 mm, corrected average penetration number, $N = 10$.

Solution

From Eq. (4.12)

$$q_{\text{all(net)}} = 11.98N\left(1 + \frac{0.33D_f}{B}\right)\left(\frac{s}{25.4}\right) \leq 15.93N\left(\frac{s}{25.4}\right)$$

Now

$$11.98N\left(1 + \frac{0.33D_f}{B}\right)\left(\frac{s}{25.4}\right) = 11.98N\left[1 + \frac{(0.33)(2)}{10}\right]\left(\frac{s}{25.4}\right)$$

$$= 12.77N\left(\frac{s}{25.4}\right)$$

So

$$q_{\text{all(net)}} = 12.77N\left(\frac{s}{25.4}\right) = 12.77(10)\left(\frac{30}{25.4}\right)$$

$$= 150.83 \text{ kN/m}^2$$

4.4   Differential Settlement of Mats

The American Concrete Institute Committee 436 (1966) has suggested the following method for calculation of the differential settlement of mat foundations. According to this method, the rigidity factor (K_r) can be calculated as

$$K_r = \frac{E'I_b}{E_s B^3} \tag{4.17}$$

where E' = modulus of elasticity of the material used in the structure
E_s = modulus of elasticity of the soil
B = width of foundation
I_b = moment of inertia of the structure per unit length at right angles to B

The term $E'I_b$ can be expressed as

$$E'I_b = E'\left(I_F + \sum I_{b'} + \sum \frac{ah^3}{12}\right) \tag{4.18}$$

where $E'I_b$ = flexural rigidity of the foundation per unit length at right angles to B
$\sum E'I_{b'}$ = flexural rigidity of the framed members
$\sum (E'ah^3/12)$ = flexural rigidity of the shear walls
a = shear wall thickness
h = shear wall height

Based on the value of K_r, the ratio (δ) of the differential settlement to the total settlement can be estimated in the following manner:

1. If $K_r > 0.5$, it can be treated as a rigid mat, and $\delta = 0$.
2. If $K_r = 0.5$, then $\delta \approx 0.1$.
3. If $K_r = 0$, $\delta = 0.35$ for square mats ($B/L = 1$), and $\delta = 0.5$ for long foundations ($B/L = 0$).

4.5   Compensated Foundations

The settlement of a mat foundation can be reduced by decreasing the net pressure increase on soil, which can be done by increasing the depth of embedment, D_f. This increase is particularly important for rafts on soft clays, where large consolidation settlements are expected. From Eq. (4.16), the net average applied pressure on soil is (Figure 4.4)

$$q = \frac{Q}{A} - \gamma D_f$$

For no increase of the net soil pressure on soil below a raft foundation, q should be equal to zero. Thus

$$D_f = \frac{Q}{A\gamma} \tag{4.19}$$

The relation for D_f given in the preceding equation is usually referred to as the depth of a *fully compensated foundation*.

The factor of safety against bearing capacity failure for partially compensated foundations (that is, $D_f < Q/A\gamma$) may be given as

$$FS = \frac{q_{u(net)}}{q} = \frac{q_{u(net)}}{\dfrac{Q}{A} - \gamma D_f} \tag{4.20}$$

For saturated clays, the factor of safety against bearing capacity failure can thus be obtained by substituting Eq. (4.10) into Eq. (4.20):

$$\boxed{FS = \frac{5.14 c_u \left(1 + \dfrac{0.195B}{L}\right)\left(1 + 0.4\dfrac{D_f}{B}\right)}{\dfrac{Q}{A} - \gamma D_f}} \tag{4.21}$$

Example 4.3

Refer to Figure 4.4. The mat has dimensions of 30 m × 40 m. The live load and dead load on the mat are 200 MN. The mat is placed over a layer of soft clay. The unit weight of the clay is 18.75 kN/m³. Find D_f for a fully compensated foundation.

Solution

From Eq. (4.19)

$$D_f = \frac{Q}{A\gamma} = \frac{200 \times 10^3 \text{ kN}}{(30 \times 40)(18.75)} = \textbf{8.89 m}$$

Example 4.4

Refer to Example 4.3. Given: c_u for the clay is 12.5 kN/m². If the required factor of safety against bearing capacity failure is 3, determine the depth of the foundation.

Solution

From Eq. (4.21)

$$FS = \frac{5.14c_u\left(1 + \frac{0.195B}{L}\right)\left(1 + 0.4\frac{D_f}{B}\right)}{\frac{Q}{A} - \gamma D_f}$$

$FS = 3$; $c_u = 12.5$ kN/m²; $B/L = 30/40 = 0.75$; $Q/A = (200 \times 10^3)/(30 \times 40) = 166.67$ kN/m². Substitution of the preceding values into Eq. (4.21) yields

$$3 = \frac{(5.14)(12.5)[1 + (0.195)(0.75)]\left[1 + 0.4\left(\frac{D_f}{30}\right)\right]}{166.67 - (18.75)D_f}$$

$$500.01 - 56.25D_f = 73.65 + 0.982D_f$$

$$426.36 = 57.23D_f$$

or

$$D_f = 7.5 \text{ m}$$

Example 4.5

Consider a mat foundation 30 m × 40 m in plan, as shown in Figure 4.5. The total dead load plus the live load on the raft is 200 MN. Estimate the consolidation settlement at the center of the foundation.

Solution

Given: $Q = 200 \times 10^3$ kN. So, net load per unit area is

$$q = \frac{Q}{A} - \gamma D_f = \frac{200 \times 10^3}{30 \times 40} - (15.7)(2) = 166.67 - 31.4 = 135.27 \text{ kN/m}^2$$

From Eq. (3.81), the average pressure increase on the clay layer below the center of the foundation is

$$\Delta p_{av} = \tfrac{1}{6}(\Delta p_t + \Delta p_m + \Delta p_b)$$

The values of $\Delta p_t, 4\,\Delta p_m$, and Δp_b can be determined by referring to Figure 3.37. At the *top of the clay layer*

$$\frac{z}{B} = \frac{15}{30} = 0.5$$

$$\frac{L}{B} = \frac{40}{30} = 1.33$$

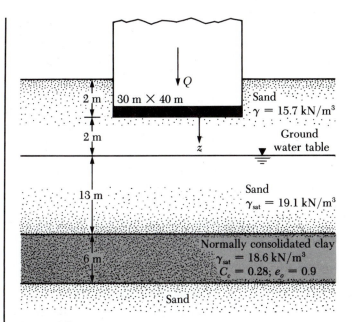

Figure 4.5

So, for $z/B = 0.5$ and $L/B = 1.33$

$$\frac{\Delta p_t}{q} = 0.75, \qquad \Delta p_t = (0.75)(135.27) = 101.45 \text{ kN/m}^2$$

Similarly, for the *middle of the clay layer*

$$\frac{z}{B} = \frac{18}{30} = 0.6$$

$$\frac{L}{B} = 1.33$$

So, $\Delta p_m/q = 0.66$, $\Delta p_m = 89.3 \text{ kN/m}^2$. At the *bottom of the clay layer*

$$\frac{z}{B} = \frac{21}{30} = 0.7$$

$$\frac{L}{B} = 1.33$$

So, $\Delta p_b/q = 0.58$, $\Delta p_b = 75.46 \text{ kN/m}^2$. Hence

$$\Delta p_{av} = \tfrac{1}{6}[101.45 + 4(89.3) + 75.46] = 89 \text{ kN/m}^2$$

From Eq. (1.65), the consolidation settlement

$$S = \frac{C_c H_c}{1 + e_o} \log \frac{p_o + \Delta p_{av}}{p_o}$$

$$p_o = 4(15.7) + 13(19.1 - 9.81) + \tfrac{6}{2}(18.6 - 9.81) = 209.94 \text{ kN/m}^2$$

So

$$S = \frac{(0.28)(6 \times 1000)}{1 + 0.9} \log \left(\frac{209.94 + 89}{209.94} \right) = \textbf{135.7 mm}$$

Note: Similar calculations can be made to obtain the settlement at the corner of the raft. The differential settlement between the center and the corner of the raft calculated in the preceding manner will, in most cases, be higher than the actual differential settlement. This is because of the stiffness of the superstructure.

The reliability of the settlement calculation will depend on several factors. If the clay is normally consolidated and uniform in thickness, the settlement calculation will yield fairly accurate results. If the compressible soil is erratic in formation and contains several types of soil, the calculation will indicate only the maximum magnitude.

4.6   Structural Design of Mat Foundations

The structural design of mat foundations can be carried out by two conventional methods: the *conventional rigid method* and the *approximate flexible method*. Finite difference and finite element methods can also be used; however, this section will cover the basic concepts of the first two design methods.

Conventional Rigid Method

The conventional rigid method of mat foundation design can be explained in a step-by-step manner with reference to Figure 4.6. The steps are as follows:

1. Figure 4.6a shows that the mat has a dimension of $L \times B$. $Q_1, Q_2, Q_3, \ldots$ are the column loads. Calculate the total column load as

$$Q = Q_1 + Q_2 + Q_3 + \cdots \tag{4.22}$$

2. Determine the pressure on the soil (q) below the mat at points A, B, C, D, ... by using the equation

$$q = \frac{Q}{A} \pm \frac{M_y x}{I_y} \pm \frac{M_x y}{I_x} \tag{4.23}$$

where $A = BL$
$I_x = (1/12)BL^3 =$ moment of inertia about the x axis
$I_y = (1/12)LB^3 =$ moment of inertia about the y axis
$M_x =$ moment of the column loads about the x axis $= Qe_y$
$M_y =$ moment of the column loads about the y axis $= Qe_x$

e_x and e_y are the load eccentricities in the direction of x and y. They can be determined by using x', y' coordinates as

$$X' = \frac{Q_1 x_1' + Q_2 x_2' + Q_3 x_3' + \cdots}{Q} \tag{4.24}$$

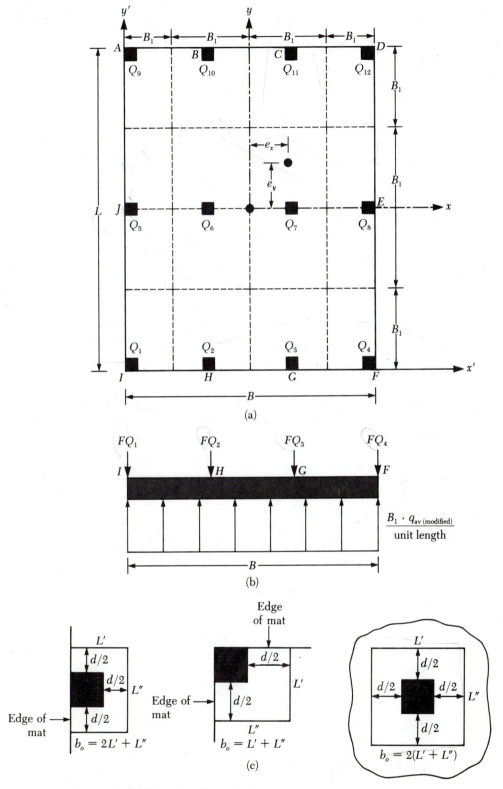

Figure 4.6 Conventional rigid mat foundation design

and

$$e_x = X' - \frac{B}{2} \tag{4.25}$$

Similarly

$$Y' = \frac{Q_1 y_1' + Q_2 y_2' + Q_3 y_3' + \cdots}{Q} \tag{4.26}$$

and

$$e_y = Y' - \frac{L}{2} \tag{4.27}$$

3. Compare the values of the soil pressures determined in Step 2 with the net allowable soil pressure to check if $q \leq q_{\text{all(net)}}$.

4. Divide the mat into several strips in x and y directions (see Figure 4.6a). Let the width of any strip be B_1.

5. Draw the shear (V) and the moment (M) diagrams for each individual strip (in x and y directions). For example, take the bottom strip in the x direction of Figure 4.6a; its average soil pressure can be given as

$$q_{\text{av}} \approx \frac{q_I + q_F}{2} \tag{4.28}$$

where q_I and q_F = soil pressures at points I and F as determined from Step 2

The total soil reaction is equal to $q_{\text{av}} B_1 B$. Now obtain the total column load on the strip as $Q_1 + Q_2 + Q_3 + Q_4$. The sum of the column loads on the strip will not be equal to $q_{\text{av}} B_1 B$ because the shear between the adjacent strips has not been taken into account. For this reason, the soil reaction and the column loads need to be adjusted, or

$$\text{average load} = \frac{q_{\text{av}} B_1 B + (Q_1 + Q_2 + Q_3 + Q_4)}{2} \tag{4.29}$$

Now, the modified average soil reaction

$$q_{\text{av(modified)}} = q_{\text{av}} \left(\frac{\text{average load}}{q_{\text{av}} B_1 B} \right) \tag{4.30}$$

Also, the column load modification factor is

$$F = \frac{\text{average load}}{Q_1 + Q_2 + Q_3 + Q_4} \tag{4.31}$$

So, the modified column loads are FQ_1, FQ_2, FQ_3, and FQ_4. This modified loading on the strip under consideration is shown in Figure 4.6b. Now the shear and the moment diagram for this strip can be drawn. This procedure can be repeated for all strips in the x and y directions.

6. Determine the depth of the mat d (refer to Figure A.1 in Appendix A for the definition of d). This can be done by checking for diagonal tension shear near various columns. According to ACI Code 318-83—revised 1986 (Section 11.11.2.1) (American Concrete Institute, 1987), for the critical section

$$U = b_o \, d[\phi(0.34)\sqrt{f'_c}] \tag{4.32}$$

where U = factored column load (MN) = (column load) $\times$ (load factor)

   ϕ = reduction factor (see Appendix A)

   f'_c = compressive strength of concrete at 28 days (MN/m^2)

The units of b_o and d in the preceding equation are in meters. In English units, Eq. (4.32) can be expressed as

$$U = b_o \, d(4\phi\sqrt{f'_c}) \tag{4.33}$$

where U is in lb, b_o and d are in in., and f'_c is in lb/in.^{2}.

The expression for b_o in terms of d, which depends on the location of the column with respect to the plan of the mat, can be obtained from Figure 4.6c.

7. From the moment diagrams of all strips *in a given direction* (that is, x or y), obtain the *maximum* positive and negative moments per unit width (that is, $M' = M/B_1$).

8. Determine the areas of steel per unit width for positive and negative reinforcement in x and y directions from the following equations [Appendix; Eqs. (A.4) and (A.6)]:

$$M_u = (M')(\text{load factor}) = \phi A_s f_y \left(d - \frac{a}{2} \right) \tag{A.6}$$

and

$$a = \frac{A_s f_y}{0.85 f'_c \, b} \tag{A.4}$$

where A_s = area of steel per unit width

   f_y = yield stress of reinforcement in tension

   M_u = factored moment

Examples 4.6 and 4.7 illustrate the use of the conventional rigid method of mat foundation design.

Example 4.6

The plan of a mat foundation with column loads is shown in Figure 4.7. Using Eq. (4.23), calculate the soil pressures at points A, B, C, D, E, and F. The size of the mat is 16.5 m $\times$ 21.5 m. All columns are 0.5 m $\times$ 0.5 m in section. Given: $q_{\text{all(net)}} = 60$

Figure 4.7

kN/m². Determine that the soil pressures are less than the net allowable soil bearing capacity.

Solution

From Eq. (4.23)

$$q = \frac{Q}{A} \pm \frac{M_y x}{I_y} \pm \frac{M_x y}{I_x}$$

A = area of the mat = $(16.5)(21.5) = 354.75$ m²

$$I_x = \frac{1}{12} BL^3 = \frac{1}{12} (16.5)(21.5)^3 = 13,665 \text{ m}^4$$

$$I_y = \frac{1}{12} LB^3 = \frac{1}{12} (21.5)(16.5)^3 \approx 8050 \text{ m}^4$$

Q = sum of the column loads = $350 + 2(400) + 450 + 2(500) + 2(1200)$
$\quad + 4(1500) = 11,000$ kN

$$M_y = Qe_x$$

$$e_x = X' - \frac{B}{2}$$

$$X' = \frac{Q_1 x_1' + Q_2 x_2' + Q_3 x_3' + \cdots}{Q} = \frac{1}{11,000}[(8.25)(500 + 1500 + 1500 + 500)$$
$$+ (16.25)(350 + 1200 + 1200 + 450) + (0.25)(400 + 1500 + 1500$$
$$+ 400)]$$
$$= 7.814 \text{ m}$$

So

$$e_x = X' - \frac{B}{2} = 7.814 - 8.25 = -0.436 \approx -0.44 \text{ m}$$

Hence, the resultant line of action is located to the left of the center of the mat. So

$$M_y = (11,000)(0.44) = 4840 \text{ kN-m}$$

Similarly

$$M_x = Qe_y$$

$$e_y = \left(Y' - \frac{L}{2}\right)$$

$$Y' = \frac{Q_1 y_1' + Q_2 y_2' + Q_3 y_3' + \cdots}{Q} = \frac{1}{11,000}[(0.25)(400 + 500 + 350)$$
$$+ (7.25)(1500 + 1500 + 1200) + (14.25)(1500 + 1500 + 1200)$$
$$+ (21.25)(400 + 500 + 450)]$$
$$= 10.85 \text{ m}$$

So

$$e_y = Y' - \frac{L}{2} = 10.85 - \frac{21.5}{2} = 0.1 \text{ m}$$

The location of the line of action of the resultant column loads is shown in Figure 4.7.

$$M_x = (11,000)(0.1) = 1100 \text{ kN-m}$$

So

$$q = \frac{11,000}{354.75} \pm \frac{(4840)x}{8050} \pm \frac{(1100)y}{13,665} = 31.0 \pm 0.6x \pm 0.086 \text{ (kN/m}^2)$$

Calculation of Soil Pressure

At A: $q = 31.0 + (0.6)(8.25) + (0.08)(10.75) = \mathbf{36.81 \text{ kN/m}^2}$
At B: $q = 31.0 + (0.6)(0) + (0.08)(10.75) = \mathbf{31.86 \text{ kN/m}^2}$
At C: $q = 31.0 - (0.6)(8.25) + (0.08)(10.75) = \mathbf{26.91 \text{ kN/m}^2}$
At D: $q = 31.0 - (0.6)(8.25) - (0.08)(10.75) = \mathbf{25.19 \text{ kN/m}^2}$
At E: $q = 31.0 + (0.6)(0) - (0.08)(10.75) = \mathbf{30.14 \text{ kN/m}^2}$
At F: $q = 31.0 + (0.6)(8.25) - (0.08)(10.75) = \mathbf{35.09 \text{ kN/m}^2}$

The soil pressures at all points are less than the given value of

$$q_{all(net)} = 60 \text{ kN/m}^2$$

Example 4.7

Divide the mat shown in Figure 4.7 into three strips, such as $AGHF$ ($B_1 = 4.25$ m), $GIJH$ ($B_1 = 8$ m), and $ICDJ$ ($B_1 = 4.25$ m). Using the results of Example 4.6, determine the reinforcement requirements in the y direction. Given: $f'_c = 20.7$ MN/m^2 and $f_y = 413.7$ MN/m^2. Use a load factor of 1.7.

(a) Strip $AGHF$

Figure 4.8

(b) Strip *GIJH*

Figure 4.8   (Continued)

Solution
Determination of Shear and Moment Diagrams for Strips

Strip AGHF: Average soil pressure $= q_{av} = (q_{(at\ A)} + q_{(at\ F)})/2 = (36.81 + 35.09)/2$ $= 35.95$ kN/m².
Total soil reaction $= q_{av} B_1 L = (35.95)(4.25)(21.50) = 3285$ kN
Total column load on this strip $= 400 + 1500 + 1500 + 400 = 3800$ kN
Average load $=$ (load soil reaction $+$ column loads)$/2 = (3285 + 3800)/2 = 3542.5$ kN
So, modified average soil pressure $= q_{av(modified)} = q_{av}(3542.5/3285) = (35.95)(3542.5/3285) = 38.768$ kN/m². The column loads can be modified in a similar manner by multiplying factor $F = 3542.5/3800 = 0.9322$.

(c) Strip *ICDJ*

Figure 4.8 (Continued)

Figure 4.8a shows the loading on the strip and the corresponding shear and moment diagrams. Note that the column loads shown in this figure have been multiplied by $F = 0.9322$. Also, the load per unit length of the beam is equal to $B_1 q_{av(modified)} = (4.25)(38.768) = 164.76$ kN/m.

Strip GIJH: In a similar manner

$$q_{av} = \frac{q_{(at\ B)} + q_{(at\ E)}}{2} = \frac{31.86 + 30.14}{2} = 31.0 \text{ kN/m}^2$$

Total soil reaction = $(31)(8)(21.5) = 5332$ kN
Total column load = 4000 kN
Average load = $(5332 + 4000)/2 = 4666$ kN
$q_{av(modified)} = 31.0\ (4666/5332) = 27.12$ kN/m^2
$F = 4666/4000 = 1.1665$
The load shear and moment diagrams are shown in Figure 4.8b.

Figure 4.8 (Continued)

Strip ICDJ: Figure 4.8c shows the load shear and moment diagrams for this strip.

Determination of the Thickness of the Mat

For this problem, the critical section for diagonal tension shear will be at the column carrying 1500 kN of load at the edge of the mat (Figure 4.8d). So

$$b_o = \left(0.5 + \frac{d}{2}\right) + \left(0.5 + \frac{d}{2}\right) + (0.5 + d) = 1.5 + 2d$$

$$U = (b_o d)[(\phi)(0.34)\sqrt{f_c'}]$$

$$U = (1.7)(1500) = 2550 \text{ kN} = 2.55 \text{ MN}$$

$$2.55 = (1.5 + 2d)(d)[(0.85)(0.34)\sqrt{20.7}]$$

or

$$(1.5 + 2d)d = 1.94$$

$$d \approx 0.68 \text{ m}$$

Assuming a minimum cover of 76 mm over the steel reinforcement and also assuming that the steel bars to be used are 25 mm in diameter, the total thickness of the slab is

$$h = 0.68 + 0.076 + 0.025 = 0.781 \text{ m} \approx \textbf{0.8 m}$$

The thickness of this mat will satisfy the wide beam shear condition across the three strips under consideration.

Determination of Reinforcement

From the moment diagrams shown in Figure 4.8a, b, and c, it can be seen that the maximum positive moment is located in strip $AGHF$, and its magnitude is

$$M' = \frac{1727.57}{B_1} = \frac{1727.57}{4.25} = 406.5 \text{ kN-m/m}$$

Similarly, the maximum negative moment is located in strip $ICDJ$, and its magnitude is $1196.19/B_1 = 1196.19/4.25 = 281.5$ kN-m/m.

Now, referring to Eq. (A.6) in Appendix A

$$M_u = (M')(\text{load factor}) = \phi A_s f_y \left(d - \frac{a}{2} \right)$$

For the positive moment

$$M_u = (406.5)(1.7) = (\phi)(A_s)(413.7 \times 1000)\left(0.68 - \frac{a}{2} \right)$$

$$\phi = 0.9$$

Also, from Eq. (A.4) in Appendix A

$$a = \frac{A_s f_y}{0.85 f'_c b} = \frac{(A_s)(413.7)}{(0.85)(20.7)(1)} = 23.51 A_s$$

or

$$A_s = 0.0425a$$

So

$$M_u = 691.22 = (0.9)(0.0425a)(413700)\left(0.68 - \frac{0.0425a}{2} \right)$$

or

$$a \approx 0.0645$$

So

$$A_s = (0.0425)(0.0645) = 0.00274 \text{ m}^2/\text{m} = 2740 \text{ mm}^2/\text{m}$$

Hence, use **25 mm-diameter bars at 175 mm center-to-center** [A_s provided $=$ $(491)(1000/175) = 2805.7$ mm^2/m].

Similarly, for negative reinforcement

$$M_u = (281.5)(1.7) = \phi A_s (413.7 \times 1000)\left(0.68 - \frac{a}{2} \right)$$

$\phi = 0.9$; $A_s = 0.0425a$. So

$$478.38 = (0.9)(0.0425a)(413.7 \times 1000)\left(0.68 - \frac{0.0425a}{2}\right)$$

Solution of the preceding equation gives $a \approx 0.045$. So

$$A_s = (0.045)(0.0425) = 0.001913 \text{ m}^2/\text{m} = 1913 \text{ mm}^2/\text{m}$$

Hence, use **25-mm diameter bars at 250 mm center-to-center** (A_s provided = 1924 mm^2).

Because negative moment occurs at midbay of strip $ICDJ$, reinforcement should be provided. This moment is $289.95/4.25 = 68.22$ kN-m/m. Hence

$$M_u = (68.22)(1.7) = (0.9)(0.0425a)(413.7 \times 1000)\left(0.68 - \frac{0.0425a}{2}\right)$$

Solution of the preceding equation gives $a \approx 0.0108$. So

$$A_s = (0.0108)(0.0425) = 0.000459 \text{ m}^2/\text{m} = 459 \text{ mm}^2/\text{m}$$

Provide **16-mm diameter bars at 400 mm center-to-center** (A_s provided = 502 mm^2).

For general arrangement of the reinforcement, see Figure 4.8e. Note that a similar type of analysis needs to be done to determine the reinforcement in the x direction of the mat.

Approximate Flexible Method

In the conventional rigid method of design, the mat is assumed to be infinitely rigid. Also, the soil pressure is distributed in a straight line, and the centroid of the soil pressure is coincidental with the line of action of the resultant column loads (see Figure 4.9a). In the approximate flexible method of design, the soil is assumed to be equivalent to an infinite number of elastic springs, as shown in Figure 4.9b. This is sometimes referred to as the *Winkler foundation*. The elastic constant of these assumed springs is referred to as the *coefficient of subgrade reaction, k*.

In order to understand the fundamental concepts behind flexible foundation design, consider a beam of width B_1 having an infinite length, as shown in Figure 4.9c. The beam is subjected to a single concentrated load Q. From the fundamentals of mechanics of materials

$$M = E_F I_F \frac{d^2 z}{dx^2} \tag{4.34}$$

where M = moment at any section
E_F = Young's modulus of the material of the beam
I_F = moment of inertia of the cross section of the beam = $(1/12)B_1 h^3$
(see Figure 4.9c)

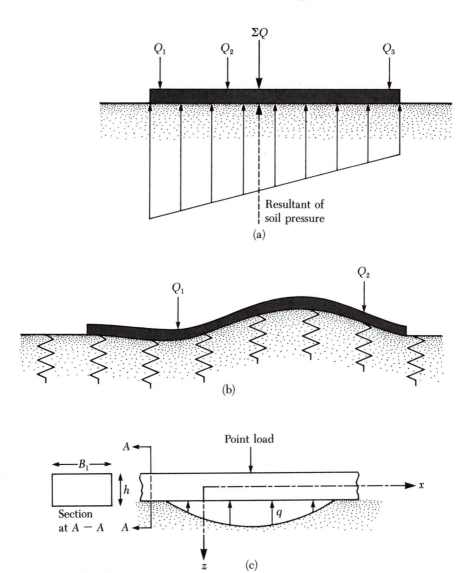

Figure 4.9 (a) Principles of design by conventional rigid method; (b) principles of approximate flexible method; (c) derivation of Eq. (4.38) for beams on elastic foundation

However

$$\frac{dM}{dx} = \text{shear force} = V$$

and

$$\frac{dV}{dx} = q = \text{soil reaction}$$

Hence

$$\frac{d^2M}{dx^2} = q \tag{4.35}$$

Now, combining Eqs. (4.34) and (4.35)

$$E_F I_F \frac{d^4z}{dx^4} = q \tag{4.36}$$

However, the soil reaction

$$q = -zk'$$

where z = deflection
$\quad\quad k' = kB_1$
$\quad\quad k$ = coefficient of subgrade reaction (unit—kN/m^3 or lb/in.^{3})

So

$$E_F I_F \frac{d^4z}{dx^2} = -zkB_1 \tag{4.37}$$

Solution of the preceding equation yields

$$z = e^{-\alpha x}(A' \cos \beta x + A'' \sin \beta x) \tag{4.38}$$

where A' and A'' are constants and

$$\beta = \sqrt[4]{\frac{B_1 k}{4E_F I_F}} \tag{4.39}$$

The unit of the term β as defined by the preceding equation is (length)$^{-1}$. This is a very important parameter in the determination of whether a mat foundation should be designed by conventional rigid method or approximate flexible method. According to the American Concrete Institute Committee 436 (1966), the design of mats should be done by the conventional rigid method if the spacings of columns in a strip are less than $1.75/\beta$. If the spacings of columns are larger than $1.75/\beta$, the approximate flexible method may be adopted. For example, determine whether or not the mat foundation in Examples 4.6 and 4.7 is rigid. Let the value of k for that foundation be 8000 kN/m^3. The strips $AFHF$ and $ICDJ$ have the lower value of B_1—that is, 4.25 m. So, according to Eq. (4.39)

$$\beta = \sqrt[4]{\frac{B_1 k}{4E_F I_F}}$$

For this section, $I_F = (1/12)B_1 h^3 = (1/12)(4.25)(0.8)^3 = 0.1813$ m^4. Also, let E_F be equal to 21×10^6 kN/m^2. So

$$\beta = \sqrt[4]{\frac{(4.25)(8000)}{(4)(21 \times 10^6)(0.1813)}} = 0.217 \text{ m}^{-1}$$

So

$$\frac{1.75}{\beta} = \frac{1.75}{0.217} = 8.07 \text{ m}$$

Because the actual spacing of columns is 7 m, which is less than $1.75/\beta$, this mat should be designed by the conventional rigid method.

To perform the analysis for the structural design of a flexible mat, one must know the principles of evaluating the *coefficient of subgrade reaction, k*. Before proceeding with the discussion of the approximate flexible design method, let us discuss this coefficient in more detail.

If a foundation of width B (Figure 4.10) is subjected to a load per unit area of q, it will undergo a settlement Δ. The coefficient of subgrade modulus k can be defined as

$$k = \frac{q}{\Delta} \tag{4.40}$$

The unit of k is in kN/m^3 (or lb/in.^{3}). The value of the coefficient of subgrade reaction is not a constant for a given soil. It depends on several factors, such as the length (L) and width (B) of the foundation and also the depth of embedment of the foundation. A comprehensive study of the parameters affecting the coefficient of subgrade reaction has been given by Terzaghi (1955). According to this study, the value of the coefficient of subgrade reaction decreases with the width of the foundation. In the field, load tests can be carried out by means of square plates measuring 0.3 m × 0.3 m (1 ft × 1 ft), and values of k can be calculated. The value of k can be related to large foundations measuring $B \times B$ as follows:

Foundations on Sandy Soils:

$$k = k_{0.3}\left(\frac{B + 0.3}{2B}\right)^2 \tag{4.41}$$

where $k_{0.3}$ and k = coefficients of subgrade reaction of footings measuring 0.3 (m) × 0.3 (m) and B (m) × B (m), respectively (unit kN/m^3).

Figure 4.10 Definition of coefficient of subgrade reaction, *k*

In English units, Eq. (4.41) can be expressed as

$$k = k_1 \left(\frac{B+1}{2B} \right)^2 \tag{4.42}$$

where k and k_1 = coefficients of subgrade reaction of footings measuring 1 ft × 1 ft and B (ft) × B (ft), respectively (unit is lb/in.^{2}).

Foundations on Clays :

$$k \ (\text{kN/m}^3) = k_{0.3} \ (\text{kN/m}^3) \left[\frac{0.3 \ (\text{m})}{B \ (\text{m})} \right] \tag{4.43}$$

The definition of k in Eq. (4.42) is the same as that given in Eq. (4.41).
In English units,

$$k \ (\text{lb/in.}^2) = k_1 \ (\text{lb/in.}^2) \left[\frac{1 \ (\text{ft})}{B \ (\text{ft})} \right] \tag{4.44}$$

The definitions of k and k_1 are the same as in Eq. (4.42).
    For rectangular foundations having dimensions of $B \times L$ (for similar soil and q)

$$k = \frac{k_{(B \times B)} \left(1 + \dfrac{B}{L} \right)}{1.5} \tag{4.45}$$

where k = coefficient of subgrade modulus of the rectangular foundation ($L \times B$)
    $k_{(B \times B)}$ = coefficient of subgrade modulus of a square foundation having dimension of $B \times B$

The preceding equation indicates that the value of k of a very long foundation with a width B is approximately equal to $0.67k_{(B \times B)}$.
    The Young's modulus of granular soils increases with depth. Because of the fact that the settlement of a foundation is dependent on the Young's modulus, the value of k increases as the depth of the foundation increases.
    Following are some typical ranges of value for the coefficient of subgrade reaction $k_{0.3}$ for sandy and clayey soils.

Sand (dry or moist)

Loose: 8–25 MN/m³ (29–92 lb/in.³)
Medium: 25–125 MN/m³ (91–460 lb/in.³)
Dense: 125–375 MN/m³ (460–1380 lb/in.³)

Sand (saturated)

Loose: 10–15 MN/m³ (38–55 lb/in.³)
Medium: 35–40 MN/m³ (128–147 lb/in.³)
Dense: 130–150 MN/m³ (478–552 lb/in.³)

Clay

Stiff $(q_u = 100–200 \text{ kN/m}^2)$: 12–25 MN/m³ (44–92 lb/in.³)
Very stiff $(q_u = 200–400 \text{ kN/m}^2)$: 25–50 MN/m³ (92–184 lb/in.³)
Hard $(q_u > 400 \text{ kN/m}^2)$:   >50 MN/m³ (>184 lb/in.³)

(*Note:* q_u = unconfined compression strength)

Scott (1981) has proposed that for sandy soils, the value of $k_{0.3}$ can be obtained from standard penetration resistance at any given depth as

$$k_{0.3} \ (\text{MN/m}^3) = 1.8N \tag{4.46}$$

where N = *corrected* standard penetration resistance

In English units,

$$k_1 \ (\text{U.S. ton/ft}^3) = 6N \tag{4.47}$$

For long beams, Vesic (1961) proposed an equation for estimation of subgrade reaction that can be expressed as

$$k' = Bk = 0.65 \sqrt[12]{\frac{E_s B^4}{E_F I_F}} \frac{E_s}{1 - \mu^2}$$

or

$$k = 0.65 \sqrt[12]{\frac{E_s B^4}{E_F I_F}} \frac{E_s}{B(1 - \mu^2)} \tag{4.48}$$

where E_s = Young's modulus of soil
   B = foundation width
   E_F = Young's modulus of foundation material
   I_F = moment of inertia of the cross section of the foundation
   μ = Poisson's ratio of soil

For most practical purposes, Eq. (4.48) can be approximated as

$$k = \frac{E_s}{B(1 - \mu^2)} \tag{4.49}$$

The coefficient of subgrade reaction is also a very useful parameter in the design of rigid highway and airfield pavements. The pavement with a concrete wearing surface is generally referred to as a *rigid pavement*, and the pavement with an asphaltic wearing surface is called a *flexible pavement*. For a surface load acting on a rigid pavement, the maximum tensile stress occurs at the base of the slab. To estimate the magnitude of the maximum horizontal tensile stress developed at the base of the rigid pavement, elastic solutions involving slabs on Winkler foundations are extremely useful. Some of the early work in this area was done by Westergaard (1926, 1939, 1947).

Now that we have discussed the coefficient of subgrade reaction, we will proceed with the discussion of the approximate flexible method of designing mat foundations. This method, as proposed by the American Concrete Institute Committee 436 (1966) is described below in a step-by-step manner. The design procedure is primarily based on the theory of plates. Its use allows the effects (that is, moment, shear, and deflection) of a concentrated column load in the area surrounding it to be evaluated. If the zones of influence of two or more columns overlap, the method of superposition can be used to obtain the net moment, shear, and deflection at any point.

1. Assume a thickness (h) for the mat. This can be done according to Step 6 as outlined in the preceding section on the conventional rigid method. (*Note:* h is the *total* thickness of the mat.)

2. Determine the flexural rigidity R of the mat as

$$R = \frac{E_F h^3}{12(1 - \mu_F^2)} \tag{4.50}$$

where E_F = Young's modulus of foundation material
μ_F = Poisson's ratio of foundation material

3. Determine the radius of effective stiffness as

$$L' = \sqrt[4]{\frac{R}{k}} \tag{4.51}$$

where k = coefficient of subgrade reaction

The zone of influence of any column load will be in the order of 3 to 4 L'.

4. Determine the moment in polar coordinate system at a point caused by a column load from the following equations (Figure 4.11a)

$$M_t = \text{tangential moment} = -\frac{Q}{4}\left[A_1 - \frac{(1 - \mu_F)A_2}{\dfrac{r}{L'}} \right] \tag{4.52}$$

$$M_r = \text{radial moment} = -\frac{Q}{4}\left[\mu_F A_1 + \frac{(1 - \mu_F)A_2}{\dfrac{r}{L'}} \right] \tag{4.53}$$

where r = radial distance from the column load
Q = column load
A_1, A_2 = functions of r/L'

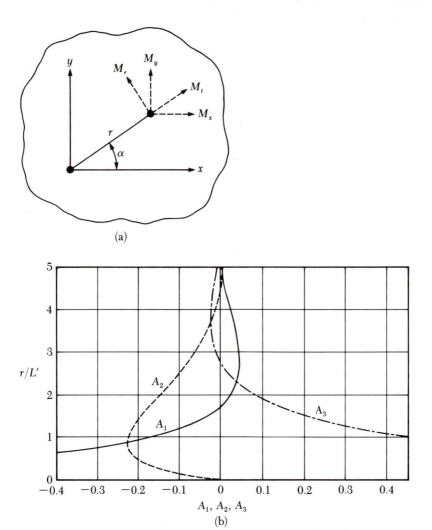

Figure 4.11 Approximate flexible method of mat design

The variations of A_1 and A_2 with r/L' are shown in Figure 4.11b (for details, see Hetenyi, 1946).

In the cartesian coordinate system (Figure 4.11a)

$$M_x = M_t \sin^2 \alpha + M_r \cos^2 \alpha \qquad (4.54)$$
$$M_y = M_t \cos^2 \alpha + M_r \sin^2 \alpha \qquad (4.55)$$

5. Determine the shear force (V) for the unit width of the mat caused by a column load as

$$V = -\frac{Q}{4L'} A_3 \qquad (4.56)$$

The variation of A_3 with r/L' is shown in Figure 4.11b.

6. If the edge of the mat is located in the zone of influence of a column, determine the moment and shear along the wedge, assuming that the mat is continuous. Moment and shear opposite in sign to those determined are applied at the edges to satisfy the known conditions.

Example 4.8

From the plate load test (plate dimension 1 ft × 1 ft) in the field, the coefficient of subgrade reaction of a sandy soil was determined to be 80 lb/in.^{3}. What will be the value of the coefficient of subgrade reaction on the same soil for a foundation with dimensions of 30 ft × 30 ft?

Solution

From Eq. (4.42)

$$k = k_1 \left(\frac{B+1}{2B} \right)^2$$

where $k_1 = 80$ lb/in.^{2}; $B = 30$ ft

So

$$k = 80 \left[\frac{30+1}{(2)(30)} \right] = \textbf{21.36 in.}^3$$

Example 4.9

Refer to Example 4.8. If the full-sized foundation has dimensions of 45 ft × 30 ft, what will be the value of the coefficient of subgrade reaction?

Solution

From Eq. (4.45)

$$k = \frac{k_{(B \times B)} \left(1 + \dfrac{B}{L} \right)}{1.5}$$

From Example 4.8

$$k_{(30 \text{ ft} \times 30 \text{ ft})} = 21.36 \text{ lb/in.}^3$$

So

$$k = \frac{(21.36) \left(1 + \dfrac{30}{45} \right)}{1.5} = \textbf{23.73 lb/in.}^3$$

PROBLEMS

4.1 Determine the net ultimate bearing capacity of a mat foundation measuring 20 m × 10 m on a saturated clay with $c_u = 100$ kN/m^2, $\phi = 0$, and $D_f = 3$ m. [Use Eq. (4.10).]

4.2 Repeat Problem 4.1 with the following: size of mat foundation = 28 ft × 20 ft, $c_u = 2300$ lb/ft^2, $\phi = 0$, and $D_f = 4.5$ ft.

4.3 Following are the results of a standard penetration test in the field (sandy soil):

Depth (m)	Field value of N
2	8
4	10
6	12
8	9
10	14

Estimate the net allowable bearing capacity of a mat foundation 6 m × 5 m in plan. Given: $D_f = 1.5$ m, allowable settlement = 50.8 mm. Assume that the unit weight of soil, $\gamma = 17.5$ kN/m^3.

4.4 Repeat Problem 4.3 for an allowable settlement of 30 mm.

4.5 Consider a mat foundation with dimensions of 50 ft × 30 ft. The dead and live load on the mat is 10^7 lb. The mat is to be placed on a clay with $c_u = 700$ lb/ft^2. The unit weight of the clay is 120 lb/ft^3. Find the depth, D_f, of the mat for a fully compensated foundation.

4.6 Repeat Problem 4.5 with the following: dimension of mat = 20 m × 20 m, dead and live load = 48 MN, $c_u = 30$ kN/m^2, $\gamma_{clay} = 18.8$ kN/m^3.

4.7 For the mat considered in Problem 4.6, what will be the depth of the mat (D_f) for a factor of safety of 2 against bearing capacity failure?

4.8 Repeat Problem 4.6 assuming that the undrained cohesion of the clay is 48.5 kN/m^2.

4.9 Consider the mat foundation shown in Figure P4.9. Given: $L = 12$ m, $B = 10$ m, $Q = 30$ MN, $D_f = 2.2$ m, $x_1 = 2$ m, $x_2 = 2$ m, $x_3 = 5.2$ m. The clay is normally consolidated. Estimate the consolidation settlement under the center of the mat.

4.10 For the mat foundation described in Problem 4.9, estimate the consolidation settlement under the corner of the mat.

4.11 Redo Problem 4.9 assuming that the preconsolidation pressure of the clay is 120 kN/m^2 and the swelling index is about $\frac{1}{5}C_c$.

4.12 Refer to Figure P4.12. For the mat, given: $Q_1, Q_3 = 40$ tons, $Q_4, Q_5, Q_6 = 60$ tons, $Q_2, Q_9 = 45$ tons, $Q_7, Q_8 = 50$ tons. All columns are 20 in. × 20 in. in cross section. Using the procedure outlined in Section 4.6, determine the pressure on the soil at A, B, C, D, F, G, and H.

4.13 Refer to the mat foundation described in Problem 4.12. Divide the mat into three strips in the y direction $(B_1 = 8$ ft, 16 ft, and 8 ft). Determine the reinforcement requirement according to the conventional rigid method. Given: $f'_c = 3000$ lb/in.^{2} and $f_y = 60,000$ lb/in.^{2}. Use a load factor of 1.6.

4.14 Refer to Example 4.6. Divide the mat shown in Figure 4.7 into four strips and determine the reinforcement requirement according to the conventional rigid method.

Figure P4.9

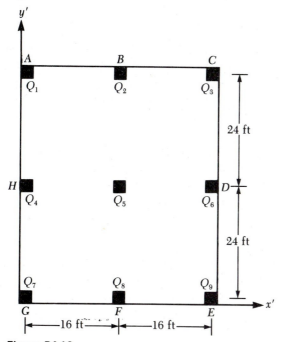

Figure P4.12

4.15 From the plate load test (plate dimension 1 ft × 1 ft) in the field, the coefficient of subgrade reaction of a sandy soil was determined to be 65 lb/in.^{3}. What will be the value of the coefficient of subgrade reaction on the same soil for a foundation with dimensions of 25 ft × 25 ft?

4.16 Refer to Problem 4.15. If the full-sized foundation has dimensions of 60 ft × 25 ft, what will be the value of the coefficient of the subgrade reaction?

4.17 The subgrade reaction of a sandy soil obtained from a plate load test (plate dimension 1 m × 0.7 m) is 22 kN/m^2. What will be the value of k on the same soil for a foundation measuring 4 m × 3 m?

REFERENCES

American Concrete Institute (1987). *ACI Standard—Building Code Requirements for Reinforced Concrete*, ACI 318-83—revised 1986, Detroit.

American Concrete Institute Committee 436 (1966). "Suggested Design Procedures for Combined Footings and Mats," *Journal of the American Concrete Institute*, Vol. 63, No. 10, pp. 1041–1057.

Hetenyi, M. (1946). *Beams on Elastic Foundations*, University of Michigan Press, Ann Arbor.

Scott, R. F. (1981). *Foundations Analysis*, Prentice-Hall, Englewood Cliffs, N.J.

Terzaghi, K. (1955). "Evaluation of the Coefficient of Subgrade Reactions," *Geotechnique*, Institute of Engineers, London, Vol. 5, No. 4, pp. 197–326.

Vesic, A. S. (1961). "Bending of Beams Resting on Isotropic Solid," *Journal of the Engineering Mechanics Division*, American Society of Civil Engineers, Vol. 87, No. EM2, pp. 35–53.

Westergaard, H. M. (1926). "Stresses in Concrete Pavements Computed by Theoretical Analysis," *Public Roads*, Vol. 7, No. 12, pp. 23–35.

Westergaard, H. M. (1939). "Stresses in Concrete Runways of Airports," *Proceedings*, Highway Research Board, Vol. 19, pp. 197–205.

Westergaard, H. M. (1947). "New Formulas for Stresses in Concrete Pavements of Airfields," *Proceedings*, American Society of Civil Engineers, Vol. 73, pp. 687–701.

LATERAL EARTH PRESSURE AND RETAINING WALLS

5.1   Introduction

A *retaining wall* is a wall that provides lateral support for a *vertical* or *near-vertical* slope of soil. It is a common structure used in many construction projects. The most common types of retaining wall may be classified as follows:

1. Gravity retaining walls
2. Semigravity retaining walls
3. Cantilever retaining walls
4. Counterfort retaining walls

Gravity retaining walls (Figure 5.1a) are constructed with plain concrete or stone masonry. These walls depend on their own weight and any soil resting on the masonry for their stability. This type of construction is not very economical for high walls.

In many cases, a small amount of steel may be used for the construction of gravity walls, thereby minimizing the size of the wall sections. Such walls are generally referred to as *semigravity walls* (Figure 5.1b).

Cantilever retaining walls (Figure 5.1c) are made of reinforced concrete that consists of a thin stem and a base slab. This type of wall is economical up to a height of about 8 m.

Counterfort retaining walls (Figure 5.1d) are similar to cantilever walls except for the fact that, at regular intervals, they have thin vertical concrete slabs known as *counterforts* that tie the wall and the base slab together. The purpose of the counterforts is to reduce the shear and the bending moments.

To design retaining walls properly, one must know the basic soil parameters—that is, the *unit weight, angle of friction,* and *cohesion*—for the

Plain
concrete
or
stone
masonry

(a) Gravity wall

←Reinforcement

(b) Semigravity wall

←Reinforcement

(c) Cantilever wall

←Counterfort

(d) Counterfort wall

Figure 5.1   Types of retaining wall

soil retained behind the wall and the soil below the base slab. Knowing the properties of the soil behind the wall enables an engineer to determine the lateral pressure distribution for which the design has to be made.

There are two phases in the design of a retaining wall. First, with the lateral earth pressure known, the structure as a whole is checked for *stability*. This includes checking for possible *overturning*, *sliding*, and *bearing capacity* failures. Second, each component of the structure is checked for *adequate strength*, and the *steel reinforcement* of each component is determined.

This chapter presents the procedures for determination of lateral earth pressure and retaining-wall stability. Checks for adequate strength of each component of the structures are covered in Appendix A.

LATERAL EARTH PRESSURE

5.2   Lateral Earth Pressure At Rest

Consider a vertical wall of height H, as shown in Figure 5.2, retaining a soil having a unit weight of γ. A uniformly distributed load, q/unit area, is also applied at the ground surface. The shear strength, s, of the soil may be given by the equation

$$s = c + \sigma' \tan \phi$$

where c = cohesion
       ϕ = angle of friction
       σ' = effective normal stress

At any depth z below the ground surface, the vertical stress below the ground surface can be given as

$$\sigma_v = q + \gamma z \tag{5.1}$$

If the *wall is not allowed to move at all* either way from the soil mass or into the soil mass (for example, zero horizontal strain), the lateral pressure at a depth z can be given as

$$\sigma_h = K_o \sigma_v' + u \tag{5.2}$$

where u = pore water pressure
       K_o = coefficient of at-rest earth pressure

Figure 5.2 At-rest earth pressure

For normally consolidated granular soil, the relation for K_o is

$$\boxed{K_o \approx 1 - \sin \phi} \tag{5.3}$$

Note that the preceding equation is an empirical approximation.

For normally consolidated clays, the coefficient of earth pressure at rest can be approximated as follows (Brooker and Ireland, 1965):

$$\boxed{K_o \approx 0.95 - \sin \phi} \tag{5.4}$$

where ϕ = drained friction angle.

Based on the experimental results of Brooker and Ireland (1965), the value of K_o for normally consolidated clays can be approximately correlated with plasticity index (PI) as

$$K_o = 0.4 + 0.007(PI) \qquad \text{(for } PI \text{ between 0 and 40)} \tag{5.5}$$

and

$$K_o = 0.64 + 0.001(PI) \qquad \text{(for } PI \text{ between 40 and 80)} \tag{5.6}$$

For overconsolidated clays

$$\boxed{K_{o(\text{overconsolidated})} \approx K_{o(\text{normally consolidated})}\sqrt{OCR}} \tag{5.7}$$

where OCR = overconsolidation ratio

With a properly selected value of the at-rest earth pressure coefficient, Eq. (5.2) can be used to determine the variation of lateral earth pressure with depth, z. Figure 5.2b shows the variation of σ_h with depth for the wall shown in Figure 5.2a. Note that if the surcharge $q = 0$ and the pore water pressure $u = 0$, the pressure diagram will be a triangle. The total force, P_o, *per unit length* of the wall given in Figure 5.2a can now be obtained from the area of the pressure diagram given in Figure 5.2b as

$$P_o = P_1 + P_2 = qK_o H + \tfrac{1}{2}\gamma H^2 K_o \tag{5.8}$$

where P_1 = area of rectangle 1
P_2 = area of triangle 2

The location of the line of action of the resultant force, P_o, can be obtained by taking the moment about the bottom of the wall. Thus

$$\bar{z} = \frac{P_1\left(\dfrac{H}{2}\right) + P_2\left(\dfrac{H}{3}\right)}{P_o} \tag{5.9}$$

If the water table is located at a certain depth, $z < H$, the at-rest pressure diagram shown in Figure 5.2b will have to be somewhat modified. This is shown in Figure 5.3. Let the effective unit weight of soil below the water be equal to γ' (that is, $\gamma_{sat} - \gamma_w$). For this condition

At $z = 0$, $\sigma'_h = K_o \sigma'_v = K_o q$

At $z = H_1$, $\sigma'_h = K_o \sigma'_v = K_o(q + \gamma H_1)$

At $z = H_2$, $\sigma'_h = K_o \sigma'_v = K_o(q + \gamma H_1 + \gamma' H_2)$

Note that in the preceding equations, σ'_v and σ'_h are effective vertical and horizontal pressures. In order to determine the total pressure distribution on the wall, the hydrostatic pressure has to be added. The hydrostatic pressure, u, is equal to zero from $z = 0$ to $z = H_1$; at $z = H_2$, $u = H_2 \gamma_w$. The variation of σ'_h and u with depth is shown in Figure 5.3b. Hence the total force per unit length of the wall can be determined from the area of the pressure diagram. Thus

$$P_o = A_1 + A_2 + A_3 + A_4 + A_5$$

where A = area of the pressure diagram

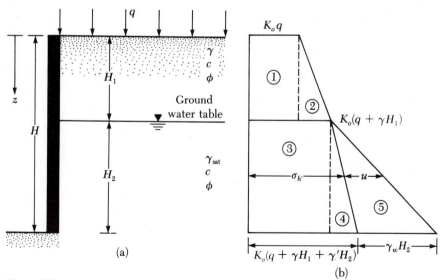

(a)   (b)

Figure 5.3

So

$$P_o = K_o qH_1 + \tfrac{1}{2}K_o \gamma H_1^2 + K_o(q + \gamma H_1)H_2 + \tfrac{1}{2}K_o \gamma' H_2^2 + \tfrac{1}{2}\gamma_w H_2^2$$

(5.10)

Sherif, Fang, and Sherif (1984) have shown by several laboratory model tests that Eq. (5.3) gives good results for estimating the lateral earth pressure at rest for loose sands. However, for compacted dense sand, it grossly underestimates the value of K_o. For that reason, they proposed the following modified relationship for K_o:

$$K_o = (1 - \sin \phi) + \left(\frac{\gamma_d}{\gamma_{d(min)}} - 1\right)5.5$$

(5.11)

where γ_d = *in situ* unit weight of sand
   $\gamma_{d(min)}$ = minimum possible dry unit weight of sand (see Chapter 1)

Example 5.1

A retaining wall is 10 ft high. The backfill is dense sand. Given: $\phi = 40°$, $\gamma_d = 108$ lb/ft^3, and $\gamma_{d(min)} = 97$ lb/ft^3. Determine the lateral earth force at rest per unit length of the wall. [Use Eq. (5.11).]

Solution

Refer to Figure 5.4. From Eq. (5.11)

$$K_o = (1 - \sin \phi) + \left(\frac{\gamma_d}{\gamma_{d(min)}} - 1\right)5.5$$

$$= (1 - \sin 40) + \left(\frac{108}{97} - 1\right)5.5$$

$$= 0.357 + 0.634 = 0.991$$

10 ft

Sand
$\phi = 40°$
$\gamma_d = 108$ lb/ft^3
$\gamma_{d(min)} = 97$ lb/ft^3

5351.5 lb/ft

1070.3 lb/ft^2

(a)   (b)

Figure 5.4

At $z = 0$, $\sigma'_v = 0$, $\sigma'_h = 0$

At $z = 10$ ft, $\sigma'_v = (10)(108) = 1080$ lb/ft^2

$$\sigma'_h = K_o \sigma'_v = (0.991)(1080) = 1070.3 \text{ lb/ft}^2$$

The at-rest force per unit length of the wall is the area of the triangle shown in Figure 5.4b. So

$$P_o = \left(\frac{1}{2}\right)(1070.3)(10) = 5351.5 \text{ lb/ft}$$

Example 5.2

For the retaining wall shown in Figure 5.5a, determine the lateral earth force at rest per unit length of the wall. Also determine the location of the resultant earth pressure.

(a)

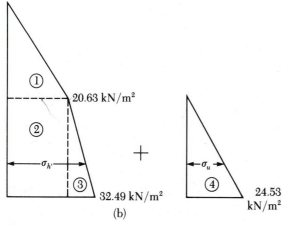

(b)

Figure 5.5

Solution

$$K_o = 1 - \sin \phi = 1 - \sin 30° = 0.5$$

At $z = 0$, $\sigma'_v = 0$; $\sigma'_h = 0$

At $z = 2.5$ m, $\sigma'_v = (16.5)(2.5) = 41.25$ kN/m^2;

$$\sigma'_h = K_o \sigma'_v = (0.5)(41.25) = 20.63 \text{ kN/m}^2$$

At $z = 5$ m, $\sigma'_v = (16.5)(2.5) + (19.3 - 9.81)2.5 = 64.98$ kN/m^2;

$$\sigma'_h = K_o \sigma'_v = (0.5)(64.98) = 32.49 \text{ kN/m}^2$$

The hydrostatic pressure distribution is as follows:

From $z = 0$ to $z = 2.5$ m, $u = 0$. At $z = 5$ m, $u = \gamma_w(2.5) = (9.81)(2.5) = 24.53$ kN/m^2. The pressure distribution for the wall is shown in Figure 5.5b.

The total force per unit length of the wall can be determined from the area of the pressure diagram, or

$$P_o = \text{Area 1} + \text{Area 2} + \text{Area 3} + \text{Area 4}$$
$$= \tfrac{1}{2}(2.5)(20.63) + (2.5)(20.63) + \tfrac{1}{2}(2.5)(32.49 - 20.63)$$
$$+ \tfrac{1}{2}(2.5)(24.53) = \mathbf{122.85 \text{ kN/m}}$$

The location of the center of pressure measured from the bottom of the wall (point O) =

$$\bar{z} = \frac{(\text{Area 1})\left(2.5 + \dfrac{2.5}{3}\right) + (\text{Area 2})\left(\dfrac{2.5}{2}\right) + (\text{Area 3} + \text{Area 4})\left(\dfrac{2.5}{3}\right)}{P_o}$$

$$= \frac{(25.788)(3.33) + (51.575)(1.25) + (14.825 + 30.663)(0.833)}{122.85}$$

$$= \frac{85.87 + 64.47 + 37.89}{122.85} = \mathbf{1.53 \text{ m}}$$

5.3   Rankine Active Earth Pressure

The lateral earth pressure condition described in Section 5.2 involves walls that do not yield at all. However, if a wall tends to move away from the soil a distance Δx, as shown in Figure 5.6a, the soil pressure on the wall at any given depth will decrease. For a wall that is *frictionless*, the horizontal stress, σ_h, at a depth z will be equal to $K_o \sigma_v$ ($= K_o \gamma z$) when Δx is equal to zero. However, with $\Delta x > 0$, σ_h will be less than $K_o \sigma_v$.

The Mohr's circles corresponding to wall displacements of $\Delta x = 0$ and $\Delta x > 0$ are shown as circles *a* and *b*, respectively, in Figure 5.6b. If the displacement of the wall, Δx, continues to increase, there will be a time when the corresponding Mohr's circle will just touch the Mohr-Coulomb failure envelope defined by the equation

$$s = c + \sigma \tan \phi$$

This circle is marked *c* in Figure 5.6b. It represents the failure condition in the soil mass; the horizontal stress at this time is equal to σ_a. This horizontal stress, σ_a, is referred to as the *Rankine active pressure*. The *slip lines*

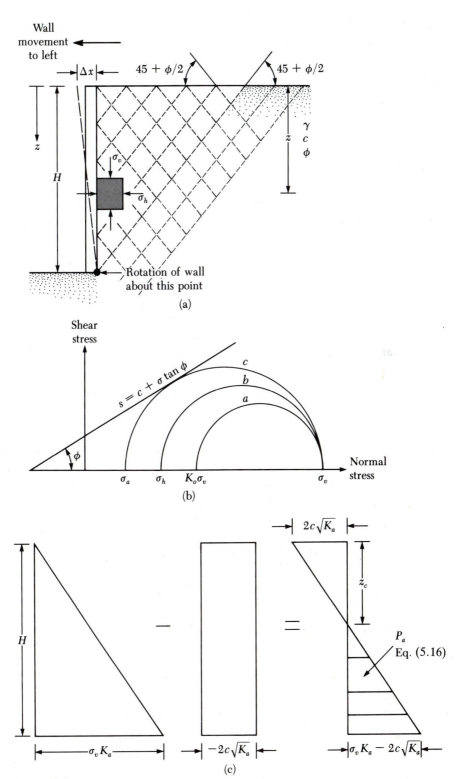

Figure 5.6 Rankine active pressure

(failure planes) in the soil mass at this time will make angles of $\pm(45 + \phi/2)$ with the horizontal. This is shown in Figure 5.6a.

Referring to Eq. (1.83), the equation relating the principal stresses for a Mohr's circle that touches the Mohr-Coulomb failure envelope can be given by

$$\sigma_1 = \sigma_3 \tan^2 \left(45 + \frac{\phi}{2} \right) + 2c \tan \left(45 + \frac{\phi}{2} \right)$$

For the Mohr's circle c in Figure 5.6b,

Major principal stress, $\sigma_1 = \sigma_v$

and

Minor principal stress, $\sigma_3 = \sigma_a$

Thus

$$\sigma_v = \sigma_a \tan^2 \left(45 + \frac{\phi}{2} \right) + 2c \tan \left(45 + \frac{\phi}{2} \right)$$

$$\sigma_a = \frac{\sigma_v}{\tan^2 \left(45 + \dfrac{\phi}{2} \right)} - \frac{2c}{\tan \left(45 + \dfrac{\phi}{2} \right)}$$

or

$$\boxed{\begin{aligned} \sigma_a &= \sigma_v \tan^2 \left(45 - \frac{\phi}{2} \right) - 2c \tan \left(45 - \frac{\phi}{2} \right) \\ &= \sigma_v K_a - 2c\sqrt{K_a} \end{aligned}}$$

(5.12)

where $K_a = \tan^2 (45 - \phi/2) =$ Rankine active pressure coefficient (Table 5.1).

The variation of the active pressure with depth for the wall shown in Figure 5.6a is given in Figure 5.6c. Note that σ_v is equal to zero at $z = 0$, and $\sigma_v = \gamma H$ at $z = H$. The pressure distribution shows that at $z = 0$, the active pressure is equal to $-2c\sqrt{K_a}$, indicating tensile stress. This tensile stress decreases with depth and becomes zero at a depth $z = z_c$, or

$$\gamma z_c K_a - 2c\sqrt{K_a} = 0$$

$$\boxed{z_c = \frac{2c}{\gamma\sqrt{K_a}}}$$

(5.13)

The depth z_c is usually referred to as the *depth of tensile crack*, because the tensile stress in the soil will eventually cause a crack along the soil-wall

Table 5.1   Variation of Rankine K_a

Soil friction angle, ϕ (deg)	$K_a = \tan^2(45 - \phi/2)$
20	0.490
21	0.472
22	0.455
23	0.438
24	0.422
25	0.406
26	0.395
27	0.376
28	0.361
29	0.347
30	0.333
31	0.320
32	0.307
33	0.295
34	0.283
35	0.271
36	0.260
37	0.249
38	0.238
39	0.228
40	0.217
41	0.208
42	0.198
43	0.189
44	0.180
45	0.172

interface. Thus the total Rankine active force per unit length of the wall before the tensile crack occurs is equal to

$$P_a = \int_0^H \sigma_a \, dz = \int_0^H \gamma z \cdot K_a \, dz - \int_0^H 2c\sqrt{K_a} \, dz$$
$$= \tfrac{1}{2}\gamma H^2 K_a - 2cH\sqrt{K_a} \tag{5.14}$$

After the occurrence of the tensile crack, the force on the wall will be caused only by the pressure distribution between depths $z = z_c$ to $z = H$, as shown by the hatched area in Figure 5.6c, and it can be expressed as

$$P_a = \frac{1}{2}(H - z_c)(\gamma H K_a - 2c\sqrt{K_a}) \tag{5.15}$$

or

$$P_a = \frac{1}{2}\left(H - \frac{2c}{\gamma\sqrt{K_a}}\right)(\gamma H K_a - 2c\sqrt{K_a}) \tag{5.16}$$

For calculation purposes in some retaining wall design problems, a cohesive soil backfill is replaced by an assumed granular soil with a triangu-

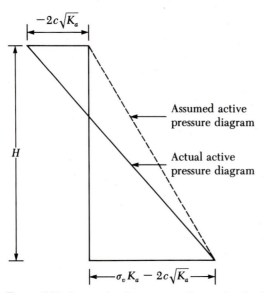

Figure 5.7 Assumed active pressure diagram for clay backfill behind a retaining wall

lar Rankine active pressure diagram with $\sigma_a = 0$ at $z = 0$, and $\sigma_a = \sigma_v K_a$ $- 2c\sqrt{K_a}$ at $z = H$ (see Figure 5.7). In such a case, the assumed active force per unit length of the wall is

$$P_a = \tfrac{1}{2}H(\gamma H K_a - 2c\sqrt{K_a}) = \tfrac{1}{2}\gamma H^2 K_a - cH\sqrt{K_a} \qquad (5.17)$$

However, the active earth pressure condition will be reached only if the wall is allowed to "yield" sufficiently. The amount of outward displacement of the wall necessary is about $0.001H$ to $0.004H$ for granular soil backfills and about $0.01H$ to $0.04H$ for cohesive soil backfills.

Example 5.3

A 6-m high retaining wall is to support a soil with unit weight $\gamma = 17.4$ kN/m³, soil friction angle $\phi = 26°$, and cohesion $c = 14.36$ kN/m². Determine the Rankine active force per unit length of the wall both before and after the tensile crack occurs, and determine the line of action of the resultant in both cases.

Solution

Given: $\phi = 26°$. So

$$K_a = \tan^2\left(45 - \frac{\phi}{2}\right) = \tan^2(45 - 13) = 0.39$$

$$\sqrt{K_a} = 0.625$$

$$\sigma_a = \gamma H K_a - 2c\sqrt{K_a}$$

Referring to Figure 5.6c,

At $z = 0$, $\sigma_a = -2c\sqrt{K_a} = -2(14.36)(0.625) = -17.95$ kN/m²
At $z = 6$ m, $\sigma_a = (17.4)(6)(0.39) - 2(14.36)(0.625)$
$\qquad = 40.72 - 17.95 = 22.77$ kN/m²

Active Force Before the Occurrence of Tensile Crack: Eq. (5.14)

$$P_a = \tfrac{1}{2}\gamma H^2 K_a - 2cH\sqrt{K_a}$$
$$= \tfrac{1}{2}(6)(40.72) - (6)(17.95) = 122.16 - 107.7 = 14.46 \text{ kN/m}$$

The line of action of the resultant can be determined by taking the moment of the area of the pressure diagrams about the bottom of the wall, or

$$P_a\bar{z} = (122.16)\left(\frac{6}{3}\right) - (107.7)\left(\frac{6}{2}\right)$$

or

$$\bar{z} = \frac{244.32 - 323.1}{14.46} = -5.448 \text{ m}$$

Active Force After the Occurrence of Tensile Crack: Eq. (5.13)

$$z_c = \frac{2c}{\gamma\sqrt{K_a}} = \frac{2(14.36)}{(17.4)(0.625)} = 2.64 \text{ m}$$

Using Eq. (5.15)

$$P_a = \tfrac{1}{2}(H - z_c)(\gamma H K_a - 2c\sqrt{K_a}) = \tfrac{1}{2}(6 - 2.64)(22.77) = 38.25 \text{ kN/m}$$

Figure 5.6c shows that the force $P_a = 38.25$ kN/m is the area of the hatched triangle. Hence, the line of action of the resultant will be located at a height of $\bar{z} = (H - z_c)/3$ above the bottom of the wall, or

$$\bar{z} = \frac{6 - 2.64}{3} = 1.12 \text{ m}$$

Note: For most retaining wall construction, a granular backfill is used and so $c = 0$. Thus Example 5.3 is an academic problem; however, it illustrates the basic principles of the Rankine active earth pressure equation.

Example 5.4

Refer to Figure 5.8a. Assume that the wall can yield sufficiently and determine the Rankine active force per unit length of the wall. Also determine the location of the resultant line of action.

Solution

If the cohesion, c, is equal to zero

$$\sigma_a = \sigma_v K_a$$

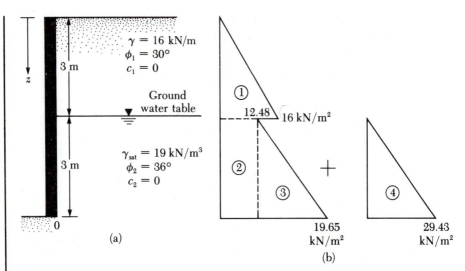

Figure 5.8

For the top soil layer, $\phi_1 = 30°$, so

$$K_{a(1)} = \tan^2 \left(45 - \frac{\phi_1}{2} \right) = \tan^2 (45 - 15) = \frac{1}{3}$$

Similarly, for the bottom soil layer, $\phi_2 = 36°$, and

$$K_{a(2)} = \tan^2 \left(45 - \frac{36}{2} \right) = 0.26$$

Because of the presence of the water table, the effective lateral pressure and the hydrostatic pressure have to be calculated separately.

At $z = 0$, $\sigma_v' = 0$, $\sigma_a' = 0$
At $z = 3$ m, $\sigma_v' = \gamma z = (16)(3) = 48$ kN/m²

At this depth, for the top soil layer

$$\sigma_a' = K_{a(1)}\sigma_v' = (\tfrac{1}{3})(48) = 16 \text{ kN/m}^2$$

Similarly, for the bottom soil layer

$$\sigma_a' = K_{a(2)}\sigma_v' = (0.26)(48) = 12.48 \text{ kN/m}^2$$

At $z = 6$ m, $\sigma_v' = (\gamma)(3) + (\gamma_{sat} - \gamma_w)(3) = (16)(3) + (19 - 9.81)(3)$
$$= 48 + 27.57 = 75.57 \text{ kN/m}^2$$
$$\sigma_a' = K_{a(2)}\sigma_v' = (0.26)(75.57) = 19.65 \text{ kN/m}^2$$

The hydrostatic pressure, u, is zero from $z = 0$ to $z = 3$ m. At $z = 6$ m, $u = 3\gamma_w = 3(9.81) = 29.43$ kN/m². The pressure distribution diagram is plotted in Figure 5.8b. The force per unit length

$$P_a = \text{Area 1} + \text{Area 2} + \text{Area 3} + \text{Area 4}$$
$$= \tfrac{1}{2}(3)(16) + (3)(12.48) + \tfrac{1}{2}(3)(19.65 - 12.48) + \tfrac{1}{2}(3)(29.43)$$
$$= 24 + 37.44 + 10.76 + 44.15 = 116.35 \text{ kN/m}$$

The distance of the line of action of the resultant from the bottom of the wall ($\bar{z}$) can be determined by taking the moments about the bottom of the wall (point O in Figure 5.8a), or

$$\bar{z} = \frac{(24)\left(3 + \dfrac{3}{3}\right) + (37.44)\left(\dfrac{3}{2}\right) + (10.76)\left(\dfrac{3}{3}\right) + (44.15)\left(\dfrac{3}{3}\right)}{116.35}$$

$$= \frac{96 + 56.16 + 10.76 + 44.15}{116.35} = 1.779 \text{ m}$$

Example 5.5

Refer to Example 5.4. Other quantities remaining the same, assume that, in the top layer, $c_1 = 24$ kN/m^2 (not zero as in Example 5.4). Determine P_a after the occurrence of the tensile crack.

Solution

From Eq. (5.13)

$$z_c = \frac{2c_1}{\gamma\sqrt{K_{a(1)}}} = \frac{(2)(24)}{(16)\sqrt{(^1\!/_3)}} = 5.2 \text{ m}$$

Since the depth of the top layer is only 3 m, the depth of the tensile crack will be only 3 m. So the pressure diagram up to $z = 3$ m will be zero. For $z > 3$ m, the pressure diagram will be the same as shown in Figure 5.8, or

$$P_a = \underbrace{\text{Area 2} + \text{Area 3} + \text{Area 4}}_{\text{Figure 5.8}}$$

$$= 37.44 + 10.76 + 44.15 = 92.35 \text{ kN/m}$$

5.4   Coulomb's Active Earth Pressure

The Rankine active earth pressure calculations discussed in the preceding section were based on the assumption that the wall is frictionless. In 1776, Coulomb proposed a theory to calculate the lateral earth pressure on a retaining wall with granular soil backfill. This theory takes wall friction into consideration.

To apply Coulomb's active earth pressure theory, let us consider a retaining wall with its back face inclined at an angle β with the horizontal as shown in Figure 5.9a. The backfill is a granular soil that slopes at an angle α with the horizontal. Also, let δ be the angle of friction between the soil and the wall (that is, angle of wall friction).

In an active pressure condition, the wall will move away from the soil mass (that is, to the left in Figure 5.9a). Coulomb assumed that, in such a case, the failure surface in the soil mass would be a plane (such as BC_1, BC_2, ...). So, to find the active force in our example, consider a possible

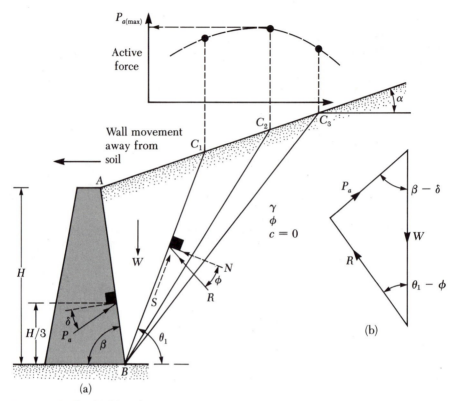

Figure 5.9 Coulomb's active pressure

soil failure wedge ABC_1. The forces acting on this wedge, ABC_1 (per unit length at right angles to the cross section shown), are as follows:

1. Weight of the wedge, W.

2. The resultant, R, of the normal and resisting shear forces along the surface, BC_1. The force R will be inclined at an angle ϕ to the normal drawn to the surface BC_1.

3. The active force per unit length of the wall, P_a. The force P_a will be inclined at an angle δ to the normal drawn to the back face of the wall.

For equilibrium purposes, a force triangle can be drawn. This is shown in Figure 5.9b. Note that θ_1 is the angle that BC_1 makes with the horizontal. Because the magnitude of W as well as the directions of all three forces are known, the value of P_a can now be determined. In a similar manner, one can determine the active forces of other trial wedges, such as ABC_2, ABC_3, ... The maximum value of P_a thus determined is Coulomb's active force (see top part of Figure 5.9a). This can be expressed as

$$P_a = \tfrac{1}{2}K_a\gamma H^2 \tag{5.18}$$

where

$$K_a = \text{Coulomb's active earth pressure coefficient}$$

$$= \frac{\sin^2 (\beta + \phi)}{\sin^2 \beta \cdot \sin (\beta - \delta)\left[1 + \sqrt{\dfrac{\sin (\phi + \delta) \cdot \sin (\phi - \alpha)}{\sin (\beta - \delta) \cdot \sin (\alpha + \beta)}}\right]^2}$$

$$H = \text{height of the wall}$$

(5.19)

The values of the active earth pressure coefficient, K_a, for a vertical retaining wall ($\beta = 90°$) with horizontal backfill ($\alpha = 0°$) are given in Table 5.2. Note that the line of action of the resultant (P_a) will act at a distance of $H/3$ above the base of the wall and will be inclined at an angle δ to the normal drawn to the back of the wall.

Table 5.2   Values of K_a [Eq. (5.19)] for $\beta = 90°$, $\alpha = 0°$

ϕ (deg)	δ (deg)					
	0	5	10	15	20	25
28	0.3610	0.3448	0.3330	0.3251	0.3203	0.3186
30	0.3333	0.3189	0.3085	0.3014	0.2973	0.2956
32	0.3073	0.2945	0.2853	0.2791	0.2755	0.2745
34	0.2827	0.2714	0.2633	0.2579	0.2549	0.2542
36	0.2596	0.2497	0.2426	0.2379	0.2354	0.2350
38	0.2379	0.2292	0.2230	0.2190	0.2169	0.2167
40	0.2174	0.2098	0.2045	0.2011	0.1994	0.1995
42	0.1982	0.1916	0.1870	0.1841	0.1828	0.1831

In the actual design of retaining walls, the value of the wall friction angle, δ, is assumed to be between $\phi/2$ and $\frac{2}{3}\phi$. The active earth pressure coefficients for various values of ϕ, α, and β with $\delta = \frac{2}{3}\phi$ are given in Table 5.3. This is a very useful table for design considerations.

If a uniform surcharge of intensity q is located above the backfill as shown in Figure 5.10, the active force, P_a, can be calculated as

$$P_a = \frac{1}{2} K_a \gamma_{eq} H^2$$
$$\uparrow$$
$$\text{Eq. (5.19)}$$

(5.20)

where

$$\gamma_{eq} = \gamma + \left[\frac{\sin \beta}{\sin (\beta + \alpha)} \right]\left(\frac{2q}{H} \right)$$

(5.21)

Table 5.3   Values of K_a [Eq. (5.19)] (note: $\delta = \frac{2}{3}\phi$ in all cases)

α (deg)	φ (deg)	β (deg)					
		90	85	80	75	70	65
0	28	0.3213	0.3588	0.4007	0.4481	0.5026	0.5662
	30	0.2973	0.3349	0.3769	0.4245	0.4794	0.5435
	32	0.2750	0.3125	0.3545	0.4023	0.4574	0.5220
	34	0.2543	0.2916	0.3335	0.3813	0.4367	0.5017
	36	0.2349	0.2719	0.3137	0.3615	0.4170	0.4825
	38	0.2168	0.2535	0.2950	0.3428	0.3984	0.4642
	40	0.1999	0.2361	0.2774	0.3250	0.3806	0.4468
	42	0.1840	0.2197	0.2607	0.3081	0.3638	0.4303
5	28	0.3431	0.3845	0.4311	0.4843	0.5461	0.6191
	30	0.3165	0.3578	0.4043	0.4575	0.5194	0.5926
	32	0.2919	0.3329	0.3793	0.4324	0.4943	0.5678
	34	0.2691	0.3097	0.3558	0.4088	0.4707	0.5443
	36	0.2479	0.2881	0.3338	0.3866	0.4484	0.5222
	38	0.2282	0.2679	0.3132	0.3656	0.4273	0.5012
	40	0.2098	0.2489	0.2937	0.3458	0.4074	0.4814
	42	0.1927	0.2311	0.2753	0.3271	0.3885	0.4626
10	28	0.3702	0.4164	0.4686	0.5287	0.5992	0.6834
	30	0.3400	0.3857	0.4376	0.4974	0.5676	0.6516
	32	0.3123	0.3575	0.4089	0.4683	0.5382	0.6220
	34	0.2868	0.3314	0.3822	0.4412	0.5107	0.5942
	36	0.2633	0.3072	0.3574	0.4158	0.4849	0.5682
	38	0.2415	0.2846	0.3342	0.3921	0.4607	0.5438
	40	0.2214	0.2637	0.3125	0.3697	0.4379	0.5208
	42	0.2027	0.2441	0.2921	0.3487	0.4164	0.4990
15	28	0.4065	0.4585	0.5179	0.5869	0.6685	0.7671
	30	0.3707	0.4219	0.4804	0.5484	0.6291	0.7266
	32	0.3384	0.3387	0.4462	0.5134	0.5930	0.6895
	34	0.3091	0.3584	0.4150	0.4811	0.5599	0.6554
	36	0.2823	0.3306	0.3862	0.4514	0.5295	0.6239
	38	0.2578	0.3050	0.3596	0.4238	0.5006	0.5949
	40	0.2353	0.2813	0.3349	0.3981	0.4740	0.5672
	42	0.2146	0.2595	0.3119	0.3740	0.4491	0.5416
20	28	0.4602	0.5205	0.5900	0.6715	0.7690	0.8810
	30	0.4142	0.4728	0.5403	0.6196	0.7144	0.8303
	32	0.3742	0.4311	0.4968	0.5741	0.6667	0.7800
	34	0.3388	0.3941	0.4581	0.5336	0.6241	0.7352
	36	0.3071	0.3609	0.4233	0.4970	0.5857	0.6948
	38	0.2787	0.3308	0.3916	0.4637	0.5587	0.6580
	40	0.2529	0.3035	0.3627	0.4331	0.5185	0.6243
	42	0.2294	0.2784	0.3360	0.4050	0.4889	0.5931

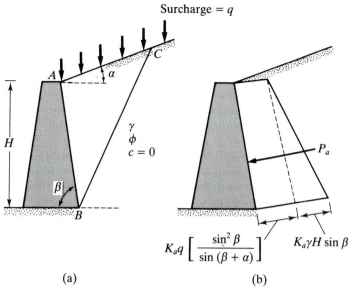

Figure 5.10 Coulomb's active pressure with a surcharge on the backfill

The derivations of Eqs. (5.20) and (5.21) have been shown in other soil mechanics texts (for example, Das, 1987).

Example 5.6

Consider the retaining wall shown in Figure 5.9a. Given: $H = 4.6$ m; unit weight of soil = 16.5 kN/m^3; angle of friction of soil = 30°; wall friction angle, $\delta = \frac{2}{3}\phi$; soil cohesion, $c = 0$; $\alpha = 0$, and $\beta = 90°$. Calculate the Coulomb's active force per unit length of the wall.

Solution

From Eq. (5.18)

$$P_a = \frac{1}{2}\gamma H^2 K_a$$

From Table 5.2, for $\alpha = 0°$, $\beta = 90°$, $\phi = 30°$, and $\delta = \frac{2}{3}\phi = 20°$, $K_a = 0.297$. Hence

$$P_a = \frac{1}{2}(16.5)(4.6)^2(0.297) = \mathbf{51.85\ kN/m}$$

5.5   Rankine Passive Pressure

Figure 5.11a shows a vertical frictionless retaining wall with a horizontal backfill. At a depth z, the vertical pressure on a soil element is equal to $\sigma_v = \gamma z$. Initially, if the wall does not yield at all, then the lateral stress at that depth will be equal to $\sigma_h = K_o \sigma_v$. This state of stress is illustrated by the Mohr's circle a in Figure 5.11b. Now, if the wall is pushed into the soil mass by an amount Δx as shown in Figure 5.11a, the vertical stress at depth

z will stay the same; however, the horizontal stress will increase. Thus σ_h will be greater than $K_o \sigma_v$. The state of stress can now be represented by the Mohr's circle b in Figure 5.11b. If the wall moves farther inward (that is, Δx is increased still more), the stresses at a depth z will ultimately reach the state represented by Mohr's circle c (Figure 5.11b). Note that this Mohr's circle touches the Mohr-Coulomb failure envelope. This implies that the soil behind the wall will fail by being pushed upward. The horizontal stress, σ_h, at this point is referred to as the *Rankine passive pressure*, or $\sigma_h = \sigma_p$.

For Mohr's circle c in Figure 5.11b, the major principal stress is σ_p, and the minor principal stress is σ_v. Substituting them into Eq. (1.83) yields

$$\sigma_p = \sigma_v \tan^2 \left(45 + \frac{\phi}{2} \right) + 2c \tan \left(45 + \frac{\phi}{2} \right) \qquad (5.22)$$

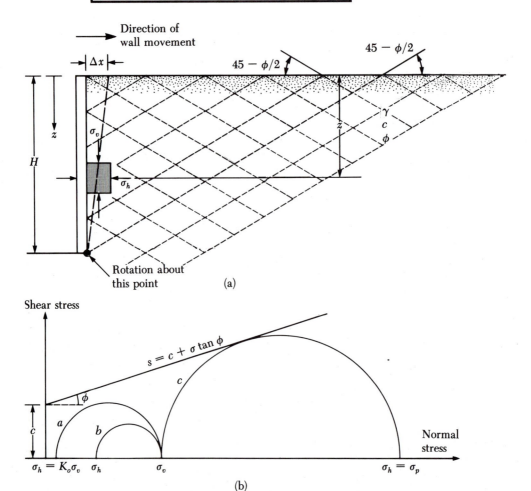

Figure 5.11 Rankine passive pressure

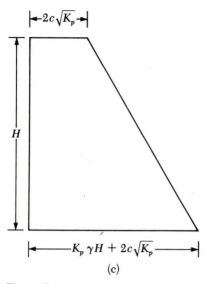

(c)

Figure 5.11 (Continued)

Now, let

$$K_p = \text{Rankine passive earth pressure coefficient}$$
$$= \tan^2\left(45 + \frac{\phi}{2}\right)$$

(5.23)

(see Table 5.4). Hence, from Eq. (5.22)

$$\sigma_p = \sigma_v K_p + 2c\sqrt{K_p}$$

(5.24)

Using the preceding equation, Figure 5.11c shows the passive pressure diagram for the wall shown in Figure 5.11a. Note that

at $z = 0$, $\sigma_v = 0$, $\sigma_p = 2c\sqrt{K_p}$

and

at $z = H$, $\sigma_v = \gamma H$, $\sigma_p = \gamma H K_p + 2c\sqrt{K_p}$

The passive force per unit length of the wall can be determined from the area of the pressure diagram, or

$$P_p = \tfrac{1}{2}\gamma H^2 K_p + 2cH\sqrt{K_p}$$

(5.25)

Table 5.4   Variation of Rankine K_p

Soil friction angle, ϕ (deg)	$K_p = \tan^2 (45 + \phi/2)$
20	2.040
21	2.117
22	2.198
23	2.283
24	2.371
25	2.464
26	2.561
27	2.663
28	2.770
29	2.882
30	3.000
31	3.124
32	3.255
33	3.392
34	3.537
35	3.690
36	3.852
37	4.023
38	4.204
39	4.395
40	4.599
41	4.815
42	5.045
43	5.289
44	5.550
45	5.828

The magnitudes of the wall movements, Δx, required to develop failure under passive conditions are as follows:

Soil type	Wall movement for passive condition, Δx
Dense sand	$0.005H$
Loose sand	$0.01H$
Stiff clay	$0.01H$
Soft clay	$0.05H$

Example 5.7

A 3-m high wall is shown in Figure 5.12a. Determine the Rankine passive force per unit length of the wall.

Solution

For the top layer

$$K_{p(1)} = \tan^2 \left(45 + \frac{\phi_1}{2} \right) = \tan^2 (45 + 15) = 3$$

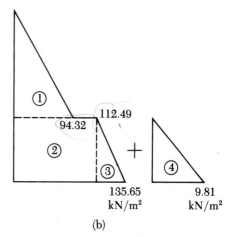

Figure 5.12

From the bottom soil layer

$$K_{p(2)} = \tan^2\left(45 + \frac{\phi_2}{2}\right) = \tan^2(45 + 13) = 2.56$$

$$\sigma_p = \sigma'_v K_p + 2c\sqrt{K_p}$$

where σ'_v = effective vertical stress

at $z = 0$, $\sigma'_v = 0$, $c_1 = 0$, $\sigma_p = 0$

at $z = 2$ m, $\sigma'_v = (15.72)(2) = 31.44$ kN/m^2, $c_1 = 0$

So, for the top soil layer

$$\sigma_p = 31.44K_{p(1)} + 2(0)\sqrt{K_p} = 31.44(3) = 94.32 \text{ kN/m}^2$$

At this depth, that is, $z = 2$ m, for the bottom soil layer

$$\sigma_p = \sigma'_v K_{p(2)} + 2c\sqrt{K_{p(2)}} = 31.44(2.56) + 2(10)\sqrt{2.56}$$
$$= 80.49 + 32 = 112.49 \text{ kN/m}^2$$

Again, at $z = 3$ m, $\sigma'_v = (15.72)(2) + (\gamma_{sat} - \gamma_w)(1)$
$$= 31.44 + (18.86 - 9.81)(1) = 40.49 \text{ kN/m}^2$$

Hence

$$\sigma_p = \sigma'_v K_{p(2)} + 2c\sqrt{K_{p(2)}} = 40.49(2.56) + (2)(10)(1.6)$$
$$= 135.65 \text{ kN/m}^2$$

Note that, because a water table is present, the hydrostatic stress, u, also has to be taken into consideration. For $z = 0$ to 2 m, $u = 0$; at $z = 3$ m, $u = (1)(\gamma_w) = 9.81$ kN/m^2.

The passive pressure diagram is plotted in Figure 5.12b. The passive force per unit length of the wall can be determined from the area of the pressure diagram as follows:

Area no.	Area	
1	$(^1/_2)(2)(94.32)$	$= 94.32$
2	$(112.49)(1)$	$= 112.49$
3	$(^1/_2)(1)(135.65 - 112.49)$	$= 11.58$
4	$(^1/_2)(9.81)(1)$	$= 4.905$

$$P_p \approx 223.3 \text{ kN/m}$$

5.6   Coulomb's Passive Earth Pressure

Coulomb (1776) also presented an analysis for determination of the passive earth pressure (that is, when the wall moves *into* the soil mass) for walls possessing friction (δ = angle of wall friction) and retaining a granular backfill material similar to that discussed in Section 5.4.

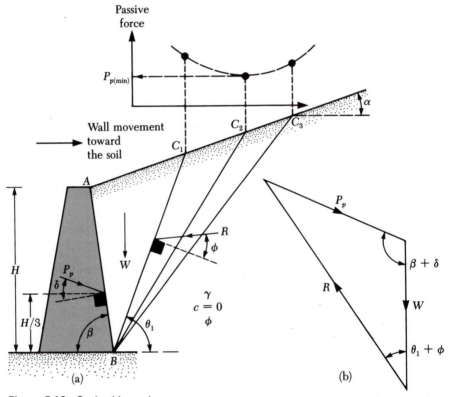

Figure 5.13 Coulomb's passive pressure

To understand the determination of Coulomb's passive force, P_p, consider the wall shown in Figure 5.13a. As in the case of active pressure, Coulomb assumed that the potential failure surface in soil is a plane. If one pictures a trial failure wedge of soil, such as ABC_1, the forces per unit length of the wall acting on the wedge are as follows:

1. The weight of the wedge, W
2. The resultant, R, of the normal and shear forces on the plane BC_1
3. The passive force, P_p

Figure 5.13b shows the force triangle, for equilibrium, for the trial wedge ABC_1. From this force triangle, the value of P_p can be determined, because the direction of all three forces as well as the magnitude of one force are known.

Similar force triangles for several trial wedges, such as ABC_1, ABC_2, ABC_3, ..., can be constructed, and the corresponding values of P_p can be determined. The top part of Figure 5.13a shows the nature of variation of the P_p values for different wedges. The *minimum value of P_p* in this diagram is *Coulomb's passive force*. Mathematically, this can be expressed as

$$P_p = \tfrac{1}{2}\gamma H^2 K_p \tag{5.26}$$

where

$$
\begin{aligned}
K_p &= \text{Coulomb's passive pressure coefficient} \\[6pt]
&= \frac{\sin^2 (\beta - \phi)}{\sin^2 \beta \cdot \sin (\beta + \delta) \left[1 - \sqrt{\dfrac{\sin (\phi + \delta) \sin (\phi + \alpha)}{\sin (\beta + \delta) \sin (\beta + \alpha)}} \right]^2}
\end{aligned} \tag{5.27}
$$

The values of the passive pressure coefficient, K_p, for various values of ϕ and δ are given in Table 5.5 ($\beta = 90°$, $\alpha = 0°$).

Note that the resultant passive force, P_p, will act at a distance of $H/3$ measured from the bottom of the wall and will be inclined at an angle δ to the normal drawn to the back face of the wall.

Table 5.5 Values of K_p [Eq. (5.27)] for $\beta = 90°$ and $\alpha = 0°$

ϕ (deg)	δ (deg)				
	0	5	10	15	20
15	1.698	1.900	2.130	2.405	2.735
20	2.040	2.313	2.636	3.030	3.525
25	2.464	2.830	3.286	3.855	4.597
30	3.000	3.506	4.143	4.977	6.105
35	3.690	4.390	5.310	6.854	8.324
40	4.600	5.590	6.946	8.870	11.772

Table 5.6 General Range of Wall Friction Angles for Masonry or Mass Concrete Walls

Backfill material	Range of δ (deg)
Gravel	27–30
Coarse sand	20–28
Fine sand	15–25
Stiff clay	15–20
Silty clay	12–16

5.7   Range of Wall Friction Angle, δ

Retaining walls are generally constructed with masonry or mass concrete. It is always useful for a designer to have a general idea of the range of the wall friction angle, δ, that may be encountered. Table 5.6 shows the general range of the values of δ for various backfill materials.

5.8   Comments on the Failure Surface Assumption for Coulomb Pressure Calculation

Coulomb's pressure calculation methods for active and passive pressure have been discussed in Sections 5.4 and 5.6. The fundamental assumption for these analyses is the acceptance of *plane failure surfaces*. However, for walls with friction, this assumption does not apply in practice. The nature of *actual* failure surfaces in the soil mass for active and passive pressure is shown in Figure 5.14a and b, respectively (for a vertical wall with a horizontal backfill). Note that the failure surfaces marked *BC* are curved. The failure surfaces marked *CD* are planes.

Although the actual failure surface in soil for the case of active pressure is somewhat different from that assumed in the calculation of the Coulomb pressure, the results are not greatly different. However, this is not so in the case of passive pressure, when, as the value of δ increases, Coulomb's method of calculation gives increasingly erroneous values of P_p. This factor of error could lead to an unsafe condition, because the values of P_p would become higher than the soil could resist. To remedy this situation, several investigations have been done to calculate the value of P_p using the type of failure surface shown in Figure 5.14b. Some of those investigations are listed in the following table.

Investigator	Assumption for the curved part, *BC* (Figure 5.14b)
1. Packshaw (1969)	Circle
2. Caquot and Kerisel (1948)	Ellipse
3. James and Bransby (1971)	Log spiral
4. Shields and Tolunay (1972, 1973)	Log spiral
5. Terzaghi and Peck (1967)	Log spiral

The details of the calculation procedures involved in the studies listed in the table are beyond the scope of this text. However, the results of Shields and Tolunay's work (1973) appear to be reasonable. They show that

$$\boxed{P_p = \tfrac{1}{2}\gamma H^2 K_p}$$

(5.28)

where K_p = passive pressure coefficient

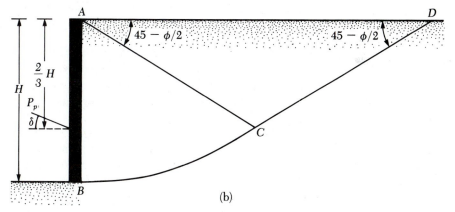

Figure 5.14   Nature of failure surface in soil with wall friction for (a) active pressure case; (b) passive pressure case

Table 5.7   Values of K_p [Eq. (5.28)] for a Vertical Wall with Horizontal Cohesionless Soil as Backfill[a]

δ (deg)	ϕ (deg)					
	20	25	30	35	40	45
0	2.04	2.46	3.00	3.69	4.60	5.83
5	2.27	2.78	3.44	4.31	5.46	4.09
10	2.47	3.08	3.86	4.92	6.36	8.43
15	2.64	3.34	4.28	5.53	7.30	9.89
20	2.87	3.61	4.68	6.17	8.30	11.49
25		4.0	5.12	6.85	9.39	13.20
30			5.81	7.61	10.60	15.31
35				8.85	12.00	17.65
40					14.40	20.48
45						25.47

[a] After Shields and Tolunay (1973)

Figure 5.15 Variation of K_p obtained from Terzaghi and Peck's method [vertical backface of wall ($\beta = 90°$) and horizontal granular backfill]

The values of K_p for a vertical wall with a horizontal granular backfill (that is, $c = 0$) are given in Table 5.7.

For comparison purposes, the variation of K_p [Eq. (5.28)] as obtained by using the procedure outlined by Terzaghi and Peck (1967), in which the curved part BC in Figure 5.14b is a log spiral, is shown in Figure 5.15.

Example 5.8

Refer to Figure 5.14b, which shows a vertical retaining wall with a horizontal backfill. Given: $H = 13.5$ ft, $\gamma = 105$ lb/ft³, $\phi = 35°$, and $\delta = 10°$. Based on Shields and Tolunay's work (Table 5.7), what would be the passive force per meter of the wall?

Solution

From Eq. (5.28)

$$P_p = \tfrac{1}{2}\gamma H^2 K_p$$

For $\phi = 35°$ and $\delta = 10°$, $K_p = 4.92$ (Table 5.7):

$$P_p = (\tfrac{1}{2})(105)(13.5)^2(4.92) = \mathbf{47{,}047 \ lb/ft}$$

5.9 Rankine Active and Passive Earth Pressure for Inclined Granular Backfill

If the backfill of a frictionless retaining wall is a granular soil (that is, $c = 0$) and rises at an angle α with respect to the horizontal (Figure 5.16), the *active*

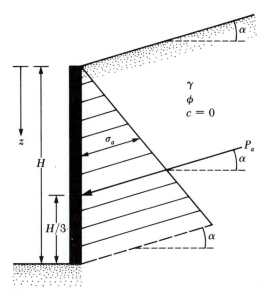

Figure 5.16 Notations for active pressure—Eqs. (5.29), (5.30), (5.31)

earth pressure coefficient, K_a, can be expressed in the form

$$K_a = \cos \alpha \frac{\cos \alpha - \sqrt{\cos^2 \alpha - \cos^2 \phi}}{\cos \alpha + \sqrt{\cos^2 \alpha - \cos^2 \phi}} \qquad (5.29)$$

where ϕ = angle of friction of soil

At any given depth, z, the *Rankine active pressure* can be expressed as

$$\sigma_a = \gamma z K_a \qquad (5.30)$$

Also, the total force per unit length of the wall (P_a) is equal to

$$P_a = \tfrac{1}{2}\gamma H^2 K_a \qquad (5.31)$$

Note that, in this case, the direction of the resultant force, P_a, is *inclined at an angle α with the horizontal* and intersects the wall at a distance of $H/3$ measured from the base of the wall. Table 5.8 presents the values of K_a (active earth pressure) for various values of α and ϕ.

In a similar manner, the *Rankine passive earth pressure* for a wall of height H with a granular sloping backfill can be represented by the equation

$$P_p = \tfrac{1}{2}\gamma H^2 K_p \qquad (5.32)$$

Table 5.8 Active Earth Pressure Coefficient, K_a [Eq. (5.29)]

	ϕ (deg) →						
↓ α (deg)	28	30	32	34	36	38	40
0	0.361	0.333	0.307	0.283	0.260	0.238	0.217
5	0.366	0.337	0.311	0.286	0.262	0.240	0.219
10	0.380	0.350	0.321	0.294	0.270	0.246	0.225
15	0.409	0.373	0.341	0.311	0.283	0.258	0.235
20	0.461	0.414	0.374	0.338	0.306	0.277	0.250
25	0.573	0.494	0.434	0.385	0.343	0.307	0.275

Note: With $\alpha = \phi$, $K_a = \cos \alpha$. So, $\alpha = \phi = 28°$, $K_a = 0.883$
$\alpha = \phi = 30°$, $K_a = 0.866$
$\alpha = \phi = 32°$, $K_a = 0.848$
$\alpha = \phi = 34°$, $K_a = 0.829$
$\alpha = \phi = 36°$, $K_a = 0.809$
$\alpha = \phi = 38°$, $K_a = 0.788$
$\alpha = \phi = 40°$, $K_a = 0.866$

where K_p = passive earth pressure coefficient

$$K_p = \cos \alpha \cdot \frac{\cos \alpha + \sqrt{\cos^2 \alpha - \cos^2 \phi}}{\cos \alpha - \sqrt{\cos^2 \alpha - \cos^2 \phi}} \tag{5.33}$$

As in the case of the active force, the resultant force, P_p, is inclined at an angle α with the horizontal and intersects the wall at a distance of $H/3$ measured from the bottom of the wall. The values of K_p (passive earth pressure coefficient) for various values of α and ϕ are given in Table 5.9.

Table 5.9 Passive Earth Pressure Coefficient, K_p [Eq. (5.33)]

	ϕ (deg) →						
↓ α (deg)	28	30	32	34	36	38	40
0	2.770	3.000	3.255	3.537	3.852	4.204	4.599
5	2.715	2.943	3.196	3.476	3.788	4.136	4.527
10	2.551	2.775	3.022	3.295	3.598	3.937	4.316
15	2.284	2.502	2.740	3.003	3.293	3.615	3.977
20	1.918	2.132	2.362	2.612	2.886	3.189	3.526
25	1.434	1.664	1.894	2.135	2.394	2.676	2.987

Note: With $\alpha = \phi$, $K_p = \cos \alpha$. So, $\alpha = \phi = 28°$, $K_p = 0.883$
$\alpha = \phi = 30°$, $K_p = 0.866$
$\alpha = \phi = 32°$, $K_p = 0.848$
$\alpha = \phi = 34°$, $K_p = 0.829$
$\alpha = \phi = 36°$, $K_p = 0.809$
$\alpha = \phi = 38°$, $K_p = 0.788$
$\alpha = \phi = 40°$, $K_p = 0.766$

5.10   Lateral Earth Pressure Due to Surcharge

In several instances, the theory of elasticity is used to determine the lateral earth pressure on retaining structures caused by various types of surcharge loading, such as *line loading* (Figure 5.17a) and *strip loading* (Figure 5.17b).

According to the theory of elasticity, the stress at any depth, z, on a retaining structure caused by a line load of intensity q/unit length (Figure 5.17a) may be given as

$$\sigma = \frac{2q}{\pi H} \cdot \frac{a^2 b}{(a^2 + b^2)^2} \tag{5.34}$$

where σ = horizontal stress at a depth $z = bH$

(See Figure 5.17a for explanations of the terms a and b.)

However, because soil is not a perfectly elastic medium, some deviations from Eq. (5.34) may be expected. The modified forms of this equation generally accepted for use with soils are as follows:

$$\boxed{\sigma = \frac{4q}{\pi H} \frac{a^2 b}{(a^2 + b^2)^2} \quad \text{for } a > 0.4} \tag{5.35}$$

and

$$\boxed{\sigma = \frac{q}{H} \frac{0.203b}{(0.16 + b^2)^2} \quad \text{for } a \leq 0.4} \tag{5.36}$$

Figure 5.17b shows a strip load with an intensity of q/unit area located at a distance b' from a wall of height H. Based on the theory of elasticity, the horizontal stress, σ, at any depth z on a retaining structure can be given by the equation

$$\sigma = \frac{q}{H} (\beta - \sin \beta \cdot \cos 2\alpha) \tag{5.37}$$

(The angles α and β are defined in Figure 5.17b.)

However, in the case of soils, the right side of Eq. (5.37) is doubled to account for the yielding soil continuum, or

$$\boxed{\sigma = \frac{2q}{H} (\beta - \sin \beta \cdot \cos 2\alpha)} \tag{5.38}$$

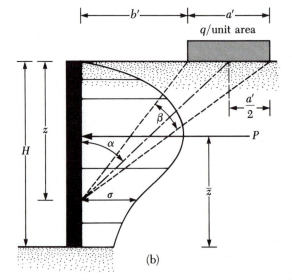

Figure 5.17 Lateral earth pressure caused by (a) line load; (b) strip load

The total force per unit length (P) and the location of the resultant force, $\bar{z}$, due to the *strip loading only* can be expressed as follows (Jarquio, 1981):

$$P = \frac{q}{90}\left[H(\theta_2 - \theta_1)\right] \tag{5.39}$$

where $\theta_1 = \tan^{-1}\left(\dfrac{b'}{H}\right)$ in degrees   (5.40)

$$\theta_2 = \tan^{-1}\left(\dfrac{a' + b'}{H}\right) \text{ in degrees}$$ (5.41)

$$\bar{z} = H - \dfrac{H^2(\theta_2 - \theta_1) + (R - Q) - 57.30a'H}{2H(\theta_2 - \theta_1)}$$

$$= \dfrac{H^2(\theta_2 - \theta_1) - (R - Q) + 57.30a'H}{2H(\theta_2 - \theta_1)}$$ (5.42)

where $R = (a' + b')^2(90 - \theta_2)$ (5.43)

$Q = b'^2(90 - \theta_1)$ (5.44)

Example 5.9

Refer to Figure 5.17b. Given: $a' = 2$ m, $b' = 1$ m, $q = 40$ kN/m^2, and $H = 6$ m.

a. Determine the total pressure on the wall caused by the strip loading only.
b. Determine the location of the center of pressure, $\bar{z}$, measured from the bottom of the wall.

Solution
Part a

Using Eqs. (5.40) and (5.41)

$$\theta_1 = \tan^{-1}\left(\dfrac{1}{6}\right) = 9.46°$$

$$\theta_2 = \tan^{-1}\left(\dfrac{2 + 1}{6}\right) = 26.57°$$

From Eq. (5.39)

$$P = \dfrac{q}{90}\left[H(\theta_2 - \theta_1)\right] = \dfrac{40}{90}\left[6(26.57 - 9.46)\right] = \mathbf{45.63 \ kN/m}$$

Part b

Again, from Eq. (5.42)

$$H - \bar{z} = \dfrac{H^2(\theta_2 - \theta_1) - (R - Q) + 57.30a'H}{2H(\theta_2 - \theta_1)}$$

$$R = (a' + b')^2(90 - \theta_2) = (2 + 1)^2(90 - 26.57) = 570.87$$

$$Q = b'^2(90 - \theta_1) = 1^2(90 - 9.46) = 80.54$$

So

$$H - \bar{z} = \frac{6^2(26.57 - 9.46) - (570.87 - 80.54) + 57.30(2)(6)}{(2)(6)(26.57 - 9.46)}$$

$$= \frac{615.96 - 490.33 + 687.6}{205.32} = 3.96 \text{ m}$$

Hence

$$\bar{z} = H - 3.96 = 6.0 - 3.96 = \textbf{2.04 m}$$

Example 5.10

Refer to Figure 5.17a. Given: $H = 16$ ft, $a = 0.3$, and $q = 1600$ lb/ft. Determine the variation of σ on the face of the wall at $z = 0, 4, 8, 12$, and 16 ft. Plot the variation of σ with depth.

Solution

Since $a < 0.4$, Eq. (5.36) will be used:

$$\sigma = \frac{q}{H} \frac{0.203b}{(0.16 + b^2)^2}$$

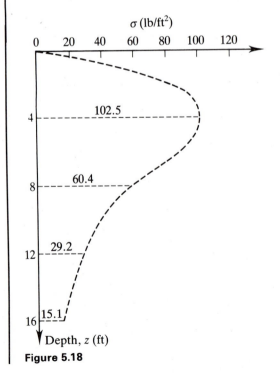

Figure 5.18

Now the following table can be prepared.

z (ft)	$b = z/H$	σ (lb/ft^2)
0	0	0
4	0.25	102.5
8	0.50	60.4
12	0.75	29.2
16	1.0	15.1

The variation of σ with depth is shown in Figure 5.18.

5.11   Active Earth Pressure with Earthquake Forces

Coulomb's active earth pressure theory (see Section 5.4) can be extended to take into account the forces caused by an earthquake. Figure 5.19 shows a condition of active pressure with a granular backfill ($c = 0$). Note that the forces acting on the soil failure wedge in Figure 5.19 are essentially the same as those shown in Figure 5.9a, with the addition of $k_h W$ and $k_v W$ in the horizontal and vertical directions, respectively; k_h and k_v may be defined as follows:

$$k_h = \frac{\text{horizontal earthquake acceleration component}}{\text{acceleration due to gravity, } g} \tag{5.45}$$

$$k_v = \frac{\text{vertical earthquake acceleration component}}{\text{acceleration due to gravity, } g} \tag{5.46}$$

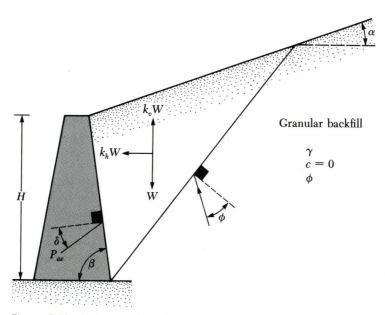

Figure 5.19   Derivation of Eq. (5.47)

Proceeding in a manner similar to that described in Section 5.4, the relation for the active force per unit length of the wall (P_{ae}) can be determined as

$$P_{ae} = \tfrac{1}{2}\gamma H^2(1 - k_v)K_{ae}$$

(5.47)

where

K_{ae} = active earth pressure coefficient

$$= \frac{\sin^2 (\phi + \beta - \theta')}{\cos \theta' \sin^2 \beta \sin (\beta - \theta' - \delta)\left[1 + \sqrt{\dfrac{\sin (\phi + \delta) \sin (\phi - \theta' - \alpha)}{\sin (\beta - \delta - \theta') \sin (\alpha + \beta)}}\right]^2}$$

(5.48)

$$\theta' = \tan^{-1}\left[\frac{k_h}{1 - k_v}\right]$$

(5.49)

Note that, for no earthquake condition

$$k_h = 0$$
$$k_v = 0$$
$$\theta' = 0$$

Hence, $K_{ae} = K_a$ [as given by Eq. (5.19)].

The variation of $K_{ae} \cos \delta$ with k_h for the case of $k_v = 0$, $\beta = 90°$, $\alpha = 0°$, and $\delta = \phi/2$ is shown in Figure 5.20.

Equation (5.47) is usually referred to as the *Mononobe-Okabe* solution. Unlike the case shown in Figure 5.9a, the resultant earth pressure in this situation, as calculated by Eq. (5.47), *does not act* at a distance of $H/3$ measured from the bottom of the wall. In order to obtain the location of the resultant earth pressure, the following procedure may be used.

1. Calculate P_{ae} by using Eq. (5.47).
2. Calculate P_a by using Eq. (5.18).
3. Calculate

$$\Delta P_{ae} = P_{ae} - P_a$$

(5.50)

4. Assume that P_a acts at a distance of $H/3$ from the bottom of the wall (Figure 5.21).
5. Assume that ΔP_{ae} acts at a distance of $0.6H$ from the bottom of the wall (Figure 5.21).

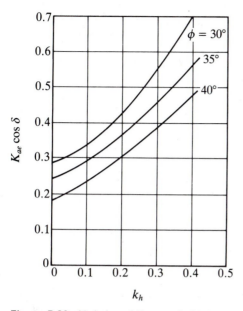

Figure 5.20 Variation of K_{ae} cos δ with k_h, (*note*: $k_v = 0$, $\beta = 90°$, $a = 0°$, and $\delta = \phi/2$)

6. Calculate the location of the resultant as

$$\bar{z} = \frac{(0.6H)(\Delta P_{ae}) + \left(\dfrac{H}{3}\right)(P_a)}{P_{ae}} \tag{5.51}$$

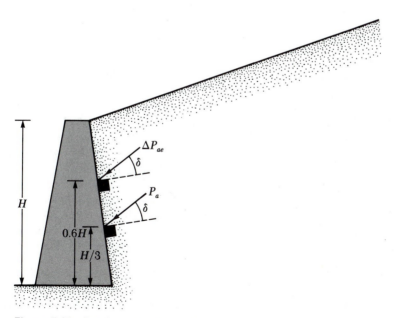

Figure 5.21 Determination of line of action of P_{ae}

Example 5.11

Refer to Figure 5.19. Given: $H = 10$ ft, $\beta = 90°$, $\alpha = 0°$, $k_v = 0$, $k_h = 0.3$, $\gamma = 105$ lb/ft^3, $\phi = 35°$, and $\delta = \phi/2$. Determine the following:

   a. P_{ae}
   b. The location of the resultant, $\bar{z}$, measured from the bottom of the wall

Solution
Part a

From Eq. (5.47)

$$P_{ae} = \tfrac{1}{2}\gamma H^2(1 - k_v)K_{ae}$$

Given: $\gamma = 105$ lb/ft^3, $H = 10$ ft, and $k_v = 0$. Since $\delta = \phi/2$, we can use Figure 5.20 to determine K_{ae}. For $k_h = 0.3$, $K_{ae} \approx 0.472$. So

$$P_{ae} = (\tfrac{1}{2})(105)(10)^2(1 - 0)(0.472) = \textbf{2478 lb/ft}$$

Part b

From Eq. (5.18)

$$P_a = \tfrac{1}{2}\gamma H^2 K_a$$

From Eq. (5.19), with $\delta = 17.5°$, $\beta = 90°$, and $\alpha = 0°$, $K_a \approx 0.24$. So

$$P_a = (\tfrac{1}{2})(105)(10)^2(0.24) = 1260 \text{ lb/ft}$$

$$\Delta P_{ae} = P_{ae} - P_a = 2478 - 1260 = 1218 \text{ lb/ft}$$

From Eq. (5.51)

$$\bar{z} = \frac{(0.6H)(\Delta P_{ae}) + \left(\dfrac{H}{3}\right)(P_a)}{P_{ae}}$$

$$= \frac{[(0.6)(10)](1218) + \left(\dfrac{10}{3}\right)(1260)}{2478} = \textbf{4.64 ft}$$

5.12   Passive Earth Pressure with Earthquake Forces

Coulomb's passive pressure theory as given in Section 5.6 can be extended to estimate the passive earth pressure, taking into consideration the earthquake forces. Figure 5.22 shows the forces acting on a soil failure wedge on a retaining wall with a granular backfill. The passive force per unit length of the wall, P_{pe}, can be taken as follows (Kapila, 1962):

$$\boxed{P_{pe} = \tfrac{1}{2}\gamma H^2(1 - k_v)K_{pe}} \qquad (5.52)$$

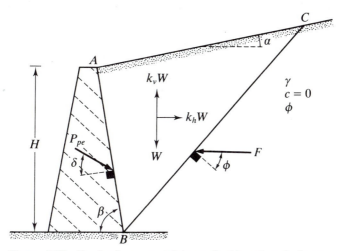

Figure 5.22   Passive force on retaining wall with earthquake force

where

$$
\boxed{
\begin{aligned}
K_{pe} &= \text{passive earth pressure coefficient} \\[2mm]
&= \frac{\sin^2\,(\beta + \theta' - \phi)}{\cos\,\theta' \cdot \sin^2\,\beta \cdot \sin\,(\delta + \beta + \theta' - 90)} \\[2mm]
&\quad \cdot \left[1 - \sqrt{\frac{\sin\,(\phi + \delta) \cdot \sin\,(\phi + \alpha - \theta')}{\sin\,(\beta + \delta + \theta') \cdot \sin\,(\alpha + \beta)}} \right]^2
\end{aligned}
}
\qquad (5.53)
$$

The definition of θ' is as given by Eq. (5.49). The variation of K_{pe} with the soil friction angle ϕ and k_h (for $\beta = 90°$, $\alpha = 0°$, $\delta = 0$, and $k_v = 0$) is shown in Figure 5.23.

STABILITY OF RETAINING WALLS

5.13   Proportioning Retaining Walls

When designing retaining walls, one must assume some of the dimensions. This is referred to as *proportioning*, which allows the engineer to check trial sections for stability. If the stability checks yield undesirable results, the sections can be changed and rechecked. Figure 5.24 shows the general proportions of various retaining-wall components that can be used for initial checks.

Note that the top of the stem of any retaining wall should not be less than about 0.3 m (≈ 12 in.) for proper placement of concrete. The depth, D,

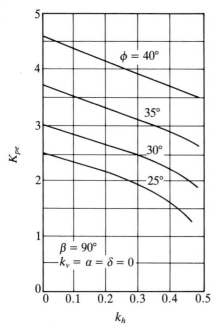

Figure 5.23 Variation of K_{pe} with k_h for $k_v = \alpha = \delta = 0$ and $\beta = 90°$ (after Davies, Richards, and Chen, 1986)

(a)

Figure 5.24 Approximate dimensions for various components of retaining wall for initial stability checks: (a) gravity wall; (b) cantilever wall (*note:* minimum dimension of D is 0.6 m) (continued on page 289)

(b)

Figure 5.24 (Continued)

to the bottom of the base slab should be kept at a minimum of 0.6 m (≈ 2 ft). However, the bottom of the base slab should be kept below the seasonal frost line.

For counterfort retaining walls, the general proportion of the stem and the base slab is the same as for cantilever walls. However, the counterfort slabs may be about 0.3 m (≈ 12 in.) thick and spaced at center-to-center distances of $0.3H$ to $0.7H$.

5.14   Application of Lateral Earth Pressure Theories to Design

The preceding sections have presented the fundamental theories of calculation of lateral earth pressure. To use these theories in design, one must make several simple assumptions. In the case of cantilever walls, if the Rankine earth pressure theory is to be used for stability checks, we need to draw a vertical line AB through point A, as shown in Figure 5.25a (which is located at the edge of the heel of the base slab). It is assumed that the Rankine active condition exists along the vertical plane AB. Rankine active earth pressure equations may then be used to calculate the lateral pressure on the face AB. In the analysis of stability for the wall, the force $P_{a(\text{Rankine})}$, the weight of soil above the heel (W_s), and the weight of the concrete (W_c)

(a)

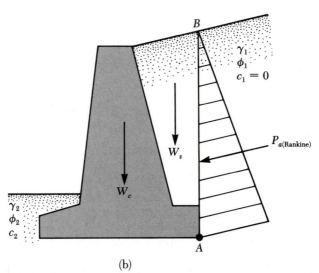

(b)

Figure 5.25 Assumption for the determination of lateral earth pressure: (a) cantilever wall; (b) and (c) gravity wall (continued on page 291)

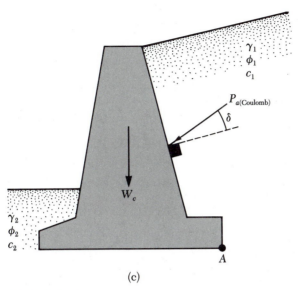

(c)

Figure 5.25   (Continued)

should all be taken into consideration. The assumption for the development of Rankine active pressure along the soil face AB is theoretically correct if the shear zone bounded by the line AC is not obstructed by the stem of the wall. The angle, η, that the line AC makes with the vertical can be given by the equation

$$\eta = 45 + \frac{\alpha}{2} - \frac{\phi}{2} - \sin^{-1}\left(\frac{\sin \alpha}{\sin \phi}\right) \qquad (5.54)$$

For gravity walls, a similar type of analysis can be used, as shown in Figure 5.25b. However, Coulomb's theory can also be used, as shown in Figure 5.25c. If *Coulomb's active pressure theory* is used, then the forces to be considered are only $P_{a(Coulomb)}$ and the weight of the wall, W_c.

In the case of ordinary retaining walls, the problems of water tables and, hence, hydrostatic pressure are not encountered. Facilities for drainage from the soils retained are always provided.

In several instances, for small retaining walls, *semiempirical charts* are used for evaluation of lateral earth pressure. Figures 5.26 and 5.27 show two semiempirical charts given by Terzaghi and Peck (1967). Figure 5.26 is for backfills with plane surfaces, and Figure 5.27 is for backfills that slope upward from the crest of the wall for a limited distance and then become horizontal. Note that, in these figures, $\frac{1}{2}K_v H'^2$ is the vertical component of the active force on the plane AB; similarly, $\frac{1}{2}K_h H'^2$ is the horizontal force. The numerals on the curves indicate the type of soil described in Table 5.10.

Table 5.10 Types of Backfill for Retaining Walls[a]

1. Coarse-grained soil without admixture of fine soil particles, very permeable (clean sand or gravel).
2. Coarse-grained soil of low permeability due to admixture of particles of silt size.
3. Residual soil with stones, fine silty sand, and granular materials with conspicuous clay content.
4. Very soft or soft clay, organic silts, or silty clays.
5. Medium or stiff clay, deposited in chunks and protected in such a way that a negligible amount of water enters the spaces between the chunks during floods or heavy rains. If this condition of protection cannot be satisfied, the clay should not be used as backfill material. With increasing stiffness of the clay, danger to the wall due to infiltration of water increases rapidly.

[a] From *Soil Mechanics in Engineering Practice*, Second Edition, by K. Terzaghi and R. B. Peck. Copyright 1967 by John Wiley and Sons. Reprinted with permission.

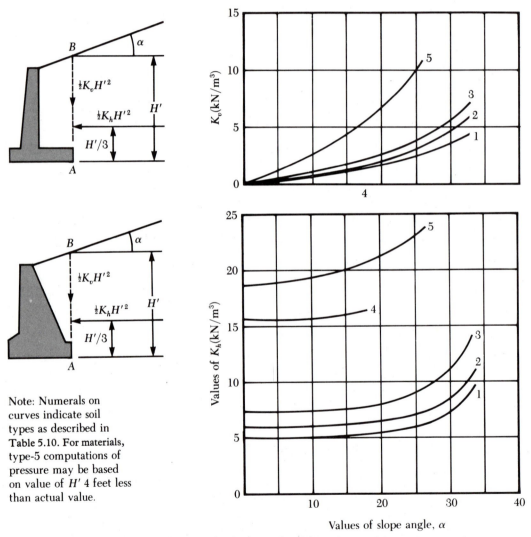

Note: Numerals on curves indicate soil types as described in Table 5.10. For materials, type-5 computations of pressure may be based on value of H' 4 feet less than actual value.

Figure 5.26 Chart for estimating pressure of backfill against retaining walls supporting backfills with plane surface (after *Soil Mechanics in Engineering Practice*, Second Edition, by K. Terzaghi and R. B. Peck. Copyright 1967 by John Wiley and Sons. Reprinted with permission)

Figure 5.26   (Continued)

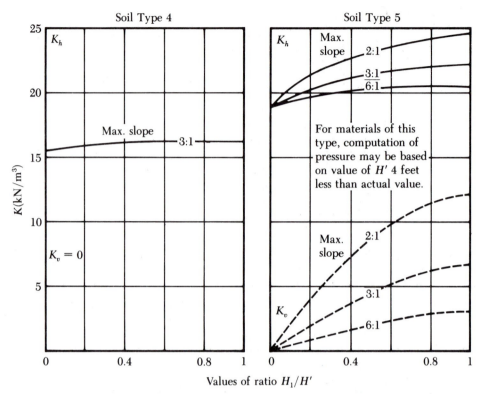

Figure 5.27   Chart for estimating pressure of backfill against retaining walls supporting backfills with surface that slopes upward from crest of wall for limited distance and then becomes horizontal (after *Soil Mechanics in Engineering Practice*, Second Edition, by K. Terzaghi and R. B. Peck. Copyright 1967 by John Wiley and Sons. Reprinted with permission)

5.15   Stability Checks

To check the stability of a retaining wall, the following steps are necessary:

   1. Check for *overturning* about its toe
   2. Check for *sliding failure* along its base
   3. Check for *bearing capacity failure* of the base
   4. Check for *settlement*
   5. Check for *overall stability*

This section will describe the procedure for checking for overturning and sliding and bearing capacity failure. The principles of investigation for settlement have been covered in Chapter 3 and will not be repeated here. Some problems regarding the overall stability of retaining walls have been discussed in Section 5.17.

Check for Overturning

Figure 5.28 shows the forces acting on a cantilever and gravity retaining wall with the assumption that the Rankine active pressure is acting along a vertical plane AB drawn through the heel. P_p is the Rankine passive pressure; its magnitude can be given as

$$P_p = \tfrac{1}{2}K_p \gamma_2 D^2 + 2c_2\sqrt{K_p}\,D \tag{5.25}$$

where γ_2 = unit weight of soil in front of the heel and under the base slab

K_p = Rankine passive earth pressure coefficient

= $\tan^2(45 + \phi_2/2)$

c_2, ϕ_2 = cohesion and soil friction angle, respectively

The factor of safety against overturning about the toe—that is, about point C in Figure 5.28—can be expressed as

$$FS_{(\text{overturning})} = \frac{\sum M_R}{\sum M_O} \tag{5.55}$$

where $\sum M_O$ = sum of the moments of forces tending to overturn about point C

$\sum M_R$ = sum of the moments of forces tending to resist overturning about point C

The overturning moment can be given as

$$\sum M_O = P_h\left(\frac{H'}{3}\right) \tag{5.56}$$

where $P_h = P_a \cos \alpha$

For calculation of the resisting moment, $\sum M_R$ (neglecting P_p), a table (such as Table 5.11) can be prepared. The weight of the soil above the heel

Table 5.11   Procedure for Calculation of $\sum M_R$

Section (1)	Area (2)	Weight/unit length of wall (3)	Moment arm measured from C (4)	Moment about C (5)
1	A_1	$W_1 = \gamma_1 \times A_1$	X_1	M_1
2	A_2	$W_2 = \gamma_2 \times A_2$	X_2	M_2
3	A_3	$W_2 = \gamma_c \times A_3$	X_3	M_3
4	A_4	$W_2 = \gamma_c \times A_4$	X_4	M_4
5	A_5	$W_2 = \gamma_c \times A_5$	X_5	M_5
6	A_6	$W_2 = \gamma_c \times A_6$	X_6	M_6
		P_v	B	M_v
		$\sum V$		$\sum M_R$

Note: γ_1 = unit weight of backfill
γ_c = unit weight of concrete

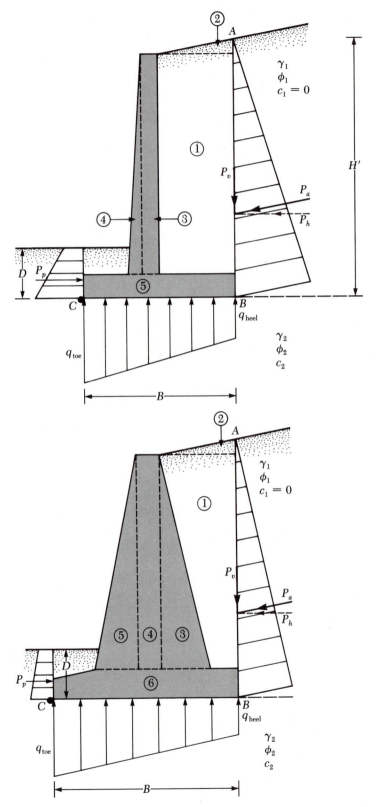

Figure 5.28 Check for overturning; assume Rankine pressure is valid

and the weight of the concrete (or masonry) are both forces that contribute to the resisting moment. Note that the force P_v also contributes to the resisting moment. P_v is the vertical component of the active force P_a, or

$$P_v = P_a \sin \alpha \tag{5.57}$$

The moment of the force P_v about C is equal to

$$M_v = P_v B = P_a \sin \alpha \cdot B \tag{5.58}$$

where B = width of the base slab

Once $\sum M_R$ is known, the factor of safety can be calculated as

$$FS_{\text{(overturning)}} = \frac{M_1 + M_2 + M_3 + M_4 + M_5 + M_6 + M_v}{P_a \cos \alpha \dfrac{H'}{3}} \tag{5.59}$$

The usual minimum desirable value of the factor of safety with respect to overturning is 1.5 to 2.

Some designers prefer to use the following equation for determination of the factor of safety against overturning:

$$FS_{\text{(overturning)}} = \frac{M_1 + M_2 + M_3 + M_4 + M_5 + M_6}{(P_a \cos \alpha) \dfrac{H'}{3} - M_v} \tag{5.60}$$

Check for Sliding Along the Base

The factor of safety against sliding may be expressed by the equation

$$FS_{\text{(sliding)}} = \frac{\sum F_{R'}}{\sum F_d} \tag{5.61}$$

where $\sum F_{R'}$ = sum of the horizontal resisting forces
$\sum F_d$ = sum of the horizontal driving forces

Referring to Figure 5.29, one sees that the shear strength of the soil below the base slab can be represented as

$$s = \sigma \tan \phi_2 + c_2$$

Thus, the maximum resisting force that can be derived from the soil per unit length of the wall along the bottom of the base slab is

$$R' = s \,(\text{area of cross section}) = s(B \times 1) = B\sigma \tan \phi_2 + Bc_2$$

However

$$B\sigma = \text{sum of the vertical force} = \sum V \text{ (see Table 5.11)}$$

So

$$R' = \left(\sum V\right) \tan \phi_2 + Bc_2$$

Figure 5.29 Check for sliding along the base

As can be seen in Figure 5.29, the passive force P_p is also a horizontal resisting force. The expression for P_p is given in Eq. (5.25). Hence

$$\sum F_{R'} = \left(\sum V\right) \tan \phi_2 + Bc_2 + P_p \tag{5.62}$$

The only horizontal force that will tend to cause sliding of the wall (*driving force*) is the horizontal component of the active force P_a, so

$$\sum F_d = P_a \cos \alpha \tag{5.63}$$

Combining Eqs. (5.61), (5.62), and (5.63)

$$FS_{\text{(sliding)}} = \frac{\left(\sum V\right) \tan \phi_2 + Bc_2 + P_p}{P_a \cos \alpha} \tag{5.64}$$

A minimum factor of safety of 1.5 against sliding is generally required.

In many cases, the passive force P_p is ignored for calculation of the factor of safety with respect to sliding. The friction angle, ϕ_2, is also reduced in several instances for safety. The reduced soil friction angle may be in the order of one-half to two-thirds of the angle ϕ_2. In a similar manner, the cohesion c_2 may be reduced to the value of 0.5 to $0.67c_2$. Thus

$$FS_{\text{(sliding)}} = \frac{\left(\sum V\right) \tan (k_1\phi_2) + Bk_2 c_2 + P_p}{P_a \cos \alpha} \tag{5.65}$$

where k_1 and k_2 are in the range of $\frac{1}{2}$ to $\frac{2}{3}$

In some instances, certain walls may not yield a desired factor of safety of 1.5. In order to improve their resistance to sliding, a base key may be used. Base keys are illustrated by broken lines in Figure 5.29. This figure indicates that the passive force at the toe *without the key* can be given as

$$P_p = \tfrac{1}{2}\gamma_2 D^2 K_p + 2c_2 D\sqrt{K_p}$$

However, if a key is included, the passive force per unit length of the wall will be

$$P_p = \tfrac{1}{2}\gamma_2 D_1^2 K_p + 2c_2 D'\sqrt{K_p}$$

where $K_p = \tan^2 (45 + \phi_2/2)$

Because $D_1 > D$, it is obvious that a key will help increase the passive resistance at the toe and, hence, the factor of safety against sliding. Usually the base key is constructed below the stem, and some main steel is run into the key.

Another possible way to increase the value of $FS_{\text{(sliding)}}$ is to consider reducing the value of P_a [see Eq. (5.65)]. One possible way this can be achieved is described below (Elman and Terry, 1988). This discussion is limited to the case in which the retaining wall has a horizontal granular backfill (Figure 5.30). In Figure 5.30a, the active force, P_a, is horizontal ($\alpha = 0$). This means that

$$P_a \cos \alpha = P_h = P_a$$

and

$$P_a \sin \alpha = P_v = 0$$

However

$$P_a = P_{a(1)} + P_{a(2)} \qquad\qquad (5.66)$$

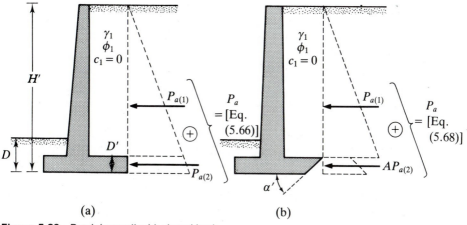

(a)   (b)

Figure 5.30   Retaining wall with sloped heel

The magnitude of $P_{a(2)}$ can be reduced if the heel of the retaining wall is sloped as shown in Figure 5.30b. For this case

$$P_a = P_{a(1)} + AP_{a(2)} \tag{5.67}$$

Elman and Terry (1988) have provided the magnitude of A, shown in Figure 5.31, which is valid for $\alpha' = 45°$. However, note that, in Figure 5.30a

$$P_{a(1)} = \tfrac{1}{2}\gamma K_a(H' - D')^2$$

and

$$P_a = \tfrac{1}{2}\gamma K_a H'^2$$

Hence

$$P_{a(2)} = \tfrac{1}{2}\gamma K_a[H'^2 - (H' - D')^2]$$

So, for the active pressure diagram shown in Figure 5.30b

$$\boxed{P_a = \frac{1}{2}\gamma K_a(H' - D')^2 + \frac{A}{2}\gamma K_a[H^2 - (H - D')^2]} \tag{5.68}$$

This procedure of sloping the heel can be extremely helpful in some cases.

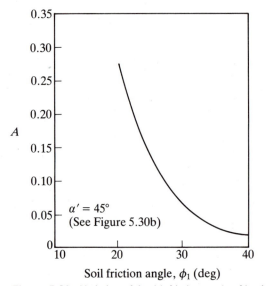

Figure 5.31   Variation of A with friction angle of backfill [Eq. (5.68)]

Check for Bearing Capacity Failure

The vertical pressure as transmitted to the soil by the base slab of the retaining wall should be checked against the ultimate bearing capacity of the soil. The nature of variation of the vertical pressure transmitted by the base slab into the soil is shown in Figure 5.32. Note that q_{toe} and q_{heel} are the *maximum* and the *minimum* pressures occurring at the ends of the toe and heel sections, respectively. The magnitudes of q_{toe} and q_{heel} can be determined in the following manner:

The sum of the vertical forces acting on the base slab is equal to $\sum V$ (see Col. 3, Table 5.11), and the horizontal force is $P_a \cos \alpha$. Let R be the resultant force, or

$$\vec{R} = \sum \vec{V} + \overrightarrow{(P_a \cos \alpha)} \tag{5.69}$$

The net moment of these forces about the point C (Figure 5.32) is equal to

$$M_{net} = \sum M_R - \sum M_O \tag{5.70}$$

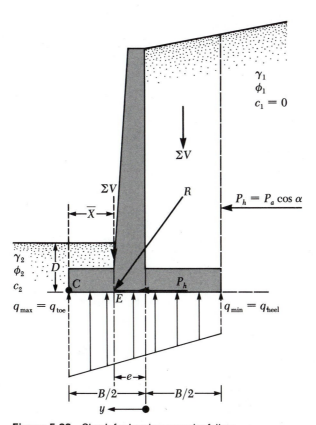

Figure 5.32 Check for bearing capacity failure

Note that the values of $\sum M_R$ and $\sum M_O$ have been previously determined [see Col. 5, Table 5.11 and Eq. (5.56)]. Let the line of action of the resultant, R, intersect the base slab at E, as shown in Figure 5.32. The distance CE can now be determined as

$$\overline{CE} = \overline{X} = \frac{M_{net}}{\sum V} \tag{5.71}$$

Hence, the eccentricity of the resultant, R, can be expressed as

$$e = \frac{B}{2} - \overline{CE} \tag{5.72}$$

The pressure distribution under the base slab can be determined by using the simple principles of mechanics of materials:

$$q = \frac{\sum V}{A} \pm \frac{M_{net} y}{I} \tag{5.73}$$

where M_{net} = moment = $(\sum V)e$
I = moment of inertia per unit length of the base section
$= \frac{1}{12}(1)(B^2)$

For maximum and minimum pressures, the value of y in Eq. (5.73) is equal to $B/2$. Substituting the preceding values into Eq. (5.73)

$$q_{max} = q_{toe} = \frac{\sum V}{(B)(1)} + \frac{e(\sum V)\dfrac{B}{2}}{\left(\dfrac{1}{12}\right)(B^3)} = \frac{\sum V}{B}\left(1 + \frac{6e}{B}\right) \tag{5.74}$$

Similarly

$$q_{min} = q_{heel} = \frac{\sum V}{B}\left(1 - \frac{6e}{B}\right) \tag{5.75}$$

Note that $\sum V$ includes the soil weight, as shown in Table 5.11, and that when the value of the eccentricity, e, becomes greater than $B/6$, q_{min} becomes negative [Eq. (5.75)]. This indicates that there will be some tensile stress at the end of the heel section. This stress is not desirable because the tensile strength of soil is very small. If the analysis of a given design shows that $e > B/6$, then the design should be reproportioned, and calculations should be redone.

The relationships for the ultimate bearing capacity of a shallow foundation were discussed in Chapter 3 and can be expressed as

$$q_u = c_2 N_c F_{cd} F_{ci} + q N_q F_{qd} F_{qi} + \frac{1}{2}\gamma_2 B' N_\gamma F_{\gamma d} F_{\gamma i} \tag{5.76}$$

where $q = \gamma_2 D$

$B' = B - 2e$

$F_{cd} = 1 + 0.4 \dfrac{D}{B'}$

$F_{qd} = 1 + 2 \tan \phi_2 (1 - \sin \phi_2)^2 \dfrac{D}{B'}$

$F_{\gamma d} = 1$

$F_{ci} = F_{qi} = \left(1 - \dfrac{\psi^\circ}{90^\circ}\right)^2$

$F_{\gamma i} = \left(1 - \dfrac{\psi^\circ}{\phi^\circ}\right)^2$

$\psi^\circ = \tan^{-1}\left(\dfrac{P_a \cos \alpha}{\sum V}\right)$

Note that the shape factors F_{cs}, F_{qs}, and $F_{\gamma s}$, which were given in Table 3.4, are all equal to one, because they can be treated as a continuous foundation. For this reason, the shape factors are not shown in Eq. (5.76).

Once the ultimate bearing capacity of the soil has been calculated by using Eq. (5.76), the factor of safety against bearing capacity failure can be determined, or

$$FS_{(\text{bearing capacity})} = \frac{q_u}{q_{max}} \qquad (5.77)$$

Generally, a factor of safety of 3 is required. In Chapter 3 we noted that the ultimate bearing capacity of shallow foundations occurs at a settlement of about 10% of the foundation width. In the case of retaining walls, the width B is large. Hence, the ultimate load q_u will occur at a fairly large foundation settlement. A factor of safety of 3 against bearing capacity failure may not insure, in all cases, that settlement of the structure will be within the tolerable limit. Thus this situation needs further investigation.

Example 5.12

The cross section of a cantilever retaining wall is shown in Figure 5.33. Calculate the factors of safety with respect to overturning and sliding and bearing capacity.

Solution

Referring to Figure 5.33

$H' = H_1 + H_2 + H_3 = 2.6 \tan 10^\circ + 6 + 0.7$

$= 0.458 + 6 + 0.7 = 7.158$ m

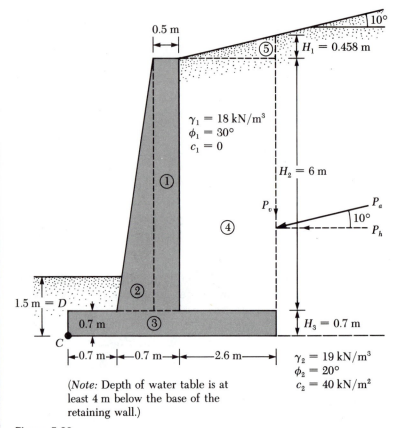

Figure 5.33

(*Note:* Depth of water table is at least 4 m below the base of the retaining wall.)

Rankine active force per unit length of wall $= P_p = \frac{1}{2}\gamma_1 H'^2 K_a$. For $\phi_1 = 30°$, $\alpha = 10°$, K_a is equal to 0.350 (Table 5.8). Thus,

$$P_a = \frac{1}{2}(18)(7.158)^2(0.35) = 161.4 \text{ kN/m}$$
$$P_v = P_a \sin 10° = 161.4(\sin 10°) = 28.03 \text{ kN/m}$$
$$P_h = P_a \cos 10° = 161.4(\cos 10°) = 158.95 \text{ kN/m}$$

Factor of Safety Against Overturning

The following table can now be prepared for determination of the resisting moment. The overturning moment, M_O

$$M_O = P_h\left(\frac{H'}{3}\right) = 158.95\left(\frac{7.158}{3}\right) = 379.25 \text{ kN-m}$$

$$FS_{(\text{overturning})} = \frac{\sum M_R}{M_O} = \frac{1128.98}{379.25} = 2.98 > 2\text{—O.K.}$$

Section no.[a]	Area (m^2)	Weight/unit length (kN/m)	Moment arm from point C (m)	Moment (kN-m)
1	$6 \times 0.5 = 3$	70.74	1.15	81.35
2	$\frac{1}{2}(0.2)6 = 0.6$	15.15	0.833	11.79
3	$4 \times 0.7 = 2.8$	66.02	2.0	132.04
4	$6 \times 2.6 = 15.6$	280.80	2.7	758.16
5	$\frac{1}{2}(2.6)(0.458) = 0.595$	10.71	3.13	33.52
		$P_v = 28.03$	4.0	112.12
		$\sum V = 470.45$		$\sum 1128.98$ $= \sum M_R$

[a] For section numbers, refer to Figure 5.33
$\gamma_{concrete} = 23.58$ kN/m^3

Factor of Safety Against Sliding

From Eq. (5.65)

$$FS_{(sliding)} = \frac{(\sum V)\tan(k_1\phi_1) + Bk_2c_2 + P_p}{P_a \cos\alpha}$$

Let $k_1 = k_2 = \frac{2}{3}$
Also

$$P_p = \frac{1}{2}K_p\gamma_2 D^2 + 2c_2\sqrt{K_p}D$$

$$K_p = \tan^2\left(45 + \frac{\phi_2}{2}\right) = \tan^2(45 + 10) = 2.04$$

$$D = 1.5 \text{ m}$$

So

$$P_p = \frac{1}{2}(2.04)(19)(1.5)^2 + 2(40)(\sqrt{2.04})(1.5)$$

$$= 43.61 + 171.39 = 215 \text{ kN/m}$$

Hence

$$FS_{(sliding)} = \frac{(470.45)\tan\left(\dfrac{2 \times 20}{3}\right) + (4)\left(\dfrac{2}{3}\right)(40) + 215}{158.95}$$

$$= \frac{111.5 + 106.67 + 215}{158.95} = 2.73 > 1.5\text{---O.K.}$$

Note: For some designs, the depth D for passive pressure calculation may be taken to be *equal to the thickness of the base slab.*

Factor of Safety Against Bearing Capacity Failure

Combining Eqs. (5.70), (5.71), and (5.72)

$$e = \frac{B}{2} - \frac{\sum M_R - \sum M_O}{\sum V} = \frac{4}{2} - \frac{1128.98 - 379.25}{470.45}$$

$$= 0.406 \text{ m} < \frac{B}{6} = \frac{4}{6} = 0.666 \text{ m}$$

Again, from Eqs. (5.74) and (5.75)

$$q_{\substack{toe \\ heel}} = \frac{\sum V}{B}\left(1 \pm \frac{6e}{B}\right) = \frac{470.45}{4}\left(1 \pm \frac{6 \times 0.406}{4}\right) = 189.2 \text{ kN/m}^2 \text{ (toe)}$$

$$= 45.99 \text{ kN/m}^2 \text{ (heel)}$$

The ultimate bearing capacity of the soil can be determined from Eq. (5.76):

$$q_u = c_2 N_c F_{cd} F_{ci} + q N_q F_{qd} F_{qi} + \tfrac{1}{2}\gamma_2 B' N_\gamma F_{\gamma d} F_{\gamma i}$$

For $\phi_2 = 20°$ (Table 3.2), $N_c = 14.83$, $N_q = 6.4$, and $N_\gamma = 5.39$. Also

$$q = \gamma_2 D = (19)(1.5) = 28.5 \text{ kN/m}^2$$

$$B' = B - 2e = 4 - 2(0.406) = 3.188 \text{ m}$$

$$F_{cd} = 1 + 0.4\left(\frac{D}{B'}\right) = 1 + 0.4\left(\frac{1.5}{3.188}\right) = 1.188$$

$$F_{qd} = 1 + 2 \tan \phi_2 (1 - \sin \phi_2)^2\left(\frac{D}{B'}\right) = 1 + 0.315\left(\frac{1.5}{3.188}\right) = 1.148$$

$$F_{\gamma d} = 1$$

$$F_{ci} = F_{qi} = \left(1 - \frac{\psi°}{90°}\right)^2$$

$$\psi = \tan^{-1}\left(\frac{P_a \cos \alpha}{\sum V}\right) = \tan^{-1}\left(\frac{158.95}{470.45}\right) = 18.67°$$

So

$$F_{ci} = F_{qi} = \left(1 - \frac{18.67}{90}\right)^2 = 0.628$$

$$F_{\gamma i} = \left(1 - \frac{\psi}{\phi}\right)^2 = \left(1 - \frac{18.67}{20}\right)^2 \approx 0$$

Hence

$$q_u = (40)(14.83)(1.188)(0.628) + (28.5)(6.4)(1.148)(0.628)$$
$$+ \tfrac{1}{2}(19)(5.93)(3.188)(1)(0)$$
$$= 442.57 + 131.50 + 0 = 574.07 \text{ kN/m}^2$$

$$FS_{\text{(bearing capacity)}} = \frac{q_u}{q_{toe}} = \frac{574.07}{189.2} = 3.03 > 3\text{—O.K.}$$

Example 5.13

A concrete gravity retaining wall is shown in Figure 5.34. Determine the following:

   a. The factor of safety against overturning
   b. The factor of safety against sliding
   c. The pressure on the soil at the toe and heel

(*Note:* Unit weight of concrete $= \gamma_c = 150 \ \text{lb/ft}^3$.)

Solution

$$H' = 15 + 2.5 = 17.5 \ \text{ft}$$

$$K_a = \tan^2\left(45 - \frac{\phi_1}{2}\right) = \tan^2\left(45 - \frac{30}{2}\right) = \frac{1}{3}$$

$$P_a = \frac{1}{2}\gamma(H')^2 K_a = \frac{1}{2}(121)(17.5)^2\left(\frac{1}{3}\right) = 6176 \ \text{lb/ft}$$

$$= 6.176 \ \text{kips/ft}$$

Since $\alpha = 0$

$$P_h = P_a = 6.176 \ \text{kips/ft}$$
$$P_v = 0$$

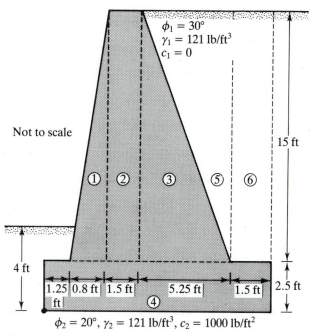

$\phi_1 = 30°$
$\gamma_1 = 121 \ \text{lb/ft}^3$
$c_1 = 0$

Not to scale

15 ft

4 ft

1.25 ft | 0.8 ft | 1.5 ft | 5.25 ft | 1.5 ft | 2.5 ft

$\phi_2 = 20°$, $\gamma_2 = 121 \ \text{lb/ft}^3$, $c_2 = 1000 \ \text{lb/ft}^2$

Figure 5.34

Part a: Factor of Safety Against Overturning

The following table can now be prepared to obtain $\sum M_R$.

Area (from Figure 5.34)	Weight (kips)	Moment arm from C (ft)	Moment about C (kip-ft)
1	$\frac{1}{2}(0.8)(15)(\gamma_c)$ $= 0.9$	$1.25 + \frac{2}{3}(0.8) = 1.783$	1.605
2	$(1.5)(15)(\gamma_c)$ $= 3.375$	$1.25 + 0.8 + 0.75 = 2.8$	9.45
3	$\frac{1}{2}(5.25)(15)(\gamma_c)$ $= 5.906$	$1.25 + 0.8 + 1.5 + 5.25/3 = 5.3$	31.30
4	$(10.3)(2.5)(\gamma_c)$ $= 3.863$	$10.3/2 = 5.15$	19.89
5	$\frac{1}{2}(5.25)(15)(0.121)$ $= 4.764$	$1.25 + 0.8 + 1.5 + \frac{2}{3}(5.25) = 7.05$	33.59
6	$(1.5)(15)(0.121)$ $= 2.723$ $\overline{21.531}$	$1.25 + 0.8 + 1.5 + 5.25 + 0.75 = 9.55$	26.0 $\overline{121.84}$ $= M_R$

The overturning moment

$$M_O = \frac{H'}{3} P_a = \left(\frac{17.5}{3}\right)(6.176) = 36.03 \text{ kip-ft}$$

$$FS_{(\text{overturning})} = \frac{121.84}{36.03} = \textbf{3.38}$$

Part b: Factor of Safety Against Sliding

From Eq. (5.65) with $k_1 = k_2 = \frac{2}{3}$, and assuming that $P_p = 0$

$$FS_{(\text{sliding})} = \frac{\sum V \tan\left(\frac{2}{3}\right)\phi_2 + B\left(\frac{2}{3}\right)c_2}{P_a}$$

$$= \frac{21.531 \tan\left(\frac{2 \times 20}{3}\right) + 10.3\left(\frac{2}{3}\right)(1.0)}{6.176}$$

$$= \frac{5.1 + 6.87}{6.176} = \textbf{1.94}$$

Part c: Pressure on the Soil at the Toe and Heel

From Eqs. (5.70), (5.71), and (5.72)

$$e = \frac{B}{2} - \frac{\sum M_R - \sum M_O}{\sum V} = \frac{10.3}{2} - \frac{121.84 - 36.03}{21.531} = 5.15 - 3.99 = 1.16 \text{ ft}$$

$$q_{toe} = \frac{\sum V}{B}\left[1 + \frac{6e}{B}\right] = \frac{21.531}{10.3}\left[1 + \frac{(6)(1.16)}{10.3}\right] = 3.5 \text{ kips/ft}^2$$

$$q_{heel} = \frac{\sum V}{B}\left[1 - \frac{6e}{B}\right] = \frac{21.531}{10.3}\left[1 - \frac{(6)(1.16)}{10.3}\right] = 0.678 \text{ kip/ft}^2$$

Example 5.14

Repeat Example 5.13 using Coulomb's active pressure for calculation and $\delta = 2\phi/3$.

Solution

Refer to Figure 5.35 for pressure calculation.

$$\delta = \frac{2}{3}\phi = \left(\frac{2}{3}\right)(30) = 20°$$

From Table 5.3, $K_a = 0.4794$ ($\alpha = 0°, \beta = 70°$)

$P_a = \frac{1}{2}(0.121)(17.5)^2(0.4794) = 8.882 \text{ kips/ft}$
$P_h = P_a \cos 40 = (8.882)(\cos 40) = 6.8 \text{ kips/ft}$
$P_v = P_a \sin 40 = 5.71 \text{ kips/ft}$

Not to scale

Figure 5.35

Part a: Factor of Safety Against Overturning

Refer to Figures 5.36 and 5.34.

Figure 5.36

Area (from Figures 5.34 and 5.36)	Weight (kips)	Moment arm from C (ft)	Moment about c (kip-ft)
1	0.9[a]	1.783[a]	1.605
2	3.375[a]	2.8[a]	9.46
3	5.906[a]	5.3[a]	31.30
4	3.863[a]	5.15[a]	19.89
$P_v = 5.71$		$1.25 + 0.8 + 1.5 +$ $5.25 - 1.21 = 7.59$	43.34
	19.75		105.6

[a] Same as in Example 5.13

Overturning moment:

$$M_O = P_h \frac{H'}{3} = (6.8)\left(\frac{17.5}{3}\right) = 39.67 \text{ kip-ft}$$

$$FS_{(overturning)} = \frac{105.6}{39.67} = \textbf{2.66}$$

Part b: Factor of Safety Against Sliding

$$FS_{(sliding)} = \frac{\Sigma V \tan\left(\frac{2}{3}\right)\phi_2 + B\left(\frac{2}{3}\right)c_2}{P_h}$$

$$= \frac{19.75 \tan\left(\frac{2}{3}\right)(20) + 10.3\left(\frac{2}{3}\right)(1.0)}{6.8} = \textbf{1.7}$$

Part c: Pressure on the Soil at the Toe and Heel

$$e = \frac{B}{2} - \frac{\sum M_R - \sum M_o}{\sum V} = \frac{10.3}{2} - \frac{(105.6 - 39.67)}{19.67} = 1.8 \text{ ft}$$

$$q_{\text{toe}} = \frac{19.75}{10.3} \left[1 + \frac{(6)(1.8)}{10.3} \right] = 3.93 \text{ kips/ft}^2$$

$$q_{\text{heel}} = \frac{19.75}{10.3} \left[1 - \frac{(6)(1.8)}{10.3} \right] = -0.093 \text{ kips/ft}^2 \approx 0$$

5.16 Gravity Retaining Wall Design for Earthquake Conditions

In Sections 5.11 and 5.12, the lateral earth pressures (passive and active) on retaining walls, taking earthquake forces into consideration, have been presented. Even in mild earthquakes, most retaining walls undergo limited lateral displacement. Richards and Elms (1979) have proposed a procedure for designing gravity retaining walls for earthquake conditions that allows limited lateral displacement. This procedure takes into consideration the wall inertia effect. Figure 5.37 shows a retaining wall with various forces acting on it. Note that W_w is equal to the weight of the retaining wall per unit length. All other notations in this figure are the same as in Section 5.11. Also note that the cohesion, c_2, of the soil on which the retaining wall is resting is assumed to be zero.

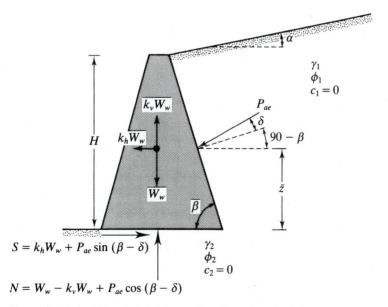

$$S = k_h W_w + P_{ae} \sin (\beta - \delta)$$

$$N = W_w - k_v W_w + P_{ae} \cos (\beta - \delta)$$

Figure 5.37 Stability of a retaining wall under earthquake forces

Considering the equilibrium of the wall, it can be shown as

$$\boxed{W_w = [\tfrac{1}{2}\gamma_1 H^2(1 - k_v)K_{ae}]C_{IE}} \tag{5.78}$$

where γ_1 = unit weight of the backfill;

$$\boxed{C_{IE} = \frac{\sin(\beta - \delta) - \cos(\beta - \delta)\tan\phi_2}{(1 - k_v)(\tan\phi_2 - \tan\theta')}} \tag{5.79}$$

where $\theta' = \tan^{-1}\left(\dfrac{k_h}{1 - k_v}\right)$

For detailed derivation of Eq. (5.79), readers may refer to Das (1983).

Based on Eqs. (5.78) and (5.79), the following procedure may be adopted to determine the weight of the retaining wall, W_w, for tolerable displacement that may take place during an earthquake.

1. Determine the tolerable displacement of the wall, Δ.
2. Obtain a design value of k_h by using the following equation:

$$\boxed{k_h = A_a\left(\frac{0.2A_v^2}{A_a\Delta}\right)^{0.25}} \tag{5.80}$$

In the preceding equation, A_a and A_v are effective acceleration coefficients and Δ is displacement in inches. The magnitudes of A_a and A_v are given by the Applied Technology Council (1978) for various regions of the United States.

3. Assume that $k_v = 0$, and, with the above value of k_h, calculate K_{ae} from Eq. (5.48).
4. Using the value of K_{ae} determined in Step 3, obtain the weight of the wall (W_w).
5. Apply a factor of safety to the value of W_w obtained in Step 4.

Example 5.15

Refer to Example 5.11. Determine the required weight of the wall for

a. Static condition
b. Zero displacement during an earthquake
c. Lateral displacement of 1.5 in. during an earthquake

Assume that $k_h = 0.25$, $A_a = 0.2$, and $A_v = 0.2$. Also assume that $\gamma_1 = \gamma_2 = 105$ lb/ft^3 and $\phi_1 = \phi_2 = 35° = \phi$ (see Figure 5.38). Use a factor of safety of 1.5.

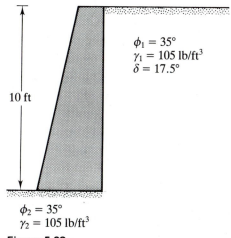

$\phi_1 = 35°$
$\gamma_1 = 105$ lb/ft^3
$\delta = 17.5°$

10 ft

$\phi_2 = 35°$
$\gamma_2 = 105$ lb/ft^3
Figure 5.38

Solution
Part a

For static condition, Eq. (5.79) becomes (since $\theta' = 0$)

$$C_{IE} = \frac{\sin(\beta - \delta) - \cos(\beta - \delta)\tan\phi}{\tan\phi}$$

From Example 5.11, $\beta = 90°$, $\delta = 17.5°$, and $\phi = 35°$. So

$$C_{IE} = \frac{\sin(90 - 17.5) - \cos(90 - 17.5)\tan 35}{\tan 35}$$

$$= \frac{0.954 - (0.3)(0.7)}{0.7} = 1.06$$

For static condition, $K_{ae} = K_a$. So

$$W_w = \tfrac{1}{2}\gamma H^2 K_a C_{IE}$$

$K_a \approx 0.24$ [Eq. (5.19)].

$$W_w = (\tfrac{1}{2})(105)(10)^2(0.24)(0.16) = 1169 \text{ lb/ft}$$

With a factor of safety of 1.5, the required

$$W_w = (1169)(1.5) = \textbf{1753.5 lb/ft}$$

Part b

For zero displacement condition, $k_v = 0$.

$$C_{IE} = \frac{\sin(\beta - \delta) - \cos(\beta - \delta) \tan \phi_2}{\tan \phi_2 - \tan \theta'}$$

$$\tan \theta' = \frac{k_h}{1 - k_v} = \frac{0.25}{1 - 0} = 0.25$$

$$C_{IE} = \frac{\sin(90 - 17.5) - \cos(90 - 17.5) \tan 35}{\tan 35 - 0.25}$$

$$= \frac{0.954 - 0.21}{0.7 - 0.25} = 1.653$$

$$W_w = \underbrace{[\tfrac{1}{2}\gamma_1 H^2 (1 - k_v) K_{ae}]}_{2478 \text{ lb/ft}} C_{IE}$$

$$= (2478)(1.653) = 4096 \text{ lb/ft}$$

With a factor of safety of 1.5, $W_w = \mathbf{6144 \ lb/ft}$

Part c

For a lateral displacement of 1.5 in.

$$k_h = A_a \left(\frac{0.2 A_v^2}{A_a \Delta}\right)^{0.25} = (0.2) \left[\frac{(0.2)(0.2)^2}{(0.2)(1.5)}\right]^{0.25} = 0.081$$

$$\tan \theta' = \frac{k_h}{1 - k_v} = \frac{0.081}{1 - 0} = 0.081$$

$$C_{IE} = \frac{\sin(90 - 17.5) - \cos(90 - 17.5) \tan 35}{\tan 35 - 0.081} = 1.2$$

$$W_w = \tfrac{1}{2}\gamma_1 H^2 K_{ae} \, C_{IE}$$
$$\uparrow$$
$$\approx 0.294 \text{ [Eq. (5.48)]}$$

$$W_w = (\tfrac{1}{2})(105)(10)^2(0.294)(1.2) = 1852 \text{ lb/ft}$$

With a factor of safety of 1.5, $W_w = \mathbf{2778 \ lb/ft.}$

5.17   Other Types of Possible Failure of Retaining Walls

In addition to the three types of possible failure mode for retaining walls discussed in Section 5.15, two other types of failure could occur: *shallow shear failure* and *deep shear failure*.

Shallow shear failure in soil below the base of a retaining wall takes place along a cylindrical surface *abc* passing through the heel, as shown in Figure 5.39a. The center of the arc of the circle *abc* is located at *O*, which is found by trial and error (corresponds to the minimum factor of safety). This type of failure can occur as the result of excessive induced shear stress along the cylindrical surface in soil. In general, the factor of safety against horizontal sliding is lower than that obtained by shallow shear failure. So, if

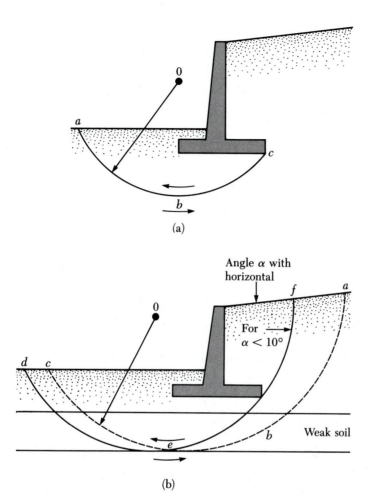

Figure 5.39   (a) Shallow shear failure; (b) deep shear failure

$FS_{(sliding)}$ is greater than about 1.5, shallow shear failure under the base may not occur.

Deep shear failure can occur along a cylindrical surface *abc*, as shown in Figure 5.39b, as the result of the existence of a weak layer of soil underneath the wall at a depth of about 1.5 times the height of the retaining wall. In such cases, the critical cylindrical failure surface *abc* has to be determined by trial and error with various centers, such as *O* (Figure 5.39b). The failure surface along which the minimum factor of safety is obtained is the *critical surface of sliding*. For the backfill slope with α less than about 10°, it has been observed that the critical failure circle passes through the edge of the heel slab (such as *def* in Figure 5.39b). In this situation, the minimum factor of safety also has to be determined by trial and error by changing the center of the trial circle.

An approximate procedure for determination of the factor of safety against a deep-seated shear failure for a gently sloping backfill ($\alpha < 10°$) is outlined as follows in a step-by-step manner (Teng, 1962) (refer to Figure 5.40):

1. Draw the retaining wall and the underlying soil layer to a convenient scale.

2. For a trial center O, draw an arc of a circle *abcd*. For all practical purposes, the weight of the soil in the area *abcde* may be taken to be symmetrical about a vertical line drawn through point O. Let the radius of the trial circle be equal to r.

3. To determine the driving force on the failure surface causing instability (Figure 5.40a), divide the area in the zone *efgh* into several slices. These slices can be treated as rectangles or triangles, as the case may be.

4. Determine the area of each of these slices, and then determine the weight W of the soil (and/or concrete) contained inside each of these slices (per unit length of the wall).

5. Draw a vertical line through the centroid of each slice, and locate the point of intersection of each vertical line with the trial failure circle.

6. Join point O (that is, the center of the trial circles) with the points of intersection as determined in Step 5.

7. Determine the angle, ω, that each vertical line makes with the radial line.

8. Calculate $W \sin \omega$ for each slice.

9. Determine the active force P_a on the face *df*, $\frac{1}{2}\gamma_1 H'^2 K_a$.

10. The total driving force can now be calculated as

$$\sum (W \sin \omega) + \frac{P_a \overline{X}}{r} \tag{5.81}$$

where $\overline{X}$ = perpendicular distance between the line of action of P_a and the center O

11. To determine the resisting force on the failure surface (Figure 5.40b), divide the area in the zones *abk* and *idefj* into several slices, and determine the weight of each slice, W_1 (per unit length of the wall). Note that the points *b* and *i* are on the top of the soft clay layer; the weight of each slice shown in Figure 5.40b is W_1 in contrast to the weight of each slice W, as shown in Figure 5.40a.

12. Draw a vertical line through the centroid of each slice, and locate the point of intersection of each line with the trial failure circle.

13. Join point O with the points of intersection as determined in Step 12. Determine the angles, ω_1, that the vertical lines make with the radial lines.

14. For each slice, obtain

$$W_1 \tan \phi_2 \cos \omega_1$$

(a)

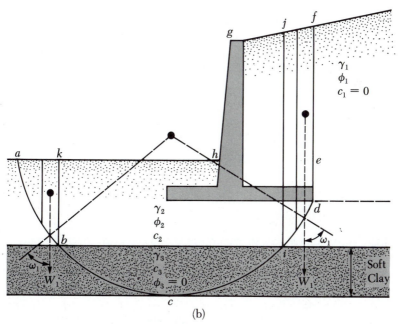

(b)

Figure 5.40 Deep shear failure analysis

15. Calculate

$$c_2 l_1 + c_3 l_2 + c_2 l_3$$

where l_1, l_2, and l_3 are the lengths of the arcs ab, bi, and id

16. The maximum resisting force that can be derived along the failure surface is

$$\sum (W_1 \tan \phi_2 \cos \omega_1) + c_2 l_1 + c_3 l_2 + c_2 l_3 \qquad (5.82)$$

17. To determine the factor of safety against deep shear failure for this trial failure surface:

$$FS_{(\text{deep shear failure})} = \frac{\sum (W_1 \tan \phi_2 \cos \omega_1) + c_2 l_1 + c_3 l_2 + c_2 l_3}{\sum (W \sin \omega) + \dfrac{P_a \overline{X}}{r}}$$

$$(5.83)$$

Several other trial failure surfaces may be drawn, and the factor of safety can be determined in a similar manner. The lowest value of the factor of safety obtained from all trial surfaces is the desired factor of safety.

5.18   Comments Relating to Stability

When a weak soil layer is located at a shallow depth—that is, within a depth of about 1.5 times the height of the retaining wall—the bearing capacity of the weak layer should be carefully investigated. One should also look into the possibility of excessive settlement. In some cases, the use of lightweight backfill material behind the retaining wall may solve the problem.

In many instances, piles are used to transmit the foundation load to a firmer layer. However, often the thrust of the sliding wedge of soil, in the case of deep shear failure, may bend the piles and eventually cause them to fail. Careful attention should be given to this possibility when considering the option of pile foundations for retaining walls. (Pile foundations may sometimes be required for bridge abutments to avoid the problem of scouring.)

5.19   Drainage from the Backfill
of the Retaining Wall

As the result of rainfall or other wet conditions, the backfill material for a retaining wall may become saturated. Saturation will, in effect, increase the pressure on the wall and create an unstable condition. For this reason, adequate drainage must be provided by means of *weepholes* and/or *perforated drainage pipes* (see Figure 5.41).

The *weepholes*, if provided, should have a minimum diameter of about 0.1 m and be adequately spaced. Note that there is always a possibility that the backfill material may be washed into weepholes or drainage pipes. This will ultimately clog up the drainage facilities. Thus a filter material needs to be placed behind the weepholes or around the drainage pipes, as the case

Figure 5.41 Drainage provisions for the backfill of a retaining wall

may be. Nowadays, geotextiles are used to serve the same purpose. Whenever granular soil is used as a filter, the principles outlined in Section 1.11 should be followed. Example 5.16 gives the procedure for designing a filter.

Example 5.16

Figure 5.42 shows the grain-size distribution of a backfill material. Using the conditions outlined in Section 1.11, determine the range of the grain-size distribution for the filter material.

Figure 5.42

Solution

From the grain-size distribution curve given in Figure 5.42, the following values can be determined:

$$D_{15(B)} = 0.04 \text{ mm}$$
$$D_{85(B)} = 0.25 \text{ mm}$$
$$D_{50(B)} = 0.13 \text{ mm}$$

Conditions of Filter

1. $D_{15(F)}$ should be less than $5D_{85(F)}$—that is, $5 \times 0.25 = 1.25$ mm.
2. $D_{15(F)}$ should be greater than $4D_{15(B)}$—that is, $4 \times 0.04 = 0.16$ mm.
3. $D_{50(F)}$ should be less than $25D_{50(B)}$—that is, $25 \times 0.13 = 3.25$ mm.
4. $D_{15(F)}$ should be less than $20D_{15(B)}$—that is, $20 \times 0.04 = 0.8$ mm.

These limiting points are plotted in Figure 5.42. Through these points, two curves can be drawn that are similar in nature to the grain-size distribution curve of the backfill material. These curves define the range for the filter material to be used.

5.20   Provision of Joints in the Construction of Walls

A retaining wall may be constructed with all or any of the following joints:

1. *Construction Joints (Figure 5.43a):* These are vertical and horizontal joints that are used between two successive pours of concrete. In order to

Figure 5.43   (a) Construction joints; (b) contraction joint; (c) expansion joint

increase the shear at the joints, keys may be used. If keys are not used, the surface of the first pour is cleaned and roughened before the next placement of concrete.

2. *Contraction Joints (Figure 5.43b):* These are vertical joints (grooves) placed in the face of the wall (from the top of the base slab to the top of the wall) that allow the concrete to shrink without noticeable harm. The grooves may be about 6–8 mm (≈ 0.25 to 0.3 in.) wide and 12–16 mm (≈ 0.5 to 0.6 in.) deep.

3. *Expansion Joints (Figure 5.43c):* To take into account the expansion of concrete caused by temperature changes, vertical expansion joints going from the base to the top of the wall may also be used. These joints may be filled with flexible joint fillers. In most cases, horizontal reinforcing steel bars running across the stem are continuous through all joints. The steel is greased to allow the concrete to expand.

PROBLEMS

5.1 Refer to Figure 5.2a. Given: $H = 12$ ft, $q = 0$, $\gamma = 108$ lb/ft^3, $c = 0$, and $\phi = 30°$. Determine the at-rest lateral earth force per foot length of the wall. Also find the location of the resultant. Use Eq. (5.3).

5.2 Repeat Problem 5.1 with the following: $H = 3.5$ m, $q = 20$ kN/m^2, $\gamma = 18.2$ kN/m^3, $c = 0$, and $\phi = 35°$.

5.3 Refer to Figure 5.2a. Given: $H = 15$ ft, $q = 400$ lb/ft^2, $\gamma = 110$ lb/ft^3, $c = 0$, $\phi = 38°$, and $\gamma_{min} = 96$ lb/ft^3. Determine the at-rest lateral earth force per foot length of the wall. Use Eq. (5.11).

5.4 Refer to Figure 5.3a. Given: $H = 4$ m, $H_1 = 1.5$ m, $q = 0$, $\gamma = 16.8$ kN/m^3, $\gamma_{sat} = 19.6$ kN/m^3, $\phi = 32°$, $c = 0$. Determine the at-rest lateral earth force per foot length of the wall. Also find the location of the resultant. Use Eq. (5.3).

5.5 Refer to Figure 5.3a. Given: $H_1 = 4.5$ m, $H_2 = 0$, $q = 0$, and $\gamma = 17$ kN/m^3. The backfill is an overconsolidated clay with a plasticity index of 23. If the overconsolidation ratio is 2.2, determine the at-rest lateral earth force per meter of the wall. Also find the location of the resultant. Use Eqs. (5.5) and (5.7).

5.6 Redo Problem 5.5 assuming that the surcharge, $q = 30$ kN/m^2.

5.7 Refer to Figure 5.5a. Given: the height of the retaining wall, $H = 18$ ft; the backfill is a saturated clay with $\phi = 0°$, $c = 500$ lb/ft^2, $\gamma_{sat} = 120$ lb/ft^3.
   a. Determine the Rankine active pressure distribution diagram behind the wall.
   b. Determine the depth of the tensile crack, z_c.
   c. Estimate the Rankine active force per foot of the wall before and after the occurrence of tensile crack.

5.8 Refer to Problem 5.7.
   a. Draw the Rankine passive pressure distribution diagram behind the wall.
   b. Estimate the Rankine passive force per foot of the wall and also the location of the resultant.

5.9 A vertical retaining wall (Figure 5.5a) is 6.3 m high with a horizontal backfill. For the backfill, assume that $\gamma = 17.9$ kN/m^3, $\phi = 26°$, and $c = 15$ kN/m^2. Determine the

Rankine active force per unit length of the wall after the occurrence of the tensile crack.

5.10 Repeat Problem 5.9 for the Rankine passive case.

5.11 Refer to Figure P5.11. Given: $H_1 = 8$ ft, $H_2 = 16$ ft, $\gamma_1 = 110$ lb/ft³, $\phi_1 = 38°$, $c_1 = 0$, $\gamma_2 = 140$ lb/ft³, $\phi_2 = 25°$, and $c_2 = 209$ lb/ft². Determine the Rankine active force per unit length of the wall.

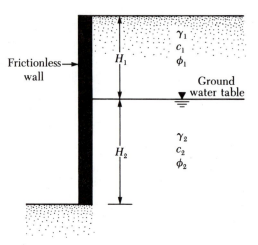

Figure P5.11

5.12 Repeat Problem 5.11 with the following changes: $\phi_1 = 28°$, $c_1 = 418$ lb/ft². Determine P_a after the occurrence of the tensile crack.

5.13 Repeat Problem 5.11 for the Rankine passive case.

5.14 Repeat Problem 5.12 for the Rankine passive case.

5.15 Refer to Figure 5.9a. Given: $H = 12$ ft, $\gamma = 105$ lb/ft³, $\phi = 30°$, $\delta = 20°$, $c = 0$, $\alpha = 10°$, and $\beta = 85°$. Determine the Coulomb's active force per foot of the wall and the location and direction of the resultant.

5.16 For the retaining wall described in Problem 5.15, determine the Coulomb's passive force per foot of the wall and the location and direction of the resultant.

5.17 Refer to Figure 5.10. Given: $H = 5$ m, $\gamma = 17.5$ kN/m³, $q = 20$ kN/m², $\phi = 36°$, $\delta = 24°$, $c = 0$, $\alpha = 15°$, $\beta = 85°$. Determine Coulomb's active force per meter of the wall.

5.18 A vertical retaining wall has a horizontal granular backfill. Given:
   Wall height, $H = 10$ ft
       Backfill: $\gamma = 112$ lb/ft³
           $\phi = 40°$
           $\delta = 30°$
   Determine the passive force, P_p, per foot length of the wall. Use Eq. (5.28) and Table 5.7.

5.19 Refer to Figure 5.16. For the retaining wall, $H = 8.2$ m, $\phi = 35°$, $\alpha = 10°$, $\gamma = 18.2$ kN/m³, and $c = 0$.

a. Determine the intensity of the Rankine active force at $z = 2, 4, 6$, and 7.5 m.
b. Determine the Rankine active force per meter of the wall and also the location and direction of the resultant.

5.20 Redo Problem 5.15 assuming that earthquake forces have to be taken into consideration, that is, $k_h = 0.25$ and $k_v = 0$. Determine the location of the resultant, P_{ae}.

5.21 Redo Problem 5.16 assuming that earthquake forces have to be taken into consideration, that is, $k_h = 0.3$ and $k_v = 0$.

5.22 Refer to Figure 5.17. Given: $H = 12$ ft, $a' = 3$ ft, $b' = 4.5$ ft, and $q = 525$ lb/ft^2.
a. Determine the lateral force per unit length of the wall caused by surcharge loading only.
b. Determine the location of the center of pressure, $\bar{z}$.

5.23 For the cantilever retaining wall shown in Figure P5.23, given:
Wall dimensions: $H = 8$ m, $x_1 = 0.4$ m, $x_2 = 0.6$ m, $x_3 = 1.5$ m, $x_4 = 3.5$ m, $x_5 = 0.96$ m, $D = 1.75$ m, $\alpha = 10°$
Soil properties: $\gamma_1 = 16.8$ kN/m^3, $\phi_1 = 32°$, $\gamma_2 = 17.6$ kN/m^3, $\phi_2 = 28°$, $c_2 = 30$ kN/m^2

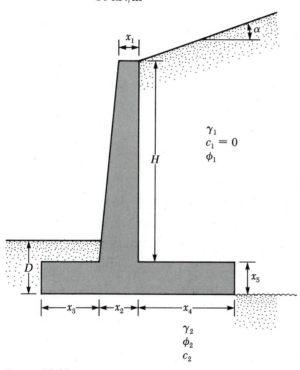

Figure P5.23

Calculate the factor of safety with respect to overturning and sliding and bearing capacity.

5.24 Repeat Problem 5.23 with the following:
Wall dimensions: $H = 18$ ft, $x_1 = 18$ in., $x_2 = 30$ in., $x_3 = 4$ ft, $x_4 = 6$ ft, $x_5 = 2.75$ ft, $D = 4$ ft, $\alpha = 10°$
Soil properties: $\gamma_1 = 117$ lb/ft^3, $\phi_1 = 34°$, $\gamma_2 = 107$ lb/ft^3, $\phi_2 = 18°$, $c_2 = 900$ lb/ft^2

5.25 Repeat Problem 5.23 with the following:

Wall dimensions: $H = 6.5$ m, $x_1 = 0.3$ m, $x_2 = 0.6$ m, $x_3 = 0.8$ m, $x_4 = 2$ m, $x_5 = 0.8$ m, $D = 1.5$ m, $\alpha = 0°$

Soil properties: $\gamma_1 = 18.08$ kN/m³, $\phi_1 = 36°$, $\gamma_2 = 19.65$ kN/m³, $\phi_2 = 15°$, $c_2 = 30$ kN/m²

5.26 Refer to Problem 5.25. What will be the factor of safety against sliding if the heel of the wall is sloped with $\alpha' = 45°$? (See Figure 5.30 for the definition of α'.)

5.27 A gravity retaining wall is shown in Figure P5.27. Calculate the factor of safety with respect to overturning and sliding. Given:

Wall dimensions: $H = 6$ m, $x_1 = 0.6$ m, $x_2 = 0.2$ m, $x_3 = 2$ m, $x_4 = 0.5$ m, $x_5 = 0.75$ m, $x_6 = 0.8$ m, $D = 1.5$ m

Soil properties: $\gamma_1 = 16.5$ kN/m³, $\phi_1 = 32°$, $\gamma_2 = 18$ kN/m³, $\phi_2 = 22°$, $c_2 = 40$ kN/m²

Use Rankine active pressure for calculation (see Figure 5.25).

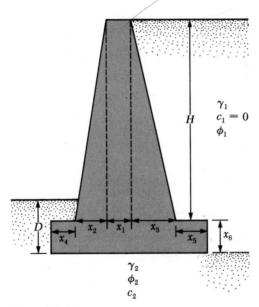

Figure P5.27

5.28 Repeat Problem 5.27 using Coulomb's active pressure for calculation and $\delta = \tfrac{2}{3}\phi_1$.

REFERENCES

Applied Technology Council (1978). "Tentative Provisions for the Development of Seismic Regulations for Buildings," *Publication ATC-3-06.*

Brooker, E. W., and Ireland, H. O. (1965). "Earth Pressure at Rest Related to Stress History," *Canadian Geotechnical Journal,* Vol. 2, No. 1, pp. 1–15.

Caquot, A., and Kerisel, J. (1948). *Tables for Calculation of Passive Pressure, Active Pressure, and Bearing Capacity of Foundations,* Gauthier-Villars, Paris, France.

Coulomb, C. A. (1776). *Essai sur une Application des Regles des Maximis et Minimum à quelques Problemes de Statique Relatifs à l'Architecture*, Mem. Acad. Roy. des Sciences, Paris, Vol. 3, p. 38.

Das, B. M. (1983). *Fundamentals of Soil Dynamics*, Elsevier, New York.

Das, B. M. (1987). *Theoretical Foundation Engineering*, Elsevier, Amsterdam.

Davies, T. G., Richards, R., and Chen, K. H. (1986). "Passive Pressure During Seismic Loading," *Journal of Geotechnical Engineering*, American Society of Civil Engineers, Vol. 112, No. GT4, pp. 479–483.

Elman, M. T., and Terry, C. F. (1988). "Retaining Walls with Sloped Heel," *Journal of Geotechnical Engineering*, American Society of Civil Engineers, Vol. 114, No. GT10, pp. 1194–1199.

James, R. G., and Bransby, P. L. (1971). "A Velocity Field for Some Passive Earth Pressure Problems," *Geotechnique*, London, England, Vol. 21, No. 1, pp. 61–84.

Jarquio, R. (1981). "Total Lateral Surcharge Pressure Due to Strip Load," *Journal of the Geotechnical Engineering Division*, American Society of Civil Engineers, Vol. 107, No. GT10, pp. 1424–1428.

Kapila, J. P. (1962). "Earthquake Resistant Design of Retaining Walls," *Proceedings*, Second Earthquake Symposium, Roorkee, India, pp. 97–108.

Packshaw, S. (1969). "Earth Pressure and Earth Resistance," *A Century of Soil Mechanics*, The Institution of Engineers, London, England, pp. 409–435.

Richards, R., and Elms, D. G. (1979). "Seismic Behavior of Gravity Retaining Walls," *Journal of the Geotechnical Engineering Division*, American Society of Civil Engineers, Vol. 105, No. GT4, pp. 449–464.

Sherif, M. A., Fang, Y. S., and Sherif, R. I. (1984). "k_a and k_0 Behind Rotating and Non-Yielding Walls." *Journal of Geotechnical Engineering*, American Society of Civil Engineers, Vol. 110, No. GT1, pp. 41–56.

Shields, D. H., and Tolunay, A. Z. (1972). "Passive Pressure Coefficients for Sand by the Terzaghi and Peck Method," *Canadian Geotechnical Journal*, Vol. 9, No. 4.

Shields, D. H., and Tolunay, A. Z. (1973). "Passive Pressure by Method of Slices," *Journal of the Soil Mechanics and Foundations Division*, American Society of Civil Engineers, Vol. 99, No. SM12, pp. 1043–1053.

Teng, W. C. (1962). *Foundation Design*, Prentice-Hall, Englewood Cliffs, N.J.

Terzaghi, K., and Peck, R. B. (1967). *Soil Mechanics in Engineering Practice*, Wiley, New York.

Chapter 6

SHEET PILE WALLS

6.1   Introduction

Connected or semiconnected sheet piles are often used to build continuous walls for waterfront structures that may range from small waterfront pleasure-boat launching facilities to large dock facilities (Figure 6.1a). In contrast to the construction of other types of retaining wall, the building of sheet pile walls does not usually require dewatering of the site. Sheet piles are also used for some temporary structures, such as braced cuts (Figure 6.1b). The principles of sheet pile design used in braced cuts are given in Chapter 7.

There are several types of sheet pile that are in common use: (a) *wooden sheet piles*, (b) *precast concrete sheet piles*, and (c) *steel sheet piles*. Aluminum sheet piles are also marketed.

Wooden sheet piles are used only for temporary light structures located above the water table. The most common types are ordinary wooden planks and *Wakefield piles*. The wooden planks are about 50 mm × 300 mm in cross section and are driven edge to edge (Figure 6.2a). The Wakefield piles are made by nailing three planks together with the middle plank offset by 50–75 mm (Figure 6.2b). Wooden planks can also be milled to form *tongue-and-groove piles*. This is shown in Figure 6.2c. Figure 6.2d shows another type of wooden sheet pile that has grooves that are cut in the mills. Metal *splines* are driven into the grooves of the adjacent sheetings to hold them together after they are driven into the ground.

Precast concrete sheet piles are heavy and are designed with reinforcements to withstand the permanent stresses to which the structure will be subjected after construction and also to handle the stresses produced during construction. In cross section, these piles are about 500–800 mm wide and 150–250 mm thick. Figure 6.2e shows schematic diagrams of the elevation and the cross section of a reinforced concrete sheet pile.

The *steel sheet piles* in the United States are about 10–13 mm thick. European sections may be thinner and wider. Sheet pile sections may be Z,

326

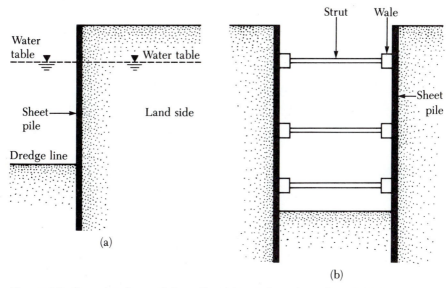

Figure 6.1 Examples of uses of sheet piles: (a) waterfront sheet pile wall; (b) braced cut

Figure 6.2 Various types of wooden and concrete sheet pile

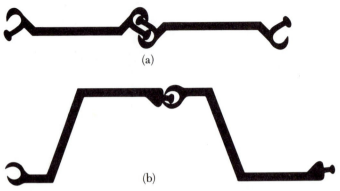

(a)

(b)

Figure 6.3   Nature of sheet pile connections: (a) thumb-and-finger type; (b) ball-and-socket type

deep arch, low arch, or *straight web* sections. The interlocks of the sheet pile sections are shaped like a *thumb-and-finger* or a *ball-and-socket* type for watertight connections. Figure 6.3a shows schematic diagrams of the thumb-and-finger type of interlocking for straight web sections. The ball-and-socket type of interlocking for Z section piles is shown in Figure 6.3b. Table 6.1 shows the properties of the sheet pile sections produced by the U.S. Steel Corporation. The allowable design flexural stress for the steel sheet piles is as follows:

Type of steel	Allowable stress (MN/m^2)
ASTM A-328	170 ($\approx 25{,}000$ lb/in.^{2})
ASTM A-572	210 ($\approx 30{,}000$ lb/in.^{2})
ASTM A-690	210 ($\approx 30{,}000$ lb/in.^{2})

Steel sheet piles are convenient to use because of their resistance to high-driving stress developed when being driven into hard soils. They are also lightweight and reusable.

This chapter primarily discusses the design principles of waterfront retaining structures built with sheet piles.

6.2 Construction Methods for Sheet Pile Walls—General

Sheet pile walls may be divided into two basic categories: (a) cantilever sheet pile walls and (b) anchored sheet pile walls. The principles of anchoring sheet pile walls are discussed in more detail in Section 6.12.

In the construction of sheet pile walls, sheet piles may be driven into the ground and then the backfill placed on the land side, or the sheet pile may first be driven into the ground and the soil in front of the sheet pile dredged. In any case, the soil used for backfill behind the sheet pile wall is usually granular. The soil below the dredge line may be sandy or clayey soil.

Table 6.1 Properties of Sheet Pile Sections (Produced by United States Steel Corporation)

Section designation	Sketch of section	Driving distance (mm)	Section modulus		Moment of inertia		Comments
			m^3/m of wall $\times 10^5$	in.^{3}/ft of wall	m^4/m of wall $\times 10^6$	in.^{4}/ft of wall	
PZ-38	12.7 mm, 9.53 mm, 12.7 mm, 304.8 mm, Driving distance	457.2	251.32	46.8	383.29	280.8	Interlock with each other and also with PSA-23 or PSA-28
PZ-32	12.7 mm, 9.53 mm, 12.7 mm, 292.1 mm	533.4	205.67	38.3	300.85	220.4	Interlock with each other and also with PSA-23 or PSA-28
PZ-27	12.7 mm, 9.53 mm, 12.7 mm, 304.8 mm	457.2	162.17	30.2	251.43	184.2	Interlock with each other and also with PSA-23 or PSA-28
PDA-27	9.53 mm, 12.7 mm	406.4	57.46	10.7	54.33	39.8	Interlock with each other

Table 6.1 (Continued)

Section designation	Sketch of section	Driving distance (mm)	Section modulus		Moment of inertia		Comments
			m³/m of wall × 10⁵	in.³/ft of wall	m⁴/m of wall × 10⁶	in.⁴/ft of wall	
PMA-22	(82.55 mm, 9.53 mm)	498.48	29.00	5.4	18.70	13.7	Interlock with each other
PSA-28	(34.13 mm, 12.7 mm)	406.4	13.43	2.5	6.14	4.5	Interlock with each other
PSA-23	(34.13 mm, 9.53 mm)	406.4	12.89	2.4	5.6	4.1	Interlock with each other
PSX-32	(11.51 mm)	419.1	12.89	2.4	5.05	3.7	Interlock with each other
PS-32	(12.7 mm)	381	10.20	1.9	3.96	2.9	Interlock with each other
PS-28	(9.53 mm)	381	10.20	1.9	3.82	2.8	Interlock with each other

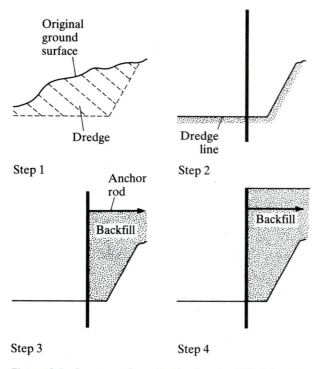

Figure 6.4   Sequence of construction for a backfilled structure

The surface of soil on the water side is referred to as the *mud line* or *dredge line*. Based on this, the construction method can be generally divided into two categories (Tsinker, 1983):

1. *Backfilled structure*
2. *Dredged structure*

The sequence of construction for a backfilled structure is as follows (Figure 6.4):

Step 1. Dredge the *in situ* soil in front and back of the proposed structure.

Step 2. Drive the sheet piles.

Step 3. Backfill up to the level of the anchor and place the anchor system.

Step 4. Backfill up to the top of the wall.

For a cantilever type of wall, only Steps 1, 2, and 4 apply.

The sequence of construction for a dredged structure is as follows (Figure 6.5):

Step 1. Drive the sheet piles.

Step 2. Backfill up to the anchor level and place the anchor system.

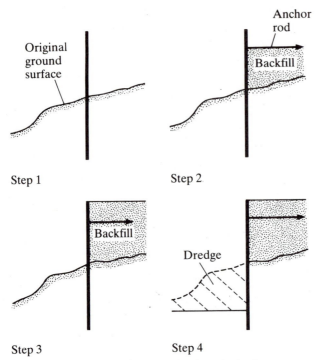

Figure 6.5   Sequence of construction for a dredged structure

Step 3. Backfill up to the top of the wall.
Step 4. Dredge the front side of the wall.

For cantilever sheet pile walls, Step 2 is not required.

CANTILEVER SHEET PILE WALL

Cantilever sheet pile walls are usually recommended for walls of moderate height—about 6 m ($\approx$20 ft) or less measured above the dredge line. In these walls, the sheet piles act like a wide cantilever beam above the dredge line. The basic principles for the estimation of net lateral pressure distribution on a cantilever sheet pile wall can be explained with the aid of Figure 6.6, which shows the nature of lateral yielding of a cantilever wall penetrating a sand layer below the dredge line. The wall rotates about a point O. Because the hydrostatic pressures at any depth from both sides of the wall will cancel each other, we will consider only the effective lateral soil pressures. In Zone A, the lateral pressure is only the active pressure from the land side. In Zone B, because of the nature of yielding of the wall, there will be active pressure from the land side and passive pressure from the water side. The condition is reversed in Zone C—that is, below the point of rotation, O. The net actual pressure distribution on the wall is like that shown in Figure 6.6b. However, for design purposes, Figure 6.6c shows a simplified version.

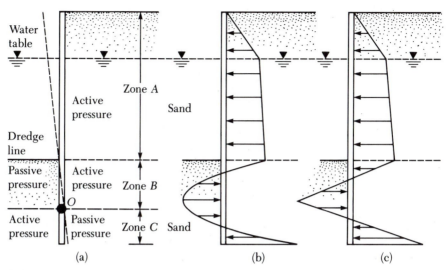

Figure 6.6   Cantilever sheet pile penetrating sand

The following sections (Sections 6.3 through 6.6) present the mathematical formulation of the analysis of cantilever sheet pile walls. Note that, in some waterfront structures, the water level may fluctuate as the result of tidal effects. Care should be taken in determining the water level that will affect the net pressure diagram.

6.3   Cantilever Sheet Piling Penetrating Sandy Soils

To develop the relationships for the proper depth of embedment of sheet piles driven into a granular soil, we refer to Figure 6.7a. The soil retained by the sheet piling above the dredge line is also sand. The water table is located at a depth of L_1 below the top of the wall. Let the angle of friction of the sand be ϕ. The intensity of the active pressure at a depth $z = L_1$ can be given as

$$p_1 = \gamma L_1 K_a \tag{6.1}$$

where K_a = Rankine active pressure coefficient = $\tan^2 (45 - \phi/2)$
$\quad\quad\;\gamma$ = unit weight of soil above the water table

Similarly, the active pressure at a depth of $z = L_1 + L_2$ (that is, at the level of the dredge line) is equal to

$$p_2 = (\gamma L_1 + \gamma' L_2) K_a \tag{6.2}$$

where γ' = effective unit weight of soil = $\gamma_{sat} - \gamma_w$

Note that, at the level of the dredge line, the hydrostatic pressures from both sides of the wall are of the same magnitude and cancel each other.

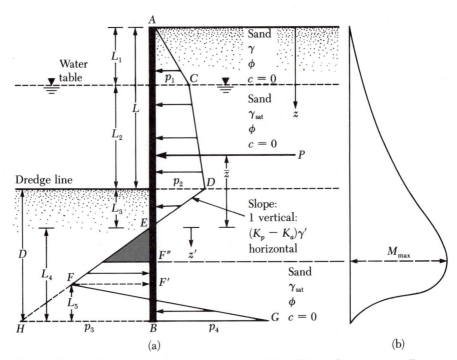

Figure 6.7 Cantilever sheet pile penetrating sand: (a) variation of net pressure diagram; (b) variation of moment

In order to determine the net lateral pressure below the dredge line up to the point of rotation O, as shown in Figure 6.6a, one has to consider the passive pressure acting from the left side (water side) toward the right side (land side) and also the active pressure acting from the right side toward the left side of the wall. For such cases, ignoring the hydrostatic pressure from both sides of the wall, the active pressure at a depth z can be given as

$$p_a = [\gamma L_1 + \gamma' L_2 + \gamma'(z - L_1 - L_2)]K_a \qquad (6.3)$$

Also, the passive pressure at that depth z is equal to

$$p_p = \gamma'(z - L_1 - L_2)K_p \qquad (6.4)$$

where K_p = Rankine passive pressure coefficient = $\tan^2(45 + \phi/2)$

Hence, combining Eqs. (6.3) and (6.4), the net lateral pressure can be obtained as

$$p = p_a - p_p = (\gamma L_1 + \gamma' L_2)K_a - \gamma'(z - L_1 - L_2)(K_p - K_a)$$
$$= p_2 - \gamma'(z - L)(K_p - K_a) \qquad (6.5)$$

where $L = L_1 + L_2$

The net pressure, p, becomes equal to zero at a depth L_3 below the dredge line; or

$$p_2 - \gamma'(z - L)(K_p - K_a) = 0$$

or

$$(z - L) = L_3 = \frac{p_2}{\gamma'(K_p - K_a)} \tag{6.6}$$

From the preceding equation, it is apparent that the slope of the net pressure distribution line DEF is 1 vertical to $(K_p - K_a)\gamma'$ horizontal. So, in the pressure diagram

$$\overline{HB} = p_3 = L_4(K_p - K_a)\gamma' \tag{6.7}$$

At the bottom of the sheet pile, passive pressure (p_p) acts from the right toward the left side, and active pressure acts from the left toward the right side of the sheet pile. So, at $z = L + D$

$$p_p = (\gamma L_1 + \gamma' L_2 + \gamma' D)K_p \tag{6.8}$$

At the same depth

$$p_a = \gamma' D K_a \tag{6.9}$$

Hence, the net lateral pressure at the bottom of the sheet pile is equal to

$$
\begin{aligned}
p_p - p_a = p_4 &= (\gamma L_1 + \gamma' L_2)K_p + \gamma' D(K_p - K_a) \\
&= (\gamma L_1 + \gamma' L_2)K_p + \gamma' L_3(K_p - K_a) + \gamma' L_4(K_p - K_a) \\
&= p_5 + \gamma' L_4(K_p - K_a)
\end{aligned} \tag{6.10}
$$

$$\text{where } p_5 = (\gamma L_1 + \gamma' L_2)K_p + \gamma' L_3(K_p - K_a) \tag{6.11}$$

$$D = L_3 + L_4 \tag{6.12}$$

For the stability of the wall, the principles of statics can now be applied; or

$$\sum \text{horizontal forces per unit length of wall} = 0$$

and

$$\sum \text{moment of the forces per unit length of wall about point } B = 0$$

For summation of the horizontal forces,

area of the pressure diagram $ACDE$ – area of $EFHB$

+ area of $FHBG = 0$

or

$$p - \tfrac{1}{2}p_3 L_4 + \tfrac{1}{2}L_5(p_3 + p_4) = 0 \tag{6.13}$$

where P = area of the pressure diagram $ACDE$

Summing the moment of all the forces about point B

$$P(L_4 + \bar{z}) - \left(\frac{1}{2}L_4 p_3\right)\left(\frac{L_4}{3}\right) + \frac{1}{2}L_5(p_3 + p_4)\left(\frac{L_5}{3}\right) = 0 \tag{6.14}$$

From Eq. (6.13)

$$L_5 = \frac{p_3 L_4 - 2P}{p_3 + p_4} \tag{6.15}$$

Combining Eqs. (6.7), (6.10), (6.14), and (6.15) and simplifying them further, one obtains the following fourth-degree equation in terms of L_4.

$$\boxed{L_4^4 + A_1 L_4^3 - A_2 L_4^2 - A_3 L_4 - A_4 = 0} \tag{6.16}$$

where

$$A_1 = \frac{p_5}{\gamma'(K_p - K_a)} \tag{6.17}$$

$$A_2 = \frac{8P}{\gamma'(K_p - K_a)} \tag{6.18}$$

$$A_3 = \frac{6P[2\bar{z}\gamma'(K_p - K_a) + p_5]}{\gamma'^2(K_p - K_a)^2} \tag{6.19}$$

$$A_4 = \frac{P(6\bar{z}p_5 + 4P)}{\gamma'^2(K_p - K_a)^2} \tag{6.20}$$

Step-by-Step Procedure for Obtaining the Pressure Diagram

Based on the preceding theory, the step-by-step procedure for obtaining the pressure diagram for a cantilever sheet pile wall penetrating a granular soil is as follows:

1. Calculate K_a and K_p.
2. Calculate p_1 [Eq. (6.1)] and p_2 [Eq. (6.2)]. *Note:* L_1 and L_2 will be given.
3. Calculate L_3 [Eq. (6.6)].
4. Calculate P.
5. Calculate $\bar{z}$ (that is, the center of pressure for the area $ACDE$) by taking the moment about E.
6. Calculate p_5 [Eq. (6.11)].
7. Calculate A_1, A_2, A_3, and A_4 [Eqs. (6.17) to (6.20)].
8. Solve Eq. (6.16) by trial and error to determine L_4.
9. Calculate p_4 [Eq. (6.10)].

10. Calculate p_3 [Eq. (6.7)].

11. Obtain L_5 from Eq. (6.15).

12. Now the pressure distribution diagram as shown in Figure 6.7a can easily be drawn.

13. Obtain the theoretical depth [Eq. (6.12)] of penetration as $L_3 + L_4$. The actual depth of penetration is increased by about 20–30%.

Note: Some designers prefer to use a factor of safety on the passive earth pressure coefficient at the beginning. In that case, in Step 1

$$K_{p(design)} = \frac{K_p}{FS}$$

where FS = factor of safety (usually between 1.5 to 2)

For this type of analysis, follow Steps 1 through 12 with the value of $K_a = \tan^2 (45 - \phi/2)$ and $K_{p(design)}$ (instead of K_p). The actual depth of penetration can now be determined by adding L_3, obtained from Step 3, and L_4, obtained from Step 8.

Calculation of Maximum Bending Moment

The nature of variation of the moment diagram for a cantilever sheet pile wall is shown in Figure 6.7b. The maximum moment will occur between the points E and F'. To obtain the maximum moment (M_{max}) per unit length of the wall, one must determine the point of zero shear. Adopting a new axis z' (with origin at point E) for zero shear

$$P = \tfrac{1}{2}(z')^2(K_p - K_a)\gamma'$$

or

$$z' = \sqrt{\frac{2P}{(K_p - K_a)\gamma'}} \tag{6.21}$$

Once the point of zero shear force is determined (point F'' in Figure 6.7a), the magnitude of the maximum moment can be obtained as

$$\boxed{M_{max} = P(\bar{z} + z') - [\tfrac{1}{2}\gamma'z'^2(K_p - K_a)](\tfrac{1}{3})z')} \tag{6.22}$$

The sizing of the necessary profile of the sheet piling is then made according to the allowable flexural stress of the sheet pile material, or

$$\boxed{S = \frac{M_{max}}{\sigma_{all}}} \tag{6.23}$$

where S = section modulus of the sheet pile required per unit length of the structure

σ_{all} = allowable flexural stress of the sheet pile

Example 6.1

Refer to Figure 6.7. For a cantilever sheet pile wall penetrating a granular soil, given: $L_1 = 2$ m, $L_2 = 3$ m. The granular soil has the following properties:

$\phi = 32°$

$c = 0$

$\gamma = 15.9$ kN/m^3

$\gamma_{sat} = 19.33$ kN/m^3

Make the necessary calculations to determine the theoretical and actual depth of penetration. Also determine the miminum size of sheet pile (section modulus) necessary.

Solution

The step-by-step procedure given in Section 6.3 will be followed here.

Step 1

$$K_a = \tan^2\left(45 - \frac{\phi}{2}\right) = \tan^2\left(45 - \frac{32}{2}\right) = 0.307$$

$$K_p = \tan^2\left(45 + \frac{\phi}{2}\right) = 3.25$$

Step 2

$$p_1 = \gamma L_1 K_a = (15.9)(2)(0.307) = 9.763 \text{ kN/m}^2$$
$$p_2 = (\gamma L_1 + \gamma' L_2)K_a = [(15.9)(2) + (19.33 - 9.81)3]0.307$$
$$= 18.53 \text{ kN/m}^2$$

Step 3

$$L_3 = \frac{p_2}{\gamma'(K_p - K_a)} = \frac{18.53}{(19.33 - 9.81)(3.25 - 0.307)} = 0.66 \text{ m} \quad \checkmark$$

Step 4

$$P = \tfrac{1}{2}p_1 L_1 + p_1 L_2 + \tfrac{1}{2}(p_2 - p_1)L_2 + \tfrac{1}{2}p_2 L_3$$
$$= \tfrac{1}{2}(9.763)(2) + (9.763)(3) + \tfrac{1}{2}(18.53 - 9.763)3 + \tfrac{1}{2}(18.53)(0.66)$$
$$= 9.763 + 29.289 + 13.151 + 6.115 = 58.32 \text{ kN/m}$$

Step 5. Taking the moment about E

$$\bar{z} = \frac{1}{58.32}\left[9.763\left(0.66 + 3 + \frac{2}{3}\right) + 29.289\left(0.66 + \frac{3}{2}\right)\right.$$
$$\left. + 13.151\left(0.66 + \frac{3}{3}\right) + 6.115\left(0.66 \times \frac{2}{3}\right)\right] = 2.23 \text{ m}$$

Step 6

$$p_5 = (\gamma L_1 + \gamma'L_2)K_p + \gamma'L_3(K_p - K_a)$$
$$= [(15.9)(2) + (19.33 - 9.81)3]3.25 + (19.33 - 9.81)(0.66)(3.25 - 0.307)$$
$$= 196.17 + 18.49 = 214.66 \text{ kN/m}^2$$

Step 7

$$A_1 = \frac{p_5}{\gamma'(K_p - K_a)} = \frac{214.66}{(9.52)(2.943)} = 7.66$$

$$A_2 = \frac{8P}{\gamma'(K_p - K_a)} = \frac{(8)(58.32)}{(9.52)(2.943)} = 16.65$$

$$A_3 = \frac{6P[2\bar{z}\gamma'(K_p - K_a) + p_5]}{\gamma'^2(K_p - K_a)^2}$$

$$= \frac{(6)(58.32)[(2)(2.23)(9.52)(2.943) + 214.66]}{(9.52)^2(2.943)^2} = 151.93$$

$$A_4 = \frac{P(6\bar{z}p_5 + 4P)}{\gamma'^2(K_p - K_a)^2}$$

$$= \frac{58.32[(6)(2.23)(214.66) + (4)(58.32)]}{(9.52)^2(2.943)^2} = 230.72$$

Step 8. From Eq. (6.16)

$$L_4^4 + 7.66L_4^3 - 16.65L_4^2 - 151.39L_4 - 230.72 = 0$$

The following table shows the solution of the preceding equation by trial and error.

Assumed L_4 (m)	Left side of Eq. (6.16)
4	-356.44
5	+178.58
4.8	+36.96

So, $L_4 \approx 4.8$ m

Step 9

$$p_4 = p_5 + \gamma'L_4(K_p - K_a)$$
$$= 214.66 + (9.52)(4.8)(2.943) = 349.14 \text{ kN/m}^2$$

Step 10

$$p_3 = \gamma'(K_p - K_a)L_4 = (9.52)(2.943)(4.8) = 134.48 \text{ kN/m}^2$$

Step 11

$$L_5 = \frac{p_3L_4 - 2P}{p_3 + p_4} = \frac{(134.48)(4.8) - 2(58.32)}{134.48 + 349.14} = 1.09 \text{ m}$$

Step 12. The net pressure distribution diagram can now be drawn, as shown in Figure 6.7a.

Step 13. The actual depth of penetration = $1.3(L_3 + L_4) = 1.3(0.66 + 4.8) =$ **7.1 m.**
The theoretical depth of penetration = $0.66 + 4.8 =$ **5.46 m.**

Size of Sheet Piling

Using Eq. (6.21)

$$z' = \sqrt{\frac{2P}{\gamma'(K_p - K_a)}} = \sqrt{\frac{(2)(58.32)}{9.52(2.943)}} = 2.04 \text{ m}$$

From Eq. (6.22)

$$
\begin{aligned}
M_{max} &= P(\bar{z} + z') - \left[\frac{1}{2}\gamma'z'^2(K_p - K_a)\right]\left(\frac{z'}{3}\right) \\
&= (58.32)(2.23 + 2.04) - \frac{1}{2}(9.52)(2.04)^2(2.943)\left(\frac{2.04}{3}\right) \\
&= 249.03 - 39.64 = 209.39 \text{ kN-m}
\end{aligned}
$$

The required section modulus of the sheet pile

$$S = \frac{M_{max}}{\sigma_{all}}$$

With $\sigma_{all} = 172.5 \text{ MN/m}^2$

$$S = \frac{209.39 \text{ kN-m}}{172.5 \times 10^3 \text{ kN/m}^2} = 1.214 \times 10^{-3} \text{ m}^3/\text{m of wall}$$

6.4   Special Cases for Cantilever Wall (Penetrating a Sandy Soil)

Following are two special cases of the mathematical formulation shown in Section 6.3.

Case 1: Sheet Pile Wall with the Absence of Water Table

In the absence of the water table, the net pressure diagram on the cantilever sheet pile wall will be as shown in Figure 6.8, which is a modified version of Figure 6.7. For this figure

$$p_2 = \gamma L K_a \tag{6.24}$$

$$p_3 = L_4(K_p - K_a)\gamma \tag{6.25}$$

$$p_4 = p_5 + \gamma L_4(K_p - K_a) \tag{6.26}$$

$$p_5 = \gamma L K_p + \gamma L_3(K_p - K_a) \tag{6.27}$$

$$L_3 = \frac{p_2}{\gamma(K_p - K_a)} = \frac{LK_a}{(K_p - K_a)} \tag{6.28}$$

$$P = \frac{1}{2}p_2 L + \frac{1}{2}p_2 L_3 \tag{6.29}$$

$$\bar{z} = L_3 + \frac{L}{3} = \frac{LK_a}{K_p - K_a} + \frac{L}{3} = \frac{L(2K_a + K_p)}{3(K_p - K_a)} \tag{6.30}$$

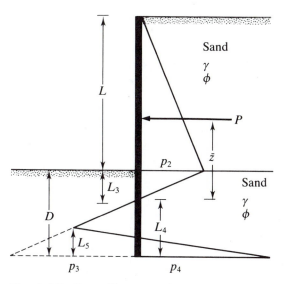

Figure 6.8 Sheet piling penetrating a sandy soil with the absence of water table

For this case, Eq. (6.16) transforms to the form

$$L_4^4 + A_1' L_4^3 - A_2' L_4^2 - A_3' L_4 - A_4' = 0 \tag{6.31}$$

where

$$A_1' = \frac{p_5}{\gamma(K_p - K_a)} \tag{6.32}$$

$$A_2' = \frac{8P}{\gamma(K_p - K_a)} \tag{6.33}$$

$$A_3' = \frac{6P[2\bar{z}\gamma(K_p - K_a) + p_5]}{\gamma^2(K_p - K_a)^2} \tag{6.34}$$

$$A_4' = \frac{P(6\bar{z}p_5 + 4P)}{\gamma^2(K_p - K_a)^2} \tag{6.35}$$

Case 2: Free Cantilever Sheet Piling

Figure 6.9 shows a free cantilever sheet pile wall penetrating a sandy soil and subjected to a line load of P per unit length of the wall. For this case

$$D^4 - \left[\frac{8P}{\gamma(K_p - K_a)}\right]D^2 - \left[\frac{12PL}{\gamma(K_p - K_a)}\right]D - \left[\frac{2P}{\gamma(K_p - K_a)}\right]^2 = 0 \tag{6.36}$$

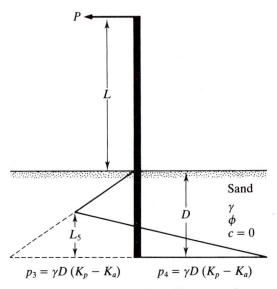

$$p_3 = \gamma D (K_p - K_a) \qquad\qquad p_4 = \gamma D (K_p - K_a)$$

Figure 6.9 Free cantilever sheet piling penetrating a sand layer

and

$$L_5 = \frac{\gamma(K_p - K_a)D^2 - 2P}{2D(K_p - K_a)\gamma} \qquad\qquad (6.37)$$

$$M_{max} = P(L + z') - \frac{\gamma z^3(K_p - K_a)}{6} \qquad\qquad (6.38)$$

$$z' = \sqrt{\frac{2P}{\gamma(K_p - K_a)}} \qquad\qquad (6.39)$$

Example 6.2

Refer to Figure 6.9. Given: $L = 15$ ft, $\gamma = 110$ lb/ft^3, $\phi = 30°$, and $P = 2000$ lb/ft. Determine the following:

a. The theoretical depth of penetration, D
b. The maximum moment, M_{max} (lb-ft/ft)

Solution

$$K_p = \tan^2\left(45 + \frac{\phi}{2}\right) = \tan^2\left(45 + \frac{30}{2}\right) = 3$$

$$K_a = \tan^2\left(45 - \frac{\phi}{2}\right) = \tan^2\left(45 - \frac{30}{2}\right) = \frac{1}{3}$$

$$K_p - K_a = 3 - 0.333 = 2.667$$

Part a

From Eq. (6.36)

$$D^4 - \left[\frac{8P}{\gamma(K_p - K_a)}\right]D^2 - \left[\frac{12PL}{\gamma(K_p - K_a)}\right]D - \left[\frac{2P}{\gamma(K_p - K_a)}\right]^2 = 0$$

$$\frac{8P}{\gamma(K_p - K_a)} = \frac{(8)(2000)}{(110)(2.667)} = 54.54$$

$$\frac{12PL}{\gamma(K_p - K_a)} = \frac{(12)(2000)(15)}{(110)(2.667)} = 1227.1$$

$$\frac{2P}{\gamma(K_p - K_a)} = \frac{(2)(2000)}{(110)(2.667)} = 13.63$$

So

$$D^4 - 54.54D^2 - 1227.1D - (13.63)^2 = 0$$

From the above equation, $D \approx$ **13 ft**

Part b

From Eq. (6.39)

$$z' = \sqrt{\frac{2P}{\gamma(K_p - K_a)}} = \sqrt{\frac{(2)(2000)}{(110)(2.667)}} = 3.69 \text{ ft}$$

From Eq. (6.35)

$$M_{\max} = P(L + z') - \frac{\gamma z'^3(K_p - K_a)}{6}$$

$$= (2000)(15 + 3.69) - \frac{(110)(3.69)^3(2.667)}{6}$$

$$= 37,387 - 2456.65 \approx \textbf{34,923 lb-ft/ft}$$

6.5   Cantilever Sheet Piling Penetrating Clay

In several cases, cantilever sheet piles must be driven into a clay layer possessing an undrained cohesion, c ($\phi = 0$ concept). The net pressure diagram will be somewhat different from that shown in Figure 6.7a. Figure 6.10 shows a cantilever sheet pile wall driven into clay with a backfill of granular soil above the level of the dredge line. Let the water table be located at a depth of L_1 below the top of the wall. As before, using Eqs. (6.1) and (6.2), the intensity of the net pressures p_1 and p_2 can be calculated, and the

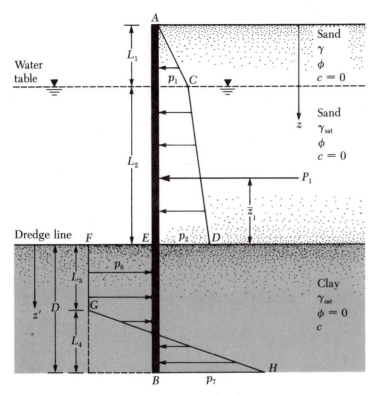

Figure 6.10 Cantilever sheet pile penetrating clay

diagram for pressure distribution above the level of the dredge line can be drawn.

The diagram for net pressure distribution below the dredge line can now be determined as follows.

At any depth z greater than $L_1 + L_2$ and above the point of rotation (point O in Figure 6.6a), the active pressure (p_a) from right to left can be expressed as

$$p = [\gamma L_1 + \gamma' L_2 + \gamma_{sat}(z - L_1 - L_2)]K_a - 2c\sqrt{K_a} \qquad (6.40)$$

where K_a = Rankine active earth pressure coefficient; with $\phi = 0$, it is
        equal to one

Similarly, the passive pressure (p_p) from left to right can be given as

$$p_p = \gamma_{sat}(z - L_1 - L_2)K_p + 2c\sqrt{K_p} \qquad (6.41)$$

where K_p = Rankine passive earth pressure coefficient; with $\phi = 0$, K_p is
        equal to one

Thus, the net pressure

$$
\begin{aligned}
p_6 = p_p - p_a &= [\gamma_{\text{sat}}(z - L_1 - L_2) + 2c] \\
&\quad - [\gamma L_1 + \gamma' L_2 + \gamma_{\text{sat}}(z - L_1 - L_2)] + 2c \\
&= 4c - (\gamma L_1 + \gamma' L_2)
\end{aligned}
\tag{6.42}
$$

At the bottom of the sheet pile, the passive pressure from right to left is

$$
p_p = (\gamma L_1 + \gamma' L_2 + \gamma_{\text{sat}} D) + 2c
\tag{6.43}
$$

Similarly, the active pressure from left to right is

$$
p_a = \gamma_{\text{sat}} D - 2c
\tag{6.44}
$$

Hence, the net pressure

$$
p_7 = p_p - p_a = 4c + (\gamma L_1 + \gamma' L_2)
\tag{6.45}
$$

For equilibrium analysis, $\sum F_H = 0$—that is, area of pressure diagram $ACDE$ − area of $EFIB$ + area of $GIH = 0$, or

$$
\begin{aligned}
P_1 - [4c - (\gamma L_1 + \gamma' L_2)]D + \tfrac{1}{2}L_4[4c - (\gamma L_1 + \gamma' L_2) \\
+ 4c + (\gamma L_1 + \gamma' L_2)] = 0
\end{aligned}
$$

where P_1 = area of the pressure diagram $ACDE$

Simplifying the preceding equation

$$
L_4 = \frac{D[4c - (\gamma L_1 + \gamma' L_2)] - P_1}{4c}
\tag{6.46}
$$

Now, taking the moment about point B, $\sum M_B = 0$, or

$$
P_1(D + \bar{z}_1) - [4c - (\gamma L_1 + \gamma' L_2)]\frac{D^2}{2} + \frac{1}{2}L_4(8c)\left(\frac{L_4}{3}\right) = 0
\tag{6.47}
$$

where $\bar{z}_1$ = distance of the center of pressure of the pressure diagram $ACDE$ measured from the level of the dredge line

Combining Eqs. (6.46) and (6.47) yields

$$
\boxed{D^2[4c - (\gamma L_1 + \gamma' L_2)] - 2DP_1 - \frac{P_1(P_1 + 12c\bar{z}_1)}{(\gamma L_1 + \gamma' L_2) + 2c} = 0}
\tag{6.48}
$$

The preceding equation can now be solved to obtain D, the theoretical depth of penetration of the clay layer by the sheet pile.

Step-by-Step Procedure to Obtain the Pressure Diagram

1. Calculate $K_a = \tan^2(45 - \phi/2)$ for the granular soil (backfill).
2. Obtain p_1 and p_2 [Eqs. (6.1) and (6.2)].
3. Calculate P_1 and $\bar{z}_1$.
4. Use Eq. (6.48) to obtain the theoretical value of D.
5. Using Eq. (6.46), calculate L_4.
6. Calculate p_6 and p_7 [Eqs. (6.42) and (6.45)].
7. Draw the pressure distribution diagram as shown in Figure 6.10.
8. The actual depth of penetration

$$D_{actual} = 1.4 \text{ to } 1.6(D_{theoretical})$$

Maximum Bending Moment

According to Figure 6.10, the maximum moment (that is, zero shear) will occur between $L_1 + L_2 < z < L_1 + L_2 + L_3$. Using a new coordinate system z' ($z' = 0$ at dredge line) for zero shear

$$P_1 - p_6 z' = 0$$

or

$$z' = \frac{P_1}{p_6} \tag{6.49}$$

The magnitude of the maximum moment can now be obtained as

$$M_{max} = P_1(z' + \bar{z}_1) - \frac{p_6 z'^2}{2} \tag{6.50}$$

Knowing the maximum bending moment, one can determine the section modulus of the sheet pile section from Eq. (6.23).

Example 6.3

Redo Example 6.1. Assume that the properties of the backfill material are the same. However, the soil below the dredge line is clay. The value of the unconsolidated undrained shear strength of this clay is 47 kN/m².

Solution

We will follow the step-by-step procedure given in Section 6.5; thus

Step 1

$$K_a = 0.307$$

Step 2

$$p_1 = 9.763 \text{ kN/m}^2$$
$$p_2 = 18.53 \text{ kN/m}^2$$

Step 3. Referring to the net pressure distribution diagram given in Figure 6.10

$$P_1 = \frac{1}{2} p_1 L_1 + p_1 L_2 + \frac{1}{2}(p_2 - p_1)L_2$$

$$= 9.763 + 29.289 + 13.151 = 52.2 \text{ kN/m}$$

$$\bar{z}_1 = \frac{1}{52.2}\left[9.763\left(3 + \frac{2}{3}\right) + 29.289\left(\frac{3}{2}\right) + 13.151\left(\frac{3}{3}\right)\right]$$

$$= 1.78 \text{ m}$$

Step 4. From Eq. (6.48)

$$D^2[4c - (\gamma L_1 + \gamma' L_2)] - 2DP_1 - \frac{P_1(P_1 + 12c\bar{z}_1)}{(\gamma L_1 + \gamma' L_2) + 2c} = 0$$

Substituting proper values

$$D^2\{(4)(47) - [(2)(15.9) + (19.33 - 9.81)3]\} - 2D(52.2)$$
$$- \frac{52.2[52.2 + (12)(47)(1.78)]}{[(15.9)(2) + (19.33 - 9.81)3] + (2)(47)} = 0$$

or

$$127.64D^2 - 104.4D - 357.15 = 0$$

Solving the preceding equation, $D = 2.13$ m.

Step 5. From Eq. (6.46)

$$L_4 = \frac{D[4c - (\gamma L_1 + \gamma' L_2)] - P_1}{4c}$$

$$4c - (\gamma L_1 + \gamma' L_2) = (4)(47) - [(15.9)(2) + (19.33 - 9.81)3]$$
$$= 127.64 \text{ kN/m}^2$$

So

$$L_4 = \frac{2.13(127.64) - 52.2}{(4)(47)} = 1.17 \text{ m}$$

Step 6

$$p_6 = 4c - (\gamma L_1 + \gamma' L_2) = 127.64 \text{ kN/m}^2$$
$$p_7 = 4c + (\gamma L_1 + \gamma' L_2) = 248.36 \text{ kN/m}^2$$

Step 7. The net pressure distribution diagram can now be drawn, as shown in Figure 6.10.

Step 8. $D_{\text{actual}} \approx 1.5 D_{\text{theoretical}} = 1.5(2.13) \approx \mathbf{3.2}$ **m**

Maximum Moment Calculation

From Eq. (6.49)

$$z' = \frac{P_1}{p_6} = \frac{52.2}{127.64} \approx 0.41 \text{ m}$$

Again, from Eq. (6.50)

$$M_{max} = P_1(z' + \bar{z}_1) - \frac{p_6 z'^2}{2}$$

So

$$M_{max} = 52.2(0.41 + 1.78) - \frac{127.64(0.41)^2}{2}$$

$$= 114.32 - 10.73 = 103.59 \text{ kN-m}$$

The minimum required section modulus (assuming $\sigma_{all} = 172.5 \text{ MN/m}^2$)

$$S = \frac{103.59 \text{ kN-m}}{172.5 \times 10^3 \text{ kN/m}^2} = \textbf{0.6} \times \textbf{10}^{-3} \textbf{ m}^3/\textbf{m of the wall}$$

6.6   Special Cases for Cantilever Wall (Penetrating Clay)

As in Section 6.4, relationships for special cases for cantilever walls penetrating clay can also be derived. They are as follows.

Case 1: Sheet Pile Wall with the Absence of Water Table

Referring to Figure 6.11, we can write that

$$p_2 = \gamma L K_a \tag{6.51}$$
$$p_6 = 4c - \gamma L \tag{6.52}$$
$$p_7 = 4c + \gamma L \tag{6.53}$$
$$P_1 = \tfrac{1}{2}Lp_2 = \tfrac{1}{2}\gamma L^2 K_a \tag{6.54}$$
$$L_4 = \frac{D(4c - \gamma L) - \tfrac{1}{2}\gamma L^2 K_a}{4c} \tag{6.55}$$

The theoretical depth of penetration, D, can be calculated by solving the following equation [similar to Eq. (6.48)]:

$$D^2(4c - \gamma L) - 2DP_1 - \frac{P_1(P_1 + 12c\bar{z}_1)}{\gamma L + 2c} = 0 \tag{6.56}$$

where $\bar{z}_1 = \dfrac{L}{3}$ $\tag{6.57}$

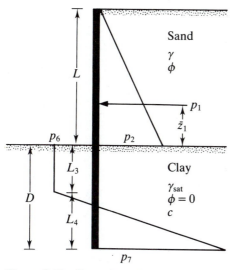

Figure 6.11   Sheet pile wall penetrating clay

The magnitude of the maximum moment in the wall is

$$M_{\max} = P_1(z' + \bar{z}_1) - \frac{p_6 z'^2}{2} \tag{6.58}$$

where $z' = \dfrac{P_1}{p_6} = \dfrac{\frac{1}{2}\gamma L^2 K_a}{4c - \gamma L}$ (6.59)

Case 2: Free Cantilever Sheet Pile Wall Penetrating Clay

Figure 6.12 shows a free cantilever sheet pile wall penetrating a clay layer. The wall is being subjected to a line load of P per unit length. For this case

$$p_6 = p_7 = 4c \tag{6.60}$$

The depth of penetration, D, can be obtained from the following equation:

$$4D^2 c - 2PD - \frac{P(P + 12cL)}{2c} = 0 \tag{6.61}$$

Also note that, for pressure diagram construction

$$L_4 = \frac{4cD - P}{4c} \tag{6.62}$$

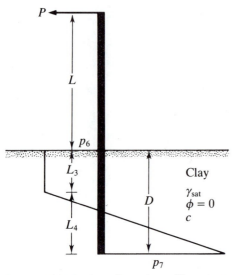

Figure 6.12 Free cantilever sheet piling penetrating clay

The maximum moment in the wall

$$M_{\text{max}} = P(L + z') - \frac{4cz'^2}{2} \tag{6.63}$$

where $z' = \dfrac{P}{4c}$ $\qquad\qquad$ (6.64)

Example 6.4

Refer to Figure 6.12. For the free cantilever sheet pile wall, the following are given: $P = 2200$ lb/ft, $L = 10$ ft, and $c = 200$ lb/ft^2. Calculate the theoretical depth of penetration, D.

Solution

From Eq. (6.61)

$$4D^2c - 2PD - \frac{P(P + 12cL)}{2c} = 0$$

or

$$(4)(D^2)(200) - (2)(2200)(D) - \frac{2200[2200 + (12)(200)(10)]}{(2)(200)} = 0$$

$$800D^2 - 4400D - 144{,}100 = 0$$

Solving for D

$$D \approx 16.5 \text{ ft}$$

ANCHORED SHEET PILE WALL

When the height of the backfill material behind a cantilever sheet pile wall exceeds about 6 m ($\approx$ 20 ft), it becomes more economical to tie the sheet pile wall near the top to anchor plates, anchor walls, or anchor piles. This is referred to as *anchored sheet pile wall* or an *anchored bulkhead*. Anchors minimize the depth of required penetration by the sheet piles and also reduce the cross section area and weight of the sheet piles needed for construction. However, the tie rods and anchors must be carefully designed.

There are two basic methods of constructing anchored sheet pile walls: (a) the *free earth support* method and (b) the *fixed earth support* method. Figure 6.13 shows the assumed nature of deflection of the sheet piles designed by the two methods.

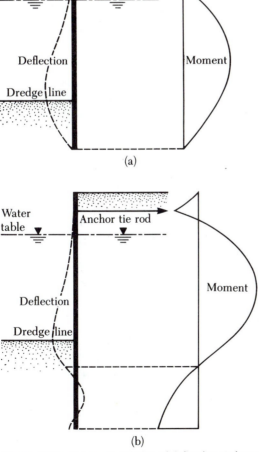

Figure 6.13 Nature of variation of deflection and moment for anchored sheet piles: (a) free earth support method; (b) fixed earth support method

The free earth support method is the method of minimum penetration depth. Below the dredge line, no pivot point exists for the statical system. The nature of variation of the bending moment with depth for both of the methods is also shown in Figure 6.13.

6.7   Free Earth Support Method for Penetration of Sandy Soil

Figure 6.14 shows an anchor sheet pile wall with a granular soil backfill; the wall has been driven into a granular soil. The tie rod connecting the sheet pile and the anchor is located at a depth of l_1 below the top of the sheet pile wall.

The diagram of net pressure distribution above the dredge line will be similar to that shown in Figure 6.7. At a depth $z = L_1$, $p_1 = \gamma L_1 K_a$; and, at $z = L_1 + L_2$, $p_2 = (\gamma L_1 + \gamma' L_2) K_a$. Below the dredge line, the net pressure will be equal to zero at a depth of $z = L_1 + L_2 + L_3$. The relation for L_3 can be given by Eq. (6.6), or

$$L_3 = \frac{p_2}{\gamma'(K_p - K_a)}$$

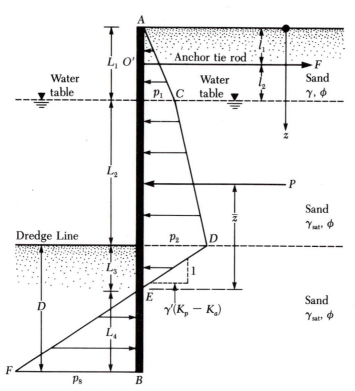

Figure 6.14 Anchored sheet pile wall penetrating sand

At a depth $z = (L_1 + L_2 + L_3 + L_4)$, the net pressure can be given by

$$p_8 = \gamma'(K_p - K_a)L_4 \tag{6.65}$$

Note that the slope of the line DEF is 1 vertical to $\gamma'(K_p - K_a)$ horizontal.

For equilibrium of the sheet pile, Σ horizontal forces $= 0$, and Σ moment about $O' = 0$. (*Note:* Point O' is located at the level of the tie rod.)

Summing the forces in the horizontal direction (per unit length of the wall)

area of the pressure diagram $ACDE$ − area of $EBF − F = 0$

where $F =$ tension in the tie rod/unit length of the wall, or

$$P - \tfrac{1}{2}p_8 L_4 - F = 0$$

or

$$\boxed{F = P - \tfrac{1}{2}[(\gamma'(K_p - K_a)]L_4^2} \tag{6.66}$$

where $P =$ area of the pressure diagram $ACDE$

Now, taking the moment about point O'

$$-P[(L_1 + L_2 + L_3) - (\bar{z} + l_1)] + \tfrac{1}{2}[\gamma'(K_p - K_a)] \cdot L_4^2$$
$$\cdot \left(l_2 + L_2 + L_3 + \tfrac{2}{3}L_4\right) = 0$$

or

$$\boxed{L_4^3 + 1.5L_4^2(l_2 + L_2 + L_3) - \frac{3P[(L_1 + L_2 + L_3) - (\bar{z} + l_1)]}{\gamma'(K_p - K_a)} = 0}$$

$$\tag{6.67}$$

The preceding equation can be solved by trial and error to determine the theoretical depth, L_4. Thus, the theoretical depth of penetration is equal to

$$D_{\text{theoretical}} = L_3 + L_4$$

The theoretical depth is increased by about 30−40% in the actual construction work.

$$D_{\text{actual}} = 1.3 \text{ to } 1.4D_{\text{theoretical}} \tag{6.68}$$

The step-by-step procedure in Section 6.3 pointed out that a factor of safety can be applied to K_p at the beginning [that is, $K_{p(\text{design})} = K_p/FS$]. If this is done, there is no need to increase the theoretical depth by 30−40%. This approach is often more conservative.

The maximum theoretical moment to which the sheet pile will be subjected will occur at a depth between $z = L_1$ to $z = L_1 + L_2$. The depth, z, for zero shear and, hence, maximum moment can be evaluated from the equation

$$\tfrac{1}{2}p_1 L_1 - F + p_1(z - L_1) + \tfrac{1}{2}K_a \gamma'(z - L_1)^2 = 0 \tag{6.69}$$

Once the value of z is determined, the magnitude of the maximum moment can be obtained easily. The procedure of determination of the holding capacity of anchors is treated in Section 6.12.

Example 6.5

Redo Example 6.1 for an anchored sheet pile wall to

    a. Determine the theoretical and actual depth of penetration
    b. Find the anchor force per unit length of the wall
    c. Determine M_{max}

The anchor line is located at a depth of 1 m below the top of the wall.

Solution

Refer to Figure 6.14 for the nature of variation of the net lateral pressure diagram. From Example 6.1

$$p_1 = 9.763 \text{ kN/m}^2$$
$$p_2 = 18.53 \text{ kN/m}^2$$
$$L_3 = 0.66 \text{ m}$$
$$P = 58.32 \text{ kN/m}$$
$$\bar{z} = 2.23 \text{ m}$$

Part a: Depth of Penetration

Referring to Eq. (6.67)

$$L_3^4 + 1.5L_4^2(l_2 + L_2 + L_3) - \frac{3P[(L_1 + L_2 + L_3) - (\bar{z} + l_1)]}{\gamma'(K_p - K_a)} = 0$$

Noting that

$$l_1 = 1 \text{ m} \qquad K_p = 3.25$$
$$l_2 = 1 \text{ m} \qquad K_a = 0.307$$

$$L_4^3 + 1.5L_4^2(1 + 3 + 0.66) - \frac{3(58.32)[(2 + 3 + 0.66) - (2.33 + 1)]}{9.52(3.25 - 0.307)} = 0$$

or

$$L_4^3 + 6.99L_4^2 - 14.55 = 0 \tag{a}$$

The value of L_4 can be obtained by trial and error, as shown in the following table.

Assumed L_4 (m)	Left side of Eq. (a)
2.0	+21.41
1.5	+3.55
1.4	+2.89
1.3	−0.54

The preceding table shows that L_4 is approximately equal to 1.4 m. Hence,

$$D_{\text{theoretical}} = L_3 + L_4 = 0.66 + 1.4 = 2.06 \text{ m}$$
$$D_{\text{actual}} \approx 1.4 D_{\text{theory}} = (1.4)(2.06) = 2.88 \text{ m (rounded off to } \textbf{2.9 m)}$$

Part b: Anchor Force

From Eq. (6.66)

$$F = P - \tfrac{1}{2}[\gamma'(K_p - K_a)]L_4^2$$
$$= 58.32 - \tfrac{1}{2}[9.52(3.25 - 0.307)](1.4)^2 = \textbf{30.86 kN/m}$$

Part c: Maximum Moment ($M_{\max}$)

Referring to Eq. (6.69), for zero shear

$$\tfrac{1}{2}p_1 L_1 - F + p_1(z - L_1) + \tfrac{1}{2}K_a \gamma'(z - L_1)^2 = 0$$

or

$$(\tfrac{1}{2})(9.763)(2) - 30.86 + (9.763)(z - 2) + \tfrac{1}{2}(0.307)(9.52)(z - 2)^2 = 0$$

Let $z - 2 = x$. So

$$9.763 - 30.86 + 9.763x + 1.461x^2 = 0$$
$$x^2 + 6.682x - 14.44 = 0$$
$$x = 1.72 \text{ m}$$

or

$$z = x + 2 = 1.72 + 2 = 3.72 \text{ m}$$
$$(L_1 + L_2 < z < L_1 \text{—checks})$$

Taking the moment about the point to zero shear force (that is, $z = 3.72$ m or $x = 1.72$ m)

$$M_{\max} = -\left(\frac{1}{2}p_1 L_1\right)\left[x + \left(\frac{1}{3}\right)(2)\right] + F(x + 1) - (p_1 x)\left(\frac{x}{2}\right) - \frac{1}{2}K_a \gamma'(x)^2\left(\frac{x}{3}\right)$$

or

$$M_{\max} = -(9.763)(2.387) + (30.86)(2.72) - \frac{9.763(1.72)^2}{2} - \frac{(0.307)(9.52)(1.72)^3}{6}$$
$$= -23.3 + 83.94 - 14.44 - 2.48 = \textbf{43.72 kN-m/m}$$

6.8 Free Earth Support Method for Penetration of Clay

Figure 6.15 shows an anchor sheet pile wall that has penetrated a clay soil and that has a granular soil backfill. The diagram of pressure distribution above the dredge line is similar to that shown in Figure 6.10. The net pressure distribution below the dredge line (from $z = L_1 + L_2$ to $z = L_1 + L_2 + D$) can be given as [Eq. (6.42)]

$$p_6 = 4c - (\gamma L_1 + \gamma' L_2)$$

For static equilibrium, summing the forces in the horizontal direction

$$\boxed{P_1 - p_6 D = F} \tag{6.70}$$

where P_1 = area of the pressure diagram ACD
F = anchor force per unit length of the sheet pile wall

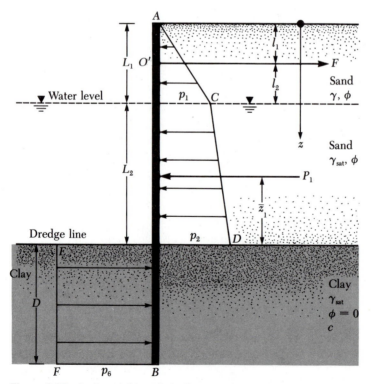

Figure 6.15 Anchored sheet pile wall penetrating clay

Again, taking the moment about O'

$$P_1(L_1 + L_2 - l_1 - \bar{z}_1) - p_6 D\left(l_2 + L_2 + \frac{D}{2}\right) = 0$$

Simplification of the preceding equation yields

$$\boxed{p_6 D^2 + 2p_6 D(L_1 + L_2 - l_1) - 2P_1(L_1 + L_2 - l_1 - \bar{z}_1) = 0} \qquad (6.71)$$

The theoretical depth of penetration, D, can be determined from the preceding equation.

As in Section 6.7, the maximum moment in this case will occur at a depth $L_1 < z < L_1 + L_2$. The depth of zero shear (and thus the maximum moment) can be determined by using Eq. (6.69).

Example 6.6

Refer to Figure 6.15. Given: $L_1 = 10.8$ ft, $L_2 = 21.6$ ft, $l_1 = 5.4$ ft, $\gamma = 108$ lb/ft³, $\gamma_{sat} = 127.2$ lb/ft³, $\phi = 35°$, and $c = 850$ lb/ft².

a. Determine the theoretical depth of embedment.
b. Calculate the anchor force per unit length of the sheet pile wall.

Solution
Part a

Given: $\phi = 35°$. So

$$K_a = \tan^2\left(45 - \frac{\phi}{2}\right) = 0.271$$

$$K_p = \tan^2\left(45 + \frac{\phi}{2}\right) = 3.69$$

Refer to the pressure diagram shown in Figure 6.16.

$p_1 = \gamma L_1 K_a = (0.108)(10.8)(0.271) = 0.316$ kip/ft²
$p_2 = (\gamma L_1 + \gamma' L_2)K_a$
$\quad = [(10.8)(0.108) + (0.0648)(21.6)](0.217)$
$\quad = (1.1664 + 1.4)(0.271) = 0.695$ kip/ft²
$P_1 = $ areas of $1 + 2 + 3 = 1.706 + 6.826 + 4.093$
$\quad = 12.625$ kips/ft

$$\bar{z}_1 = \frac{(1.706)\left(21.6 + \frac{10.8}{3}\right) + (6.826)(10.8) + (4.093)\left(\frac{21.6}{3}\right)}{12.625}$$

$$= \frac{42.99 + 73.72 + 29.47}{12.625} = 11.58 \text{ ft}$$

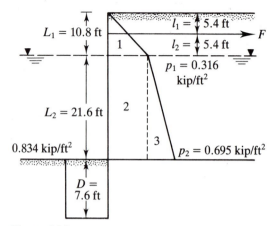

Figure 6.16

From Eq. (6.71)

$$p_6 D^2 + 2p_6 D(L_1 + L_2 - l_1) - 2P_1(L_1 + L_2 - l_1 - \bar{z}_1) = 0$$
$$p_6 = 4c - (\gamma L_1 + \gamma' L_2)$$
$$= 4(0.850) - (1.1664 + 1.4) \approx 0.834 \text{ kip/ft}^2$$

So

$$0.834D^2 - (2)(0.834)(D)(27) - (2)(12.625)(15.42) = 0$$
$$D^2 - 54D - 466.85 = 0$$
$$D = \textbf{7.6 ft}$$

Part b

From Eq. (6.70)

$$F = P_1 - p_6 D = 12.625 - (0.834)(7.6) = \textbf{6.29 kips/ft}$$

6.9   Moment Reduction for Anchored Sheet Pile Walls

Sheet piles are flexible. Because of this flexibility, sheet pile walls will yield (that is, displace laterally). Their yielding results in a redistribution of lateral earth pressure. This change will tend to reduce the maximum bending moment, M_{max}, as calculated by the procedure outlined in Sections 6.7 and 6.8. For that reason, Rowe (1952, 1957) suggested a procedure to reduce the maximum design moment on the sheet pile walls as obtained via the free earth support method. This section discusses the procedure of moment reduction as proposed by Rowe.

In Figure 6.17, which is valid for the case of a sheet pile penetrating sand, the following notations have been used:

1. H' = total height of pile drive (that is, $L_1 + L_2 + D_{actual}$)

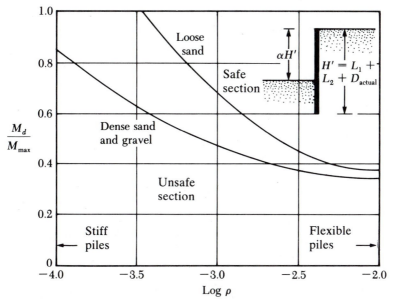

Figure 6.17 Plot of log ρ against M_d/M_{max} for sheet pile walls penetrating sand (after Rowe, 1952)

2. Relative flexibility of pile $= \rho = 10.91 \times 10^{-7}\left(\dfrac{H'^4}{EI}\right)$ (6.72)

where H is in meters

$\quad\quad E =$ Young's modulus of the pile material (MN/m^2)
$\quad\quad I =$ moment of inertia of the pile section per foot of the wall
$\quad\quad\quad$ (m^4/m of wall)

3. $M_d =$ design moment
4. $M_{max} =$ maximum theoretical moment

In English units, Eq. (6.70) takes the form

$\rho = \dfrac{H'^4}{EI}$ (6.73)

where H' is in ft, E is in $lb/in.^2$, and I is in $in.^4/ft$ of the wall.

The procedure for the use of the moment reduction diagram (Figure 6.17) is as follows:

Step 1. Choose a sheet pile section (such as those given in Table 6.1).

Step 2. Find the section modulus (S) of the selected section (Step 1) per unit length of the wall.

Step 3. Determine the moment of inertia of the section (Step 1) per unit length of the wall.

Step 4. Obtain H' and calculate ρ [Eqs. (6.72) or Eq. (6.73)].

Step 5. Find log ρ.

Step 6. Find the moment capacity of the pile section chosen in Step 1 as $M_d = \sigma_{all} \cdot S$.

Step 7. Determine M_d/M_{max}. Note that M_{max} is the maximum theoretical moment determined before.

Step 8. Plot log ρ (Step 5) and M_d/M_{max} in Figure 6.17.

Step 9. Repeat Steps 1 through 8 for several sections. The points that fall above the curve (loose sand or dense sand, as the case may be) are all *safe sections*. Those points that fall below the curve are the *unsafe sections*. The cheapest section can now be chosen from those points that fall above the proper curve. Note that the section chosen will have an $M_d < M_{max}$.

For piles penetrating clay soils, the notations in Figure 6.18 are as follows:

1. Stability number =

$$S_n = 1.25 \frac{c}{(\gamma L_1 + \gamma' L_2)} \tag{6.74}$$

where c = undrained cohesion ($\phi = 0$ condition)

For the definition of γ, γ', L_1, and L_2, see Figure 6.15.

2.

$$\alpha = \frac{L_1 + L_2}{L_1 + L_2 + D_{actual}} \tag{6.75}$$

3. Flexibility number, ρ [see Eqs. (6.72) or Eq. (6.73)]

4. M_d = design moment

M_{max} = maximum theoretical moment

The procedure for moment reduction using Figure 6.18 is as follows:

Step 1. Obtain H'.

Step 2. Determine $\alpha = (L_1 + L_2)/H'$.

Step 3. Determine S_n [Eq. (6.74)].

Step 4. For a given value of α and S_n (Steps 2 and 3), determine M_d/M_{max} for various values of log ρ from Figure 6.18 and plot a graph of M_d/M_{max} against log ρ.

Step 5. Follow Steps 1 through 9 as outlined for the case of moment reduction of sheet pile walls penetrating granular soil.

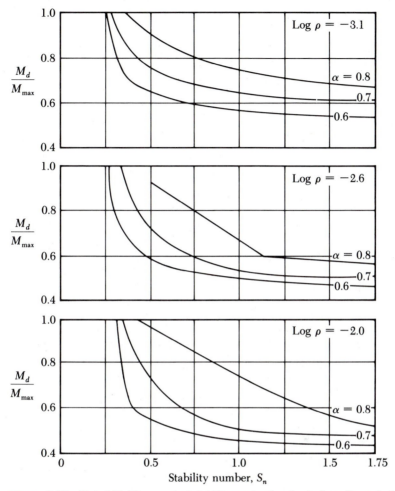

Figure 6.18 Plot of M_d/M_{max} against stability number for sheet pile wall penetrating clay (after Rowe, 1957)

Example 6.7

Refer to Example 6.5, in which $L_1 = 2$ m, $L_2 = 3$ m, and $D_{actual} = 2.9$ m. Use Rowe's moment reduction diagram (Figure 6.17) and find the most appropriate sheet pile section.

Solution

$$H' = L_1 + L_2 + D_{actual} = 2 + 3 + 2.9 = 7.9 \text{ m}$$

For moment reduction, the following table can be prepared.

Section (1)	I (m^4/m) (2)	H' (m) (3)	$\rho = \dfrac{10.91 \times 10^{-7}\, H'^4}{EI}$ (4)	$\log \rho$ (5)	S (m^3/m) (6)	$M_d = S \cdot \sigma_{all}$ (kN-m) (7)	$\dfrac{M_d}{M_{max}}$ (8)
PDA-27	54.33×10^{-6}	7.9	0.000378	-3.42	57×10^{-5}	98.32	2.250
PMA-22	18.7×10^{-6}	7.9	0.001098	-2.96	29×10^{-5}	50.02	1.140
PS-32	3.96×10^{-6}	7.9	0.005184	-2.29	10.20×10^{-5}	17.60	0.403
PS-28	3.82×10^{-6}	7.9	0.005370	-2.27	10.2×10^{-5}	17.60	0.403

(1), (2), (6) from Table 6.1
(4) Eq. (6.72), $E = 207 \times 10^3$ MN/m^2
(7) $\sigma_{all} = 172{,}500$ kN/m^2
(8) $M_{max} = 43.72$ kN-m/m

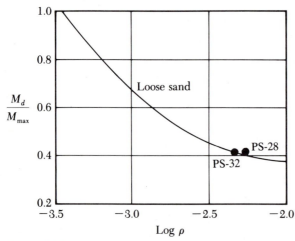

Figure 6.19

Figure 6.19 shows the calculated values of $\log \rho$ and the corresponding values of M_d/M_{max} (it is assumed that the soil behaves like loose sand). Also shown in Figure 6.19 is the design curve of $\log \rho$ against M_d/M_{max}, as given by Rowe (from Figure 6.17). Note that all points plot above the curve, so they are all safe for design use. The point corresponding to **Section PS-28** is closest to the curve: this is the section to be adopted.

6.10   Computational-Pressure-Diagram Method for Penetration of Sandy Soil

The computational-pressure-diagram method (CPD method) for sheet pile penetrating a sandy soil is a simplified method of design and an alternative to the free earth method described in Sections 6.7 and 6.9 (Nataraj and Hoadley, 1984). In this method, the net pressure diagram shown in Figure 6.14 is replaced by rectangular pressure diagrams, as shown in Figure 6.20.

Figure 6.20 Computational-pressure-diagram method (*note:* $L_1 + L_2 = L$)

Note that $\bar{p}_a$ is the width of the net active pressure diagram above the dredge line and $\bar{p}_p$ is the width of the net passive pressure diagram below the dredge line. The magnitudes of $\bar{p}_a$ and $\bar{p}_p$ can be expressed as

$$\bar{p}_a = CK_a\gamma_{av}L \tag{6.76}$$

$$\bar{p}_p = RCK_a\gamma_{av}L = R\bar{p}_a \tag{6.77}$$

where

γ_{av} = average effective unit weight of sand

$$\approx \frac{\gamma L_1 + \gamma' L_2}{L_1 + L_2} \tag{6.78}$$

C = coefficient

$$R = \text{coefficient} = \frac{L(L - 2l_1)}{D(2L + D - 2l_1)} \tag{6.79}$$

The range of values for C and R is given in Table 6.2.

Table 6.2 Range of Values for C and R
[Eqs. (6.76) and (6.77)]

Soil type	C^a	R
Loose sand	0.8–0.85	0.3–0.5
Medium sand	0.7–0.75	0.55–0.65
Dense sand	0.55–0.65	0.60–0.75

[a] Valid for the case in which there is no surcharge above the granular backfill, i.e., on the right side of the wall as shown in Figure 6.20

The depth of penetration (D), anchor force per unit length of the wall (F), and maximum moment in the wall (M_{max}) can be obtained from the following relationships.

Depth of penetration

$$D^2 + 2DL\left[1 - \left(\frac{l_1}{L}\right)\right] - \left(\frac{L^2}{R}\right)\left[1 - 2\left(\frac{l_1}{L}\right)\right] = 0$$

(6.80)

Anchor force

$$F = \bar{p}_a(L - RD)$$

(6.81)

Maximum moment

$$M_{max} = 0.5\bar{p}_a L^2\left[\left(1 - \frac{R}{D}\right)^2 - \left(\frac{2l_1}{L}\right)\left(1 - \frac{RD}{L}\right)\right]$$

(6.82)

It is important to note the following:

1. The magnitude of D obtained from Eq. (6.80) is about 1.25 to 1.5 times the value of $D_{theoretical}$ obtained by the conventional free earth support method (Section 6.7). So

$$D \approx D_{actual}$$
$$\uparrow \quad \uparrow$$
Eq. (6.80)   Eq. (6.68)

2. The magnitude of F obtained by using Eq. (6.81) is about 1.2 to 1.6 times the value obtained by using Eq. (6.66). So an additional factor of safety for actual design of anchors need not be used.

3. The magnitude of M_{max} obtained by using Eq. (6.82) is about 0.6 to 0.75 times the value of M_{max} obtained by using the conventional free earth support method. So this value of M_{max} can be used as the actual design value, and Rowe's moment reduction may not be applied.

Example 6.8

Refer to Figure 6.20. Given: $L_1 = 9$ ft, $L_2 = 26$ ft, $l_1 = 5$ ft, $\gamma = 108.5$ lb/ft^3, $\gamma_{sat} = 128.5$ lb/ft^3, and $\phi = 35°$. Using the CPD method, determine the following:

a. D
b. F
c. M_{max}

Solution
Part a

$$\gamma' = \gamma_{\text{sat}} - \gamma_w = 128.5 - 62.4 = 66.1 \text{ lb/ft}^3$$

From Eq. (6.78)

$$\gamma_{av} = \frac{\gamma L_1 + \gamma' L_2}{L_1 + L_2} = \frac{(108.5)(9) + (66.1)(26)}{9 + 26} = 77 \text{ lb/ft}^3$$

$$K_a = \tan^2\left(45 - \frac{\phi}{2}\right) = \tan^2\left(45 - \frac{35}{2}\right) = 0.271$$

$$\bar{p}_a = CK_a\gamma_{av}L = (0.65)(0.271)(77)(35) = 474.7 \text{ lb/ft}^2$$

Note that it is assumed that it is a *medium sand* and that $C = 0.65$ (from Table 6.2).

$$\bar{p}_p = R\bar{p}_a$$

Assume that $R = 0.6$ (using the range from Table 6.2). So

$$\bar{p}_p = (0.6)(474.7) = 284.82 \text{ lb/ft}^2$$

From Eq. (6.80)

$$D^2 + 2DL\left[1 - \left(\frac{l_1}{L}\right)\right] - \frac{L^2}{R}\left[1 - 2\left(\frac{l_1}{L}\right)\right] = 0$$

or

$$D^2 + 2(D)(35)\left[1 - \left(\frac{5}{35}\right)\right] - \frac{(35)^2}{0.6}\left[1 - 2\left(\frac{5}{35}\right)\right] = 0$$

$$D^2 + 60D - 1458.3 = 0$$

Solving the above equation

$$D \approx \textbf{18.6 ft}$$

Check for the assumption of R. Or

$$R = \frac{L(L - 2l_1)}{D(2L + D - 2l_1)} = \frac{35[35 - (2)(5)]}{18.6[(2)(35) + 18.6 - (2)(5)]}$$

$$= \textbf{0.599} \approx \textbf{0.6—O.K.}$$

Part b

From Eq. (6.81):

$$F = \bar{p}_a(L - RD) = 474.7[35 - (0.6)(18.6)]$$

$$= \textbf{11,316.8 lb/ft}$$

Part c

From Eq. (6.82):

$$M_{\text{max}} = 0.5\bar{p}_a L^2\left[\left(1 - \frac{RD}{L}\right)^2 - \left(\frac{2l_1}{L}\right)\left(1 - \frac{RD}{L}\right)\right]$$

$$1 - \frac{RD}{L} = 1 - \frac{(0.6)(18.6)}{35} = 0.681$$

So

$$M_{max} = (0.5)(474.7)(35)^2 \left[(0.681)^2 - \frac{(2)(5)(0.681)}{35} \right]$$

$$= \textbf{78,260 lb-ft/ft}$$

Note: The following table compares the values of D, F, and M_{max} obtained from the CPD method with those that would be obtained if the problem were solved by the conventional free earth support method (Section 6.7).

Item	CPD method	Conventional free earth method	
D	18.6 ft	11.23 ft	$\dfrac{D_{(CPD)}}{D_{actual-free\ earth}} = 1.66$
F	11,316.8 lb/ft	8070 lb/ft	$\dfrac{D_{(CPD)}}{D_{actual-free\ earth}} = 1.4$
M_{max}	78,260 lb-ft/ft	92,790 lb-ft/ft	$\dfrac{M_{max(CPD)}}{M_{max-free\ earth}} = 0.84$

6.11   Fixed Earth Support Method for Penetration of Sandy Soil

When using the fixed earth support method, one assumes that the toe of the pile is restrained from rotating, as shown in Figure 6.21a. The net lateral pressure distribution diagram for this condition is also shown in Figure 6.21a. In the fixed earth support solution, the lower portion of the pressure distribution diagram—that is, *HFH'GB*—is replaced by a concentrated force, P'. In order to calculate L_4, a simplified solution called the *equivalent beam solution* is generally used. In order to understand the equivalent beam solution, refer to point I, which is the point of inflection on the deflected shape of the sheet pile. At this point, the pile can be assumed to be hinged and the bending moment to be equal to zero (Figure 6.21b). The vertical distance between the point I and the dredge line is equal to L_5. Blum (1931) has given a mathematical solution between L_5 and $L_1 + L_2$. Figure 6.21d is a plot of $L_5/(L_1 + L_2)$ against the soil friction angle, ϕ.

With the known values of ϕ and $L_1 + L_2$, the magnitude of L_5 can now be obtained. The portion of sheet pile (Figure 6.21c) above the point I can now be treated as a beam that resists the net lateral earth pressure via the anchor force F (kN/m) and the shear P'' (kN/m). The shear force P'' can be calculated by taking the moment about O' (that is, at the anchor level).

Once the value of P'' is known, the length L_4 can be obtained by taking the moment about the point H (see the bottom diagram of Figure 6.21c). The depth of penetration, D, can now be given as 1.2 to 1.4 $(L_3 + L_4)$.

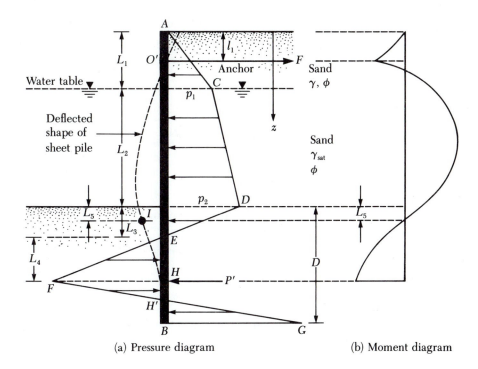

(a) Pressure diagram   (b) Moment diagram

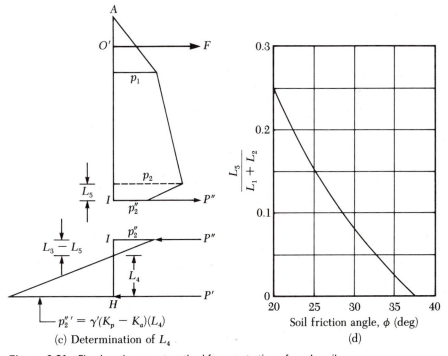

(c) Determination of L_4 (d)

Figure 6.21   Fixed earth support method for penetration of sandy soil

Step-by-Step Procedure for Obtaining D

The following step-by-step procedure for calculating the depth of embedment of sheet piles is based on the procedure just described.

Step 1. Obtain K_a and K_p.
Step 2. Calculate p_1 and p_2 from Eqs. (6.1) and (6.2), respectively.
Step 3. Calculate L_3 using Eq. (6.6).
Step 4. Determine L_5 from Figure 6.21d.
Step 5. Calculate p_2'' (Figure 6.21c):

$$p_2'' = \frac{p_2(L_3 - L_5)}{L_3} \tag{6.83}$$

Step 6. Draw the pressure distribution for the portion of the sheet pile located above I, as shown in Figure 6.21c.
Step 7. For the diagram drawn in Step 6, take the moment about O' to calculate P''.
Step 8. Knowing P'', draw the pressure distribution diagram for the portion of the sheet pile located between points I and H, as shown in Figure 6.21c. Note that in this diagram p_2''' is equal to $\gamma'(K_p - K_a)(L_4)$.
Step 9. For the diagram drawn in Step 8, take the moment about H to calculate L_4.
Step 10. Calculate $D = 1.2$ to $1.4(L_3 + L_4)$.

6.12   Anchors

Sections 6.7 through 6.11 presented the analysis of anchored sheet pile walls. These sections also discussed the expression for obtaining the force, F, per unit length of the sheet pile wall that has to be taken by the anchors. This section covers in more detail the various types of anchor generally used and the procedures for evaluating their ultimate holding capacity.

The general types of anchor used in sheet pile walls are as follows:

1. Anchor plates and beams (deadman)
2. Tie backs
3. Vertical anchor piles
4. Anchor beams supported by batter (compression and tension) piles

Anchor plates and beams are generally made of cast-concrete blocks (Figure 6.22a). The anchors are attached to the sheet pile by *tie rods*. A *wale* is placed at the front or back face of a sheet pile for the purpose of conveniently attaching the tie rod to the wall. To protect the tie rod from corrosion, it is generally coated with paint or asphaltic materials.

In the construction of *tie backs*, bars or cables are placed in predrilled holes (Figure 6.22b) with concrete grout (cables are commonly high-strength, prestressed steel tendons). Figures 6.22c and 6.22d show a vertical anchor pile and an anchor beam with batter piles.

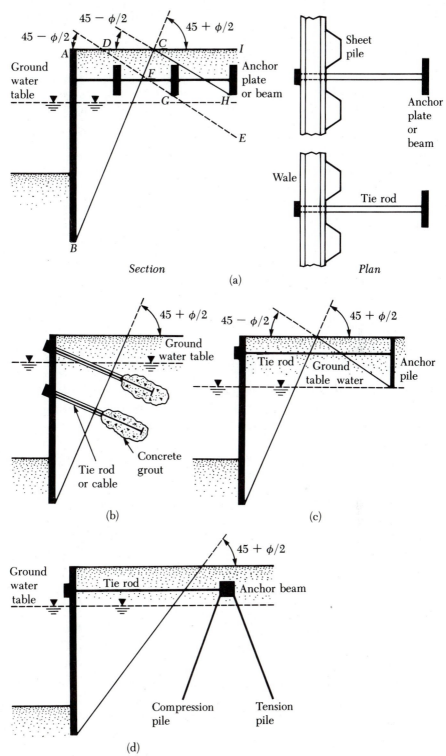

Figure 6.22 Various types of anchoring for sheet pile walls: (a) anchor plate or beam; (b) tie back; (c) vertical anchor pile; (d) anchor beam with batter piles

369

Placement of Anchors

The resistance offered by anchor plates and beams is primarily derived from the passive force of the soil located in front of them. To learn how to determine the best location for an anchor plate (for maximum efficiency), refer to Figure 6.22a, in which *AB* is the sheet pile wall. If the anchor is placed inside the wedge *ABC*, which is the Rankine active zone, it would not provide any resistance to failure. As an alternative, one could place the anchor in the zone *CFEH*. Note that line *DFG* is the slip line for the Rankine passive pressure. If part of the passive wedge is located inside the active wedge *ABC*, full passive resistance of the anchor cannot be realized upon failure of the sheet pile wall. However, if the anchor is placed in zone *ICH*, the Rankine passive zone in front of the anchor slab or plate is located completely outside the Rankine active zone *ABC*. In this case, full passive resistance from the anchor can be realized.

Figures 6.22b, 6.22c, and 6.22d also show the proper locations for placement of tie backs, vertical anchor piles, and anchor beams supported by batter piles.

Calculation of the Ultimate Resistance Offered by Anchor Plates and Beams in Sand

Teng (1962) has proposed the following equation to determine the ultimate resistance of anchor plates or walls in granular soils located at or near the ground surface ($H/h \leq 1.5$ to 2 in Figure 6.23).

$$P_u = B(P_p - P_a) \quad \text{(for continuous plates or beams—that is, } B/h \approx \infty\text{)} \tag{6.84}$$

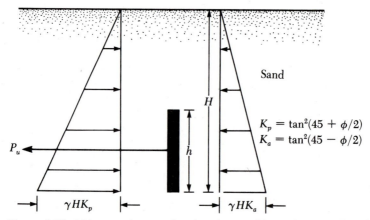

$$K_p = \tan^2(45 + \phi/2)$$
$$K_a = \tan^2(45 - \phi/2)$$

Figure 6.23 Ultimate resistance of anchor plates and beams in sand—Eqs. (6.84) and (6.87) where P_u = ultimate resistance of anchor

                B = length of anchor at right angle to the cross section shown

     P_p and P_a = Rankine passive and active force per unit length of anchor

Note that P_p acts in front of the anchor, as shown in Figure 6.23. Also

$$P_p = \frac{1}{2}\gamma H^2 \tan^2\left(45 + \frac{\phi}{2}\right) \tag{6.85}$$

and

$$P_a = \frac{1}{2}\gamma H^2 \tan^2\left(45 - \frac{\phi}{2}\right) \tag{6.86}$$

Equation (6.84) is valid for the plane-strain type of condition. For all practical cases, $B/h > 5$ may be considered as plane-strain condition.

For $B/h <$ about 5, considering the three-dimensional failure surface (that is, accounting for the friction resistance developed at the two ends of an anchor), Teng (1962) has given the following relation for the ultimate anchor resistance:

$$P_u = B(P_p - P_a) + \frac{1}{3}K_0\gamma(\sqrt{K_p} + \sqrt{K_a})H^3 \tan\phi$$

$$\left(\text{for }\frac{H}{h} \leq 1.5 \text{ to } 2\right) \tag{6.87}$$

where K_O = earth pressure coefficient at rest ≈ 0.4

More recently, Ovesen and Stromann (1972) have proposed a method to determine the ultimate anchor resistance in sand. The method is described below and is recommended for use, since this is the most rational method available at this time. The calculation is done in three steps.

Step 1. Basic case consideration. Determine the depth of embedment, H. Assume that the anchor slab has a height of H and is continuous (that is, B = length of anchor slab perpendicular to the cross section = ∞), as shown in Figure 6.24. This is referred to as the *basic case*. In Figure 6.24, the following notations have been used:

    P_p = passive force per unit length of the anchor
    P_a = active force per unit length of the anchor
    ϕ = soil friction angle
    δ = friction angle between the anchor slab and the soil
    P'_u = ultimate resistance per unit length of the anchor
    W = weight per unit length of the anchor slab

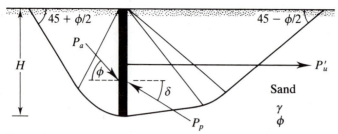

Figure 6.24 Basic case—continuous vertical anchor in granular soil

The magnitude of P'_u can be given as

$$P'_u = \tfrac{1}{2}\gamma H^2 K_p \cos\delta - P_a \cos\phi = \tfrac{1}{2}\gamma H^2 K_p \cos\delta - \tfrac{1}{2}\gamma H^2 K_a \cos\phi$$
$$= \tfrac{1}{2}\gamma H^2 (K_p \cos\delta - K_a \cos\phi)$$

(6.88)

where K_a = active pressure coefficient with $\delta = \phi$ (see Figure 6.25a)
     K_p = passive pressure coefficient

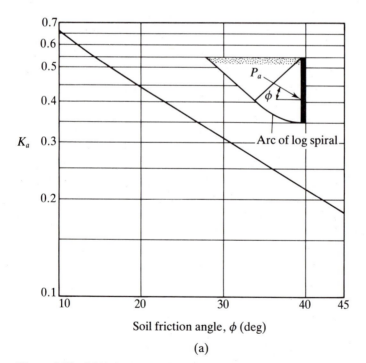

Soil friction angle, ϕ (deg)

(a)

Figure 6.25 (a) Variation of K_a (for $\delta = \phi$); (b) variation of $K_p \cos\delta$ with $K_p \sin\delta$ based on Ovesen and Stromann's analysis (continued on page 373)

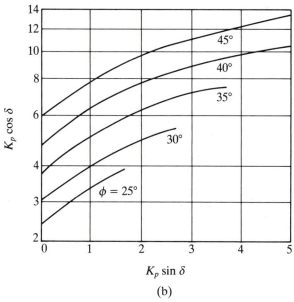

Figure 6.25   (Continued)

In order to obtain $K_p \cos \delta$, first calculate $K_p \sin \delta$ as

$$K_p \sin \delta = \frac{W + P_a \sin \phi}{\frac{1}{2}\gamma H^2} = \frac{W + \frac{1}{2}\gamma H^2 K_a \sin \phi}{\frac{1}{2}\gamma H^2} \qquad (6.89)$$

Using the value of $K_p \sin \delta$ obtained from Eq. (6.89), the magnitude of $K_p \cos \delta$ can be estimated from the plots given in Figure 6.25b.

   Step 2. Strip case. Determine the actual height of the anchor, h, to be constructed. If a continuous anchor (that is, $B = \infty$) of height h is placed in the soil such that its depth of embedment is H, as shown in Figure 6.26, the ultimate resistance per unit length can be given as

$$P'_{us} = \left[\frac{C_{ov} + 1}{C_{ov} + \left(\dfrac{H}{h}\right)}\right]\underset{\text{Eq. 6.88}}{\overset{\uparrow}{P'_u}} \qquad (6.90)$$

where P'_{us} = ultimate resistance for the *strip case*
        C_{ov} = 19 for dense sand and 14 for loose sand

Figure 6.26 Strip case for a vertical anchor

Step 3. Actual case. In practice, the anchor plates are placed in a row with center-to-center spacing, S', as shown in Figure 6.27a. For the *actual case*, the ultimate resistance of each anchor, P_u, can be calculated as

$$P_u = P'_{us} B_e \tag{6.91}$$

where B_e = equivalent length

The equivalent length is a function of S', B, H, and h. Figure 6.27b shows a plot of $(B_e - B)/(H + h)$ against $(S' - B)/(H + h)$ for the cases of loose and dense sand. With known values of S', B, H, and h, the value of B_e can be calculated and used in Eq. (6.91) to obtain P_u.

Relatively few studies have been conducted so far to determine the relation of load to displacement of anchors. Figure 6.28 shows an example of nondimensional displacement of anchors for various values of B/h and H/h as obtained by Neeley, Stuart, and Graham (1973) experimentally in medium to dense sand. Das (1975) and Das and Seeley (1975) also found essentially similar relations for anchors tested in loose sand. Based on the experimental results, Das and Seeley (1975) presented the following load-displacement relationship for anchors:

$$\bar{P} = \frac{\bar{\Delta}}{0.15 + 0.85\bar{\Delta}} \tag{6.92}$$

where $\bar{P} = \dfrac{\text{load on anchor at horizontal displacement, } \Delta}{\text{ultimate load at horizontal displacement, } \Delta_u}$ $\tag{6.93}$

$$\bar{\Delta} = \frac{\Delta}{\Delta_u} \tag{6.94}$$

The relationship given by Eq. (6.92) was found to be valid for B/h varying from 1 to 5 and H/h varying from 1 to 5.

Ultimate Resistance of Anchor Plates and Beams in Clay ($\phi = 0$ Condition)

Relatively few studies have so far been conducted to estimate the ultimate resistance of anchor plates and beams in clayey soils ($\phi = 0$ condition).

(a)

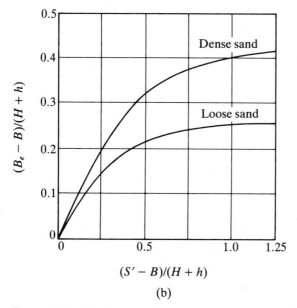

(b)

Figure 6.27   (a) Actual case for row of anchors; (b) variation of $(B_e - B)/(H + h)$ with $(S' - B)/(H + h)$ based on Ovesen and Stromann's analysis

Mackenzie (1955) and Tschebotarioff (1973) have provided the nature of variation of the ultimate resistance of strip anchors and beams as a function of H, h, and c (undrained cohesion based on $\phi = 0$ condition) in a non-dimensional form based on laboratory model test results. Das, Tarquin, and Moreno (1985) have suggested the following procedure to obtain the ultimate resistance of an anchor embedded in clay.

When an anchor plate having dimensions of $h \times B$ is embedded at a depth H, the failure surface in soil at ultimate load may extend to the ground surface, as shown in Figure 6.29a. This condition will arise when the ratio H/h is relatively small. However, for larger values of H/h, local shear failure takes place at ultimate load (Figure 6.29b). The critical value of H/h

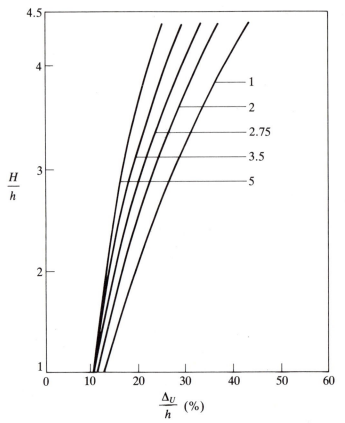

Figure 6.28 Vertical anchor plates or beams: horizontal displacement at ultimate load (redrawn after Neeley, Stuart, and Graham, 1973)

at which general shear failure changes to local shear failure in soil can be given as

$$\left(\frac{H}{h}\right)_{\text{cr-S}} = 4.7 + 2.9 \times 10^{-3} c \le 7$$
$$\text{(for } square \text{ anchors, that is, } B/h = 1) \qquad (6.95)$$

and

$$\left(\frac{H}{h}\right)_{\text{cr-R}} = \left(\frac{H}{h}\right)_{\text{cr-S}}\left[0.9 + 0.1\left(\frac{B}{h}\right)\right] \le 1.3\left(\frac{H}{h}\right)_{\text{cr-S}}$$
$$\text{(for } rectangular \text{ anchors, that is, } B/h \ge 1) \qquad (6.96)$$

In Eqs. (6.95) and (6.96), the unit of the undrained cohesion is lb/ft^2.

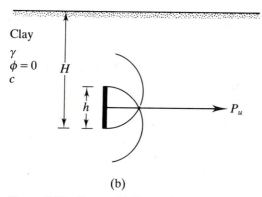

Figure 6.29 Nature of failure surface in soil around vertical anchor plate: (a) H/h relatively small; (b) $H/h > (H/h)_{cr}$

The ultimate resistance of an anchor plate can be expressed in a non-dimensional form as

$$F_c = \frac{P_u}{Bhc} \tag{6.97}$$

where F_c = breakout factor
P_u = ultimate resistance

Figure 6.30 shows the nature of variation of F_c against H/h for an anchor plate embedded in clay. Note that, for $H/h \geq (H/h)_{cr}$, the magnitude of F_c is equal to $F_{c(max)}$ which is a constant. For square anchors (that is, $B = h$), $F_{c(max)}$ is equal to 9. So, with $H/h \geq (H/h)_{cr\text{-}S}$

$$P_u = 9h^2c \quad \text{(for square anchors)} \tag{6.98}$$

For rectangular anchors with $H/h \geq (H/h)_{cr\text{-}R}$, the ultimate resistance can be given as

$$P_u = 9Bhc\left[0.825 + 0.175\left(\frac{h}{B}\right)\right]$$

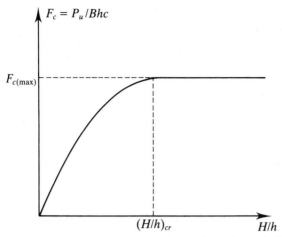

Figure 6.30 Nature of variation of F_c with H/h for vertical anchor in clay

or

$$P_u = Bch\left[7.425 + 1.575\left(\frac{h}{B}\right)\right] \tag{6.99}$$

So, for square and rectangular anchors with $H/h \leq (H/h)_{cr}$, the ultimate resistance can be calculated from the following empirical relationship:

$$\frac{\left[\dfrac{H/h}{(H/h)_{cr}}\right]}{\left[\dfrac{P_u/cBh}{7.425 + 1.575(h/B)}\right]} = 0.41 + 0.59\left[\frac{H/h}{(H/h)_{cr}}\right] \tag{6.100}$$

Factor of Safety for Anchor Plates and Beams

The allowable resistance per anchor plate, P_{all}, can be given as

$$P_{all} = \frac{P_u}{FS} \tag{6.101}$$

where FS = factor of safety

Generally, a factor of safety of 2 is suggested.

Spacing of Anchor Plates

The center-to-center spacing of anchors, S', can be obtained thus:

$$S' = \frac{P_{all}}{F} \tag{6.102}$$

where F = force per unit length of the sheet pile

Figure 6.31 Parameters for defining the ultimate resistance of tie backs

Ultimate Resistance of Tie Backs

According to Figure 6.31, the ultimate resistance offered by a tie back in sand can be given as follows:

$$P_u = \pi dl\bar{\sigma}'_v K \tan \phi \qquad (6.103)$$

where P_u = ultimate resistance
ϕ = angle of friction of soil
$\bar{\sigma}'_v$ = average effective vertical stress ($= \gamma z$ in dry sand)
K = earth pressure coefficient

The value of K can be taken to be equal to the earth pressure coefficient at rest (K_O) if the concrete grout is placed under pressure (Littlejohn, 1970). The lower limit of K can be taken to be equal to the Rankine active earth pressure coefficient.

In clays, the ultimate resistance of tie backs can be approximated as

$$P_u = \pi dl c_a \qquad (6.104)$$

where c_a = adhesion

The value of c_a can be approximated as $\frac{2}{3} c$ (where c = undrained cohesion).

A factor of safety of 1.5–2 may be used over the ultimate resistance to obtain the allowable resistance offered by each tie back.

Example 6.9

Refer to Figure 6.27a. Given: $B = h = 0.4$ m, $S' = 1.2$ m, $H = 1$ m, $\gamma = 16.51$ kN/m³, and $\phi = 35°$. Determine the ultimate resistance for each anchor plate. The anchor plates are made of concrete and have thicknesses of 0.15 m.

Solution

From Figure 6.25a for $\phi = 35°$, the magnitude of K_a is about 0.27.

$$W = Ht\gamma_{concrete} = (1 \text{ m})(0.15 \text{ m})(23.5 \text{ kN/m}^3)$$
$$= 3.525 \text{ kN/m}$$

From Eq. (6.89)

$$K_p \sin \delta = \frac{W + \frac{1}{2}\gamma H^2 K_a \sin \phi}{\frac{1}{2}\gamma H^2}$$

$$= \frac{3.535 + (0.5)(16.51)(1)^2(0.26)(\sin 35)}{(0.5)(16.51)(1)^2} = 0.576$$

From Figure 6.25b with $\phi = 35°$ and $K_p \sin \delta = 0.576$, the value of $K_p \cos \delta$ is about 4.5. Now, using Eq. (6.88)

$$P'_u = \frac{1}{2}\gamma H^2(K_p \cos \delta - K_a \cos \phi)$$
$$= (\frac{1}{2})(16.51)(1)^2[4.5 - (0.26)(\sin 35)] = 35.92 \text{ kN/m}$$

In order to calculate P'_{us}, let us assume the sand to be loose. So, C_{ov} in Eq. (6.90) is equal to 14. Hence

$$P'_{us} = \left[\frac{C_{ov}+1}{C_{ov}+\left(\dfrac{H}{h}\right)}\right]P'_u = \left[\frac{14+1}{14+\left(\dfrac{1}{0.4}\right)}\right] = 32.65 \text{ kN/m}$$

$$\frac{S'-B}{H+h} = \frac{1.2-0.4}{1+0.4} = \frac{0.8}{1.4} = 0.571$$

For $(S'-B)/(H+h) = 0.571$ and loose sand, Figure 6.27b yields

$$\frac{B_e - B}{H - h} = 0.229$$

So

$$B_e = (0.229)(H+h) + B = (0.229)(1+0.4) + 0.4$$
$$= 0.72$$

Hence, from Eq. (6.91)

$$P_u = P'_{us} B_e = (32.65)(0.72) = \textbf{23.51 kN}$$

Example 6.10

Refer to Example 6.9. Determine the anchor resistance for a maximum horizontal displacement (Δ) of 40 mm.

Solution

From Figure 6.28, for $H/h = 2.5$, $B/h = 1$, and $\Delta_u/h = 0.24$. Hence, at ultimate load, P_u, the displacement of the anchor will be

$$\Delta_u = 0.24h = 0.24 \times 0.4 = 0.096 \text{ m} = 96 \text{ mm}$$

If $\Delta = 40$ mm, $\bar{\Delta} = \Delta/\Delta_u = 40/96 = 0.417$. Referring to Eq. (6.92)

$$\bar{P} = \frac{\bar{\Delta}}{0.15 + 0.85\bar{\Delta}}$$

Hence

$$\bar{P} = \frac{0.417}{0.15 + (0.85)(0.417)} = 0.827$$

Thus

$$P_{\text{at }\Delta=40\text{ mm}} = 0.827 \times P_u = (0.827)(23.51) = \mathbf{19.44\ kN}$$

Example 6.11

For a vertical anchor plate in clay, the following parameters are given: $B = 5$ ft, $h = 2$ ft, $H = 6$ ft, and $c = 500$ lb/ft^2. Determine the ultimate resistance of the anchor.

Solution

Using Eq. (6.95)

$$\left(\frac{H}{h}\right)_{\text{cr-S}} = 4.7 + 2.9 \times 10^{-3}c$$
$$= 4.7 + (2.9 \times 10^{-3})(500) = 6.15$$

Again, from Eq. (6.96)

$$\left(\frac{H}{h}\right)_{\text{cr-R}} = \left(\frac{H}{h}\right)_{\text{cr-S}}\left[0.9 + 0.1\left(\frac{B}{h}\right)\right]$$
$$= (6.15)\left[0.9 + (0.1)\left(\frac{5}{2}\right)\right] = 7.07$$

Since $H/h = 6/2 = 3$ is less than $(H/h)_{\text{cr}}$, we will use Eq. (6.100) to obtain P_u, from which

$$\frac{\left(\frac{3}{7.07}\right)}{\left[\frac{(P_u/cBh)}{7.425 + 1.575(2/5)}\right]} = 0.41 + (0.59)\left(\frac{3}{7.07}\right)$$

or

$$\frac{0.424}{\left[\frac{(P_u/cBh)}{8.055}\right]} = 0.66$$

So

$$\frac{P_u}{cBh} = \left(\frac{0.424}{0.66}\right)(8.055) = 5.17$$
$$P_u = (5.17)cBh = (5.17)(500)(5)(2) = \mathbf{25,850\ lb}$$

PROBLEMS

6.1 Figure P6.1 shows a cantilever sheet pile wall penetrating a granular soil. Given: $L_1 = 10$ ft, $L_2 = 20$ ft, $\gamma = 120$ lb/ft^3, $\gamma_{sat} = 129.4$ lb/ft^3, and $\phi = 40°$.

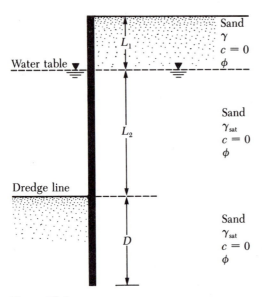

Figure P6.1

   a. What is the theoretical depth of embedment, D?
   b. Draw the net pressure distribution diagram.
   c. Assuming a 30% increase in D, what should be the total length of sheet piles to be used?

6.2 From part (b) of Problem 6.1, determine the maximum moment in the sheet pile section. Also calculate the appropriate section modulus, S, for the required sheet pile.

6.3 Redo Problem 6.1(a), (b), and (c) with the following: $L_1 = 3$ m, $L_2 = 6$ m, $\gamma = 17.3$ kN/m^3, $\gamma_{sat} = 19.4$ kN/m^3, and $\phi = 30°$.

6.4 For Problem 6.3, determine the theoretical maximum moment for the sheet pile section and then choose one.

6.5 Solve Problem 6.1 with the following: $L_1 = 4$ m, $L_2 = 8$ m, $\gamma = 16.1$ kN/m^3, $\gamma_{sat} = 18.2$ kN/m^3, and $\phi = 32°$.

6.6 For Problem 6.5, determine the theoretical maximum moment for the sheet pile section.

6.7 Refer to Figure 6.8. Given: $L = 15$ ft, $\gamma = 108$ lb/ft^3, and $\phi = 35°$. Calculate the theoretical depth of penetration, D, and the maximum moment.

6.8 Repeat Problem 6.7 with the following: $L = 3$ m, $\gamma = 16.7$ kN/m^3, and $\phi = 30°$.

6.9 Refer to Figure 6.9, which shows a free cantilever sheet pile wall penetrating a sand layer. Determine the theoretical depth of penetration and the maximum moment. Given: $\gamma = 105$ lb/ft^3, $\phi = 38°$, $L = 10$ ft, and $P = 1000$ lb/ft.

6.10 Repeat Problem 6.9 with the following: $\gamma = 18$ kN/m³, $\phi = 34°$, $L = 2.8$ m, and $P = 12$ kN/m.

6.11 Refer to Figure P6.11. Given: $L_1 = 2.4$ m, $L_2 = 4.6$ m, $\gamma = 15.7$ kN/m³, $\gamma_{sat} = 17.3$ kN/m³, $c = 29$ kN/m², and $\phi = 30°$.

Figure P6.11

a. Find the theoretical depth of penetration, D.
b. Draw the pressure distribution diagram with all dimensions.
c. Increase D by 40%. What is the length of sheet piles needed for this work?

6.12 For Problem 6.11, determine the maximum moment and choose a sheet pile section.

6.13 Redo Problem 6.11 with the following: $L_1 = 10$ ft, $L_2 = 12$ ft, $c = 1200$ lb/ft², $\gamma = 110$ lb/ft³, $\gamma_{sat} = 126$ lb/ft³, $\phi = 40°$.

6.14 For Problem 6.13, determine the maximum moment and the appropriate section.

6.15 Refer to Figure 6.11. Given: $L = 15$ ft, $\gamma = 108$ lb/ft³, $\phi = 36°$, and $c = 800$ lb/ft². Find the theoretical value of D and the maximum moment.

6.16 Refer to Figure 6.12. Given: $L = 3$ m, $P = 6$ kN/m, $c = 35$ kN/m², and $\gamma_{sat} = 19.2$ kN/m³. Find the theoretical value of D and the maximum moment.

6.17 An anchored sheet pile bulkhead is shown in Figure P6.17. Given: $L_1 = 5$ m, $L_2 = 8$ m, $l_1 = 2$ m, $\gamma = 16.5$ kN/m³, $\gamma_{sat} = 18.5$ kN/m³, and $\phi = 32°$.
a. Calculate the theoretical value of the depth of embedment, D.
b. Draw the pressure distribution diagram.
c. Determine the anchor force per unit length of the wall.
Use the free earth support method.

Figure P6.17

6.18 Refer to Problem 6.17. Assume that $D_{actual} = 1.25D_{theoretical}$.
  a. Determine the theoretical maximum moment.
  b. Choose a sheet pile section using Rowe's moment reduction technique.

6.19 Redo Problem 6.17 with the following: $L_1 = 10$ ft, $L_2 = 25$ ft, $l_1 = 4$ ft, $\gamma = 120$ lb/ft^3, $\gamma_{sat} = 129.4$ lb/ft^3, and $\phi = 40°$.

6.20 Refer to Problem 6.19. Assume that $D_{actual} = 1.3D_{theoretical}$. Using Rowe's moment reduction method, choose a sheet pile section.

6.21 An anchored sheet pile bulkhead is shown in Figure P6.21. Given: $L_1 = 2$ m, $L_2 = 6$ m, $l_1 = 1$ m, $\gamma = 16$ kN/m^3, $\gamma_{sat} = 18.86$ kN/m^3, $\phi = 32°$, and $c = 27$ kN/m^2.

Figure P6.21

a. Determine the theoretical depth of embedment, D.
b. Calculate the anchor force per unit length of the sheet pile wall.
Use the free earth support method.

6.22 Repeat Problem 6.21 with the following: $L_1 = 8$ ft, $L_2 = 20$ ft, $l_1 = 4$ ft, $c = 1500$ lb/ft², $\gamma = 115$ lb/ft³, $\gamma_{sat} = 128$ lb/ft³, and $\phi = 40°$.

6.23 Refer to Figure 6.16. Given: $L_1 = 10$ ft, $L_2 = 20$ ft, $l_1 = 5$ ft, $\gamma = 110$ lb/ft³, $\gamma_{sat} = 122.4$ lb/ft³, and $\phi = 35°$. Using the computational diagram method (Section 6.10), determine D, F, and M_{max}. Assume that $C = 0.68$ and $R = 0.6$.

6.24 Repeat Problem 6.23 with the following: $L_1 = 3.05$ m, $L_2 = 6.1$ m, $l_1 = 1.53$ m, $\gamma = 16$ kN/m³, $\gamma_{sat} = 18.5$ kN/m³, and $\phi = 30°$. Assume that $C = 0.8$ and $R = 0.5$.

6.25 Refer to Figure 6.27a. For the anchor slab in sand, $H = 6$ ft, $h = 2.5$ ft, $B = 3.5$ ft, $S' = 8$ ft, $\phi = 30°$, and $\gamma = 105$ lb/ft³. Calculate the ultimate holding capacity of each anchor. The anchor plates are made of concrete and have a thickness of 3 in.

6.26 A single anchor slab is shown in Figure P6.26. Given: $H = 0.9$ m, $h = 0.3$ m, $\gamma = 17$ kN/m³, and $\phi = 32°$. Calculate the ultimate holding capacity of the anchor slab if the width B is (a) 0.3 m, (b) 0.6 m, and (c) 0.9 m. (*Note:* center-to-center spacing, $S' = \infty$.)

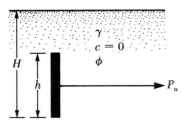

Figure P6.26

6.27 Refer to Problem 6.26. Estimate the probable horizontal displacement of the anchors at ultimate load. Also estimate the resistance of anchors if the allowable displacement is 20 mm.

6.28 A vertical anchor plate in clay has the following: $B = 3$ ft, $H = 5.5$ ft, $h = 3$ ft, and $c = 650$ lb/ft². Estimate the ultimate resistance of the anchor.

REFERENCES

Blum, H. (1931). *Einspannungsverhältnisse bei Bohlwerken*, W. Ernst und Sohn, Berlin, Germany.

Das, B. M. (1975). "Pullout Resistance of Vertical Anchors," *Journal of the Geotechnical Engineering Division*, American Society of Civil Engineers, Vol. 101, No. GT1, pp. 87–91.

Das, B. M., and Seeley, G. R. (1975). "Load-Displacement Relationships for Vertical Anchor Plates," *Journal of the Geotechnical Engineering Division*, American Society of Civil Engineers, Vol. 101, No. GT7, pp. 711–715.

Das, B. M., Tarquin, A. J., and Moreno, R. (1985). "Model Tests for Pullout Resistance of Vertical Anchor in Clay," *Journal of Civil Engineering for Practicing and Design Engineers*, Vol. 4, No. 22, pp. 191–209.

Littlejohn, G. S. (1970). "Soil Anchors," *Proceedings*, Conference on Ground Engineering, Institute of Civil Engineers, London, pp. 33–44.

Mackenzie, T. R. (1955). *Strength of Deadman Anchors in Clay*, M.S. Thesis, Princeton University, Princeton, N.J.

Nataraj, M. S., and Hoadley, P. G. (1984). "Design of Anchored Bulkheads in Sand," *Journal of Geotechnical Engineering*, American Society of Civil Engineers, Vol. 110, No. GT4, pp. 505–515.

Neeley, W. J., Stuart, J. G., and Graham, J. (1973). "Failure Loads of Vertical Anchor Plates in Sand," *Journal of the Soil Mechanics and Foundations Division*, American Society of Civil Engineers, Vol. 99, No. SM9, pp. 669–685.

Ovesen, N. K., and Stromann, H. (1972). "Design Methods for Vertical Anchor Slabs in Sand," *Proceedings*, Specialty Conference on Performance of Earth and Earth-Supported Structures, American Society of Civil Engineers, Vol. 2.1, pp. 1481–1500.

Rowe, P. W. (1952). "Anchored Sheet Pile Walls," *Proceedings*, Institute of Civil Engineers, Vol. 1, Part 1, pp. 27–70.

Rowe, P. W. (1957). "Sheet Pile Walls in Clay," *Proceedings*, Institute of Civil Engineers, Vol. 7, pp. 629–654.

Teng, W. C. (1962). *Foundation Design*, Prentice-Hall, Englewood Cliffs, N.J.

Tschebotarioff, G. P. (1973). *Foundations, Retaining and Earth Structures*, 2nd ed., McGraw-Hill, New York.

Tsinker, G. P. (1983). "Anchored Sheet Pile Bulkheads: Design Practice," *Journal of Geotechnical Engineering*, American Society of Civil Engineers, Vol. 109, No. GT8, pp. 1021–1038.

Chapter 7

BRACED CUTS

7.1 Introduction

Sometimes construction work requires ground excavations with vertical or near-vertical faces—for example, basements of buildings in developed areas or underground transportation facilities at shallow depths below the ground surface (cut-and-cover type of construction). The vertical faces of the cuts need to be protected by temporary bracing systems to avoid failure that may be accompanied by considerable settlement or by bearing capacity failure of nearby foundation(s).

Figure 7.1 shows two types of braced cut commonly used in construction work. One type uses the *soldier beam* (Figure 7.1a), which is driven into the ground before excavation. Soldier beams are vertical steel or timber beams. After the excavation is started, *laggings* are placed between the soldier beams as the excavation proceeds. Laggings are horizontal timber planks. When the excavation reaches the desired depth, *wales* and *struts* are properly installed (wales and struts are horizontal steel beams). The struts are horizontal compression members. Figure 7.1b shows another type of braced excavation. In this case, interlocking *sheet piles* are driven into the soil before excavation. As the excavation proceeds, wales and struts are inserted immediately after reaching the appropriate depth.

To design braced excavations (that is, to select wales, struts, sheet piles, and soldier beams), one must estimate the lateral earth pressure to which the braced cuts will be subjected. This topic is discussed in Section 7.2; subsequent sections cover the procedures of analysis and design of braced cuts.

7.2 Lateral Earth Pressure in Braced Cuts

In Chapter 5 we learned that a retaining wall rotates about its bottom (Figure 7.2a). With sufficient yielding of the wall, the lateral earth pressure can be approximated to be equal to that obtained by Rankine's or Coulomb's theory. In contrast to retaining walls, braced cuts show a different type of wall yielding (see Figure 7.2b). In this case, the deformation of

387

Figure 7.1 Types of braced cut: (a) use of soldier beams; (b) use of sheet piles

the wall gradually increases with the depth of excavation. The variation of the amount of deformation will depend on several factors, such as the type of soil, the depth of excavation, and the workmanship. However, one can easily visualize that, with very little wall yielding at the top of the cut, the lateral earth pressure will be close to the at-rest pressure. At the bottom of the wall, with a much larger degree of yielding, the lateral earth pressure

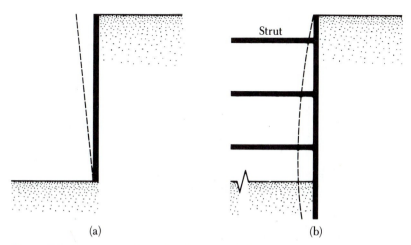

Figure 7.2 Nature of yielding of walls: (a) retaining wall; (b) braced cut

will be substantially lower than the Rankine active earth pressure. As a result, the distribution of lateral earth pressure will vary substantially in comparison to the linear distribution assumed in the case of retaining walls.

A theoretical evaluation of the total lateral force, P, imposed on a wall can be made by using Terzaghi's general wedge theory (1943a) (Figure 7.3a), in which the failure surface is assumed to be the arc of a logarithmic

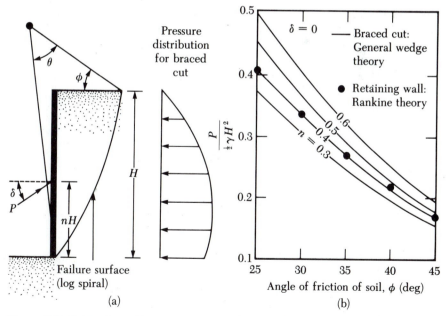

Figure 7.3 Comparison of lateral earth pressure for braced cuts and retaining walls in sand ($\delta = 0$)

spiral, defined by the equation

$$r = r_o e^{\theta \tan \phi} \tag{7.1}$$

where ϕ = angle of friction of soil

A detailed outline for the evaluation of P is beyond the scope of this text; readers should check a soil mechanics text for more information (for example, Das, 1990). However, a comparison of the lateral earth pressure for braced cuts in sand (with angle of wall friction $\delta = 0$) with that for a retaining wall ($\delta = 0$) is shown in Figure 7.3b. If $\delta = 0$, a retaining wall of height H will be subjected to a Rankine active earth pressure, and the result-ant active force will intersect the wall at a distance of nH measured from the bottom of the wall. For this case, $n = \frac{1}{3}$. In contrast, the value of n for a braced cut may vary from 0.33 to 0.5 or 0.6. The general wedge theory can also be used to analyze braced cuts in saturated clay (for example, see Das and Seeley, 1975).

In any event, when choosing a lateral soil pressure distribution for design of braced cuts, one should keep in mind that the nature of failure in braced cuts is much different from that in retaining walls. After observation of several braced cuts, Peck (1969) suggested using *design pressure envelopes* for braced cuts in sand and clay. Figures 7.4, 7.5, and 7.6 show Peck's pres-sure envelopes, to which the following guidelines apply.

Cuts in Sand

Figure 7.4 shows the pressure envelope for cuts in sand. This pressure, p_a, can be given by the relationship

$$\boxed{p_a = 0.65\gamma H K_a} \tag{7.2}$$

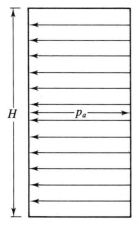

Figure 7.4   Peck's apparent pressure envelope for cuts in sand

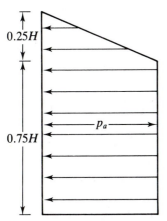

Figure 7.5   Peck's apparent pressure envelope for cuts in soft to medium clay

where γ = unit weight
    H = height of the cut
    K_a = Rankine active pressure coefficient = $\tan^2 (45 - \phi/2)$

Cuts in Soft and Medium Clay

The pressure envelope for soft to medium clay is shown in Figure 7.5. This is applicable for the condition

$$\frac{\gamma H}{c} > 4$$

where c = undrained cohesion ($\phi = 0$ condition)

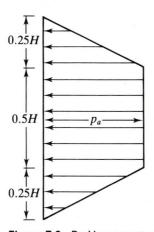

Figure 7.6   Peck's apparent pressure envelope for cuts in stiff clay

The pressure, p_a, is expressed as the larger of the following:

$$p_a = \gamma H \left[1 - \left(\frac{4c}{\gamma H} \right) \right]$$

or   (7.3)

$$p = 0.3\gamma H$$

where γ = unit weight of clay

Cuts in Stiff Clay

This is applicable for the condition of $\gamma H/c \leq 4$. The pressure envelope is shown in Figure 7.6, in which

$$p_a = 0.2\gamma H \text{ to } 0.4\gamma H \qquad \text{(with an average of } 0.3\gamma H) \qquad (7.4)$$

Peck's pressure envelopes are sometimes referred to as *apparent pressure envelopes*. However, the actual pressure distribution is a function of the construction sequence and the relative flexibility of the sheet pile wall.

Sometimes one encounters layers of both sand and clay when constructing a braced cut. In this case, Peck (1943) proposed that an equivalent value of cohesion ($\phi = 0$ concept) should be determined in the following manner (refer to Figure 7.7a):

$$c_{av} = \frac{1}{2H} [\gamma_s K_s H_s^2 \tan \phi_s + (H - H_s)n'q_u] \qquad (7.5)$$

where H = total height of the cut
γ_s = unit weight of sand
H_s = height of the sand layer
K_s = a lateral earth pressure coefficient for the sand layer (≈ 1)
ϕ_s = angle of friction of sand
q_u = unconfined compression strength of clay
n' = a coefficient of progressive failure (ranges from 0.5 to one; average value 0.75)

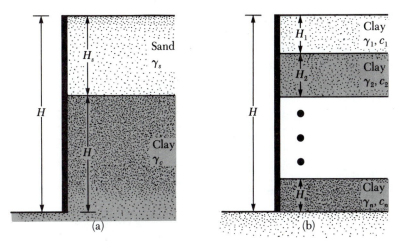

Figure 7.7 Layered soils in braced cuts

The average unit weight, γ_a, of the layers can be expressed as

$$\gamma_a = \frac{1}{H} \left[\gamma_s H_s + (H - H_s)\gamma_c \right]$$

(7.6)

where γ_c = saturated unit weight of clay layer

Once the average values of cohesion and unit weight are determined, the pressure envelopes in clay can be used to design the cuts.

In a similar manner, when a number of clay layers are encountered in the cut (Figure 7.7b), the average undrained cohesion can be expressed by the equation

$$c_{av} = \frac{1}{H} \left(c_1 H_1 + c_2 H_2 + \cdots + c_n H_n \right)$$

(7.7)

where $c_1, c_2, \ldots, c_n$ = undrained cohesion in layers $1, 2, \ldots, n$
$H_1, H_2, \ldots, H_n$ = thicknesses of layers $1, 3, \ldots, n$

The average unit weight, γ_a, can be given as

$$\gamma_a = \frac{1}{H} \left(\gamma_1 H_1 + \gamma_2 H_2 + \gamma_3 H_3 + \cdots + \gamma_n H_n \right)$$

(7.8)

7.3   Tschebotarioff's Pressure Envelopes

Similar to the pressure envelopes shown in Figures 7.4, 7.5, and 7.6, which were suggested by Peck (1969), another form of distribution of pressure has

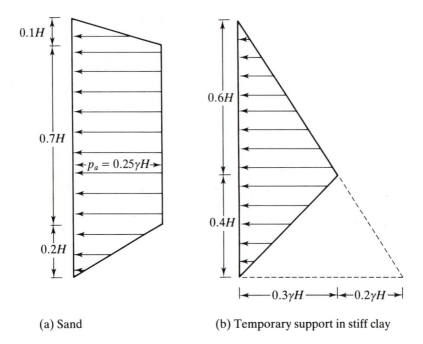

(a) Sand

(b) Temporary support in stiff clay

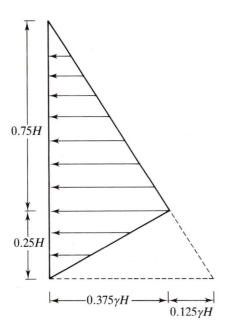

(c) Permanent support in medium clay

Figure 7.8 Tschebotarioff's pressure envelopes

been suggested by Tschebotarioff (1973). This is shown in Figure 7.8 and is sometimes used for design. In this text, however, *Figures 7.4, 7.5, and 7.6 will be used for all calculations and problems.*

7.4   Design of Various Components of a Braced Cut

Struts

In construction work, the struts should have a minimum vertical spacing of about 2.75 m (≈ 9 ft) or more. The struts are actually horizontal columns subject to bending. The load-carrying capacity of columns will depend on the *slenderness ratio*, l/r. The slenderness ratio can be reduced by providing vertical and horizontal supports at intermediate points. For cuts with large widths, it may be necessary to splice the struts. In the case of braced cuts in clayey soils, the depth of the first strut below the ground surface should be less than the depth of tensile crack, z_c. From Eq. (5.12)

$$\sigma_a = \gamma z K_a - 2c\sqrt{K_a}$$

where K_a = coefficient of Rankine active pressure

For determination of the depth of tensile crack

$$\sigma_a = 0 = \gamma z_c K_a - 2c\sqrt{K_a}$$

or

$$z_c = \frac{2c}{\sqrt{K_a}\,\gamma}$$

With $\phi = 0$, $K_a = \tan^2(45 - \phi/2) = 1$. So

$$z_c = \frac{2c}{\gamma}$$

A simplified conservative procedure can be used to determine the strut loads. This procedure will vary depending on the engineers involved in the project. Following is a step-by-step outline of it (refer to Figure 7.9).

1. Draw the pressure envelope for the braced cut (see Figures 7.4, 7.5, and 7.6). Also show the proposed strut levels. Figure 7.9a shows a pressure envelope for a sandy soil; however, it could also be for a clay. Also, in this figure, the strut levels are marked A, B, C, and D. The sheet piles (or soldier beams) can be assumed to be hinged at the strut levels, except for the top and bottom ones. In Figure 7.9a, the hinges are at the level of struts B and C. (Many designers also assume the sheet piles, or soldier beams, to be hinged at all strut levels, except for the top.)
2. Determine the reactions for the two simple cantilever beams (top and bottom) and all the simple beams in between. In Figure 7.9b, these reactions are A, B_1, B_2, C_1, C_2, and D.

Figure 7.9 Determination of strut loads: (a) section and plan of the cut; (b) method for determination of strut loads.

3. The strut loads in Figure 7.9 can now be calculated as follows:

$$
\left.
\begin{aligned}
P_A &= (A)(s)\\
P_B &= (B_1 + B_2)(s)\\
P_C &= (C_1 + C_2)(s)\\
P_D &= (D)(s)
\end{aligned}
\right\}
\tag{7.9}
$$

where

P_A, P_B, P_C, P_D = loads to be taken by the individual struts at levels *A*, *B*, *C*, and *D*, respectively

A, B_1, B_2, C_1, C_2, D = reactions calculated in Step 2 (note unit: force/unit length of the braced cut)

s = horizontal spacing of the struts (see plan in Figure 7.9a)

4. Knowing the strut loads at each level and the intermediate bracing conditions, one can now select the proper sections by using the steel construction manual.

Sheet Piles

In order to design the sheet piles, perform the following steps:

1. For each of the sections shown in Figure 7.9b, determine the maximum bending moment.

2. Determine the maximum value of the maximum bending moments (M_{max}) obtained in Step 1. Note that the unit of this moment will be, for example, kN-m/meter length of the wall.

3. Obtain the required section modulus of the sheet piles:

$$S = \frac{M_{max}}{\sigma_{all}} \tag{7.10}$$

where σ_{all} = allowable flexural stress of the sheet pile material

4. A sheet pile having a section modulus greater than or equal to the required section modulus can now be chosen from a table such as Table 6.1.

Wales

1. Wales can be treated as continuous horizontal members if they are spliced properly. Conservatively, they may also be treated as though they are pinned at the struts. For the section shown in Figure 7.9a, the maximum moments for the wales (assuming that they are pinned at the struts) are as follows:

$$\text{at level } A, M_{max} = \frac{(A)(s^2)}{8}$$

$$\text{at level } B, M_{max} = \frac{(B_1 + B_2)s^2}{8}$$

$$\text{at level } C, M_{max} = \frac{(C_1 + C_2)s^2}{8}$$

$$\text{at level } D, M_{max} = \frac{(D)(s^2)}{8}$$

where A, B_1, B_2, C_1, C_2, and D are the reactions under the struts per unit length of the wall (Step 2 of strut design).

2. Determine the section modulus of the wales:

$$S = \frac{M_{max}}{\sigma_{all}}$$

The wales are sometimes fastened to the sheet piles at points that satisfy
the lateral support requirements.

Example 7.1

The cross section of a long braced cut is shown in Figure 7.10a.
   a. Draw the earth pressure envelope.
   b. Determine the strut loads at levels A, B, and C.
   c. Determine the section of the struts subjected to the largest load.
   d. Determine the sheet pile section required.
   e. Determine a design section for the wales at level B.

Note: The struts are placed at 3 m center-to-center in the plan.

Solution
Part a

Given $\gamma = 18$ kN/m^3, $c = 35$ kN/m^2, and $H = 6$ m.

$$\frac{\gamma H}{c} = \frac{(18)(7)}{35} = 3.6 < 4$$

So, the pressure envelope will be like the one in Figure 7.6. This is plotted in Figure
7.10a with maximum pressure intensity, p_a, equal to $0.3\gamma H = 0.3(18)(7) = $ **37.8
kN/m^2**.

Part b

For determination of the strut loads, refer to Figure 7.10b. Taking the moment
about B_1, $\Sigma M_{B_1} = 0$

$$A(2.5) - \left(\frac{1}{2}\right)(37.8)(1.75)\left(1.75 + \frac{1.75}{3}\right) - (1.75)(37.8)\left(\frac{1.75}{2}\right) = 0$$

or

   $A = 54.02$ kN/m

Also, Σ vertical forces $= 0$. Thus

$$\frac{1}{2}(1.75)(37.8) + (37.8)(1.75) = A + B_1$$

   $33.08 + 66.15 - A = B_1$

So

   $B_1 = 45.2$ kN/m

Due to symmetry

   $B_2 = 45.2$ kN/m
   $C = 54.02$ kN/m

(a)

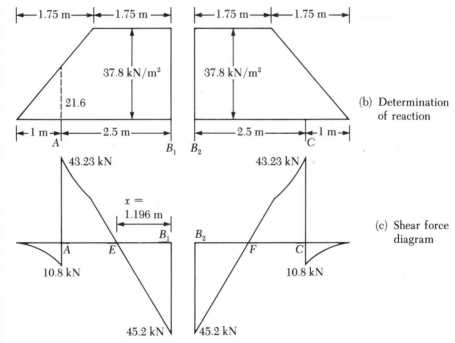

(b) Determination of reaction

(c) Shear force diagram

Figure 7.10

Strut load at level

$$P_A = 54.02 \times \text{horizontal spacing, } s = 54.02 \times 3 = \textbf{162.06 kN}$$
$$P_B = (B_1 + B_2)3 = (45.2 + 45.2)3 = \textbf{271.2 kN}$$
$$P_C = 54.02 \times 3 = \textbf{162.06 kN}$$

Part c

The struts at level B are subjected to the largest load—that is, $P_B = 271.2$ kN. For the struts, effective length (KL: refer to the American Institute of Steel Construction, *Manual of Steel Construction*, 1980, pp. 3–29) with respect to x and y axes is 6 m. Accordingly, the section **W 250 mm × 49 kg/m** (in English units, it is section **W 10 × 33**) will be more than sufficient. (*Note:* $F_y = 248.4$ MN/m^2.)

Part d

Refer to the left side of Figure 7.10b. For the maximum moment, the shear force should be zero. The nature of variation of the shear force is shown in Figure 7.10c. The location of point E can be given as

$$x = \frac{\text{reaction at } B_1}{37.8} = \frac{45.2}{37.8} = 1.196 \text{ m}$$

The magnitude of moment at $A = \frac{1}{2}\,(1)\left(\frac{37.8}{1.75} \times 1\right)\left(\frac{1}{3}\right)$

$$= 3.6 \text{ kN-m/meter of wall}$$

The magnitude of moment at $E = (45.2 \times 1.196) - (37.8 \times 1.196)\left(\frac{1.196}{2}\right)$

$$= 54.06 - 27.03$$
$$= 27.03 \text{ kN-m/meter of wall}$$

Because the loading on the left and right sections of Figure 7.10b are the same, the magnitude of moments at F and C (Figure 7.10c) will be the same as E and A, respectively. Hence, the maximum moment $= 27.03$ kN-m/meter of wall.

The section modulus of the sheet piles

$$S = \frac{M_{max}}{\sigma_{all}} = \frac{27.03 \text{ kN-m}}{170 \times 10^3 \text{ kN/m}^2} = 15.9 \times 10^{-5} \text{ m}^3/\text{m of the wall}$$

According to Table 6.1, **section PMA-22** can be used.

Part e

The reaction at level B has been calculated in Part b. Hence

$$M_{max} = \frac{(B_1 + B_2)s^2}{8} = \frac{(45.2 + 45.2)3^2}{8} = 101.7 \text{ kN-m}$$

Section modulus, $S_x = \dfrac{101.7}{\sigma_{all}} = \dfrac{101.7}{0.6F_y} = \dfrac{101.7}{0.6(248.4 \times 1000)}$

$$= 0.682 \times 10^{-3} \text{ m}^3$$

So, a section of **W 310 mm × 54 kg/m** (in English units, **W 12 × 36**) with an $S_x = 0.754 \times 10^{-3}$ m^3 can be used. (*Note:* $L_c = 2.1$ m and $L_u = 4.085$ m—AISC Manual.)

Example 7.2

Refer to the braced cut shown in Figure 7.11. Given: $\gamma = 112$ lb/ft³, $\phi = 32°$, and $c = 0$. The struts are located at 12 ft center-to-center in the plan. Draw the earth pressure envelope and determine the strut loads at levels A, B, and C.

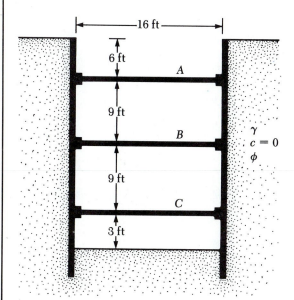

Figure 7.11

Solution

For this case, the earth pressure envelope shown in Figure 7.4 will apply.

$$K_a = \tan^2\left(45 - \frac{\phi}{2}\right) = \tan^2\left(45 - \frac{32}{2}\right) = 0.307$$

From Equation (7.2)

$$p_a = 0.65\gamma H K_a = (0.65)(112)(27)(0.307) = 603.44 \text{ lb/ft}^2$$

Figure 7.12a shows the pressure envelope. Now, referring to Figure 7.12b

$$\sum M_{B_1} = 0$$

$$A = \frac{(603.44)(15)\left(\dfrac{15}{2}\right)}{9} = 7543 \text{ lb/ft}$$

$$B_1 = (603.44)(15) - 7543 = 1508.6 \text{ lb/ft}$$

(a)

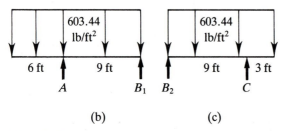

(b)   (c)

Figure 7.12

Again, referring to Figure 7.12c

$$\sum M_{B_2} = 0$$

$$C = \frac{(603.44)(12)\left(\dfrac{12}{2}\right)}{9} = 4827.5 \text{ lb/ft}$$

$$B_2 = (603.44)(12) - 4827.5 = 2413.7 \text{ lb/ft}$$

Strut load at A = (7.543)(spacing) = (7.543)(12)

$$= \textbf{90.52 kips}$$

Strut load at B = $(B_1 + B_2)$(spacing) = (1.509 + 2.414)(12)

$$= \textbf{47.07 kips}$$

Strut load at C = (4.827) = (4.827)(12)

$$= \textbf{57.93 kips}$$

Example 7.3

For the braced cut described in Example 7.2, determine the following:

    a. The sheet pile section

    b. The required section modulus of the wales at level A. Assume that $\sigma_{all} = 25$ kips/in.^{2}.

Solution
Part a

Refer to the load diagrams shown in Figures 7.12b and 7.12c. Based on the load diagrams, the shear force diagrams are as given in Figure 7.13.

$$x_1 = \frac{3.923}{0.603} = 6.5 \text{ ft}$$

$$x_2 = \frac{3.017}{0.603} = 5 \text{ ft}$$

Moment at $A = \frac{1}{2}(3.62)(6) = 10.86$ kip-ft

Moment at $C = \frac{1}{2}(1.81)(3) = 2.715$ kip-ft

Moment at $B' = \frac{1}{2}(1.509)(2.5) = 1.89$ kip-ft

Moment at $B'' = \frac{1}{2}(2.414)(4) = 4.828$ kip-ft

M_A is maximum.

$$S_x = \frac{M_{max}}{\sigma_{all}} = \frac{10.86 \text{ kip-ft} \times 12}{24 \text{ kips/in.}^2} = 5.43 \text{ in.}^3$$

Referring to Table 6.1

Use PDA-27. $S_x = 10.7$ in.^{3}/ft of wall.

(PMA-22 is just a little small.)

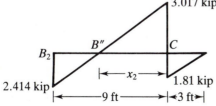

Figure 7.13

Part b

For the wale at level A

$$M_{max} = \frac{A(s^2)}{8}$$

$A = 7543$ lb/ft (from Example 7.2). So

$$M_{max} = \frac{(7.543)(12^2)}{8} = 135.77 \text{ kip-ft}$$

$$S_x = \frac{M_{max}}{\sigma_{all}} = \frac{(135.77)(12)}{25 \text{ kips/in.}^2} = \textbf{67.9 in.}^3\textbf{/ft of wall}$$

7.5   Stability of Braced Cuts

Heave of the Bottom of the Cut in Clay

Braced cuts in clay may become unstable as a result of heaving of the bottom of the excavation. Terzaghi (1943b) has analyzed the factor of safety of braced excavations against bottom heave. The failure surface for such a case is shown in Figure 7.14. The vertical load per unit length of the cut at the level of the bottom of the cut along the line bd and af is equal to

$$Q = \gamma H B_1 - cH \qquad (7.11)$$

where $B_1 = 0.7B$
$\quad\quad c$ = cohesion ($\phi = 0$ concept)

This load Q can be treated like a load per unit length on a continuous foundation at the level of bd (and af) having a width of $B_1 = 0.7B$. Based on

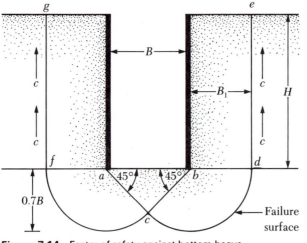

Note: cd and cf
are arcs of circles
with centers at b
and a, respectively

Figure 7.14   Factor of safety against bottom heave

Terzaghi's bearing capacity theory, the net ultimate load-carrying capacity per unit length of this foundation can be given by the equation [Chapter 3; see Eqs. (3.3) and (3.20)]

$$Q_u = cN_c B_1 = 5.7cB_1$$

Hence, from Eq. (7.11), the factor of safety against bottom heave can be given as

$$FS = \frac{Q_u}{Q} = \frac{5.7cB_1}{\gamma H B_1 - cH} = \frac{1}{H}\left(\frac{5.7c}{\gamma - \dfrac{c}{0.7B}}\right) \tag{7.12}$$

The preceding factor of safety [Eq. (7.12)] has been derived based on the assumption that the clay layer is homogeneous, at least up to a depth of 0.7B below the bottom of the cut. However, if a *hard layer of rock or rock-like material is located at a depth $D < 0.7B$*, the failure surface will be modified to some extent. In such a case, the factor of safety can be modified to the form

$$FS = \frac{1}{H}\left(\frac{5.7c}{\gamma - c/D}\right) \tag{7.13}$$

Bjerrum and Eide (1956) also studied the problem of bottom heave for braced cuts in clay, and they proposed the following equation for the factor of safety:

$$FS = \frac{cN_c}{\gamma H} \tag{7.14}$$

The bearing capacity factor, N_c, varies with the ratio of H/B and also L/B (where L = length of the cut). For infinitely long cuts ($B/L = 0$), $N_c = 5.14$ at $H/B = 0$ and increases to a value of $N_c = 7.6$ at $H/B = 4$. Beyond that—that is, for $H/B > 4$—the value of N_c remains constant. For cuts square in plan ($B/L = 1$), $N_c = 6.3$ at $H/B = 0$, and $N_c = 9$ for $H/B \geq 4$. In general, at any given H/B

$$N_{c(\text{rectangle})} = N_{c(\text{square})}\left(0.84 + 0.16\,\frac{B}{L}\right) \tag{7.15}$$

Figure 7.15 shows the variation of the value of N_c for $L/B = 1, 2, 3,$ and ∞.

Figure 7.15   Variation of N_c with L/B and H/B—based on Bjerrum and Eide's equation [Eq. (7.15)]

Equations (7.14) and (7.15) can now be combined, and the factor of safety against heave can be expressed as

$$FS = \frac{cN_{c(square)}\left(0.84 + 0.16\frac{B}{L}\right)}{\gamma H} \qquad (7.16)$$

Equation (7.16) and the variation of the bearing capacity factor, N_c, as shown in Figure 7.15 are based on the facts that the clay layer below the bottom of the cut is homogeneous and the magnitude of the undrained cohesion in the soil that contains the failure surface is equal to c (Figure 7.16). However, if a stronger clay layer is encountered at a shallow depth, as shown in Figure 7.17a, the failure surface below the cut will be controlled by the undrained cohesions c_1 and c_2. For this type of condition, the factor of safety can be given as

$$FS = \frac{c_1[N'_{c(strip)}F_d]F_s}{\gamma H} \qquad (7.17)$$

where $N'_{c(strip)}$ = bearing capacity factor of an infinitely long cut (that is, $B/L = 0$), which is a function of h'/B and c_2/c_1
       F_d = depth factor, which is a function of H/B
       F_s = shape factor

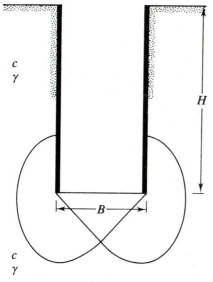

Figure 7.16 Derivation of Eq. (7.16)

The variation of $N'_{c(\text{strip})}$ is shown in Figure 7.17b, and the variation of F_d as a function of H/B is given in Figure 7.17c. The shape factor, F_s, can be given as

$$F_s = 1 + 0.2\,\frac{B}{L} \qquad\qquad (7.18)$$

In most cases, a factor of safety of about 1.5 is generally recommended. If F_s becomes less than about 1.5, the sheet pile is driven deeper (Figure 7.18). Usually the depth, d (Figure 7.18), is kept less than or equal to $B/2$. In that case, the force, P, per unit length of the buried sheet pile (that is, aa' and bb') can be expressed as follows (U.S. Department of the Navy, 1971):

$$P = 0.7(\gamma HB - 1.4cH - \pi cB) \qquad \text{for } d > 0.47B \qquad\qquad (7.19)$$

and

$$P = 1.5d\left(\gamma H - \frac{1.4cH}{B} - \pi c\right) \qquad \text{for } d < 0.47B \qquad\qquad (7.20)$$

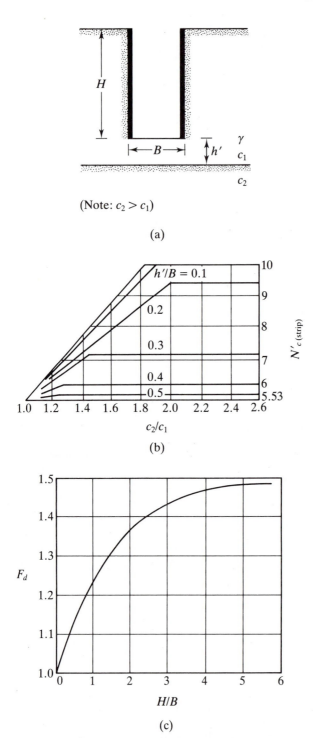

(Note: $c_2 > c_1$)

(a)

(b)

(c)

Figure 7.17   (a) Layered clay below the bottom of the cut; (b) variation of $N'_{c(strip)}$ with c_2/c_1, and h'/B (redrawn after Reddy and Srinivasan, 1967); (c) variation of F_d with H/B

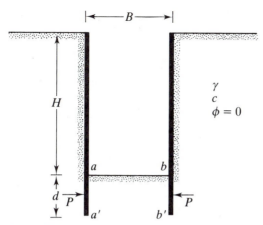

Figure 7.18   Force on the buried length of sheet pile

Stability of the Bottom of the Cut in Sand

The bottom of a cut in sand is generally stable. When the ground water table is encountered, the bottom of the cut is stable as long as the water level inside the excavation is higher than the ground water level. If the water level inside the cut is lowered below the ground water level by pumping, instability may be created as a result of the upward seepage of water into the cut. Section 7.6 discusses this problem in more detail.

Lateral Yielding of Sheet Piles

In braced cuts, some lateral movement of sheet pile walls may be expected (Figure 7.19). Of course, the lateral yield will depend on several factors, the most important of which is time elapsed after excavation that is required for the placement of wales and struts. Mana and Clough (1981) analyzed the field records of several braced cuts in clay from the San Francisco, Oslo (Norway), Boston, Chicago, and Bowline Point (New York) areas. Under ordinary construction conditions, it was found that the maximum lateral wall yield $[\delta_{H(max)}]$ has a definite relationship with the factor of safety against heave. This is shown in Figure 7.19. Note that the factor of safety against heave as plotted in Figure 7.19 has been calculated by using Eqs. (7.12) and (7.13).

As discussed before, in several instances, the sheet piles (or the soldier piles, as the case may be) are driven to a certain depth below the bottom of the excavation. This is done to reduce the lateral yielding of the walls during the last stages of excavation. The lateral yielding of the walls will cause settlement of the ground surface surrounding the cut. The degree of lateral yielding, however, depends mostly on the soil type below the bottom of the cut. If clay below the cut extends to a great depth and $\gamma H/c$ is less than about 6, extension of the sheet piles or soldier piles below the bottom of the cut will help considerably in reducing the lateral yield of the walls.

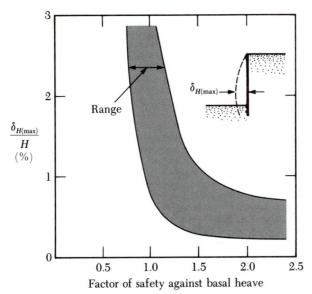

Figure 7.19 Range of variation of $\delta_{H(max)}/H$ with *FS* against basal heave from field observation (redrawn after Mana and Clough, 1981)

However, under similar circumstances, if $\gamma H/c$ is about 8, the extension of sheet piles into the clay below the cut does not help to a great extent. In such circumstances, one may expect a great degree of wall yielding that may result in the total collapse of the bracing systems. If a hard soil layer is located below a clay layer at the bottom of the cut, the piles should be embedded in the stiffer layer. This will have a great effect in reducing the lateral yield.

Ground Settlement

The lateral yielding of walls will generally induce ground settlement (δ_V) around a braced cut. This is generally referred to as *ground loss*. Based on several field observations, Peck (1969) has provided curves for prediction of ground settlement in various types of soil (see Figure 7.20). The magnitude of ground loss varies extensively; however, Figure 7.20 can be used as a general guide.

Based on the field data obtained from various cuts in the areas of San Francisco, Oslo, and Chicago, Mana and Clough (1981) have provided a correlation between the maximum lateral yield of sheet piles $[\delta_{H(max)}]$ and the maximum ground settlement $[\delta_{V(max)}]$. This is shown in Figure 7.21. It can be seen that

$$\delta_{V(max)} \approx 0.5 \text{ to } 1 \; \delta_{H(max)} \tag{7.21}$$

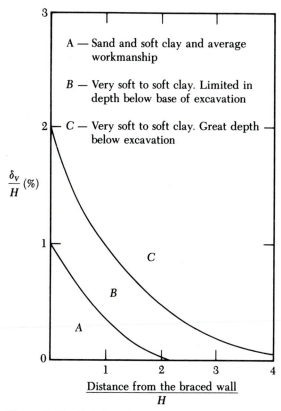

Figure 7.20 Variation of ground settlement with distance (after Peck, 1969)

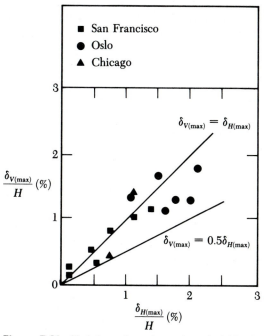

Figure 7.21 Variation of maximum lateral yield with maximum ground settlement (after Mana and Clough, 1981)

411

Example 7.4

Refer to Example 7.1. Determine the factor of safety against bottom heave using Eqs. (7.12) and (7.14).

Solution

In Example 7.1, $\gamma = 18$ kN/m³, $c = 35$ kN/m², and $H = 7$ m.

Factor of Safety from Eq. (7.12)

$$FS = \frac{1}{H}\left(\frac{5.7c}{\gamma - \dfrac{c}{0.7B}}\right) = \frac{1}{7}\left[\frac{(5.7)(35)}{18 - \dfrac{35}{(0.7)(6)}}\right] = \textbf{2.95}$$

Factor of Safety from Eq. (7.14)

$$FS = \frac{cN_c}{\gamma H}$$

According to Figure 7.15, for $H/B = 7/6 = 1.16$ and $B/L \approx 0$, the value of N_c is equal to 6.46. Thus

$$FS = \frac{(35)(6.46)}{(18)(7)} = \textbf{1.79}$$

Example 7.5

Figure 7.22 shows a braced cut. For the cut, $B = 10$ ft and $L = 30$ ft. Determine the factor of safety against bottom heave.

|←——10 ft——→|

$\gamma = 115$ lb/ft³
$c = 800$ lb/ft²

20 ft

3 ft

$c = 1120$ lb/ft²

Figure 7.22

Solution

Since layered clay is encountered below the bottom of the cut, Equation (7.17) will be used.

$$\frac{c_2}{c_1} = \frac{1120}{800} = 1.4$$

$$\frac{h'}{B} = \frac{3}{10} = 0.3$$

From Figure 7.17b for $c_2/c_1 = 1.4$ and $h'/B \approx 0.3$, the value of $N'_{c(strip)} \approx 7$. Again

$$\frac{H}{B} = \frac{20}{10} = 2$$

So, from Figure 7.17c, $F_d \approx 1.36$. Also

$$F_s = 1 + 0.2\left(\frac{B}{L}\right) = 1 + 0.2\left(\frac{10}{30}\right) = 1.067$$

So

$$FS = \frac{c_1[N'_{c(strip)}F_d]F_s}{\gamma H} = \frac{(800)[(7)(1.36)](1.067)}{(20)(115)} = \mathbf{3.53}$$

7.6   Failure of Single Wall Cofferdams by Piping

Sheet piles are sometimes driven for excavations that need dewatering (Figure 7.23). In such cases, the factor of safety against piping should be checked. [*Piping* is another term for failure by heave, as defined in Section 1.12; see Eq. (1.48)]. Piping may occur when a high hydraulic gradient is created as a result of water flow into the excavation. One can check the factor of safety by drawing flow nets and determining the maximum exit gradient [$i_{max(exit)}$] that will occur at points A and B. Figure 7.24 shows a flow net to illustrate the problem. The maximum exit gradient for this flow net can be calculated as

$$i_{max(exit)} = \frac{\dfrac{h}{N_d}}{a} = \frac{h}{N_d a} \tag{7.22}$$

where a = length of the flow element at A (or B)

N_d = number of drops (*Note:* In Figure 7.24, $N_d = 8$—also see Section 1.10.

The factor of safety against piping can be expressed as

$$\boxed{FS = \frac{i_{cr}}{i_{max(exit)}}} \tag{7.23}$$

Figure 7.23

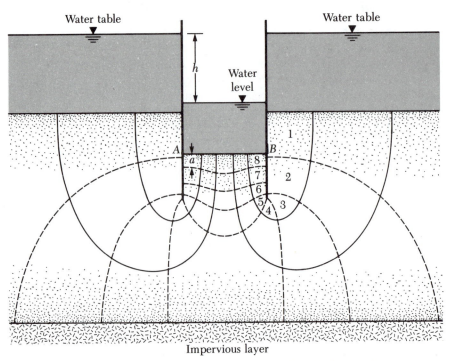

Figure 7.24 Determination of the factor of safety against piping by drawing flow net

(a)

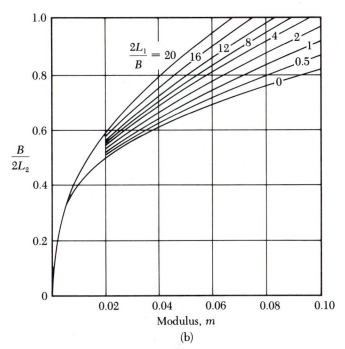

(b)

Figure 7.25   Variation of modulus (from *Groundwater and Seepage*, by M. E. Harr. Copyright © 1962 by McGraw-Hill. Used with permission.)

(a)

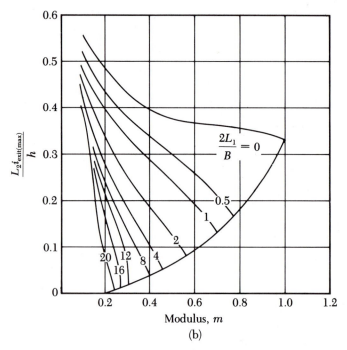

(b)

Figure 7.26 Variation of maximum exit gradient with modulus (from *Groundwater and Seepage*, by M. E. Harr. Copyright © 1962 by McGraw-Hill. Used with permission.)

where i_{cr} = critical hydraulic gradient

The relationship for i_{cr} has been given in Chapter 1 [Eq. (1.48)] as

$$i_{cr} = \frac{G_s - 1}{e + 1}$$

The value of i_{cr} varies between 0.9 and 1.1 in most soils, with an average of about one. A factor of safety of about 1.5 is desirable.

The maximum exit gradient for sheeted excavations in sands with $L_3 = \infty$ can also be theoretically evaluated (Harr, 1962). (Only the results of these mathematical derivations will be presented here. For further details, refer to the original work.) To calculate the maximum exit gradient, refer to Figures 7.25 and 7.26 and perform the following steps:

1. Determine the modulus, m, Figure 7.25 by obtaining $2L_2/B$ (or $B/2L_2$) and $2L_1/B$.

2. With the known modulus and $2L_1/B$, refer to Figure 7.26 and determine $L_2 i_{exit(max)}/h$. Because L_2 and h will be known, $i_{exit(max)}$ can be calculated.

3. The factor of safety against piping can be evaluated by using Eq. (7.23).

Marsland (1958) has provided the results of model tests conducted to study the influence of seepage on the stability of sheeted excavations in sand. These have been summarized by the U.S. Department of the Navy (1971) in *NAVFAC DM-7* and are given in Figures 7.27a, b, and c. Note

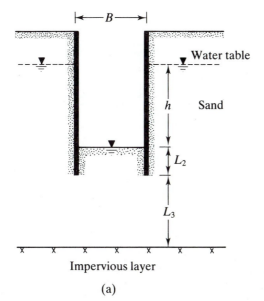

(a)

Figure 7.27 Influence of seepage on the stability of sheeted excavation (after U.S. Department of the Navy, 1971) (continued on page 418)

(b)

(c)

Figure 7.27 (Continued)

that Figure 7.27b is for the case of determining the sheet pile penetration (L_2) needed for required factor of safety against piping when the sand layer extends to a great depth below the excavation. However, Figure 7.27c represents the case in which an impervious layer is located at a depth of $L_2 + L_3$ below the bottom of the excavation.

Example 7.6

Refer to Figure 7.23. Given: $h = 4.5$ m, $L_1 = 5$ m, $L_2 = 4$ m, $B = 5$ m, and $L_3 = \infty$. Determine the factor of safety against piping.

Solution

$$\frac{2L_1}{B} = \frac{2(5)}{5} = 2$$

$$\frac{B}{2L_2} = \frac{5}{2(4)} = 0.625$$

According to Figure 7.25b, for $2L_1/B = 2$ and $B/2L_2 = 0.625$, $m \approx 0.033$. Again referring to Figure 7.26a, for $m = 0.033$ and $2L_1/B = 2$, $L_2\, i_{\text{exit(max)}}/h = 0.55$. Hence

$$i_{\text{exit(max)}} = \frac{0.55(h)}{L_2} = 0.55(4.5)/4 = 0.619$$

$$FS = \frac{i_{\text{cr}}}{i_{\text{max(exit)}}} = \frac{1}{0.619} = \textbf{1.616}$$

PROBLEMS

7.1 Refer to the braced cut shown in Figure P7.1. Given: $\gamma = 17$ kN/m³, $\phi = 35°$, and $c = 0$. The struts are located at 3 m center-to-center in the plan. Draw the earth pressure envelope and determine the strut loads at levels A, B, and C.

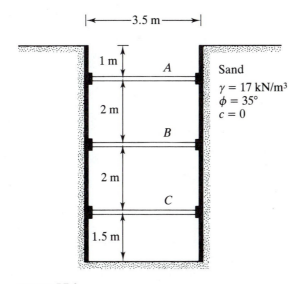

Figure P7.1

7.2 For the braced cut described in Problem 7.1, determine the following:
a. The sheet pile section
b. The section modulus of the wales at level B

Assume that $\sigma_{all} = 170$ MN/m^2.

7.3 Redo Problem 7.1 with $\gamma = 18$ kN/m^3, $\phi = 40°$, $c = 0$, and center-to-center strut spacing in the plan $= 4$ m.

7.4 Determine the sheet pile section required for the braced cut described in Problem 7.3. Given: $\sigma_{all} = 170$ MN/m^2.

7.5 Refer to Figure 7.7a. For the braced cut, given: $H = 8$ m, $H_s = 3$ m, $\gamma_s = 17.5$ kN/m^3; angle of friction of sand, $\phi_s = 34°$; $H_c = 5$ m, $\gamma_c = 18.2$ kN/m^3; and uncon-fined compression strength of clay layer, $q_u = 55$ kN/m^2.
a. Estimate the average cohesion (c_{av}) and average unit weight (γ_{av}) for the construc-tion of the earth pressure envelope.
b. Plot the earth pressure envelope.

7.6 Refer to Figure 7.7b, which shows a braced cut in clay. Given: $H = 25$ ft, $H_1 = 5$ ft, $c_1 = 2125$ lb/ft^2, $\gamma_1 = 111$ lb/ft^3, $H_2 = 10$ ft, $c_2 = 1565$ lb/ft^2, $\gamma_2 = 107$ lb/ft^3, $H_3 = 10$ ft, $c_3 = 1670$ lb/ft^2, and $\gamma_3 = 109$ lb/ft^3.
a. Determine the average cohesion (c_{av}) and average unit weight (γ_{av}) for calculation of the earth pressure envelope.
b. Plot the earth pressure envelope.

7.7 Refer to Figure P7.7. Given: $\gamma = 118$ lb/ft^3, $c = 800$ lb/ft^2, and center-to-center spacing of struts in the plan $= 12$ ft. Draw the earth pressure envelope and deter-mine the strut loads at levels A, B, and C.

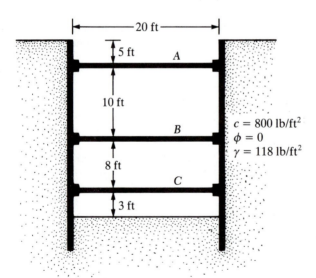

Figure P7.7

7.8 Determine the sheet pile section for the braced cut described in Problem 7.7.

7.9 Redo Problem 7.7, assuming that $c = 600$ lb/ft^2.

7.10 Determine the factor of safety against bottom heave for the braced cut described in Problem 7.7. Use Eqs. (7.12) and (7.16). For Eq. (7.16), assume the length of the cut, $L = 30$ ft.

7.11 Refer to Figure 7.17a. Given: $H = 8$ m, $B = 3$ m, $L = 8$ m, $\gamma = 17.8$ kN/m^3, $c_1 = 30$ kN/m^2, $c_2 = 45$ kN/m^2, and $h' = 2$ m. Determine the factor of safety against bottom heave.

7.12 Determine the factor of safety against bottom heave for the braced cut described in Problem 7.9. Use Eq. (7.16). Length of the cut is 40 ft.

REFERENCES

American Institute of Steel Construction (1980). *Manual of Steel Construction*, 8th ed., Chicago.

Bjerrum, L., and Eide, O. (1956). "Stability of Strutted Excavation in Clay," *Geotechnique*, Vol. 6, No. 1, pp. 32–47.

Das, B. M. (1990). *Principles of Geotechnical Engineering*, 2nd ed., PWS-Kent Publishing Company, Boston.

Das, B. M., and Seeley, G. R. (1975). "Active Thrust on Braced Cut in Clay," *Journal of the Construction Division*, American Society of Civil Engineers, Vol. 101, No. CO4, pp. 945–949.

Harr, M. E. (1962). *Ground Water and Seepage*, McGraw-Hill, New York.

Mana, A. I., and Clough, G. W. (1981). "Prediction of Movements for Braced Cuts in Clay," *Journal of the Geotechnical Engineering Division*, American Society of Civil Engineers, Vol. 107, No. GT8, pp. 759–777.

Marsland, A. (1958). "Model Experiments to Study the Influence of Seepage on the Stability of a Sheeted Excavation in Sand," *Geotechnique*, London, Vol. 3, p. 223.

Peck, R. B. (1943). "Earth Pressure Measurements in Open Cuts, Chicago (Ill.) Subway," *Transactions*, American Society of Civil Engineers, Vol. 108, pp. 1008–1058.

Peck, R. B. (1969). "Deep Excavation and Tunneling in Soft Ground," *Proceedings*, Seventh International Conference on Soil Mechanics and Foundation Engineering, Mexico City, State-of-the-Art Volume, pp. 225–290.

Reddy, A. S., and Srinivasan, R. J. (1967). "Bearing Capacity of Footing on Layered Clay," *Journal of the Soil Mechanics and Foundations Division*, American Society of Civil Engineers, Vol. 93, No. SM2, pp. 83–99.

Terzaghi, K. (1943a). "General Wedge Theory of Earth Pressure," *Transactions*, American Society of Civil Engineers, Vol. 106, pp. 68–97.

Terzaghi, K. (1943b). *Theoretical Soil Mechanics*, Wiley, New York.

Tschebotarioff, G. P. (1973). *Foundations, Retaining and Earth Structures*, 2nd ed., McGraw-Hill, New York.

U.S. Department of the Navy (1971). "Design Manual—Soil Mechanics, Foundations, and Earth Structures," NAVFAC DM-7, Washington, D.C.

Chapter **8**

PILE FOUNDATIONS

Introduction

Piles are structural members that are made of steel, concrete, and/or timber. They are used to build pile foundations, which are deep and which cost more than shallow foundations (Chapter 3). Despite the cost, the use of piles often becomes necessary to ensure that the structure under consideration is safe. Following is a list of some of the conditions that require pile foundations:

1. When the upper soil layer(s) is (are) highly compressible and too weak to support the load transmitted by the superstructure, piles are used to transmit the load to underlying bedrock or a stronger soil layer. This is shown in Figure 8.1a. When bedrock is not located at a reasonable depth below the ground surface, piles are used to transmit the structural load to the soil gradually. The resistance to the applied structural load is derived mainly from the frictional resistance developed at the soil-pile interface (Figure 8.1b).

2. When subjected to horizontal forces (see Figure 8.1c), pile foundations can resist by bending, while still supporting the vertical load transmitted by the superstructure. This type of situation is generally encountered in the design and construction of earth-retaining structures and foundations of tall structures that are subjected to high wind and/or earthquake forces.

3. In many cases, expansive and collapsible soils (Chapter 10) may be encountered at the site of a proposed structure. These soils may extend to a great depth below the ground surface. Expansive soils swell and shrink depending on the increase or decrease of moisture content. The swelling pressure of such soils can be considerably high. If shallow foundations are used in such circumstances, the structure may encounter considerable damage. However, pile foundations may be considered as an alternative when piles are extended beyond the active zone, which undergoes swelling and shrinking (Figure 8.1d).

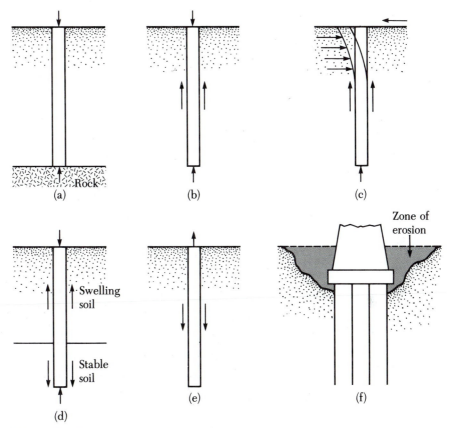

Figure 8.1 Conditions for use of pile foundations

Soils such as loess are collapsible in nature. When these soils undergo an increase of moisture content, their structures may break down. A sudden decrease of void ratio of soil induces large settlements of structures supported by shallow foundations. In such cases, pile foundations may be used in which piles are extended into stable soil layers beyond the zone of possible moisture change.

4. Foundations of some structures, such as transmission towers, offshore platforms, and basement mats below the water table, are subjected to uplifting forces. Piles are sometimes used for these foundations to resist the uplifting force (Figure 8.1e).

5. Bridge abutments and piers are usually constructed over pile foundations to avoid the possible loss of bearing capacity that a shallow foundation might suffer because of soil erosion at the ground surface (Figure 8.1f).

Although numerous investigations, both theoretical and experimental, have been conducted in the past to predict the behavior and the load-bearing capacity of piles in granular and cohesive soils, the mechanisms are not yet entirely understood and may never be. The design of pile foundations may be considered somewhat of an "art" as a result of the uncertainties involved in working with some subsoil conditions. This chapter discusses the present state of the art for design and analysis of pile foundations.

8.2   Types of Pile and Their Structural Characteristics

Different types of pile are used in construction work depending on the type of load to be carried, the subsoil conditions, and the ground water table. Piles can be divided into the following categories: (a) *steel piles*, (b) *concrete piles*, (c) *wooden (timber) piles*, and (d) *composite piles*. This section discusses each of these types in detail.

Steel Piles

The steel piles generally used are either *pipe piles* or *rolled steel H-section piles*. Pipe piles can be driven into the ground with their ends open or closed. Wide-flange and I-section steel beams can also be used as piles. However, H-section piles are usually preferred because their web and flange thicknesses are equal. In wide-flange and I-section beams, the web thicknesses are smaller than the thicknesses of the flange. Table 8.1 gives the dimensions of some standard H-section steel piles used in the United States.

Table 8.1a   Common H-Pile Sections Used in the United States (SI Units)

Designation, size (mm) × weight (kN/m)	Depth d_1 (mm)	Section area ($m^2 \times 10^{-3}$)	Flange and web thickness w (mm)	Flange width d_2 (mm)	Moment of inertia ($m^4 \times 10^{-6}$)	
					I_{xx}	I_{yy}
HP 200 × 0.52	204	6.84	11.3	207	49.5	16.8
HP 250 × 0.834	254	10.8	14.4	260	123	42
× 0.608	246	8.0	10.6	256	87.5	24
HP 310 × 1.226	312	15.9	17.5	312	271	89
× 1.079	308	14.1	15.49	310	237	77.5
× 0.912	303	11.9	13.1	308	197	63.7
× 0.775	299	10.0	11.05	306	164	62.9
HP 330 × 1.462	334	19.0	19.45	335	370	123
× 1.264	329	16.5	16.9	333	314	104
× 1.069	324	13.9	14.5	330	263	86
× 0.873	319	11.3	11.7	328	210	69
HP 360 × 1.707	361	22.2	20.45	378	508	184
× 1.491	356	19.4	17.91	376	437	158
× 1.295	351	16.8	15.62	373	374	136
× 1.060	346	13.8	12.82	371	303	109

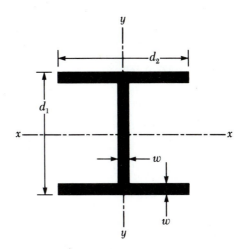

Table 8.1b   Common H-Pile Sections Used in the United States (English Units)

Designation size (in.) × weight (lb/ft)	Depth d_1 (in.)	Section area (in.^{2})	Flange and web thickness w (in.)	Flange width d_2 (in.)	Moment of inertia (in.^{4})	
					I_{xx}	I_{yy}
HP 8 × 36	8.02	10.6	0.445	8.155	119	40.3
HP 10 × 57	9.99	16.8	0.565	10.225	294	101
× 42	9.70	12.4	0.420	10.075	210	71.7
HP 12 × 84	12.28	24.6	0.685	12.295	650	213
× 74	12.13	21.8	0.610	12.215	570	186
× 63	11.94	18.4	0.515	12.125	472	153
× 53	11.78	15.5	0.435	12.045	394	127
HP 13 × 100	13.15	29.4	0.766	13.21	886	294
× 87	12.95	25.5	0.665	13.11	755	250
× 73	12.74	21.6	0.565	13.01	630	207
× 60	12.54	17.5	0.460	12.90	503	165
HP 14 × 117	14.21	34.4	0.805	14.89	1220	443
× 102	14.01	30.0	0.705	14.78	1050	380
× 89	13.84	26.1	0.615	14.70	904	326
× 73	13.61	21.4	0.505	14.59	729	262

Table 8.2 shows a list of selected pipe sections frequently used for piling purposes. In many cases, the pipe piles are filled with concrete after driving.

The allowable structural capacity for steel piles may be given as

$$Q_{all} = A_s \sigma_{all} \qquad (8.1)$$

where A_s = cross-sectional area of the steel
σ_{all} = allowable stress of steel

Table 8.2a   Selected Pipe Pile
Sections (SI Units)

Outside diameter (mm)	Wall thickness (mm)	Area of steel (cm^2)
219	3.17	21.5
	4.78	32.1
	5.56	37.3
	7.92	52.7
254	4.78	37.5
	5.56	43.6
	6.35	49.4
305	4.78	44.9
	5.56	52.3
	6.35	59.7
406	4.78	60.3
	5.56	70.1
	6.35	79.8
457	5.56	80
	6.35	90
	7.92	112
508	5.56	88
	6.35	100
	7.92	125
610	6.35	121
	7.92	150
	9.53	179
	12.70	238

Based on geotechnical considerations, once the design load for a pile is fixed, it is always advisable to check if $Q_{(design)}$ is within the allowable range as defined by Eq. (8.1).

Steel piles, when necessary, are spliced by welding or by riveting. Figure 8.2a shows a typical condition of splicing by welding for an H-pile. A typical case of splicing by welding for a pipe pile is shown in Figure 8.2b. Figure 8.2c shows a diagram of splicing an H-pile by rivets or bolts.

When hard driving conditions are expected, such as driving through dense gravel, shale, and soft rock, the steel piles can be fitted with driving points or shoes. Figures 8.2d and e are diagrams of two types of shoe used for pipe piles.

Steel piles may be subject to corrosion. For example, swamps, peats, and other organic soils are corrosive. Soils that have a pH greater than 7 are not so corrosive. In order to take into account the effect of corrosion, an additional thickness of steel (over the actual design cross-sectional area) is generally recommended. In many circumstances, factory-applied epoxy coatings on piles work satisfactorily against corrosion. These coatings are not very easily damaged by pile driving. Concrete encasement of steel piles in most corrosive zones is also practiced as a protection against corrosion.

Table 8.2b   Selected Pipe Pile
Sections (English Units)

Outside diameter (in.)	Wall thickness (in.)	Area of steel (in.^{2})
$8\frac{5}{8}$	0.125	3.34
	0.188	4.98
	0.219	5.78
	0.312	8.17
10	0.188	5.81
	0.219	6.75
	0.250	7.66
12	0.188	6.96
	0.219	8.11
	0.250	9.25
16	0.188	9.34
	0.219	10.86
	0.250	12.37
18	0.219	12.23
	0.250	13.94
	0.312	17.34
20	0.219	13.62
	0.250	15.51
	0.312	19.30
24	0.250	18.7
	0.312	23.2
	0.375	27.8
	0.500	36.9

Concrete Piles

Concrete piles can be divided into two basic categories: (a) *precast piles* and
(b) *cast-in-situ piles*. *Precast piles* can be prepared by using ordinary
reinforcement, and they can be square or octagonal in cross section (Figure
8.3). The reinforcement is provided to enable the pile to resist the bending
moment developed during pickup and transportation, the vertical load, and
the bending moment caused by lateral load. The piles are cast to desired
lengths and cured before being transported to the work sites.

    Precast piles can also be prestressed by the use of high-strength steel
prestressing cables. The ultimate strength of these steel cables is about 1800
MN/m^2 (≈ 261 ksi). In casting the piles, the cables are pretensioned up to
about 900–1300 MN/m^2 (≈ 130–188 ksi), and concrete is poured around
them. After proper curing, the cables are cut and thus produce a compressive force on the pile section. Table 8.3 gives additional information about
prestressed concrete piles with square and octagonal cross sections.

    Cast-in-situ, or *cast-in-place*, *piles* are built by making a hole in the
ground and then filling it with concrete. Various types of cast-in-place concrete pile are used in construction nowadays, and most of them have been

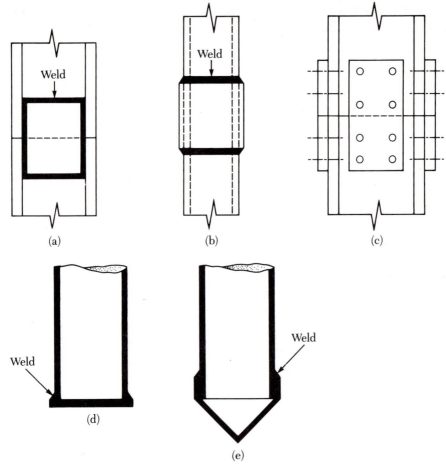

Figure 8.2 Steel piles: (a) splicing of H-pile by welding; (b) splicing of pipe pile by welding; (c) splicing of H-pile by rivets and bolts; (d) flat driving point of pipe pile; (e) conical driving point of pipe pile

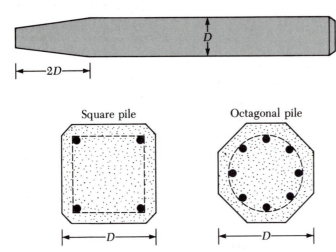

Square pile   Octagonal pile

Figure 8.3 Precast piles with ordinary reinforcement

Table 8.3a Typical Prestressed Concrete Piles in Use (SI Units)

Pile shape[a]	D (mm)	Area of cross section (cm²)	Perimeter (mm)	Number of strands 12.7-mm diameter	Number of strands 11.1-mm diameter	Minimum effective prestress force (kN)	Section modulus (m³ × 10⁻³)	Design bearing capacity (kN) Concrete strength (MN/m²) 34.5	Design bearing capacity (kN) Concrete strength (MN/m²) 41.4
S	254	645	1016	4	4	312	2.737	556	778
O	254	536	838	4	4	258	1.786	462	555
S	305	929	1219	5	6	449	4.719	801	962
O	305	768	1016	4	5	369	3.097	662	795
S	356	1265	1422	6	8	610	7.489	1091	1310
O	356	1045	1168	5	7	503	4.916	901	1082
S	406	1652	1626	8	11	796	11.192	1425	1710
O	406	1368	1346	7	9	658	7.341	1180	1416
S	457	2090	1829	10	13	1010	15.928	1803	2163
O	457	1729	1524	8	11	836	10.455	1491	1790
S	508	2581	2032	12	16	1245	21.844	2226	2672
O	508	2136	1677	10	14	1032	14.355	1842	2239
S	559	3123	2235	15	20	1508	29.087	2694	3232
O	559	2587	1854	12	16	1250	19.107	2231	2678
S	610	3658	2438	18	23	1793	37.756	3155	3786
O	610	3078	2032	15	19	1486	34.794	2655	3186

[a] S = square section; O = octagonal section

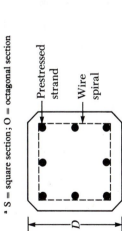

Wire spiral

Prestressed strand

Prestressed strand

Wire spiral

Table 8.3b Typical Prestressed Concrete Pile in Use (English Units)

Pile shape[a]	D (in.)	Area of cross section (in.²)	Perimeter (in.)	Number of strands 1/2-in. diameter	Number of strands 7/16-in. diameter	Minimum effective prestress force (kips)	Section modulus (in.³)	Design bearing capacity (kips) Concrete strength 5000 psi	Design bearing capacity (kips) Concrete strength 6000 psi
S	10	100	40	4	4	70	167	125	175
O	10	83	33	4	4	58	109	104	125
S	12	144	48	5	6	101	288	180	216
O	12	119	40	4	5	83	189	149	178
S	14	196	56	6	8	137	457	245	295
O	14	162	46	5	7	113	300	203	243
S	16	256	64	8	11	179	683	320	385
O	16	212	53	7	9	148	448	265	318
S	18	324	72	10	13	227	972	405	486
O	18	268	60	8	11	188	638	336	402
S	20	400	80	12	16	280	1333	500	600
O	20	331	66	10	14	234	876	414	503
S	22	484	88	15	20	339	1775	605	727
O	22	401	73	12	16	281	1166	502	602
S	24	576	96	18	23	403	2304	710	851
O	24	477	80	15	19	334	2123	596	716

[a] S = square section; O = octagonal section

Figure 8.4   Cast-in-place concrete piles (see Table 8.4 for descriptions)

patented by their manufacturers. These piles can be divided into two broad categories: (a) *cased* and (b) *uncased*. Both types may have a *pedestal* at the bottom.

Cased piles are made by driving a steel casing into the ground with the help of a mandrel placed inside the casing. When the pile reaches the proper depth, the mandrel is withdrawn and the casing is filled with concrete. Figures 8.4a, b, c, and d show some examples of cased piles without a pedestal. Table 8.4 gives additional information about these cased piles. Figure 8.4e shows a cased pile with a pedestal. The pedestal is an expanded concrete bulb that is formed by dropping a hammer on fresh concrete.

Figures 8.4f and g are two types of uncased pile, one with a pedestal and the other without. The uncased piles are made by first driving the

Table 8.4   Descriptions of the Cast-in-Place Piles Shown in Figure 8.4

Part in Figure 8.4	Name of pile	Type of casing	Maximum usual depth of pile (m)	(ft)
a	Raymond Step-Taper	Corrugated, thin cylindrical casing	30	100
b	Monotube or Union Metal	Thin fluted, tapered steel casing driven without mandrel	40	130
c	Western cased	Thin sheet casing	30–40	100–130
d	Seamless pipe or Armco	Straight steel pipe casing	50	160
e	Franki cased pedestal	Thin sheet casing	30–40	100–130
f	Western uncased without pedestal	—	15–20	50–65
g	Franki uncased pedestal	—	30–40	100–130

casing to the desired depth and then filling it with fresh concrete. The casing is gradually withdrawn in steps.

The allowable load for cast-in-place concrete piles may be given by the following equations:

Cased pile

$$Q_{all} = A_s f_s + A_c f_c \tag{8.2a}$$

where A_s = area of cross section of steel
A_c = area of cross section of concrete
f_s = allowable stress of steel
f_c = allowable stress of concrete

Uncased pile

$$Q_{all} = A_c f_c \tag{8.2b}$$

Timber Piles

Timber piles are tree trunks that have had their branches and bark carefully trimmed off. The maximum length of most timber piles is 10–20 m (30–65 ft). In order to qualify for use as a pile, the timber should be straight, sound, and without any defects. The American Society of Civil Engineers' *Manual of Practice*, No. 17 (1959), classifies timber piles into three categories:

1. *Class A Piles*: These piles carry heavy loads. Minimum diameter of the butt should be 356 mm (14 in.).
2. *Class B Piles*: These are used to carry medium loads. Minimum butt diameter should be 305–330 mm (12–13 in.).
3. *Class C Piles*: For use in temporary construction work. They can be used for structures on a permanent basis when the entire pile is below the water table. Minimum butt diameter should be 305 mm (12 in.).

In any case, pile tips should not have a diameter less than 150 mm (6 in.).

Timber piles cannot withstand hard driving stress; therefore, the pile capacity is generally limited to about 220–270 kN (25–30 tons). Steel shoes may be used to avoid damage at the pile tip (bottom). The tops of timber piles may also be damaged during the driving operation. The crushing of the wooden fibers caused by the impact of the hammer is referred to as *brooming*. To avoid damage to the pile top, a metal band or a cap may be used.

Splicing of timber piles should be avoided, particularly when they are expected to carry tensile load or lateral load. However, if splicing is necessary, it can be done by using either *pipe sleeves* (Figure 8.5a) or *metal straps and bolts* (Figure 8.5b). The length of the pipe sleeve should be at least five times the diameter of the pile. The butting ends should be cut square so that full contact can be maintained. The spliced portions should be carefully trimmed so that they fit tightly to the inside of the pipe sleeve. In the case of metal straps and bolts, the butting ends should also be cut square. Also, the sides of the spliced portion should be trimmed plane for putting the straps on.

Timber piles can stay undamaged indefinitely if they are surrounded by saturated soil. However, in a marine environment timber piles are subject to attack by various organisms and can be extensively damaged in a few

Metal sleeve

Ends cut square

Metal strap

Ends cut square

Metal strap

(a)   (b)

Figure 8.5 Splicing of timber piles: (a) use of pipe sleeves; (b) use of metal straps and bolts months. When located above the water table, the piles are subject to attack by insects. The life of the piles may be increased by treating them with preservatives such as creosote oil.

The allowable load-carrying capacity of wooden piles may be given as

$$Q_{all} = A_p f_w \qquad (8.3)$$

where A_p = average area of cross section of the pile
f_w = allowable stress for the timber

Composite Piles

The upper and lower portions of composite piles are made of different materials. For example, composite piles may be made of *steel and concrete* or *timber and concrete*. Steel and concrete piles consist of a lower portion of steel and an upper portion of cast-in-place concrete. This type of pile is the one used when the length of the pile required for adequate bearing exceeds the capacity of simple cast-in-place concrete piles. Timber and concrete piles usually consist of a lower portion of timber pile below the permanent water table and an upper portion of concrete. In any case, it is difficult to form proper joints between two dissimilar materials, and, for that reason, composite piles are not widely used.

Comparison of Pile Types

Several factors affect the selection of piles for a given structure at a given site. Table 8.5 gives a brief comparison of the advantages and disadvantages of the various types of pile based on the pile material.

8.3   Estimation of Pile Length

Selecting the type of pile to be used and estimating its necessary length are fairly difficult tasks that require good judgment. In addition to the classification given in Section 8.2, piles can be divided into three major categories depending on their lengths and the mechanisms of load transfer to the soil. These categories are (a) *point bearing piles*, (b) *friction piles*, and (c) *compaction piles*.

Point Bearing Piles

If bedrock or rock-like material is located at a given site within a reasonable depth that has been well established by soil-boring records, piles can be extended to the rock surface (Figure 8.6a). In this case, the ultimate capacity of the piles entirely depends on the load-bearing capacity of the firm material; thus the piles are called *point bearing piles*. In most of these cases, the necessary length of the pile can be fairly well established.

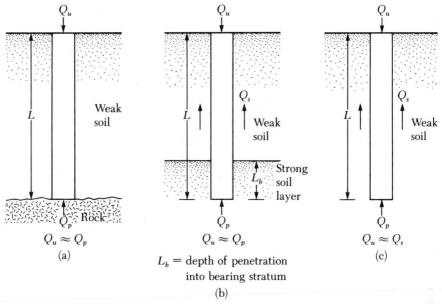

Figure 8.6   Point bearing piles [(a) and (b)]; friction piles [(c)]

Instead of bedrock, if a fairly compact and hard stratum of soil is located at a reasonable depth, piles can be extended a few meters into the hard stratum (Figure 8.6b). Piles with pedestals can be constructed on the bed of the hard stratum. For these types of pile, the ultimate pile load can be expressed as

$$Q_u = Q_p + Q_s \tag{8.4}$$

where Q_p = load carried at the pile point
       Q_s = load carried by skin friction developed at the side of the pile
              (caused by shearing resistance between the soil and the pile)

If Q_s is very small, then

$$Q_u \approx Q_p \tag{8.5}$$

In this case, necessary pile length can also be accurately estimated if proper subsoil exploration records are available.

Friction Piles

When there is no layer of rock or rock-like material located at a reasonable depth at a given site, point bearing piles becomes very long and uneconomical. For this type of subsoil condition, piles are driven through the softer material to specified depths (Figure 8.6c). The ultimate load of these

Table 8.5 Comparisons of Piles Made of Different Materials

Pile type	Usual length of piles	Maximum length of piles	Usual load	Approximate maximum load	Comments
Steel	15–60 m (50–200 ft)	Practically unlimited	300–1200 kN (67–270 kips)	Eq. (8.1)	*Advantages* a. Easy to handle with respect to cutoff and extension to the desired length b. Can stand high driving stresses c. Can penetrate hard layers such as dense gravel, soft rock d. High load-carrying capacity *Disadvantages* a. Relatively costly material b. High level of noise during pile driving c. Subject to corrosion d. H-piles may be damaged or deflected from the vertical during driving through hard layers or past major obstructions
Precast concrete	*Precast:* 10–15 m (30–50 ft) *Prestressed:* 10–35 m (30–150 ft)	*Precast:* 30 m (100 ft) *Prestressed:* 60 m (200 ft)	300–3000 kN (67–675 kips)	*Precast:* 800–900 kN (180–202 kips) *Prestressed:* 7500–8500 kN (1686–1910 kips)	*Advantages* a. Can be subjected to hard driving b. Corrosion resistant c. Can be easily combined with concrete superstructure *Disadvantages* a. Difficult to achieve proper cutoff b. Difficult to transport

Pile type	Usual length of pile	Maximum length of pile	Usual load	Approximate maximum load	Comments
Cased cast-in-place concrete	5–15 m (15–50 ft)	15–40 m (50–130 ft)	200–500 kN (45–112 kips)	800 kN (180 kips)	*Advantages* a. Relatively cheap b. Possibility of inspection before pouring concrete c. Easy to extend *Disadvantages* a. Difficult to splice after concreting b. Thin casings may be damaged during driving
Uncased cast-in-place concrete	5–15 m (15–50 ft)	30–40 m (100–130 ft)	300–500 kN (67–112 kips)	700 kN (157 kips)	*Advantages* a. Initially economical b. Can be finished at any elevation *Disadvantages* a. Voids may be created if concrete is placed rapidly b. Difficult to splice after concreting c. In soft soils, the sides of the hole may cave in, thus squeezing the concrete
Wood	10–15 m (30–50 ft)	30 m (100 ft)	100–200 kN (22–45 kips)	270 kN (60 kips)	*Advantages* a. Economical b. Easy to handle c. Permanently submerged piles are fairly resistant to decay *Disadvantages* a. Decay above water table b. Can be damaged in hard driving c. Low load-bearing capacity d. Low resistance to tensile load when spliced

piles can be expressed by Eq. (8.4). However, if the value of Q_p is relatively small

$$Q_u \approx Q_s \tag{8.6}$$

These piles are called *friction piles* because most of the resistance is derived from skin friction. However, the term *friction pile*, although used often in literature, is a misnomer: in clayey soils, the resistance to applied load is also caused by *adhesion*.

The length of friction piles depends on the shear strength of the soil, the applied load, and the pile size. To determine the necessary lengths of these piles, one needs a good understanding of the soil-pile interaction, good judgment, and experience. Theoretical procedures for the calculation of load-bearing capacity of piles are given in Section 8.6.

Compaction Piles

Under certain circumstances, piles are driven in granular soils to achieve proper compaction of soil close to the ground surface. These piles are called *compaction piles*. The length of compaction piles depends on factors such as (a) relative density of the soil before compaction, (b) desired relative density of the soil after compaction, and (c) required depth of compaction. These piles are generally short; however, some field tests are necessary to arrive at a reasonable figure.

8.4   Installation of Piles

Most piles are driven into the ground by means of *hammers* or *vibratory drivers*. In special circumstances, piles can also be inserted by *jetting* or *partial augering*. The hammers used for pile driving can be of several types, such as (a) drop hammer, (b) single-acting air or steam hammer, (c) double-acting and differential air or steam hammer, and (d) diesel hammer. In the driving operation, a cap is attached to the top of the pile. A cushion may be used between the pile and the cap. This cushion has the effect of reducing the impact force and spreading it over a longer time; however, its use is optional. A hammer cushion is placed on the pile cap. The hammer drops on the cushion.

Figure 8.7 illustrates various hammers. A drop hammer (Figure 8.7a) is raised by a winch and allowed to drop from a certain height H. This is the oldest type of hammer used for pile driving. The main disadvantage with the drop hammer is the slow rate of hammer blows. The principle of the single-acting air or steam hammer is shown in Figure 8.7b. In this case, the striking part, or ram, is raised by air or steam pressure, and then it drops by gravity. Figure 8.7c shows the operation of the double-acting and differential air or steam hammer. For these hammers, air or steam is used both to raise the ram and to push it downward. This increases the impact velocity of the ram. The diesel hammer (Figure 8.7d) essentially consists of a ram, an anvil block, and a fuel-injection system. During the operation, the ram is first raised and fuel is injected near the anvil. Then the ram is released. When the ram drops, it compresses the air-fuel mixture. This compression ignites the air-fuel mixture. This, in effect, pushes the pile downward and raises the ram. Diesel hammers work well under hard driving conditions. In soft soils, the downward movement of the pile is rather large, and the upward movement of the ram is small. This may not be sufficient to ignite the air-fuel system, so the ram may have to be lifted manually. Tables 8.6 and 8.7 give partial lists of the commercially available diesel, single-acting, double-acting, and differential hammers.

The principles of operation of a vibratory pile driver are shown in Figure 8.7e. This driver essentially consists of two counter-rotating weights. The horizontal components of the centrifugal force generated as a result of rotating masses cancel each other. As a result, this produces a sinusoidal dynamic vertical force on the pile and helps the pile to be driven downward.

Jetting is a technique sometimes used in pile driving, when the pile needs to penetrate a thin layer of hard soil (such as sand and gravel) overlying a softer soil layer. In this technique, water is discharged at the pile point

Figure 8.7 Pile-driving equipment: (a) drop hammer; (b) single-acting air or steam hammer; (c) double-acting and differential air or steam hammer; (d) diesel hammer; (e) vibratory pile driver

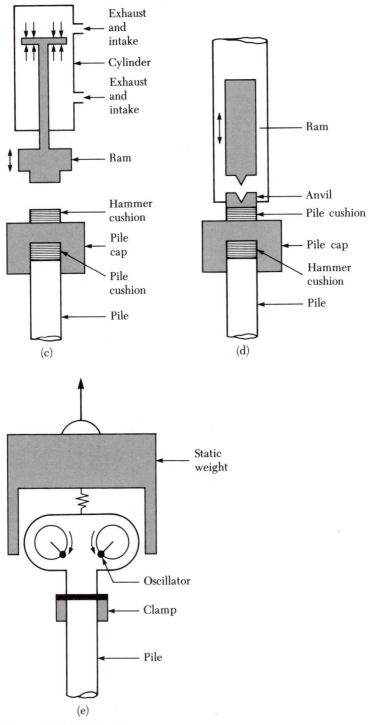

Figure 8.7 (Continued)

Table 8.6 Partial List of Typical Air and Steam Hammers

Maker of hammer[a]	Model no.	Type of hammer	Rated energy		Blows per minute	Ram weight	
			kN-m	kip-ft		kN	kips
V	3100	Single acting	406.8	300	58	448.8	100
V	540	Single acting	271.2	200	48	181.9	40.9
V	060	Single acting	244.1	180	62	266.9	60
MKT	OS-60	Single acting	244.1	180	55	266.9	60
V	040	Single acting	162.7	120	60	177.9	40
V	400C	Differential	153.9	113.5	100	177.9	40
R	8/0	Single acting	110.2	81.25	35	111.2	25
MKT	S-20	Single acting	81.4	60	60	89	20
R	5/0	Single acting	77.2	56.9	44	77.8	17.5
V	200-C	Differential	68.1	50.2	98	89	20
R	150-C	Differential	66.1	48.75	95–105	66.7	15
MKT	S-14	Single acting	50.9	37.5	60	62.3	14
V	140C	Differential	48.8	36	103	62.3	14
V	08	Single acting	35.3	26	50	35.6	8
MKT	S-8	Single acting	35.3	26	55	35.6	8
MKT	11B3	Double acting	26.1	19.2	95	22.2	5
MKT	C-5	Double acting	21.7	16.0	110	22.2	5
V	30-C	Double acting	9.9	7.3	133	13.3	3

[a] V—Vulcan Iron Works, Florida
MKT—McKiernan-Terry, New Jersey
R—Raymond International, Inc., Texas

Table 8.7 Partial List of Typical Diesel Hammers

Maker of hammer[a]	Model no.	Rated energy		Blows per minute	Piston weight	
		kN-m	kip-ft		kN	kips
K	K150	379.7	280	45–60	147.2	33.1
M	MB70	191.2–86	141–63.4	38–60	70.5	15.84
K	K-60	143.2	105.6	42–60	58.7	13.2
K	K-45	123.5	91.1	39–60	44.0	9.9
M	M-43	113.9–51.3	84–37.8	40–60	42.1	9.46
K	K-35	96	70.8	39–60	34.3	7.7
MKT	DE70B	85.4–57	63–42	40–50	31.1	7.0
K	K-25	68.8	50.7	39–60	24.5	5.51
V	N-46	44.1	32.55	50–60	17.6	3.96
L	520	35.7	26.3	80–84	22.6	5.07
M	M-14S	35.3–16.1	26–11.88	42–60	13.2	2.97
V	N-33	33.4	24.6	50–60	13.3	3.0
L	440	24.7	18.2	86–90	17.8	4.0
MKT	DE20	24.4–16.3	18.0–12.0	40–50	8.9	2.0
MKT	DE-10	11.9	8.8	40–50	4.9	1.1
L	180	11.0	8.1	90–95	7.7	1.73

[a] V—Vulcan Iron Works, Florida
M—Mitsubishi International Corporation
MKT—McKiernan-Terry, New Jersey
L—Link Belt, Cedar Rapids, Iowa
K—Kobe Diesel

by means of a pipe 50–75 mm (2–3 in.) in diameter to wash and loosen the sand and gravel.

Piles are sometimes driven at an angle to the vertical, typically 14° to 20°. These are referred to as *batter piles*. Batter piles are used in group piles when higher lateral load-bearing capacity is required. Piles can also be advanced by partial augering. This means that power augers (Chapter 2) can be used to predrill holes part of the way. The piles can then be inserted into the holes and driven to the desired depth.

Based on the nature of their placement, piles can be divided into two categories: *displacement piles* and *nondisplacement piles*. Driven piles are displacement piles, because they move some soil laterally, and, hence, there is a tendency for densification of soil surrounding them. Concrete piles and closed-ended pipe piles are high-displacement piles. However, steel H-piles displace less soil laterally during driving, and so they are low-displacement piles. In contrast, bored piles are nondisplacement type. Their placement causes very little change in the state of stress in the soil.

8.5   Load Transfer Mechanism

The load transfer mechanism from a pile to the soil is highly complicated. In order to understand it, consider a pile of length L, as shown in Figure 8.8a. Let the load on the pile be gradually increased from zero to $Q_{(z=0)}$ at the ground surface. Part of this load will be resisted by the side friction developed along the shaft (Q_1) and part by the soil below the tip of the pile (Q_2). Now, how are Q_1 and Q_2 related to the total load? If measurements are made to obtain the load carried by the pile shaft $[Q_{(z)}]$ at any depth z, the nature of variation will be like that shown in Curve 1 of Figure 8.8b. The *frictional resistance per unit area* $[f_{(z)}]$ at any given depth z can be determined as

$$f_{(z)} = \frac{\Delta Q_{(z)}}{(p)(\Delta z)} \qquad\qquad (8.7)$$

where p = perimeter of pile cross section

The nature of variation of $f_{(z)}$ with depth is shown in Figure 8.8c.

If the load Q at the ground surface is gradually increased, maximum frictional resistance along the pile shaft will be fully mobilized when the relative displacement between the soil and the pile is about 5–10 mm (0.2–0.3 in.) irrespective of pile size and length L. However, the maximum point resistance $Q_2 = Q_p$ will not be mobilized until the pile tip has gone through a movement of about 10–25% of the pile width (or diameter). The lower limit applies to driven piles, and the upper limit is for bored piles. At ultimate load (Figure 8.8d and Curve 2 in Figure 8.8b), $Q_{(z=0)} = Q_u$.

$$Q_1 = Q_s$$

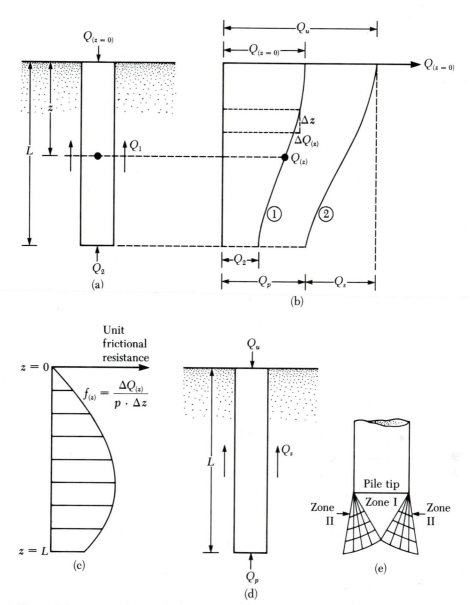

Figure 8.8   Load transfer mechanism for piles

and

$$Q_2 = Q_p$$

The preceding explanation indicates that Q_s (or the unit skin friction, f, along the pile shaft) is developed at a *much smaller pile displacement compared with the point resistance*, Q_p. This can be seen from Vesic's (1970) pile load

Figure 8.9 Relative magnitude of point load transferred at various stages of pile loading (redrawn after Vesic, 1970)

test results in granular soil, as given in Figure 8.9. Note that these results are for *pipe piles in dense sand*.

At ultimate load, the failure surface in the soil at the pile tip (bearing capacity failure caused by Q_p) is of the nature shown in Figure 8.8e. Note that pile foundations are deep foundations, and the soil fails mostly in a *punching mode*, as illustrated in Figure 3.1c and Figure 3.2. This means that a *triangular zone*, I, is developed at the pile tip, which is pushed downward without producing any other visible slip surface. In dense sands and stiff clayey soils, a *radial shear zone*, II, may partially develop. Hence, the load displacement curves of piles will resemble those shown in Figure 3.1c.

8.6   Equations for Estimation of Pile Capacity

The ultimate load-carrying capacity of a pile can be given by a simple equation as the sum of the load carried at the pile point plus the total frictional resistance (skin friction) derived from the soil-pile interface (Figure 8.10a), or

$$Q_u = Q_p + Q_s \tag{8.8}$$

where Q_u = ultimate pile capacity
Q_p = load-carrying capacity of the pile point
Q_s = frictional resistance

Numerous published studies relate to the determination of the values of Q_p and Q_s. Excellent reviews of most of the recent investigations have been

L = length of embedment
L_b = length of embedment in bearing
    stratum

(a)

(b) Open-Ended Pipe Pile Section

(c) H-Pile Section

(*Note:* A_p = area of steel + soil plug)

Figure 8.10

provided by Vesic (1977), Meyerhof (1976), and Coyle and Castello (1981). These studies provide insight into the problem of determination of ultimate pile capacity.

Point Bearing Capacity, Q_p

The ultimate bearing capacity of shallow foundations has been discussed in Chapter 3. According to Terzaghi's equations

$$q_u = 1.3cN_c + qN_q + 0.4\gamma BN_\gamma \quad \text{(for shallow square foundations)}$$

and

$$q_u = 1.3cN_c + qN_q + 0.3\gamma BN_\gamma \quad \text{(for shallow circular foundations)}$$

In a similar manner, the general bearing capacity equation for shallow foundations has been given in Chapter 3 (for vertical loading) as

$$q_u = cN_c F_{cs} F_{cd} + qN_q F_{qs} F_{qd} + \frac{1}{2}\gamma BN_\gamma F_{\gamma s} F_{\gamma d}$$

Hence, in general, the ultimate load-bearing capacity can be expressed as

$$q_u = cN_c^* + qN_q^* + \gamma BN_\gamma^* \tag{8.9a}$$

where N_c^*, N_q^*, and N_γ^* are the bearing capacity factors that include the necessary shape and depth factors.

Pile foundations are deep. However, the ultimate resistance per unit area developed at the pile tip (q_p) can be expressed by an equation similar in form to that shown in Eq. (8.9a), although the values of N_c^*, N_q^*, and N_γ^* will change. The notation used in this chapter for the width of pile is D. Hence, substitution of D for B in Eq. (8.9a) results in the form

$$q_u = q_p = cN_c^* + qN_q^* + \gamma D N_\gamma^* \tag{8.9b}$$

Because the width D of a pile is relatively small, the term $\gamma D N_\gamma^*$ can be dropped from the right side of the preceding equation without making a serious error, or

$$q_p = cN_c^* + q'N_q^* \tag{8.10}$$

Note that the term q has been replaced by q' in Eq. (8.10) to signify effective vertical stress. Hence, the point bearing of piles can be expressed as

$$\boxed{Q_p = A_p q_p = A_p(cN_c^* + q'N_q^*)} \tag{8.11}$$

where A_p = area of pile tip
 c = cohesion of the soil supporting the pile tip
 q_p = unit point resistance
 q' = effective vertical stress at the level of the pile tip
N_c^*, N_q^* = the bearing capacity factors

There are several methods for the determination of the bearing capacity factors N_c^* and N_q^*, including Meyerhof's method, Vesic's method, and Janbu's method, which we will now discuss.

Meyerhof's Method: The point bearing capacity (q_p) of a pile in sand generally increases with the depth of embedment in the bearing stratum and reaches a maximum value at an embedment ratio of $L_b/D = (L_b/D)_{cr}$. Note that in a homogeneous soil L_b is equal to the actual embedment length of the pile, L (see Figure 8.10a). However, in Figure 8.6b, where a pile has penetrated into a bearing stratum, $L_b < L$. Beyond the critical embedment ratio $(L_b/D)_{cr}$, the value of q_p remains constant (that is, $q_p = q_l$). This fact is shown in Figure 8.11 for the case of a homogeneous soil—that is, $L = L_b$. The variation of $(L_b/D)_{cr}$ with the soil friction angle is shown in Figure 8.12. In this figure, note that the broken curve is for the determination of N_c^* and the solid curve is for the determination of N_q^*. According to Meyerhof (1976), the bearing capacity factors increase with L_b/D and reach a maximum value at $L_b/D \approx 0.5(L_b/D)_{cr}$. It can be seen from Figure 8.12 that $(L_b/D)_{cr}$ for $\phi = 45°$ is about 25, and it decreases with the decrease of the friction angle, ϕ. In most practical cases, the magnitude of L_b/D for piles is greater than $0.5(L_b/D)_{cr}$. So the maximum values of N_c^* and N_q^* will apply for calculation of q_p for all piles. The variation of these maximum values of N_c^* and N_q^* with friction angle, ϕ, is shown in Figure 8.13.

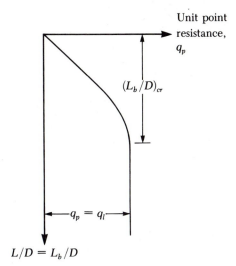

Figure 8.11   Nature of variation of unit point resistance in a homogeneous sand

For piles in sand, $c = 0$, and Eq. (8.11) simplifies to

$$Q_p = A_p q_p = A_p q' N_q^*$$ (8.12)

$\uparrow$

Figure 8.13

However, Q_p should not exceed the limiting value, which equals $A_p q_l$. So

$$Q_p = A_p q' N_q^* \leq A_p q_l$$ (8.13)

Figure 8.12   Variation of $(L_b/D)_{cr}$ with soil friction angle (after Meyerhof, 1976)

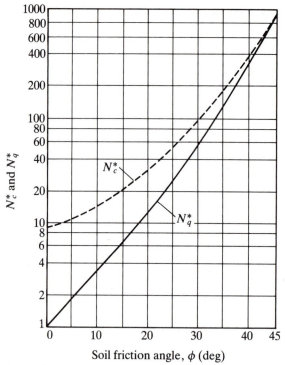

Figure 8.13 Variation of the maximum values of N_c^* and N_q^* with soil friction angle ϕ (after Meyerhof, 1976)

The limiting point resistance can be given as

$$q_l \text{ (kN/m}^2) = 50N_q^* \tan \phi \qquad (8.14)$$

where ϕ = soil friction angle in the bearing stratum

In English units, Eq. (8.14) can be expressed as

$$q_l \text{ (lb/ft}^2) = 1000N_q^* \tan \phi \qquad (8.15)$$

Based on field observations, Meyerhof (1976) also suggested that the ultimate point resistance, q_p, in a homogeneous granular soil ($L = L_b$) can be obtained from standard penetration numbers as

$$q_p \text{ (kN/m}^2) = 40NL/D \leq 400N \qquad (8.16a)$$

where N = average standard penetration number near the pile point (about $10D$ above and $4D$ below the pile point)

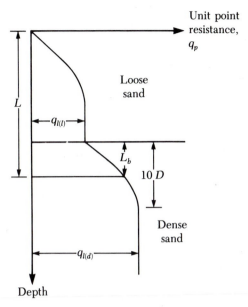

Figure 8.14 Variation of unit point resistance in layered soil

In English units

$$q_p \text{ (lb/ft}^2) = 800NL/D \le 80,000 \, N \tag{8.16b}$$

In many circumstances, a pile may initially penetrate a weak sand layer and then a dense layer, as shown in Figure 8.14. For these piles

$$q_p = q_{(l)l} + \frac{[q_{l(d)} - q_{l(l)}]L_b}{10D} \le q_{l(d)} \tag{8.17}$$

where $q_{l(l)}$ = limiting unit point resistance in the loose sand determined from Eq. (8.14) or (8.15) using the maximum value of N_q^* and ϕ value of the loose sand

$q_{l(d)}$ = limiting unit point resistance in the dense sand determined from Eq. (8.14) or (8.15) using the maximum value of N_q^* and ϕ value of the dense sand

L_b = depth of penetration into the dense sand layer

For piles in *saturated clays* in undrained conditions ($\phi = 0$)

$$\boxed{Q_p = N_c^* c_u A_p = 9c_u A_p} \tag{8.18}$$

where c_u = undrained cohesion of the soil below the pile tip

For clays with c and ϕ parameters present (effective stress basis), the ultimate point load can be given by the relation presented in Eq. (8.11).

Vesic's Method: Vesic (1977) proposed a method for estimating the pile point bearing capacity based on the theory of *expansion of cavities.* According to this theory, based on effective stress parameters,

$$Q_p = A_p q_p = A_p(c N_c^* + \sigma_o' N_\sigma^*) \qquad (8.19)$$

where σ_o' = mean normal ground stress (effective) at the level of the pile point

$$= \left(\frac{1 + 2K_o}{3}\right) q' \qquad (8.20)$$

K_o = earth pressure coefficient at rest = $1 - \sin \phi$ (8.21)
N_c^*, N_σ^* = bearing capacity factors

Note that Eq. (8.19) is a modification of Eq. (8.11) with

$$N_\sigma^* = \frac{3 N_q^*}{(1 + 2K_o)} \qquad (8.22)$$

The relation for N_c^* given in Eq. (8.19) can be expressed as

$$N_c^* = (N_q^* - 1) \cot \phi \qquad (8.23)$$

According to the theory of Vesic

$$N_\sigma^* = f(I_{rr}) \qquad (8.24)$$

where I_{rr} = reduced rigidity index for the soil

However

$$I_{rr} = \frac{I_r}{1 + I_r \Delta} \qquad (8.25)$$

where I_r = rigidity index = $\dfrac{E_s}{2(1 + \mu_s)(c + q' \tan \phi)} = \dfrac{G_s}{c + q' \tan \phi}$ (8.26)

E_s = Young's modulus of soil
μ_s = Poisson's ratio of soil
G_s = shear modulus of soil
Δ = average volumatic strain in the plastic zone below the pile point

For conditions of no volume change (that is, dense sand or saturated clay), $\Delta = 0$. So

$$I_r = I_{rr} \qquad (8.27)$$

Table 8.8 gives the values of N_c^* and N_σ^* for various values of the soil friction angle (ϕ) and I_{rr}. For $\phi = 0$ (that is, undrained condition)

$$N_c^* = \frac{4}{3}(\ln I_{rr} + 1) + \frac{\pi}{2} + 1 \tag{8.28}$$

The values of I_r can be estimated from laboratory consolidation and triaxial tests corresponding to the proper stress levels. However, for preliminary use the following values are recommended:

Soil type	I_r
Sand	70–150
Silts and clays (drained condition)	50–100
Clays (undrained condition)	100–200

Janbu's Method: Janbu (1976) has proposed a method for calculating Q_p in the following manner.

$$Q_p = A_p(cN_c^* + q'N_q^*) \tag{8.29}$$

Note that Eq. (8.29) is in the same form as Eq. (8.11). The bearing capacity factors N_c^* and N_q^* are calculated assuming a failure surface in soil at the pile tip similar to that shown in the insert of Figure 8.15. The bearing capacity factors can be given by the following relationships:

$$N_q^* = (\tan \phi + \sqrt{1 + \tan^2 \phi})^2 \cdot e^{2\eta' \tan \phi} \tag{8.30}$$

The angle η' is defined in the insert of Figure 8.15.

$$N_c^* = (N_q^* - 1) \cot \phi \tag{8.31}$$

$$\uparrow$$
$$\text{Eq. (8.30)}$$

Figure 8.15 shows the variation of N_q^* and N_c^* with ϕ and η'. The angle η' may vary from about 70° in soft clays to about 105° in dense sandy soils.

Irrespective of the theoretical procedure adopted in calculating Q_p, it should be kept in mind that full value cannot be realized until the pile tip has gone through a settlement of 10–25% of the width of the pile. This is critical in the case of sand.

Frictional Resistance, Q_s

The frictional or skin resistance of a pile can be written as

$$Q_s = \Sigma p \Delta L f \tag{8.32}$$

Table 8.8   Bearing Capacity Factors for Deep Foundations, N_c^* and N_σ^* (Eq. 8.19)

ϕ	I_{rr}									
	10	20	40	60	80	100	200	300	400	500
0	6.97	7.90	8.82	9.36	9.75	10.04	10.97	11.51	11.89	12.19
	1.00	1.00	1.00	1.00	1.00	1.00	1.00	1.00	1.00	1.00
1	7.34	8.37	9.42	10.04	10.49	10.83	11.92	12.57	13.03	13.39
	1.13	1.15	1.16	1.18	1.18	1.19	1.21	1.22	1.23	1.23
2	7.72	8.87	10.06	10.77	11.28	11.69	12.96	13.73	14.28	14.71
	1.27	1.31	1.35	1.38	1.39	1.41	1.45	1.48	1.50	1.51
3	8.12	9.40	10.74	11.55	12.14	12.61	14.10	15.00	15.66	16.18
	1.43	1.49	1.56	1.61	1.64	1.66	1.74	1.79	1.82	1.85
4	8.54	9.96	11.47	12.40	13.07	13.61	15.34	16.40	17.18	17.80
	1.60	1.70	1.80	1.87	1.91	1.95	2.07	2.15	2.20	2.24
5	8.99	10.56	12.25	13.30	14.07	14.69	16.69	17.94	18.86	19.59
	1.79	1.92	2.07	2.16	2.23	2.28	2.46	2.57	2.65	2.71
6	9.45	11.19	13.08	14.26	15.14	15.85	18.17	19.62	20.70	21.56
	1.99	2.18	2.37	2.50	2.59	2.67	2.91	3.06	3.18	3.27
7	9.94	11.85	13.96	15.30	16.30	17.10	19.77	12.46	22.71	23.73
	2.22	2.46	2.71	2.88	3.00	3.10	3.43	3.63	3.79	3.91
8	10.45	12.55	14.90	16.41	17.54	18.45	21.51	23.46	24.93	26.11
	2.47	2.76	3.09	3.31	3.46	3.59	4.02	4.30	4.50	4.67
9	10.99	13.29	15.91	17.59	18.87	19.90	23.39	25.64	27.35	28.73
	2.74	3.11	3.52	3.79	3.99	4.15	4.70	5.06	5.33	5.55
10	11.55	14.08	16.97	18.86	20.29	21.46	25.43	28.02	29.99	31.59
	3.04	3.48	3.99	4.32	4.58	4.78	5.48	5.94	6.29	6.57
11	12.14	14.90	18.10	20.20	21.81	23.13	27.64	30.61	32.87	34.73
	3.36	3.90	4.52	4.93	5.24	5.50	6.37	6.95	7.39	7.75
12	12.76	15.77	19.30	21.64	23.44	24.92	30.03	33.41	36.02	38.16
	3.71	4.35	5.10	5.60	5.98	6.30	7.38	8.10	8.66	9.11
13	13.41	16.69	20.57	23.17	25.18	26.84	32.60	36.46	39.44	41.89
	4.09	4.85	5.75	6.35	6.81	7.20	8.53	9.42	10.10	10.67
14	14.08	17.65	21.92	24.80	27.04	28.89	35.38	39.75	43.15	45.96
	4.51	5.40	6.47	7.18	7.74	8.20	9.82	10.91	11.76	12.46
15	14.79	18.66	23.35	26.53	29.02	31.08	38.37	43.32	47.18	50.39
	4.96	6.00	7.26	8.11	8.78	9.33	11.28	12.61	13.64	14.50
16	15.53	19.73	24.86	28.37	31.13	33.43	41.58	47.17	51.55	55.20
	5.45	6.66	8.13	9.14	9.93	10.58	12.92	14.53	15.78	16.83
17	16.30	20.85	26.46	30.33	33.37	35.92	45.04	51.32	56.27	60.42
	5.98	7.37	9.09	10.27	11.20	11.98	14.77	16.69	18.20	19.47
18	17.11	22.03	28.15	32.40	35.76	38.59	48.74	55.80	61.38	66.07
	6.56	8.16	10.15	11.53	12.62	13.54	16.84	19.13	20.94	22.47
19	17.95	23.26	29.93	34.59	38.30	41.42	52.71	60.61	66.89	72.18
	7.18	9.01	11.31	12.91	14.19	15.26	19.15	21.87	24.03	25.85
20	18.83	24.56	31.81	36.92	40.99	44.43	56.97	65.79	72.82	78.78
	7.85	9.94	12.58	14.44	15.92	17.17	21.73	24.94	27.51	29.67
21	19.75	25.92	33.80	39.38	43.85	47.64	61.51	71.34	79.22	85.90
	8.58	10.95	13.97	16.12	17.83	19.29	24.61	28.39	31.41	33.97
22	20.71	27.35	35.89	41.98	46.88	51.04	66.37	77.30	86.09	93.57
	9.37	12.05	15.50	17.96	19.94	21.62	27.82	32.23	35.78	38.81
23	21.71	28.84	38.09	44.73	50.08	54.66	71.56	83.68	93.47	101.83
	10.21	13.24	17.17	19.99	22.26	24.20	31.37	36.52	40.68	44.22
24	22.75	30.41	40.41	47.63	53.48	58.49	77.09	90.51	101.39	110.70
	11.13	14.54	18.99	22.21	24.81	27.04	35.32	41.30	46.14	50.29
25	23.84	32.05	42.85	50.69	57.07	62.54	82.98	97.81	109.88	120.23
	12.12	15.95	20.98	24.64	27.61	30.16	39.70	46.61	52.24	57.06

Table 8.8   (Continued)

					I_{rr}					
ϕ	10	20	40	60	80	100	200	300	400	500
26	24.98	33.77	45.42	53.93	60.87	66.84	89.25	105.61	118.96	130.44
	13.18	17.47	23.15	27.30	30.69	33.60	44.53	52.51	59.02	64.62
27	26.16	35.57	48.13	57.34	64.88	71.39	95.02	113.92	128.67	141.39
	14.33	19.12	25.52	30.21	34.06	37.37	49.88	59.05	66.56	73.04
28	27.40	37.45	50.96	60.93	69.12	76.20	103.01	122.79	139.04	153.10
	15.57	20.91	28.10	33.40	37.75	41.51	55.77	66.29	74.93	82.40
29	28.69	39.42	53.95	64.71	73.58	81.28	110.54	132.23	150.11	165.61
	16.90	22.85	30.90	36.87	41.79	46.05	62.27	74.30	84.21	92.80
30	30.03	41.49	57.08	68.69	78.30	86.64	118.53	142.27	161.91	178.98
	18.24	24.95	33.95	40.66	46.21	51.02	69.43	83.14	94.48	104.33
31	31.43	43.64	60.37	72.88	83.27	92.31	126.99	152.95	174.49	193.23
	19.88	27.22	37.27	44.79	51.03	56.46	77.31	92.90	105.84	117.11
32	32.89	45.90	63.82	77.29	88.50	98.28	135.96	164.29	187.87	208.43
	21.55	29.68	40.88	49.30	56.30	62.41	85.96	103.66	118.39	131.24
33	34.41	48.26	67.44	81.92	94.01	104.58	145.46	176.33	202.09	224.62
	23.34	32.34	44.80	54.20	62.05	68.92	95.46	115.51	132.24	146.87
34	35.99	50.72	71.24	86.80	99.82	111.22	155.51	189.11	217.21	241.84
	25.28	35.21	49.05	59.54	68.33	76.02	105.90	128.55	147.51	164.12
35	37.65	53.30	75.22	91.91	105.92	118.22	166.14	202.64	233.27	260.15
	27.36	38.32	53.67	65.36	75.17	83.78	117.33	142.89	164.33	183.16
36	39.37	55.99	79.39	97.29	112.34	125.59	177.38	216.98	250.30	279.60
	29.60	41.68	58.68	71.69	82.62	92.24	129.87	158.65	182.85	204.14
37	41.17	58.81	83.77	102.94	119.10	133.34	189.25	232.17	268.36	300.26
	32.02	45.31	64.13	78.57	90.75	101.48	143.61	175.95	203.23	227.26
38	43.04	61.75	88.36	108.86	126.20	141.50	201.78	248.23	287.50	322.17
	34.63	49.24	70.03	86.05	99.60	111.56	158.65	194.94	225.62	252.71
39	44.99	64.83	93.17	115.09	133.66	150.09	215.01	265.23	307.78	345.41
	37.44	53.50	76.45	94.20	109.24	122.54	175.11	215.78	250.23	280.71
40	47.03	68.04	98.21	121.62	141.51	159.13	228.97	283.19	329.24	370.04
	40.47	58.10	83.40	103.05	119.74	134.52	193.13	238.62	277.26	311.50
41	49.16	71.41	103.49	128.48	149.75	168.63	243.69	302.17	351.95	396.12
	43.74	63.07	90.96	112.68	131.18	147.59	212.84	263.67	306.94	345.34
42	51.38	74.92	109.02	135.68	158.41	178.62	259.22	322.22	375.97	423.74
	47.27	68.46	99.16	123.16	143.64	161.83	234.40	291.13	339.52	382.53
43	53.70	78.60	114.82	143.23	167.51	189.13	275.59	343.40	401.36	452.96
	51.08	74.30	108.08	134.56	157.21	177.36	257.99	321.22	375.28	423.39
44	56.13	82.45	120.91	151.16	177.07	200.17	292.85	365.75	428.21	483.88
	55.20	80.62	117.76	146.97	172.00	194.31	283.80	354.20	414.51	468.28
45	58.66	86.48	127.28	159.48	187.12	211.79	311.04	389.35	456.57	516.58
	59.66	87.48	128.28	160.48	188.12	212.79	312.03	390.35	457.57	517.58
46	61.30	90.70	133.97	168.22	197.67	224.00	330.20	414.26	486.54	551.16
	64.48	94.92	139.73	175.20	205.70	232.96	342.94	429.98	504.82	571.74
47	64.07	95.12	140.99	177.40	208.77	236.85	350.41	440.54	518.20	587.72
	69.71	103.00	152.19	191.24	224.88	254.99	376.77	473.42	556.70	631.25
48	66.97	99.75	148.35	187.04	220.43	250.36	371.70	468.28	551.64	626.36
	75.38	111.78	165.76	208.73	245.81	279.06	413.82	521.08	613.65	696.64
49	70.01	104.60	156.09	197.17	232.70	264.58	394.15	497.56	586.96	667.21
	81.54	121.33	180.56	227.82	268.69	305.37	454.42	573.38	676.22	768.53
50	73.19	109.70	164.21	207.83	245.60	279.55	417.82	528.46	624.28	710.39
	88.23	131.73	196.70	248.68	293.70	334.15	498.94	630.80	744.99	847.61

From "Design of Pile Foundations," by A. S. Vesic, in NCHRP *Synthesis of Highway Practice 42,* Transportation Research Board, 1977. Reprinted by permission.
Note: Upper number N_c^*, lower number N_σ^*.

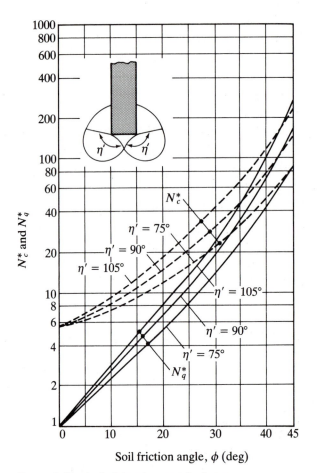

Figure 8.15   Janbu's bearing capacity factors

where p = perimeter of the pile section
   ΔL = incremental pile length over which p and f are
          taken constant (Figure 8.16a)
   f = unit friction resistance at any given depth z

Frictional Resistance in Sand: The unit frictional resistance at a given depth for a pile can be expressed as

$$f = K\sigma_v' \tan \delta \qquad (8.33)$$

where K = earth pressure coefficient
   σ_v' = effective vertical stress at the depth under consideration
   δ = soil-pile friction angle

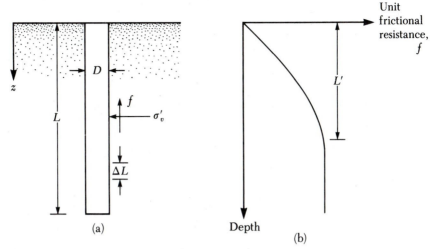

Figure 8.16 Unit frictional resistance for piles in sand

In reality, the value of K varies with depth. It is approximately equal to the Rankine passive earth pressure coefficient (K_p) at the top of the pile and may be less than the at-rest earth pressure coefficient (K_o) at the pile tip. It will also depend on the nature of pile installation. Based on presently available results, the following average values of K are recommended for use in Eq. (8.33).

For bored or jetter piles
_____

$K = K_o = 1 - \sin \phi$

For low-displacement driven piles
_____

$K = K_o$—lower limit
$K = 1.4 K_o$—upper limit

For high-displacement driven piles
_____

$K = K_o$—lower limit
$K = 1.8 K_o$—upper limit

Bhusan (1982) has recommended the following relationships for high-displacement driven piles:

$$K \tan \delta = 0.18 + 0.0065 D_r \qquad\qquad (8.34)$$

and

$$K = 0.5 + 0.008 D_r \qquad\qquad (8.35)$$

where D_r = relative density (%)

The effective vertical stress, σ_v', for use in Eq. (8.33) increases with pile depth up to a maximum limit at a depth of 15–20 pile diameters and remains constant thereafter. This is shown in Figure 8.16b. This critical depth, L', depends on several factors, such as soil friction angle and compressibility and relative density. A conservative estimate would be to assume

$$L' = 15D \tag{8.36}$$

The values of δ from various investigations appear to be in the range of 0.5ϕ to 0.8ϕ. Proper judgment should be used in choosing the value of δ.

Meyerhof (1976) has also indicated that the average unit frictional resistance (f_{av}) for driven high-displacement piles can be obtained from average standard penetration resistance values as

$$f_{av}\ (kN/m^2) = 2\overline{N} \tag{8.37a}$$

where $\overline{N}$ = average value of standard penetration resistance

In English units, the preceding equation can be stated as

$$f_{av}\ (lb/ft^2) = 40\overline{N} \tag{8.37b}$$

For driven low-displacement piles

$$f_{av}\ (kN/m^2) = \overline{N} \tag{8.38a}$$

and

$$f_{av}\ (lb/ft^2) = 20\overline{N} \tag{8.38b}$$

Thus

$$Q_s = pLf_{av} \tag{8.39}$$

Frictional (or Skin) Resistance in Clay: There are several methods presently available to obtain unit frictional (or skin) resistance of piles in clay. Some of the presently accepted procedures are briefly described below.

1. λ Method: This method was proposed by Vijayvergiya and Focht (1972). This method assumes that the displacement of soil caused by pile driving results in a passive lateral pressure at any depth, and the average unit skin resistance can be given as

$$\boxed{f_{av} = \lambda(\bar{\sigma}'_v + 2c_u)} \tag{8.40}$$

where $\bar{\sigma}'_v$ = mean effective vertical stress for the entire embedment length
$\qquad\quad c_u$ = mean undrained shear strength ($\phi = 0$ concept)

The value of λ will change with the depth of pile penetration (see Figure 8.17). Thus, the total frictional resistance can now be calculated as

$$Q_s = pLf_{av}$$

Care should be taken in obtaining the values of $\bar{\sigma}'_v$ and c_u in layered soil. This can be explained with the aid of Figure 8.18. According to

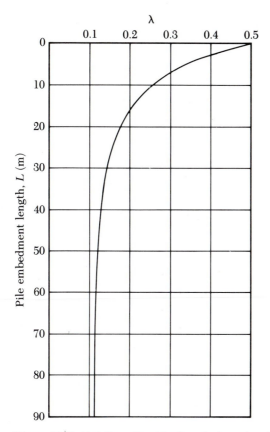

Figure 8.17   Variation of λ with pile embedment length (redrawn after McClelland, 1974)

Figure 8.18b, the mean value of c_u is $(c_{u(1)} L_1 + c_{u(2)} L_2 + \cdots)/L$. In a similar manner, Figure 8.18c shows the plot of the variation of effective stress with depth. The mean effective stress

$$\bar{\sigma}_v' = \frac{A_1 + A_2 + A_3 + \cdots}{L} \tag{8.41}$$

where $A_1, A_2, A_3, \ldots$ = areas of the vertical effective stress diagrams

2. α *Method:* According to the α method, the unit skin resistance in clayey soils can be represented by the equation

$$\boxed{f = \alpha c_u} \tag{8.42}$$

where α = empirical adhesion factor

The approximate variation of the value of α is shown in Figure 8.19. Note that for normally consolidated clays with $c_u \leq$ about 50 kN/m², the

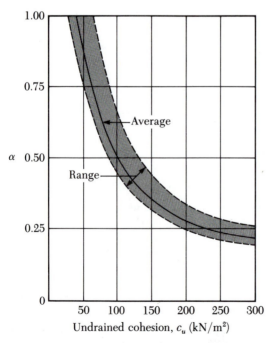

Figure 8.18 Application of λ method in layered soil

Figure 8.19 Variation of α with undrained cohesion of clay value of α is equal to one. Thus

$$\boxed{Q_s = \Sigma f p \Delta L = \Sigma \alpha c_u p \Delta L} \tag{8.43}$$

3. β Method: When piles are driven into saturated clays, the pore water pressure in the soil around the piles increases. This excess pore water pressure in normally consolidated clays may be of the magnitude of 4–6 times c_u. However, within a month or so, this pressure gradually dissipates. Hence, the unit frictional resistance for the pile can be determined on the basis of the effective stress parameters of the clay in a remolded state (that is, $c = 0$). Thus, at any given depth

$$\boxed{f = \beta \sigma'_v} \tag{8.44}$$

where σ'_v = vertical effective stress at any depth

$$\beta = K \tan \phi_R \tag{8.45}$$

ϕ_R = drained friction angle of remolded clay
K = earth pressure coefficient

The value of K can be conservatively taken as the earth pressure coefficient at rest, or

$$K = 1 - \sin \phi_R \quad \text{(for normally consolidated clays)} \tag{8.46}$$

and

$$K = (1 - \sin \phi_R)\sqrt{OCR} \quad \text{(for overconsolidated clays)} \tag{8.47}$$

where OCR = overconsolidation ratio

Combining Eqs. (8.44), (8.45), (8.46), and (8.47), for normally consolidated clays

$$f = (1 - \sin \phi_R) \tan \phi_R \sigma'_v \tag{8.48}$$

and, for overconsolidated clays

$$f = (1 - \sin \phi_R) \tan \phi_R \sqrt{OCR} \, \sigma'_v \tag{8.49}$$

Once the value of f is determined, the total frictional resistance can be evaluated as

$$Q_s = \Sigma f p \Delta L$$

Allowable Pile Capacity

Once the total ultimate load-carrying capacity of a pile is determined by summing the point bearing capacity and the frictional (or skin) resistance, a reasonable factor of safety should be used to obtain the total allowable load for each pile, or

$$Q_{\text{all}} = \frac{Q_u}{FS} \tag{8.50}$$

where Q_{all} = allowable load-carrying capacity for each pile
FS = factor of safety

The factor of safety generally used ranges from 2.5–4, depending on the uncertainties of ultimate load calculation.

General Comments

Although calculations for the ultimate load-carrying capacity of piles are made by using Eqs. (8.9a)–(8.49), one needs to keep the following points in mind:

1. For a given initial value of the soil friction angle (ϕ), driven piles in sand may show about 50–100% higher unit point resistance compared with bored piles. This is a result of the densification of soil during pile driving.

2. In sandy soils, cast-in-place piles with pedestals may show about 50–100% higher unit point resistance compared with cast-in-place piles without pedestals. The high-impact energy of the hammer building the pedestal causes substantial soil compaction and thus an increase of the soil friction angle.

3. In the calculation of the area of cross section (A_p) and the perimeter (p) of piles with developed profiles, such as H-piles and open-ended pipe piles, the effect of soil plug should be considered. According to Figure 8.10b and c, for pipe piles

$$A_p = \left(\frac{\pi}{4}\right)D^2$$

and

$$p = 2\pi D$$

Similarly, for H-piles

$$A_p = d_1 \cdot d_2$$
$$p = 2(d_1 + d_2)$$

Also, note that for H-piles, because $d_2 > d_1$, $D = d_1$.

4. The ultimate point load relations given in Eqs. (8.11), (8.19), and (8.29) are for the gross ultimate point load; that is, they include the weight of the pile. So, the net ultimate point load can approximately be given as

$$Q_{p(\text{net})} = Q_{p(\text{gross})} - q'A_p$$

In practice, for soils with $\phi > 0$, this is not done, and $Q_{p(net)}$ is assumed to be equal to $Q_{p(gross)}$.

In cohesive soils with $\phi = 0$, the value of N_q^* is equal to one (Figure 8.13). Hence, from Eq. (8.11)

$$Q_{p(gross)} = (c_u N_c^* + q')A_p$$

So

$$Q_{p(net)} = [(c_u N_c^* + q') - q']A_p = c_u N_c^* A_p = 9c_u A_p = Q_p$$

This is the relation given in Eq. (8.18).

8.7   Coyle and Castello Design Correlations

Coyle and Castello (1981) have analyzed 24 large-scale field load tests of driven piles in sand. For sands, the ultimate load can be given by the equation

$$Q_u = Q_p + Q_s = q'N_q^* A_p + f_{av} pL \tag{8.51}$$

where $q' =$ effective vertical stress at the pile tip
$f_{av} =$ average frictional resistance for the entire pile length
and can be given by the relation
$f_{av} = K\bar{\sigma}_v' \tan \delta \tag{8.52}$

where $K =$ lateral earth pressure coefficient
$\bar{\sigma}_v' =$ average effective overburden pressure
$\delta =$ soil-pile friction angle

Based on this study, the calculated values of the bearing capacity factor (N_q^*) have been correlated with the embedment ratio L/D. Figure 8.20 shows the values of N_q^* for various embedment ratios and soil friction angles. Note that N_q^* gradually increases with L/D to a maximum value and decreases thereafter.

In a similar manner, the values of deduced K for various values of ϕ and L/D ratios are given in Figure 8.21. It can be seen that, for any given soil friction angle, K decreases in a linear manner with embedment ratio. Note that, in Figure 8.21, it is assumed that

$$\delta = 0.8\phi \tag{8.53}$$

Hence, combining Eqs. (8.51), (8.52), and (8.53)

$$Q_u = q'N_q^* A_p + pLK\bar{\sigma}_v' \tan (0.8\phi) \tag{8.54}$$

Using the results of 24 pile load tests, Coyle and Castello have shown that Eq. (8.54) can predict the ultimate load with an error band of $\pm 30\%$, with a majority falling within an error band of $\pm 20\%$.

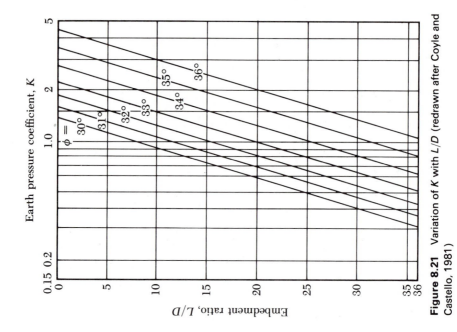

Figure 8.21 Variation of K with L/D (redrawn after Coyle and Castello, 1981)

Figure 8.20 Variation of N_q^* with L/D (redrawn after Coyle and Castello, 1981)

Example 8.1

A fully embedded precast, prestressed concrete pile is 12 m long and is driven into a homogeneous sand layer ($c = 0$). The pile is square in cross section with sides measuring 305 mm. The dry unit weight of sand (γ_d) is 16.8 kN/m^3, and the average soil friction angle is 35°. The average standard penetration resistance near the vicinity of the pile tip is 16. Calculate the ultimate point load on the pile by Coyle and Castello's method [Eq. (8.51) and Figure 8.20].

Solution

$$Q_p = q'N_q^* A_p$$

From Table 8.3a, $A_p = 929$ cm$^2 = 0.0929$ m^2

$$q' = \gamma_d L = (16.8)(12) = 201.6 \text{ kN/m}^2$$

Now, $L/D = 12/0.305 = 39.34$. For $L/D = 39.34$ and $\phi = 35°$, Figure 8.20 gives $N_q^* \approx 45$. So

$$Q_p = (201.6)(45)(0.0929) = \textbf{842.8 kN}$$

Example 8.2

Solve Example 8.1 by Meyerhof's method [Eq. (8.12) and Figure 8.13].

Solution

Since it is a homogeneous soil, $L_b = L$. Given: $\phi = 35°$. From Figure 8.13, $N_q^* \approx 120$.

$$Q_p = A_p q' N_q^* = (0.0929)(201.6)(120) = 2247.4 \text{ kN}$$

However, from Eq. (8.14)

$$q_l = 50N_q^* \tan \phi = 50(120) \tan 35° = 4201.25 \text{ kN/m}^2$$

So

$$Q_p = A_p q_l = (0.0929)(4201) = 390.3 \text{ kN} < A_p q' N_q^*$$

So

$$Q_p \approx \textbf{390 kN}$$

Example 8.3

Redo Example 8.1 by Vesic's method [Eq. (8.19)]. Use $I_r = 90 = I_{rr}$.

Solution

Given: $I_{rr} \approx 90$. With $\phi = 35°$, Table 8.8 gives $N_\sigma^* \approx 79.5$. From Eq. (8.19)

$$Q_p = A_p \sigma_o' N_\sigma^*$$

$$\sigma_o' = \frac{1 + 2K_o}{3} q'$$

$$K_o = 1 - \sin \phi = 1 - \sin 35° = 0.43$$

So

$$\sigma_o' = \left[\frac{1 + 2(0.43)}{3}\right](201.6) \approx 125 \text{ kN/m}^2$$

So

$$Q_p = (0.0929)(125)(79.5) \approx \textbf{923 kN}$$

Example 8.4

Solve Example 8.1 using Eq. (8.16).

Solution

Given: the average standard penetration resistance near the pile tip = 16. So, from Eq. (8.16a)

$$q_p = 40N\frac{L}{D} \le 400N$$

$$Q_p = A_p q_p = (0.0929)(40)(16)39.34 = 2339 \text{ kN}$$

However, the limiting value is

$$Q_p = A_p 400N = (0.0929)(400)(16) = 594.6 \text{ kN} \approx \textbf{595 kN}$$

Example 8.5

Compare the results of Examples 8.1–8.4 and estimate a design value.

Solution

For these problems Vesic's equation gives a much higher value (923 kN). The next highest value is obtained from the equation given by Coyle and Castello (842.8 kN). For a conservative estimation, one may take

$$Q_p = \frac{390.3 + 595}{2} \approx \textbf{493 kN}$$

Example 8.6

Refer to the pile given in Example 8.1. (a) Determine the total frictional resistance using $K = 1.4$ and $\delta = 0.6\phi$ [use Eqs. (8.32), (8.33), and (8.36)]. (b) Determine the total frictional resistance by using Coyle and Castello's method.

Solution
Part a

The unit skin friction at any depth is given by Eq. (8.33) as

$$f = K\sigma_v' \tan \delta$$

Also, from Eq. (8.36)

$$L' = 15D$$

So, for depth $z = 0-15D$, $\sigma'_v = \gamma z = 16.8 \times z$ (kN/m²) and beyond that—that is, $z \geq 15D$, $\sigma'_v = \gamma(15D) = (16.8)(15 \times 0.305) = 76.86$ kN/m². This is shown in Figure 8.22.

Figure 8.22

Frictional resistance from $z = 0 - 15D$:

$$Q_s = pL'f_{av} = [(4)(0.305)][15D]\left[\frac{(1.4)(76.86)\tan(0.6 \times 35)}{2}\right]$$

$$= (1.22)(4.575)(20.65) = 115.26 \text{ kN}$$

Frictional resistance from $z = 15D$ *to* 12 m:

$$Q_s = p(L - L')f_{z=15D} = [(4)(0.305)][12 - 4.575][(1.4)(76.86)\tan(0.6 \times 35)]$$

$$= (1.22)(7.425)(41.3) = 374.1 \text{ kN}$$

So, the total frictional resistance is equal to $115.26 + 374.1 = 489.35$ kN $\approx$ **490 kN**.

Part b: Coyle and Castello's Method

From Eqs. (8.52) and (8.53)

$$f_{av} = K\bar{\sigma}'_v \tan \delta$$
$$\delta = 0.8\phi$$
$$Q_s = f_{av}pL = [K\bar{\sigma}'_v \tan(0.8\phi)]pL$$

For this pile, $L/D = 39.34$. According to Figure 8.21 for determination of K, this is out of the range of the graph. By interpolation, for $L/D = 39.34$ and $\phi = 35°$, $K \approx 0.7$. Now

$$\bar{\sigma}'_v = \frac{\gamma L}{2} = \frac{(16.8)(12)}{2} = 100.8 \text{ kN/m}^2$$

So

$$Q_s = [(0.7)(100.8) \tan (0.8 \times 35)][4 \times 0.305][12] = 549.3 \text{ kN} \approx \mathbf{550 \text{ kN}}.$$

Example 8.7

Refer to Examples 8.1–8.6. Using a factor of safety of 3, estimate the allowable load for the piles.

Solution

$$Q_u = Q_p + Q_s$$

From Example 8.5, $Q_p = 493$ kN. Also, from Example 8.6, $Q_s = 490$ kN to 550 kN. So, use $Q_s = (490 + 550)/2 = 1040/2 = 520$ kN. So

$$Q_{all} = \frac{Q_s}{FS} = \frac{493 + 520}{3} = 337.7 \approx \mathbf{338 \text{ kN}}$$

Referring to Table 8.3a, the design bearing capacity of this pile is 801 kN, which is greater than 337 kN. So, $Q_{all} = \mathbf{338 \text{ kN}}$.

Example 8.8

A concrete pile having a square cross section (12 in. × 12 in.) is to be installed in a sand layer. The embedded length of the pile will be 35 ft. Assume the relative density of sand to be 40% and the unit weight of sand = 94 lb/ft³. Determine the frictional resistance of the pile. [Use Eqs. (8.32), (8.33), (8.34), and (8.36).]

Solution

From Eq. (8.33)

$$f = K\sigma'_v \tan \delta$$

Also, from Eq. (8.36)

$$L' = 15D = 15\left(\frac{12 \text{ in.}}{12}\right) = 15 \text{ ft}$$

According to Eq. (8.34)

$$K \tan \delta = 0.18 + 0.0065D_r = 0.18 + 0.0065(40) = 0.44$$

So, unit skin friction [Eq. (8.33)], $f = 0.44\sigma'_v$.

Frictional resistance from z = 0 to 15 ft:

$$Q_s = pL'f_{av} = (4 \times 1 \text{ ft})(15 \text{ ft})\left[\frac{0 + (0.44)(94 \times 15)}{2}\right]$$

$$= 18,612 \text{ lb}$$

Frictional resistance from z = 15 ft to 35 ft:

$$Q_s = p(L - L')f_{z=15\,ft}$$
$$= (4 \times 1\ \text{ft})(35 - 15)[(0.44)(94 \times 15)] = 49{,}632\ \text{lb}$$

So, the total frictional resistance is equal to

$$18{,}612 + 49{,}632 = \mathbf{68{,}244\ lb}$$

Example 8.9

An HP 310 × 1.079 steel pile is driven into sand, as shown in Figure 8.23a.

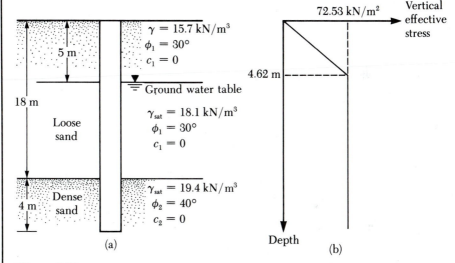

Figure 8.23

    a. Calculate the ultimate point load (1) by Meyerhof's procedure, (2) by Vesic's procedure ($I_r = 150 = I_{rr}$), and (3) using standard penetration resistance equations. (Given: average value of N in the vicinity of pile point is 45.)
    b. Estimate the value of the ultimate point load from the calculations in (a).
    c. Calculate the ultimate frictional resistance, Q_s. Use Eqs. (8.32), (8.33), and (8.36), $K = 1.4$, and $\delta = 0.6\phi$.
    d. Calculate the allowable pile load. Use $FS = 4$.

Also check the allowable load-bearing capacity of the steel section of the pile. Use σ_{all} for steel to be 62,000 kN/m².

Solution

In Table 8.1a, depth of the pile section, $d_1 = 308$ mm and flange width = 310 mm. Area of the pile section, A_p, for capacity calculation = $0.308 \times 0.310 = 0.0955$ m².

Part a: Calculation of Ultimate Point Load

Meyerhof's procedure: The variation of the unit point resistance will be like that in Figure 8.14. The depth of penetration of the lower dense sand layer, L_b, is 4 m. So $L_b/D = 4/0.308 = 12.99 > 10$. Hence, referring to Eq. (8.17)

$$q_p = q_{l(d)} = 50 N_q^* \tan \phi_2$$

For $\phi_2 = 40°$, $N_q^* \approx 350$ (Figure 8.13). Thus

$$q_p = (50)(350)(\tan 40°) \approx 14{,}684 \text{ kN/m}^2$$

So

$$Q_p = (14{,}684)(0.0955) = \textbf{1402 kN}$$

Also, check Eq. (8.12). From Eq. (8.12), $Q_p = A_p q' N_q^*$.

$$q' = 5(15.7) + 13(18.1 - 9.81) + 4(19.4 - 9.81)$$
$$= 78.5 + 107.77 + 38.36 = 224.63 \text{ kN/m}^2$$

So

$$Q_p = (0.0955)(224.63)(350) = 7508 \text{ kN}$$

Because $Q_p = 1402$ kN < 7508 kN, Eq. (8.17) controls. Thus, $Q_p = \textbf{1402 kN}$.

Vesic's procedure: Given: $I_r = 150 = I_{rr}$. From Eq. (8.19)

$$Q_p = A_p \sigma'_o N_q^*$$
$$K_o = 1 - \sin \phi = 1 - \sin 40° = 0.357$$
$$\sigma'_o = \frac{1 + 2K_o}{3} q' = \left[\frac{1 + (2)(0.357)}{3} \right] (224.63) = 128.34 \text{ kN/m}^2$$

From Table 8.8, for $\phi = 40°$ and $I_{rr} = 150$, the value of $N_\sigma^* \approx (134.52 + 193.13)/2 = 163.8$. So

$$Q_p = (0.0955)(128.34)(163.8) = \textbf{2008 kN}$$

Standard penetration resistance equations: Given: average penetration resistance, N, near the pile point is about 45. From Eq. (8.16a)

$$q_p = 40N \frac{L}{D} \leq 400N$$

So

$$q_p = 40(45)\left(\frac{22}{0.308} \right) = 128{,}571 \text{ kN/m}^2$$

or

$$q_p = (400)(N) = (400)(45) = 18{,}000 \text{ kN/m}^2$$

Hence, $q_p = 18{,}000$ kN/m² controls. So

$$Q_p = A_p q_p = (0.0955)(18000) = \textbf{1719 kN}$$

Part b: Estimation of Value for Q_p

Considering all three results, we might use

$$Q_p = \frac{1402 + 2008 + 1719}{3} \approx 1709 \text{ kN}$$

Part c: Calculation of Ultimate Frictional Resistance

According to Eq. (8.36)

$$L' = 15D = 15(0.308) = 4.62 \text{ m}$$

For ultimate frictional resistance, σ_v' will remain constant for $z > 4.62$ m. The assumed variation of σ_v' with depth is shown in Figure 8.23b.

Frictional resistance from $z = 0$ to 4.62 m $=$

$$pLf_{av} = 2(0.308 + 0.310)(4.62)\left(\frac{K\sigma_v' \tan \delta}{2}\right)$$

$$= 5.71\left[\frac{(1.4)(72.53) \tan (0.6 \times 30)}{2}\right] = \textbf{94.2 kN}$$

Frictional resistance from $z = 4.62$ m to 22 m $=$

$$pLf_{(z=4.62 \text{ m})} = 2(0.308 + 0.310)(22 - 4.62)(K\sigma_v' \tan \delta)$$

$$= 21.48[(1.4)(72.53) \tan \delta]$$

As an approximation, the value of δ may be taken as $0.6\phi_1 = (0.6)(30) = 18°$ for the entire length. Thus

$$Q_{s(z=4.62-22 \text{ m})} = (21.48)(1.4)(72.53) \tan 18° = 708.7 \text{ kN}$$

So, the total frictional resistance $=$

$$Q_s = Q_{s(z=0-4.62 \text{ m})} + Q_{s(z=4.62-22 \text{ m})}$$

$$= 94.2 + 708.7 = 802.9 \text{ kN} \approx \textbf{803 kN}$$

Part d: Calculation of Allowable Load

$Q_u = Q_p + Q_s$. From Part b, $Q_p = 1709$ kN. From Part c, $Q_s = 802.9$ kN. So $Q_u \approx 1709 + 803 = 2512$ kN.

$$Q_{all} = \frac{Q_u}{FS} = \frac{2512}{4} = 628 \text{ kN}$$

We also need to check the allowable load-bearing capacity for the steel pile section. Table 8.1a shows that the area of steel section for the pile is 14.1×10^{-3} m^2.

$$Q_{all} = (\sigma_{all})14.1 \times 10^{-3}$$

$$\sigma_{all} = 62{,}000 \text{ kN/m}^2$$

So

$$Q_{all} = (62{,}000)(14.1 \times 10^{-3}) = 874.2 \text{ kN}$$

Hence, the allowable pile load is **628 kN** (< 874.2 kN).

Example 8.10

A concrete pile 16 in. × 16 in. in cross section is shown in Figure 8.24. Determine
the allowable load that the pile can carry $(FS = 4)$. Use the α method for determination of the skin resistance.

40 ft

Saturated clay
$\phi = 0$
$c_u = 1800$ lb/ft²
$\gamma = 120$ lb/ft³ (unit weight)

16 in.

Figure 8.24

Solution

From Eq. (8.18)

$$Q_p = A_p q_p = A_p c_u N_c^* = \frac{\left(\frac{16}{12}\right)^2 (1800)(9)}{1000} = 28.78 \text{ kips}$$

Again, from Eqs. (8.42) and (8.43)

$$Q_s = \alpha c_u pL$$

From the average plot of Figure 8.19, for $c_u = 1800$ lb/ft², $\alpha = 0.55$. So

$$Q_s = \frac{(0.55)(1800)\left(4 \times \frac{16}{12}\right)(40)}{1000} = 211.1 \text{ kips}$$

$$Q_{all} = \frac{Q_p + Q_s}{FS} = \frac{28.78 + 211.2}{4} = \textbf{60 kips}$$

Example 8.11

For the pile shown in Figure 8.24, determine the ultimate skin resistance using the λ
method.

Solution

From Eq. (8.35)

$$f_{av} = \lambda(\bar{\sigma}'_v + c_u)$$

$L = 40 \text{ ft} = 12.19 \text{ m}$

From Figure 8.17, for $L = 12.19$ m, the value of λ is 0.24. So

$$f_{av} = (0.24)\left[\left(\frac{120 \times 40}{2}\right)\left(\frac{1}{1000}\right) + 2\left(\frac{1800}{1000}\right)\right] = 1.44 \text{ kips/ft}^2$$

$$Q_s = pLf_{av} = \left(4 \times \frac{16}{12}\right)(40)(1.44) = \mathbf{307.2 \text{ kips}}$$

Example 8.12

A driven pipe pile in clay is shown in Figure 8.25a. The pipe has an outside diameter of 406 mm, and the wall thickness is 6.35 mm.

Figure 8.25

a. Calculate the net point bearing capacity. Use Eq. (8.18).
b. Calculate the skin resistance (1) by using Eqs. (8.42) and (8.43) (α method), (2) by using Eq. (8.40) (λ method), and (3) by using Eq. (8.44) (β method). Given: $\phi_R = 30°$ for all clay layers. The top 10 m of clay is normally consolidated. The bottom clay layer has an *OCR* of 2.
c. Estimate the net allowable pile capacity. Use $FS = 4$.

Solution

Area of cross section of pile including the soil inside the pile =

$$A_p = \frac{\pi}{4} D^2 = \frac{\pi}{4} (0.406)^2 = 0.1295 \text{ m}^2$$

Part a: Calculation of Net Point Bearing Capacity

From Eq. (8.18)

$$Q_p = A_p q_p = A_p N_c^* c_{u(2)} = (0.1295)(9)(100) = \textbf{116.55 kN}$$

Part b: Calculation of Skin Resistance

(1) Use of Eqs. (8.42) and (8.43): From Eq. (8.43)

$$Q_s = \Sigma \alpha c_u p \Delta L$$

For the top soil layer, $c_{u(1)} = 30$ kN/m^2. According to the average plot of Figure 8.19, $\alpha_1 = 1.0$. Similarly, for the bottom soil layer, $c_{u(2)} = 100$ kN/m^2; $\alpha_2 = 0.5$. Thus

$$Q_s = \alpha_1 c_{u(1)}[(\pi)(0.406)]10 + \alpha_2 c_{u(2)}[(\pi)(0.406)]20$$
$$= (1)(30)[(\pi)(0.406)]10 + (0.5)(100)[(\pi)(0.406)]20$$
$$= 382.7 + 1275.5 = \textbf{1658.2 kN}$$

(2) Use of Eq. (8.40): $f_{av} = \lambda(\bar{\sigma}'_v + 2c_u)$

The average value of c_u is equal to

$$\frac{c_{u(1)}(10) + c_{u(2)}(20)}{30} = \frac{(30)(10) + (100)(20)}{30} = 76.7 \text{ kN/m}^2$$

To obtain the average value of $\bar{\sigma}'_v$, the diagram for vertical effective stress variation with depth is plotted in Figure 8.25b. From Eq. (8.41)

$$\bar{\sigma}'_v = \frac{A_1 + A_2 + A_3}{L} = \frac{225 + 552.38 + 4577}{30} = 178.48 \text{ kN/m}^2$$

The value of λ can be obtained from Figure 8.17 as 0.14. So

$$f_{av} = 0.14[178.48 \times (2)(76.7)] = 46.46 \text{ kN/m}^2$$

Hence

$$Q_s = pLf_{av} = \pi(0.406)(30)(46.46) = \textbf{1777.8 kN}$$

(3) Use of Eq. (8.44): The top clay layer (10 m) is normally consolidated. $\phi_R = 30°$.
For $z = 0$–5 m [Eq. (8.48)]:

$$f_{av(1)} = (1 - \sin \phi_R) \tan \phi_R \sigma'_{v(av)}$$
$$= (1 - \sin 30°)(\tan 30°)\left(\frac{0 + 90}{2}\right) = 13.0 \text{ kN/m}^2$$

Similarly, for $z = 5$–10 m:

$$f_{av(2)} = (1 - \sin 30°)(\tan 30°)\left(\frac{90 + 130.95}{2}\right) = 31.9 \text{ kN/m}^2$$

For $z = 10$–30 m [Eq. (8.49)]

$$f_{av} = (1 - \sin \phi_R) \tan \phi_R \sqrt{OCR} \, \sigma'_{v(av)}$$

Given: $OCR = 2$, so

$$f_{av(3)} = (1 - \sin 30°)(\tan 30°)\sqrt{2}\left(\frac{130.95 + 326.75}{2}\right) = 93.43 \text{ kN/m}^2$$

So

$$Q_s = p[f_{av(1)}(5) + f_{av(2)}(5) + f_{av(3)}(20)]$$
$$= (\pi)(0.406)[(13)(5) + (31.9)(5) + (93.43)(20)] = \textbf{2669.7 kN}$$

Part c: Calculation of Net Ultimate Capacity, Q_u

If we compare the three values just given, we see that the α and λ methods give close results. So use

$$Q_s = \frac{1658.1 + 1777.8}{2} \approx 1718 \text{ kN}$$

Thus

$$Q_u = Q_p + Q_s = 116.46 + 1718 = 1834.46 \text{ kN}$$
$$Q_{all} = \frac{Q_u}{FS} = \frac{1834.46}{4} = \textbf{458.6 kN}$$

8.8   Point Bearing Capacity of Piles Resting on Rocks

Sometimes piles are driven to the underlying rock layers. In such cases, one must evaluate the bearing capacity of the rocks. The ultimate unit point resistance in rocks can be approximately given as (Goodman, 1980)

$$\boxed{q_p = q_u(N_\phi + 1)} \tag{8.55}$$

where $N_\phi = \tan^2 (45 + \phi/2)$
q_u = unconfined compression strength of rock
ϕ = drained angle of friction

The unconfined compression strength of rocks can be determined by laboratory tests on rock specimens collected during field investigation. However, extreme caution should be used in obtaining the proper value of q_u, because laboratory specimens are usually small in diameter. As the diameter of the specimen increases, the unconfined compression strength decreases. This is referred to as the *scale effect*. For specimens larger than about 1 meter in diameter, the value of q_u remains approximately constant. There appears to be a 4- to 5-fold reduction of the magnitude of q_u in this process. The scale effect in rocks is primarily caused by randomly distributed large and small fractures and also by progressive ruptures along the slip lines. Hence, it is always recommended that

$$q_{u(design)} = \frac{q_{u(lab)}}{5} \qquad\qquad (8.56)$$

Table 8.9 lists some representative values of (laboratory) unconfined compression strengths of rocks. Representative values of the rock friction angle, ϕ, are given in Table 8.10.

A factor of safety of *at least 3* should be used to determine the allowable point bearing capacity of piles. Thus

$$Q_{p(all)} = \frac{[q_u(N_\phi + 1)]A_p}{FS}$$

Table 8.9 Unconfined Compressive Strengths of Specimens of Representative Rocks

	q_u	
Description	MPa	psi
Berea sandstone	73.8	10,700
Navajo sandstone	214.0	31,030
Tensleep sandstone	72.4	10,500
Hackensack siltstone	122.7	17,800
Monticello Dam s.s. (greywacke)	79.3	11,500
Solenhofen limestone	245.0	35,500
Bedford limestone	51.0	7,400
Tavernalle limestone	97.9	14,200
Oneota dolomite	86.9	12,600
Lockport dolomite	90.3	13,100
Flaming Gorge shale	35.2	5,100
Micaceous shale	75.2	10,900
Dworshak Dam gneiss		
45° to foliation	162.0	23,500
Quartz mica schist ⊥ schistocity	55.2	8,000
Baraboo quartzite	320.0	46,400
Taconic marble	62.0	8,990
Cherokee marble	66.9	9,700
Nevada Test Site granite	141.1	20,500
Pikes Peak granite	226.0	32,800
Cedar City tonalite	101.5	14,700
Palisades diabase	241.0	34,950
Nevada Test Site basalt	148.0	21,500
John Day basalt	355.0	51,500
Nevada Test Site tuff	11.3	1,639

From *Introduction to Rock Mechanics*, by R. E. Goodman. Copyright 1980 by John Wiley and Sons. Reprinted by permission.

Table 8.10 Representative Values for Angle of
Internal Friction (ϕ) for Selective Rocks

Description	Porosity (%)	ϕ (deg)
Berea sandstone	18.2	27.8
Bartlesville sandstone		37.2
Pottsville sandstone	14.0	45.2
Repetto siltstone	5.6	32.1
Muddy shale	4.7	14.4
Stockton shale		22.0
Edmonton bentonitic shale (water content 30%)	44.0	7.5
Sioux quartzite		48.0
Texas slate; loaded		
30° to cleavage		21.0
90° to cleavage		26.9
Georgia marble	0.3	25.3
Wolf Camp limestone		34.8
Indiana limestone	19.4	42.0
Chalk	40.0	31.5
Hasmark dolomite	3.5	35.5
Blaine anhydrite		29.4
Inada biotite granite	0.4	47.7
Stone Mountain granite	0.2	51.0
Nevada Test Site basalt	4.6	31.0
Schistose gneiss		
90° to schistocity	0.5	28.0
30° to schistocity	1.9	27.6

From *Introduction to Rock Mechanics*, by R. E.
Goodman. Copyright 1980 by John Wiley and Sons.
Reprinted by permission.

Example 8.13

A prestressed concrete pile is 50 ft long and is fully embedded. The pile is driven through a soft clay layer, and its tip rests on a layer of sandstone. Given:

For the rock:

square section $= D = 16$ in. (refer to Table 8.3b)

area of cross section $= 256$ in.^{2}

concrete strength $= 5000$ lb/in.^{2}

For the sandstone:

laboratory unconfined compressive strength $= 10,200$ lb/in.^{2}

friction angle, $\phi = 28°$

Estimate the allowable point bearing capacity of the pile. $FS = 4$.

Solution

From Eq. (8.56)

$$q_{u(design)} = \frac{q_{u(lab)}}{5} = \frac{10,200}{5} = 2040 \text{ lb/in.}^2$$

$$Q_{p(all)} = \frac{[q_u(N_\phi + 1)]A_p}{FS}$$

$$N_\phi = \tan^2\left(45 + \frac{\phi}{2}\right) = \tan^2\left(45 + \frac{28}{2}\right) = 2.77$$

So

$$Q_{p(all)} = \frac{[(2040)(2.77 + 1)](256)}{(4)(1000)} \approx 492 \text{ kips}$$

From Table 8.3b, the bearing design capacity of the pile should not exceed 320 kips. **Since this is smaller than $Q_{p(all)}$ calculated above, use 320 kips.**

8.9   Settlement of Piles

The settlement of a pile under a vertical working load (Q_w) is caused by three factors:

$$s = s_1 + s_2 + s_3 \tag{8.57}$$

where s = total pile settlement
s_1 = settlement of pile shaft
s_2 = settlement of pile caused by the load at the pile point
s_3 = settlement of pile caused by the load transmitted along the pile shaft

The procedures for estimating the preceding three elements of pile settlement are as follows:

Determination of s_1

If one assumes the pile material to be elastic, then the deformation of the pile shaft can be evaluated using the fundamental principles of mechanics of materials:

$$s_1 = \frac{(Q_{wp} + \xi Q_{ws})L}{A_p E_p} \tag{8.58}$$

where Q_{wp} = load carried at the pile point under working load condition
Q_{ws} = load carried by frictional (skin) resistance under working load condition
A_p = area of pile cross section
L = length of pile
E_p = Young's modulus of the pile material

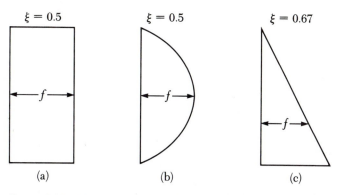

Figure 8.26 Various types of unit friction (skin) resistance distribution along the pile shaft

The magnitude of ξ will depend on the nature of unit frictional (skin) resistance distribution along the pile shaft. If the distribution of f is uniform or parabolic in nature, as shown in Figure 8.26a and b, ξ is equal to 0.5. However, for triangular distribution of f (Figure 8.26c), the value of ξ is about 0.67 (Vesic, 1977).

Determination of s_2

The settlement of a pile caused by the load carried at the pile point can be expressed in a form similar to that given for shallow foundations [Eq. (3.67)]:

$$s_2 = \frac{q_{wp}D}{E_s}(1 - \mu_s^2)I_{wp}$$

(8.59)

where D = width or diameter of pile

$\quad q_{wp}$ = point load per unit area at the pile point = Q_{wp}/A_p
$\quad E_s$ = Young's modulus of soil at or below the pile point
$\quad \mu_s$ = Poisson's ratio of soil
$\quad I_{wp}$ = influence factor

For all practical purposes, I_{wp} may be taken to be equal to α_r as given in Eq. (3.67) and can be evaluated from Figure 3.23. In the absence of any experimental results, representative values of Poisson's ratio can be obtained from Table 3.6.

Vesic (1977) has also proposed a semiempirical method to obtain the magnitude of the settlement, s_2. This can be expressed by the following equation:

$$s_2 = \frac{Q_{wp}C_p}{Dq_p}$$

(8.60)

Table 8.11 Typical Values of C_p [Eq. (8.60)]

Soil type	Driven pile	Bored pile
Sand (dense to loose)	0.02–0.04	0.09–0.18
Clay (stiff to soft)	0.02–0.03	0.03–0.06
Silt (dense to loose)	0.03–0.05	0.09–0.12

From "Design of Pile Foundations," by A. S. Vesic, in NCHRP *Synthesis of Highway Practice 42*, Transportation Research Board, 1977. Reprinted by permission.

where q_p = ultimate point resistance of the pile
$\quad\quad C_p$ = an empirical coefficient

Representative values of C_p for various soils are given in Table 8.11.

Determination of s_3

The settlement of a pile caused by the load carried by the pile shaft can be given by a relation similar to Eq. (8.59), or

$$s_3 = \left(\frac{Q_{ws}}{pL}\right)\frac{D}{E_s}(1 - \mu_s^2)I_{ws} \tag{8.61}$$

where p = perimeter of the pile
$\quad\quad L$ = embedded length of pile
$\quad\quad I_{ws}$ = influence factor

Note that the term Q_{ws}/pL in the preceding equation is the average value of f along the pile shaft. The influence factor, I_{ws}, can be expressed by a simple empirical relation as follows (Vesic, 1977):

$$I_{ws} = 2 + 0.35\sqrt{\frac{L}{D}} \tag{8.62}$$

Vesic (1977) has also proposed a simple empirical relation similar to Eq. (8.60) for obtaining s_3 as

$$s_3 = \frac{Q_{ws}C_s}{Lq_p} \tag{8.63}$$

where C_s = an empirical constant = $(0.93 + 0.16\sqrt{L/D})C_p$ $\quad\quad(8.64)$

The values of C_p for use in Eq. (8.63) can be estimated from Table 8.11.

Example 8.14

Consider the pile described in Examples 8.1–8.7. The allowable working load is 338 kN. If 240 kN is contributed by the frictional resistance and 98 kN is from the point load, determine the elastic settlement of the pile. Use $E_p = 21 \times 10^6$ kN/m², $E_s = 30{,}000$ kN/m², and $\mu_s = 0.3$.

Solution

We will use Eq. (8.57):

$$s = s_1 + s_2 + s_3$$

From Eq. (8.58)

$$s_1 = \frac{(Q_{wp} + \xi Q_{ws})L}{A_p E_p}$$

Let $\xi = 0.6$ and $E_p = 21 \times 10^6$ kN/m². So

$$s_1 = \frac{[97 + (0.6)(240)]12}{(0.305)^2(21 \times 10^6)} = 0.00148 \text{ m} = 1.48 \text{ mm}$$

From Eq. (8.58)

$$s_2 = \frac{q_{wp} D}{E_s}(1 - \mu_s^2)I_{wp}$$

From Figure 3.23, $I_{wp} = 0.82$

$$q_{wp} = \frac{Q_{wp}}{A_p} = \frac{98}{(0.305)^2} = 1042.7 \text{ kN/m}^2$$

So

$$s_2 = \left[\frac{(1042.7)(0.305)}{30{,}000}\right](1 - 0.3^2)(0.82) = 0.0079 \text{ m} = 7.9 \text{ mm}$$

Again, from Eq. (8.61)

$$s_3 = \left(\frac{Q_{wp}}{pL}\right)\frac{D}{E_s}(1 - \mu_s^2)I_{ws}$$

$$I_{ws} = 2 + 0.35\sqrt{\frac{L}{D}} = 2 + 0.35\sqrt{\frac{12}{0.305}} = 4.2$$

So

$$s_3 = \frac{240}{(\pi \times 0.305)(12)}\left(\frac{0.305}{30{,}000}\right)(1 - 0.3^2)(4.2) = 0.00081 \text{ m} = 0.81 \text{ mm}$$

Hence, the total settlement

$$s = 1.48 + 7.9 + 0.81 = \textbf{10.19 mm}$$

8.10   Pullout Resistance of Piles

Section 8.1 noted that, under certain construction conditions, piles are subjected to uplifting forces. The ultimate resistance of piles subjected to such force has not received much attention among researchers until recently. The gross ultimate resistance of a pile subjected to uplifting force can be written as follows (Figure 8.27):

$$T_{ug} = T_{un} + W \tag{8.65}$$

where T_{ug} = gross uplift capacity
   T_{un} = net uplift capacity
   W = effective weight of the pile

Piles in Clay

The net ultimate uplift capacity of piles embedded in saturated clays has been studied by Das and Seeley (1982). According to this study

$$\boxed{T_{un} = Lp\alpha'c_u} \tag{8.66}$$

where L = length of the pile
   p = perimeter of pile section
   α' = adhesion coefficient at soil-pile interface
   c_u = undrained cohesion of clay

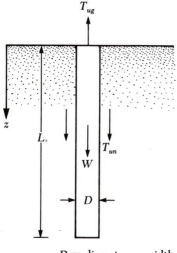

D = diameter or width
of pile

Figure 8.27   Uplift capacity of piles

For cast-*in-situ* concrete piles

$$\alpha' = 0.9 - 0.00625c_u \qquad \text{(for } c_u \leq 80 \text{ kN/m}^2) \tag{8.67}$$

and

$$\alpha' = 0.4 \qquad \text{(for } c_u > 80 \text{ kN/m}^2) \tag{8.68}$$

Similarly, for pipe piles

$$\alpha' = 0.715 - 0.0191c_u \qquad \text{(for } c_u \leq 27 \text{ kN/m}^2) \tag{8.69}$$

and

$$\alpha' = 0.2 \qquad \text{(for } c_u > 27 \text{ kN/m}^2) \tag{8.70}$$

Piles in Sand

When piles are embedded in granular soils ($c = 0$), the net ultimate uplift capacity can be expressed as follows (Das and Seeley, 1975):

$$T_{un} = \int_0^L (f_u p) \, dz \tag{8.71}$$

where f_u = unit skin friction during uplift
p = perimeter of pile cross section

The unit skin friction during uplift (f_u) usually varies in the manner shown in Figure 8.28a. It increases linearly up to a depth of $z = L_{cr}$; beyond that it remains constant. For $z \leq L_{cr}$

$$f_u = K_u \sigma'_v \tan \delta \tag{8.72}$$

where K_u = uplift coefficient
σ'_v = effective vertical stress at a depth of z
δ = soil-pile friction angle

The variation of the uplift coefficient with soil friction angle ϕ is given in Figure 8.28b. Based on the experience of the author, it appears that the values of L_{cr} and δ are dependent on the relative density of soil. Figure 8.28c shows the approximate nature of these variations with the relative density of soil. For calculation of the net ultimate uplift capacity of piles, the following procedure is suggested.

1. Determine the relative density of the soil and, using Figure 8.28c, obtain the value of L_{cr}.
2. If the length of the pile L is less than or equal to L_{cr}

$$T_{un} = p \int_0^L f_u \, dz = p \int_0^L (\sigma'_v K_u \tan \delta) \, dz \tag{8.73}$$

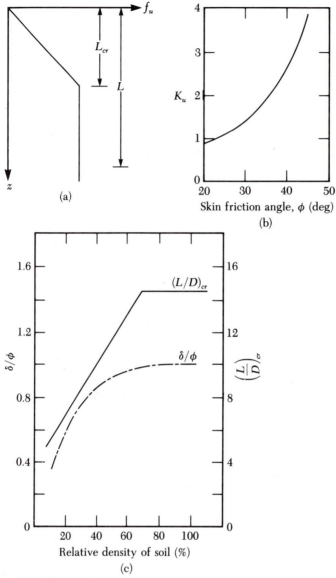

Figure 8.28 (a) Nature of variation of f_u; (b) uplift coefficient K_u; (c) variation of δ/ϕ and $(L/D)_{cr}$ with relative density of sand

In dry soils, $\sigma'_v = \gamma z$ (where γ = unit weight of soil). So

$$T_{un} = p \int_0^L (\sigma'_v K_u \tan \delta)\, dz = p \int_0^L \gamma z K_u \tan \delta\, dz$$

$$= \frac{1}{2} p\gamma L^2 K_u \tan \delta \qquad (8.74)$$

The values of K_u and δ can be obtained from Figure 8.28b and c.

3. For the case in which $L > L_{cr}$

$$T_{un} = p \int_0^L f_u \, dz = p \left[\int_0^{L_{cr}} f_u \, dz + \int_{L_{cr}}^L f_u \, dz \right]$$

$$= p \left\{ \int_0^{L_{cr}} [\sigma'_v K_u \tan \delta] \, dz + \int_{L_{cr}}^L \left[\sigma'_{v(\text{at } z = L_{cr})} K_u \tan \delta \right] dz \right\} \qquad (8.75)$$

In dry soils, Eq. (8.75) changes to the following simplified form:

$$T_{un} = \frac{1}{2} p\gamma L_{cr}^2 K_u \tan \delta + p\gamma L_{cr} K_u \tan \delta (L - L_{cr}) \qquad (8.76)$$

The values of K_u and δ can be determined from Figure 8.28b and c.

For estimation of the net allowable uplift capacity, a factor of safety of 2–3 is recommended. Thus

$$T_{u(\text{all})} = \frac{T_{ug}}{FS}$$

where $T_{u(\text{all})}$ = allowable uplift capacity

Example 8.15

For the pipe pile given in Example 8.12, determine the net ultimate uplift capacity.

Solution

We will use Eq. (8.66) for this problem. Because the top and bottom layers have clays with $c_u > 27 \text{ kN/m}^2$, the value of α' for both layers is 0.2. So, from Eq. (8.66)

$$T_{un} = p\alpha' \Sigma c_u \Delta L = \pi(0.406)(0.2)[(30)(10) + (100)(20)]$$
$$= 586.7 \text{ kN}$$

Example 8.16

Refer to Example 8.1. For the concrete pile, determine the net ultimate pullout capacity. Assume the relative density of soil to be 60%.

Solution

From Figure 8.28c, for a relative density of 60%, $(L/D)_{cr} \approx 12.7$. So

$$L_{cr} = (12.7)(0.305) = 3.87 \text{ m}$$

So, $L = 12 \text{ m} > L_{cr}$. For this case, Eq. (8.76) is to be used. Or

$$T_{un} = \frac{1}{2} p\gamma L_{cr}^2 K_u \tan \delta + p\gamma L_{cr} K_u \tan \delta (L - L_{cr})$$

From Figure 8.28b, for $\phi = 35°$. $K_u = 1.9$. Similarly, from Figure 8.28c, for a relative density = 60%, $(\delta/\phi) \approx 0.97$. So, $\delta = (0.97)(35) = 33.95°$. Substituting these values into Eq. (8.76)

$$T_{un} = \left(\frac{1}{2}\right)(4 \times 0.305)(16.8)(3.87)^2(1.9) \tan (33.95)$$

$$+ (4 \times 0.305)(16.8)(3.87)(1.9) \tan (33.95)(12 - 3.87)$$

$$= 1021.2 \text{ kN}$$

8.11   Laterally Loaded Vertical Piles

Granular Soils

A general solution for determination of moments and displacements of a vertical pile subjected to lateral load and moment at the ground surface has been given by Matlock and Reese (1960). Consider a pile of length L subjected to a lateral force Q_g and a moment M_g at the ground surface (that is, at $z = 0$), as shown in Figure 8.29a. Figure 8.29b shows the general nature of the deflected shape of the pile and the soil resistance caused by the applied load and the moment.

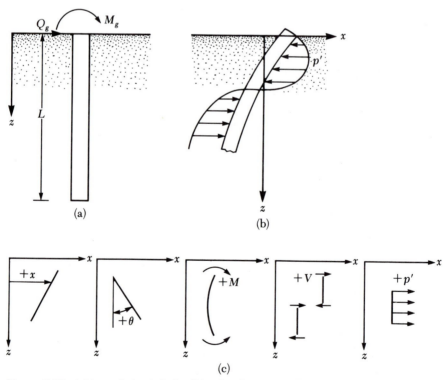

Figure 8.29   (a) Laterally loaded pile; (b) soil resistance on pile caused by lateral load; (c) sign conventions for displacement, slope, moment, shear, and soil reaction

According to a simpler Winkler's model, an elastic medium (which is soil in this case) can be replaced by a series of infinitely close independent elastic springs. With this assumption, one can write that

$$k = \frac{p' \ (kN/m)}{x \ (m)}$$

(8.77)

where k = modulus of subgrade reaction
  p' = pressure on soil
  x = deflection

The subgrade modulus for granular soils at a depth z can be defined as

$$k_z = n_h z$$

(8.78)

where n_h = constant of modulus of horizontal subgrade reaction

Referring to Figure 8.29b and using the theory of beams on an elastic foundation, one can write

$$E_p I_p \frac{d^4 x}{dz^4} = p'$$

(8.79)

where E_p = Young's modulus of the pile material
  I_p = moment of inertia of the pile section

Based on Winkler's model

$$p' = -kx$$

(8.80)

The sign in the preceding equation is negative because the soil reaction is in the opposite direction to the pile deflection.

Combining Eqs. (8.79) and (8.80)

$$E_p I_p \frac{d^4 x}{dz^4} + kx = 0$$

(8.81)

Solution of the preceding equation results in the following expressions:

Pile deflection at any depth $[x_z(z)]$:

$$\boxed{x_z(z) = A_x \frac{Q_g T^3}{E_p I_p} + B_x \frac{M_g T^2}{E_p I_p}}$$

(8.82)

Slope of pile at any depth $[\theta_z(z)]$:

$$\boxed{\theta_z(z) = A_\theta \frac{Q_g T^2}{E_p I_p} + B_\theta \frac{M_g T}{E_p I_p}}$$

(8.83)

Moment of pile at any depth $[M_z(z)]$:

$$\boxed{M_z(z) = A_m Q_g T + B_m M_g}$$
(8.84)

Shear force on pile at any depth $[V_z(z)]$:

$$\boxed{V_z(z) = A_v Q_g + B_v \frac{M_g}{T}}$$
(8.85)

Soil reaction at any depth $[p'_z(z)]$:

$$p'_z(z) = A_{p'} \frac{Q_g}{T} + B_{p'} \frac{M_g}{T^2}$$
(8.86)

where A_x, B_x, A_θ, B_θ, A_m, B_m, A_v, B_v, $A_{p'}$, and $B_{p'}$ are coefficients and

T = characteristic length of the soil-pile system

$$= \sqrt[5]{\frac{E_p I_p}{n_h}}$$
(8.87)

n_h has been defined in Eq. (8.78).

When the pile length, $L \geq 5T$, it is considered to be a *long pile*. For $L \leq 2T$, the pile is considered to be a *rigid pile*. Table 8.12 gives the values

Table 8.12 Coefficients for Long Piles, $k_z = n_h z$

z	A_x	A_θ	A_m	A_v	$A_{p'}$	B_x	B_θ	B_m	B_v	$B_{p'}$
0.0	2.435	−1.623	0.000	1.000	0.000	1.623	−1.750	1.000	0.000	0.000
0.1	2.273	−1.618	0.100	0.989	−0.227	1.453	−1.650	1.000	−0.007	−0.145
0.2	2.112	−1.603	0.198	0.956	−0.422	1.293	−1.550	0.999	−0.028	−0.259
0.3	1.952	−1.578	0.291	0.906	−0.586	1.143	−1.450	0.994	−0.058	−0.343
0.4	1.796	−1.545	0.379	0.840	−0.718	1.003	−1.351	0.987	−0.095	−0.401
0.5	1.644	−1.503	0.459	0.764	−0.822	0.873	−1.253	0.976	−0.137	−0.436
0.6	1.496	−1.454	0.532	0.677	−0.897	0.752	−1.156	0.960	−0.181	−0.451
0.7	1.353	−1.397	0.595	0.585	−0.947	0.642	−1.061	0.939	−0.226	−0.449
0.8	1.216	−1.335	0.649	0.489	−0.973	0.540	−0.968	0.914	−0.270	−0.432
0.9	1.086	−1.268	0.693	0.392	−0.977	0.448	−0.878	0.885	−0.312	−0.403
1.0	0.962	−1.197	0.727	0.295	−0.962	0.364	−0.792	0.852	−0.350	−0.364
1.2	0.738	−1.047	0.767	0.109	−0.885	0.223	−0.629	0.775	−0.414	−0.268
1.4	0.544	−0.893	0.772	−0.056	−0.761	0.112	−0.482	0.688	−0.456	−0.157
1.6	0.381	−0.741	0.746	−0.193	−0.609	0.029	−0.354	0.594	−0.477	−0.047
1.8	0.247	−0.596	0.696	−0.298	−0.445	−0.030	−0.245	0.498	−0.476	0.054
2.0	0.142	−0.464	0.628	−0.371	−0.283	−0.070	−0.155	0.404	−0.456	0.140
3.0	−0.075	−0.040	0.225	−0.349	0.226	−0.089	0.057	0.059	−0.213	0.268
4.0	−0.050	0.052	0.000	−0.106	0.201	−0.028	0.049	−0.042	0.017	0.112
5.0	−0.009	0.025	−0.033	0.015	0.046	0.000	−0.011	−0.026	0.029	−0.002

of the coefficients for long piles $(L/T \geq 5)$ in Eqs. (8.82) to (8.86). Note that, in the first column of Table 8.12, Z is the nondimensional depth, or

$$Z = \frac{z}{T} \qquad\qquad (8.88)$$

The positive sign conventions for $x_z(z)$, $\theta_z(z)$, $M_z(z)$, $V_z(z)$, and $p'_z(z)$ assumed in the derivations in Table 8.12 are shown in Figure 8.29c. Also, Figure 8.30 shows the variation of A_x, B_x, A_m, and B_m for various values of $L/T = Z_{\max}$. These figures indicate that when L/T is greater than about 5, the coefficients do not change. This is true of long piles only.

To calculate the characteristic length T for the pile, one needs to assume a proper value of n_h. Some representative values of n_h are given in Table 8.13.

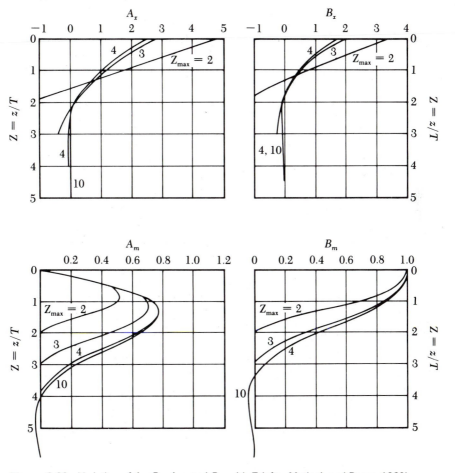

Figure 8.30   Variation of A_x, B_x, A_m, and B_m with Z (after Matlock and Reese, 1960)

Table 8.13 Representative Values of n_h

Soil type	n_h (kN/m³)
Dry or moist sand	Loose: 1800–2200 Medium: 5500–7000 Dense: 15000–18000
Submerged sand	Loose: 1000–1400 Medium: 3500–4500 Dense: 9000–12000

Note: 1 kN/m³ = 6.36 lb/ft³

Cohesive Soils

Solutions similar to those given in Eqs. (8.82)–(8.86) have been given by Davisson and Gill (1963) for the case of piles embedded in clay. According to these solutions

$$x_z(z) = A'_x \frac{Q_g R^3}{E_p I_p} + B'_x \frac{M_g R^2}{E_p I_p}$$

(8.89)

and

$$M_z(z) = A'_m Q_g R + B'_m M_g$$

(8.90)

where A'_x, B'_x, A'_m, and B'_m are coefficients and

$$R = \sqrt[4]{\frac{E_p I_p}{k}}$$

(8.91)

The values of the A and B coefficients are given in Figure 8.31. Note that, in this figure,

$$Z = \frac{z}{R}$$

(8.92)

and

$$Z_{max} = \frac{L}{R}$$

(8.93)

To use Eqs. (8.89) and (8.90), one must know the magnitude of the characteristic length, R. This can be calculated from Eq. (8.91) provided the coefficient of the subgrade reaction is known. For sands, the coefficient of

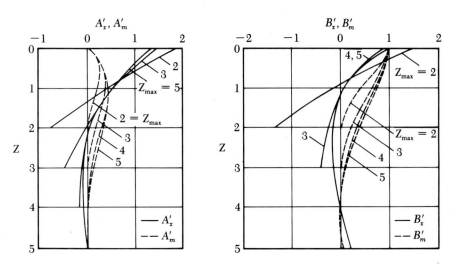

Figure 8.31 Variation of A'_x, B'_x, A'_m, and B'_m with Z (after Davisson and Gill, 1963)

subgrade reaction was given by Eq. (8.78), which showed a linear variation with depth. However, in cohesive soils, the subgrade reaction can be assumed to be approximately constant with depth. Vesic (1961) has proposed the following equation to estimate the value of k:

$$k = 0.65 \sqrt[12]{\frac{E_s D^4}{E_p I_p}} \frac{E_s}{1 - \mu_s^2} \qquad (8.94)$$

where E_s = Young's modulus of soil
  D = pile width (or diameter)
  μ_s = Poisson's ratio of the soil

The Young's modulus of clay, E_s, can be obtained from laboratory consolidation of the soil as

$$E_s = \frac{3(1 - \mu_s)}{m_v} \qquad (8.95)$$

where m_v = volume coefficient of compressibility—see Chapter 1

$$m_v = \frac{\Delta e}{\Delta p (1 + e_{av})}$$

The value of μ_s can be assumed to vary between 0.3–0.4.

Example 8.17

Consider a steel H-pile (HP 250 × 0.834) 25 m long embedded fully in a granular soil. Assume that n_h = 12,000 kN/m³. The allowable displacement at the top of the pile is 8 mm. Determine the allowable lateral load, Q_g. Assume that M_g is equal to zero.

Solution

From Table 8.1a, for an HP 250 × 0.834 pile,

$$I_p = 123 \times 10^{-6} \text{ m}^4 \qquad \text{(about the strong axis)}$$

$$E_p = 207 \times 10^6 \text{ kN/m}^2$$

From Eq. (8.87)

$$T = \sqrt[5]{\frac{E_p I_p}{n_h}} = \sqrt[5]{\frac{(207 \times 10^6)(123 \times 10^{-6})}{12,000}} = 1.16 \text{ m}$$

$L/T = 25/1.16 = 21.55 > 5$. So, this is a long pile. Because $M_g = 0$, Eq. (8.82) takes the form

$$x_z(z) = A_x \frac{Q_g T^3}{E_p I_p}$$

So

$$Q_g = \frac{x_z(z) E_p I_p}{A_x T^3}$$

Given: $x_z(z = 0) = 8 \text{ mm} = 0.008 \text{ m}$. At $z = 0$, $A_x = 2.435$ (Table 8.12). So,

$$Q_g = \frac{(0.008)(207 \times 10^6)(123 \times 10^{-6})}{(2.435)(1.16^3)} = 53.59 \text{ kN}$$

The preceding value of $Q_g = 53.59$ kN has been determined based on the *limiting displacement condition only*. However, the value of Q_g based on the *moment capacity* of the pile also needs to be determined. For that, referring to Eq. (8.84) (with $M_g = 0$),

$$M_z(z) = A_m Q_g T$$

According to Table 8.12, the maximum value of A_m at any depth is equal to 0.772. The maximum allowable moment that the pile can carry is equal to

$$M_{z(\text{max})} = \sigma_{\text{all}} \frac{I_p}{\dfrac{d_1}{2}}$$

Let $\sigma_{\text{all}} = 125,000$ kN/m². From Table 8.1a, $I_p = 123 \times 10^{-6}$ m⁴ and $d_1 = 0.254$ m. So

$$\frac{I_p}{\left(\dfrac{d_1}{2}\right)} = \frac{123 \times 10^{-6}}{\left(\dfrac{0.254}{2}\right)} = 968.5 \times 10^{-6} \text{ m}^3$$

Now

$$Q_g = \frac{M_{z(\text{max})}}{A_m T} = \frac{(968.5 \times 10^{-6})(125,000)}{(0.772)(1.16)} = 135.2 \text{ kN}$$

This value of $Q_g = 135.2$ kN is greater than 53.59 kN. So the deflection criteria apply. Hence, $Q_g = $ **53.59 kN**.

This is only the first approximation. The validity of the assumption of $n_h = 12{,}000$ kN/m^3 may now be checked using $Q_g = 53.59$ kN.

8.12 Pile-Driving Formulas

To develop the desired load-carrying capacity, a point bearing pile must sufficiently penetrate the dense soil layer or have sufficient contact with a layer of rock. This requirement cannot always be satisfied by driving a pile to a predetermined depth, because soil profiles vary. For that reason, several equations have been developed to calculate the ultimate capacity of a pile during driving. These dynamic equations are widely used in the field to determine if the pile has reached satisfactory bearing value at the predetermined depth. One of the earliest of these dynamic equations—commonly referred to as the *Engineering News Record (ENR) formula*—is derived on the basis of the work-energy theory. This means that

energy imparted by the hammer per blow =

(pile resistance) × (penetration per hammer blow)

According to the ENR formula, the pile resistance is the ultimate load Q_u and can be expressed as

$$Q_u = \frac{W_R h}{S + C} \qquad (8.96)$$

where W_R = weight of the ram (for example, see Table 8.6)
$\quad\ h$ = height of fall of the ram
$\quad\ S$ = penetration of pile per hammer blow
$\quad\ C$ = a constant

The pile penetration, S, is usually based on the average value obtained from the last few driving blows. In the equation's original form, the following values of C were recommended:

For drop hammers:

$C = 1$ in. (if the units of S and h are in inches)

For steam hammers:

$C = 0.1$ in. (if the units of S and h are in inches)

Also, a factor of safety, $FS = 6$, was recommended to estimate the allowable pile capacity. Note that, for single- and double-acting hammers, the term $W_R h$ can be replaced by EH_E (where E = hammer efficiency and H_E = rated energy of hammer). Thus

$$Q_u = \frac{EH_E}{S + C}$$ (8.97)

The ENR pile-driving formula has gone through several revisions over the years. A recent form—the *modified ENR formula*—can be given as

$$Q_u = \frac{EW_R h}{S + C} \cdot \frac{W_R + n^2 W_p}{W_R + W_p}$$ (8.98)

where E = hammer efficiency

$C = 0.1$ in. if the units of S and h are in inches

W_p = weight of the pile

n = coefficient of restitution between the ram and the pile cap

The efficiencies of various pile driving hammers, E, are in the following ranges:

Hammer type	Efficiency, E
Single- and double-acting hammers	0.7 to 0.85
Diesel hammers	0.8 to 0.9
Drop hammers	0.7 to 0.9

Representative values of the coefficient of restitution, n, are given in the following table.

Pile material	Coefficient of restitution, n
Cast iron hammer and concrete piles (without cap)	0.4 to 0.5
Wood cushion on steel piles	0.3 to 0.4
Wooden piles	0.25 to 0.3

A factor of safety (FS) of 4–6 may be used in Eq. (8.98) to obtain the allowable load-bearing capacity of a pile.

The Michigan State Highway Commission (1965) undertook a study for obtaining a rational pile-driving equation. At three diverse sites, a total of 88 piles were driven. Based on these tests, Michigan adopted a modified ENR formula:

$$Q_u = \frac{1.25 H_E}{S + C} \frac{W_R + n^2 W_p}{W_R + W_p}$$ (8.99)

where H_E = manufacturer's maximum rated hammer energy (lb-in.)
$\quad\ C = 0.1$ in.

The unit of S is in inches in the preceding equation. A factor of safety of 6 is recommended.

Another equation referred to as the *Danish formula* also yields results as reliable as any other equation's, and it is expressed as

$$Q_u = \frac{EH_E}{S + \sqrt{\dfrac{EH_E L}{2A_p E_p}}} \tag{8.100}$$

where E = hammer efficiency
$\quad H_E$ = rated hammer energy
$\quad E_p$ = Young's modulus of the pile material
$\quad L$ = length of pile
$\quad A_p$ = area of the pile cross section

(*Note:* The area of the soil plug as shown in Figure 8.10b and c should not be included.)

One has to use units consistently in the preceding equation. A factor of safety varying from 3 to 6 is recommended to estimate the allowable load-bearing capacity of piles.

Other frequently used equations for pile driving are given by the Pacific Coast Uniform Building Code (International Conference of Building Officials, 1982) and by Janbu (1953). These relations are as follows.

Pacific Coast Uniform Building Code formula:

$$Q_u = \frac{(EH_E)\left(\dfrac{W_R + nW_p}{W_R + W_p}\right)}{S + \dfrac{Q_u L}{AE}} \tag{8.101}$$

The value of n in the preceding equation should be taken to be 0.25 for steel piles and 0.1 for all other piles. A factor of safety of 4 is generally recommended.

Janbu's formula:

$$Q_u = \frac{EH_E}{K_u S} \tag{8.102}$$

where $K_u = C_d(1 + \sqrt{1 + \lambda/C_d})$ (8.103)

$C_d = 0.75 + 0.15(W_p/W_R)$ (8.104)

$\lambda = (EH_E L/A_p E_p S^2)$ (8.105)

A factor of safety of about 4–5 is generally recommended.

Example 8.18

A precast concrete pile 12 in. × 12 in. in cross section is driven by a Vulcan hammer (model no. 08). Given:

> Maximum rated hammer energy = 26 kips · ft (Table 8.6)
> Weight of ram = 8 kips
> Total length of pile = 65 ft
> Hammer efficiency = 0.8
> Coefficient of restitution = 0.45
> Weight of pile cap = 0.72 kip
> Number of blows for last 1 in. of penetration = 5

Estimate the allowable pile capacity by using the following:

    a. Equation (8.97) (use $FS = 6$)
    b. Equation (8.98) (use $FS = 4$)
    c. Equation (8.100) (use $FS = 3$)

Solution

Part a

Use of Eq. (8.97):

$$Q_u = \frac{EH_E}{S + C}$$

Given:

$$E = 0.8, \quad H_E = 26 \text{ kips} \cdot \text{ft}$$

$$S = \frac{1}{5} = 0.2 \text{ in.}$$

So

$$\overset{\text{kips} \cdot \text{in.}}{\downarrow}$$

$$Q_u = \frac{(0.8)\overbrace{(26)(12)}}{0.2 + 0.1} = 832 \text{ kips}$$

Hence

$$Q_{all} = \frac{Q_u}{FS} = \frac{832}{6} = \textbf{138.7 kips}$$

Part b

Use of Eq. (8.98):

$$Q_u = \frac{EW_R h}{S + C} = \frac{W_R + n^2 W_p}{W_R + W_p}$$

Weight of piles $= WA_p \gamma_c = (65 \text{ ft})(1 \text{ ft} \times 1 \text{ ft})(150 \text{ lb/ft}^3)$
$$= 9750 \text{ lb} = 9.75 \text{ kips}$$

W_p = weight of pile + weight of cap
$$= 9.75 + 0.72 = 10.47 \text{ kips}$$

So

$$Q_u = \left[\frac{(0.8)(26)(12)}{0.2 + 0.1}\right]\left[\frac{8 + (0.45)^2(9.75)}{8 + 9.75}\right]$$
$$= (832)(0.562) = 467.6 \text{ kips}$$

$$Q_{all} = \frac{Q_u}{FS} = \frac{467.6}{4} = \textbf{116.9 kips}$$

Part c

Use of Eq. (8.100):

$$Q_u = \frac{EH_E}{S + \sqrt{\dfrac{EH_E L}{2A_p E_p}}}$$

$E_p = 3 \times 10^6 \text{ lb/in.}^2$. So

$$\sqrt{\frac{EH_E L}{2A_p E_p}} = \sqrt{\frac{(0.8)(26 \times 12)(65 \times 12)}{(2)(12 \times 12)\left(\dfrac{3 \times 10^2}{1000}\right)}} = 0.475 \text{ in.}$$

$$\uparrow$$
$$\text{kips/in.}^2$$

Hence

$$Q_u = \frac{(0.8)(26)(12)}{0.2 + 0.475} = 369.8 \text{ kips}$$

$$Q_{all} = \frac{Q_u}{FS} = \frac{369.8}{3} = \textbf{123.3 kips}$$

8.13   **Stress on Pile During Pile Driving**

The maximum stress developed on a pile during the driving operation can be estimated from the pile-driving formulas presented in the preceding section. This can be shown, for example, by taking the modified ENR formula given in Eq. (8.98)

$$Q_u = \frac{EW_R h}{S + C} \frac{W_R + n^2 W_p}{W_R + W_p}$$

Note that, in the preceding equation, S is equal to the average penetration per hammer blow. This can also be expressed as

$$S = \frac{1}{N} \qquad (8.106)$$

where S is in inches and N = number of hammer blows per 1 in. of penetration. Thus

$$Q_u = \frac{EW_R h}{\frac{1}{N} + 0.1} \frac{W_R + n^2 W_p}{W_R + W_p} \qquad (8.107)$$

One can now assume different values of N for a given hammer and pile and calculate Q_u. The driving stress can then be calculated for each value of N as Q_u/A_p. This procedure can be demonstrated with a set of numerical values. Assume that a prestressed concrete pile 80 ft in length has to be driven by an 11B3 (MKT) hammer. The pile sides measure 10 in. From Table 8.3b, for this pile

$$A_p = 100 \text{ in.}^2$$

The weight of the pile

$$W_p = A_p L \gamma_c = \left(\frac{100 \text{ in.}^2}{144}\right)(80 \text{ ft})(150 \text{ lb/ft}^3)$$

$$= 8.33 \text{ kips}$$

Let the weight of the cap = 0.67 kip. So

$$W_p = 8.33 + 0.67 = 9 \text{ kips}$$

Again, from Table 8.6, for an 11B3 hammer

Rated energy = 19.2 kips · ft = $H_E = W_R h$
Weight of ram = 5 kips

Assume that the hammer efficiency = E = 0.85 and that n = 0.35. Substituting these values in Eq. (8.107)

$$Q_u = \left[\frac{(0.85)(19.2 \times 12)}{\frac{1}{N} + 0.1}\right]\left[\frac{5 + (0.35)^2(9)}{5 + 9}\right] = \frac{85.37}{\frac{1}{N} + 0.2} \text{ kips}$$

Now the following table can be prepared.

N	Q_u (kips)	A_p (in.^{2})	Q_u/A_p (kips/in.^{2})
0	0	100	0
2	142.3	100	1.42
4	243.9	100	2.44
6	320.1	100	3.20
8	379.4	100	3.79
10	426.9	100	4.27
12	465.7	100	4.66
20	569.1	100	5.69

Both the number of hammer blows per 1 in. and the stress can now be plotted in a graph, as shown in Figure 8.32. If such a curve is prepared, the number of blows per 1 in. of pile penetration corresponding to the allowable pile-driving stress can be easily determined.

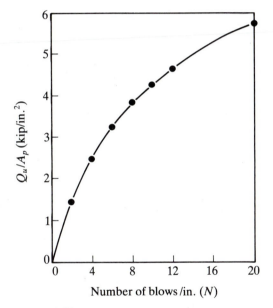

Figure 8.32

In practice, the driving stresses in wooden piles are limited to about $0.7f_u$. Similarly, for concrete and steel piles the driving stresses are limited to about $0.6f_c'$ and $0.85\sigma_y$, respectively.

In most cases, wooden piles are driven with a hammer energy of less than 60 kN-m (≈ 45 kips-ft). The driving resistances are mostly limited to 4–5 blows per 1 in. of pile penetration. For concrete and steel piles, the usual N values adopted are 6–8 and 12–14, respectively.

8.14   Pile Load Tests

In most large projects, a specific number of load tests must be conducted on piles. This is primarily because of the unreliability of prediction methods. Vertical and lateral load-bearing capacity of a given pile can be tested in the field. Figure 8.33a shows a schematic diagram of the pile load test arrangement in the field. This arrangement is for testing in *axial compression*. The load to the pile is applied by a hydraulic jack. Step loads are applied to the pile, and sufficient time is allowed to elapse after each step load so that the

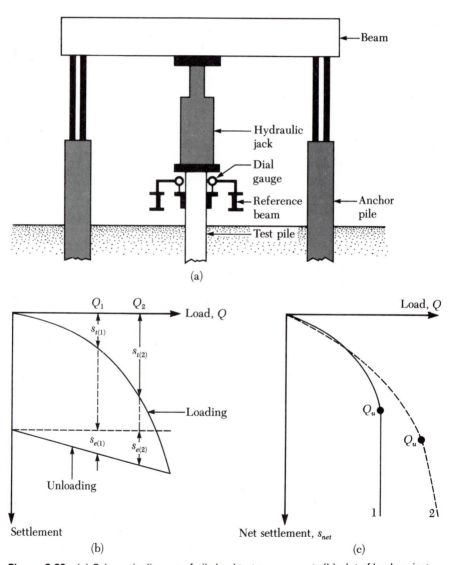

Figure 8.33   (a) Schematic diagram of pile load test arrangement; (b) plot of load against total settlement; (c) plot of load against net settlement

settlement rate reaches a small value. The settlement of the pile is recorded
by means of dial gauges. The amount of load to be applied for each step will
vary depending on local building codes. Most building codes require that
each step load be about one-fourth of the proposed working load and should
be carried out to at least a total load of two times the proposed working load.
After reaching the desired pile load, the pile is gradually unloaded.

Load tests on piles in sand can be carried out immediately after the
piles are driven. However, care should be taken in deciding the time lapse
between driving and starting the load test when piles are embedded in clay.
This time lapse can range from 30–60 days or more, because the soil
requires some time to gain its *thixotropic strength*.

Figure 8.33b shows the nature of the load settlement diagram for
loading and unloading that is obtained from the field. For any given load
(Q), the net pile settlement can be calculated as follows: When $Q = Q_1$

Net settlement, $s_{net(1)} = s_{t(1)} - s_{e(1)}$

When $Q = Q_2$

Net settlement, $s_{net(2)} = s_{t(2)} - s_{e(2)}$

$$\vdots$$

where s_{net} = net settlement
s_e = elastic settlement of the pile itself
s_t = total settlement

These values of Q can be plotted in a graph against the corresponding net
settlement (s_{net}). This is shown in Figure 8.33c. The ultimate load of the pile
can be determined from this graph. The settlement of the pile may increase
with load up to a certain point, beyond which the load-settlement curve
becomes vertical. The load corresponding to the point where the Q-s_{net}
curve becomes vertical is the ultimate load (Q_u) for the pile. This is shown
by Curve 1 in Figure 8.33c. In many cases, the latter stage of the load-
settlement curve is almost linear, showing a large degree of settlement for a
small increment of load. This is shown by Curve 2 in Figure 8.33c. The
ultimate load (Q_u) for such a case is determined from the point of the Q-s_{net}
curve where this steep linear portion starts.

The load test procedure just described requires application of step loads
on the piles and measurement of settlement. This is a *load-controlled* mode
of test. Another technique used for pile load test is the *constant-rate-of-
penetration* test. In this type of test, the load on the pile is continuously
increased to maintain a constant *rate of penetration* that can vary in the range
of 0.25–2.5 mm/min (0.01 to 0.1 in./min). This test gives a type of load-
settlement plot similar to that obtained from the load-controlled test. Other
modes of pile load tests include cyclic loading, in which an incremental load
is repeatedly applied and removed.

8.15   Group Piles—Efficiency

In most cases, piles are used in groups, as shown in Figure 8.34, to transmit the structural load to the soil. A *pile cap* is constructed over *group piles*. The pile cap can be in contact with the ground, as in most of the cases (Figure

Number of piles in group $= n_1 \times n_2$

Note: $L_g \geq B_g$

$L_g = (n_1 - 1)d + 2(D/2)$

$B_g = (n_2 - 1)d + 2(D/2)$

Figure 8.34   Pile groups

8.34a), or it may be well above the ground, as in the case of construction of offshore platforms (Figure 8.34b).

The preceding sections have discussed the load-bearing capacity of single piles. Determination of the load-bearing capacity of group piles is an extremely complicated problem and has not yet been fully resolved. When the piles are placed close to each other, it is reasonable to assume that the stresses transmitted by the piles to the soil medium will overlap (Figure 8.34c), and this may reduce the load-bearing capacity of the piles. Ideally, the piles in a group should be spaced in such a way that the load-bearing capacity of the group should not be less than the sum of the bearing capacity of the individual piles. In practice, the center-to-center pile spacings (d) are kept to a minimum of $2.5D$. However, in ordinary situations, they are kept at about 3–$3.5D$.

The efficiency of the load-bearing capacity of a group pile may be defined as

$$\boxed{\eta = \frac{Q_{g(u)}}{\Sigma Q_u}} \tag{8.108}$$

where η = group efficiency
$\quad Q_{g(u)}$ = ultimate load-bearing capacity of the group pile
$\quad\quad Q_u$ = ultimate load-bearing capacity of each pile without the group effect

Piles in Sand

Many structural engineers use a simplified analysis to obtain the group efficiency for friction piles in sand. This can be explained with the aid of Figure 8.34a. Depending on their spacing within the group, the piles may act in one of the two following ways: (1) as a *block* with dimensions $L_g \times B_g \times L$, or (2) as *individual piles*. If the piles act as a block, the frictional capacity can be given as $f_{av} p_g L \approx Q_{g(u)}$. [*Note:* p_g = perimeter of the cross section of block $= 2(n_1 + n_2 - 2)d + 4D$, and f_{av} = average unit frictional resistance.] Similarly, for each pile acting individually, $Q_u \approx pLf_{av}$. (*Note:* p = perimeter of the cross section of each pile.) Thus

$$\boxed{\begin{aligned} \eta &= \frac{Q_{g(u)}}{\Sigma Q_u} = \frac{f_{av}[2(n_1 + n_2 - 2)d + 4D]L}{n_1 n_2 p L f_{av}} \\[2mm] &= \frac{2(n_1 + n_2 - 2)d + 4D}{p n_1 n_2} \end{aligned}} \tag{8.109}$$

Hence

$$Q_{g(u)} = \left[\frac{2(n_1 + n_2 - 2)d + 4D}{pn_1n_2}\right]\Sigma Q_u \qquad (8.110)$$

From Eq. (8.110), if the center-to-center spacings (d) are large, one may obtain $\eta > 1$. In that case, the piles will behave as individual piles. Thus, in practice, if $\eta < 1$

$$Q_{g(u)} = \eta\Sigma Q_u$$

and, if $\eta \geq 1$

$$Q_{g(u)} = \Sigma Q_u$$

Another equation that is quoted often among design engineers is the *Converse-Labarre equation*, which can be stated as

$$\boxed{\eta = 1 - \left[\frac{(n_1 - 1)n_2 + (n_2 - 1)n_1}{90n_1n_2}\right]\theta} \qquad (8.111)$$

where θ (deg) $= \tan^{-1}(D/d)$ \qquad\qquad\qquad\qquad\qquad\qquad (8.112)

Figure 8.35 shows a set of the author's laboratory model test results for round piles driven into dense sand. Note that the group efficiency, in reality, can be larger than one. This is because of the soil compaction zones created around the piles during driving. Based on the experimental observa-

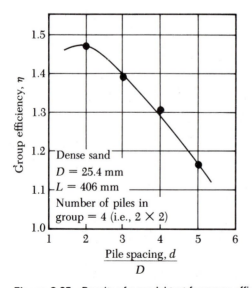

Figure 8.35   Results of a model test for group efficiency of piles in dense sand

tions for the behavior of group piles in sand made so far, the following general conclusions can be drawn:

1. For *driven* group piles in *sand* with $d \geq 3D$, $Q_{g(u)}$ may be taken to be equal to ΣQ_u. This includes the frictional and the point bearing capacity of individual piles.

2. For *bored* group piles in *sand* at conventional spacings (that is, $d \approx 3D$), $Q_{g(u)}$ may be taken to be equal to $\frac{2}{3}$ to $\frac{3}{4}$ times ΣQ_u (frictional and point bearing capacity of individual piles).

Piles in Clay

The ultimate load-bearing capacity of group piles in clay can be estimated in the following manner:

1. Determine $\Sigma Q_u = n_1 n_2 (Q_p + Q_s)$. From Eq. (8.18)

$$Q_p = A_p[9c_{u(p)}]$$

where $c_{u(p)}$ = undrained cohesion of the clay at the pile tip

Also, from Eq. (8.43)

$$Q_s = \Sigma \alpha p c_u \Delta L$$

So

$$\Sigma Q_u = n_1 n_2 [9A_p c_{u(p)} + \Sigma \alpha p c_u \Delta L] \tag{8.113}$$

2. Determine the ultimate capacity assuming that the piles in the group act as a block with dimensions of $L_g \times B_g \times L$. The skin resistance of the block =

$$\Sigma p_g c_u \Delta L = \Sigma 2(L_g + B_g) c_u \Delta L$$

The point bearing capacity can be calculated:

$$A_p q_p = A_p c_{u(p)} N_c^* = (L_g B_g) c_{u(p)} N_c^*$$

The value of the bearing capacity factor (N_c^*) may be taken from Figure 7.15. Note the change of notations: along the abscissa, the term H/B in Figure 7.15 is equivalent to L/B_g for this problem; also, L/B in Figure 7.15 is equivalent to L_g/B_g. Thus, the ultimate load is equal to

$$\Sigma Q_u = L_g B_g c_{u(p)} N_c^* + \Sigma 2(L_g + B_g) c_u \Delta L \tag{8.114}$$

3. Compare the two values of Eqs. (8.113) and (8.114). The *lower* of the two values is equal to $Q_{g(u)}$.

Piles in Rock

For point bearing piles resting on rocks, most building codes specify that $Q_{g(u)} = \Sigma Q_u$, provided that the minimum center-to-center spacing of piles is equal to $D + 300$ mm. For H-piles and piles with square cross sections, the value of D is equal to the diagonal dimension of the pile cross section.

General Comments

A pile cap resting on soil, as shown in Figure 8.34a, will contribute to the load-bearing capacity of a pile group. However, this contribution may be neglected for design purposes, because the support may be lost as a result of soil erosion or excavation during the lifetime of the project.

Example 8.19

Refer to Figure 8.34a. Given: $n_1 = 4$, $n_2 = 3$, $D = 305$ mm, and $d = 2.5D$. The piles are square in cross section and are embedded in sand. Use Eq. (8.109) to obtain the group efficiency.

Solution

From Eq. (8.109)

$$\eta = \frac{2(n_1 + n_2 - 2)d + 4D}{pn_1n_2}$$

$$d = 2.5D = (2.5)(305) = 762.5 \text{ mm}$$

$$p = 4D = (4)(305) = 1220 \text{ mm}$$

So

$$\eta = \frac{2(4 + 3 - 2)762.5 + 1220}{(1220)(4)(3)} = 0.604 = \mathbf{60.4\%}$$

Example 8.20

Redo Example 8.19 using Eq. (8.111).

Solution

According to Eq. (8.111)

$$\eta = 1 - \left[\frac{(n_1 - 1)n_2 + (n_2 - 1)n_1}{90n_1n_2}\right]\tan^{-1}\left(\frac{D}{d}\right)$$

However,

$$\tan^{-1}\left(\frac{D}{d}\right) = \tan^{-1}\left(\frac{1}{2.5}\right) = 21.8°$$

So

$$\eta = 1 - \left[\frac{(3)(3) + (2)(4)}{(90)(3)(4)}\right](21.8°) = 0.657 = \mathbf{65.7\%}$$

Example 8.21

Refer to Figure 8.34a. Given: $n_1 = 4$, $n_2 = 3$, $D = 305$ mm, $d = 1220$ mm, and $L = 15$ m. The piles are square in cross section and are embedded in a homogeneous clay with $c_u = 70$ kN/m². Using a factor of safety equal to 4, determine the allowable load-bearing capacity of the group pile.

Solution

From Eq. (8.113)

$$\Sigma Q_u = n_1 n_2 [9 A_p c_{u(p)} + \Sigma \alpha p c_u \Delta L]$$
$$A_p = (0.305)(0.305) = 0.093 \text{ m}^2$$
$$p = (4)(0.305) = 1.22 \text{ m}$$

Given: $c_u = 70$ kN/m². From Figure 8.19, for $c_u = 70$ kN/m², $\alpha = 0.63$. So

$$\Sigma Q_u = (4)(3)[(9)(0.093)(70) + (0.63)(1.22)(70)(15)]$$
$$= 12(58.59 + 807.03) \approx 10{,}387 \text{ kN}$$

Again, from Eq. (8.114), the ultimate block capacity is

$$L_g B_g c_{u(p)} N_c^* + \Sigma 2(L_g + B_g) c_u \Delta L$$
$$L_g = (n_1 - 1)d + 2\left(\frac{D}{2}\right) = (4-1)(1.22) + 0.305 = 3.965 \text{ m}$$
$$B_g = (n_2 - 1)d + 2\left(\frac{D}{2}\right) = (3-1)(1.22) + 0.305 = 2.745 \text{ m}$$
$$\frac{L}{B_g} = \frac{15}{2.745} = 5.46$$
$$\frac{L_g}{B_g} = \frac{3.965}{2.745} = 1.44$$

From Figure 7.15, $N_c^* \approx 8.6$. So

$$\text{block capacity} = (3.965)(2.745)(70)(8.6) + 2(3.965 + 2.745)(70)(15)$$
$$= 6552 + 14{,}091 = 20{,}643 \text{ kN}$$

So

$$Q_{g(u)} = 10{,}387 \text{ kN} < 20{,}643 \text{ kN}$$
$$Q_{g(\text{all})} = \frac{Q_{g(u)}}{FS} = \frac{10{,}387}{4} \approx \mathbf{2597 \text{ kN}}$$

8.16   Consolidation Settlement of Group Piles

The consolidation settlement of a group pile in clay can be approximately estimated by using the 2 : 1 stress distribution method shown in Figure 3.38. The procedure of calculation uses the following steps (refer to Figure 8.36):

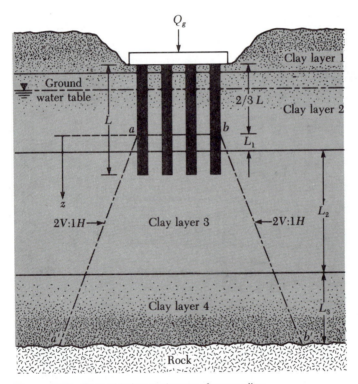

Figure 8.36   Consolidation settlement of group piles

1. Let the depth of embedment of the piles be equal to L. The group is subjected to a total load of Q_g. If the pile cap is below the original ground surface, Q_g is equal to the total load of the superstructure on the piles minus the effective weight of soil above the pile group removed by excavation.

2. Assume that the load Q_g is transmitted to the soil beginning at a depth of $2L/3$ from the top of the pile, as shown in Figure 8.36. This is depth $z = 0$ in the figure. The load Q_g spreads out along 2 vertical : 1 horizontal lines from this depth. Lines aa' and bb' are the two 2 : 1 lines.

3. Calculate the stress increase caused at the middle of each soil layer by the load Q_g:

$$\Delta p_i = \frac{Q_g}{(B_g + z_i)(L_g + z_i)} \tag{8.115}$$

where Δp_i = stress increase at the middle of layer i
L_g, B_g = length and width of the plan of pile group, respectively
z_i = distance from $z = 0$ to the middle of the clay layer, i

For example, in Figure 8.36 for layer 2, $z_i = L_1/2$. Similarly, for layer 3, $z_i = L_1 + L_2/2$; and, for layer 4, $z_i = L_1 + L_2 + L_3/2$. Note, however, that there will be no stress increase in clay layer 1, because it is above the horizontal plane $(z = 0)$ from which the stress distribution to the soil starts.

4. Calculate the settlement of each layer caused by the increased stress:

$$\Delta s_i = \left[\frac{\Delta e_{(i)}}{1 + e_{o(i)}} \right] H_i \tag{8.116}$$

where Δs_i = consolidation settlement of layer i
$\Delta e_{(i)}$ = change of void ratio caused by the stress increase in layer i
e_o = initial void ratio of layer i (before construction)
H_i = thickness of layer i (*Note:* In Figure 8.36, the value of H_i for layer 2 is equal to L_1. For layer 3, $H_i = L_2$; and for layer 4, $H_i = L_3$.)

Relations for $\Delta e_{(i)}$ are given in Chapter 1.

5. Total consolidation settlement of the pile group is then

$$\Delta s_{g(c)} = \Sigma \Delta s_i \tag{8.117}$$

Note that consolidation settlement of piles may also be initiated by fills placed nearby, adjacent floor loads, and lowering of water tables.

Example 8.22

A group pile in clay is shown in Figure 8.37. Determine the consolidation settlement of the pile groups. All clays are normally consolidated.

Pile group: $L_g = 3.3$ m; $B_g = 2.2$ m

(*not to scale*)

Figure 8.37

Solution

Because the lengths of the piles are 15 m each, the stress distribution starts at a depth of 10 m below the top of the pile. Given: $Q_g = 2000$ kN.

Calculation of Settlement of Clay Layer 1

For normally consolidated clays

$$\Delta s_1 = \left[\frac{C_{c(1)} H_1}{1 + e_{o(1)}}\right] \log \left[\frac{p_{o(1)} + \Delta p_{(1)}}{p_{o(1)}}\right]$$

$$\Delta p_{(1)} = \frac{Q_g}{(L_g + z_1)(B_g + z_1)} = \frac{2000}{(3.3 + 3.5)(2.2 + 3.5)} = 51.6 \text{ kN/m}^2$$

$$p_{o(1)} = 2(16.2) + 12.5(18.0 - 9.81) = 134.8 \text{ kN/m}^2$$

So

$$\Delta s_1 = \frac{(0.3)(7)}{1 + 0.82} \log\left[\frac{134.8 + 51.6}{134.8}\right] = 0.1624 \text{ m} = \mathbf{162.4 \text{ mm}}$$

Settlement of Layer 2

$$\Delta s_2 = \frac{C_{c(2)} H_2}{1 + e_{o(2)}} \log \left[\frac{p_{o(2)} + \Delta p_{(2)}}{p_{o(2)}} \right]$$

$$p_{o(2)} = 2(16.2) + 16(18.0 - 9.81) + 2(18.9 - 9.81) = 181.62 \text{ kN/m}^2$$

$$\Delta p_{(2)} = \frac{2000}{(3.3 + 9)(2.2 + 9)} = 14.52 \text{ kN/m}^2$$

Hence

$$\Delta s_2 = \frac{(0.2)(4)}{1 + 0.7} \log \left[\frac{181.62 + 14.52}{181.62} \right] = 0.0157 \text{ m} = \textbf{15.7 mm}$$

Settlement of Layer 3

$$p_{o(3)} = 181.62 + 2(18.9 - 9.81) + 1(19 - 9.81) = 208.99 \text{ kN/m}^2$$

$$\Delta p_{(3)} = \frac{2000}{(3.3 + 12)(2.2 + 12)} = 9.2 \text{ kN/m}^2$$

$$\Delta s_3 = \frac{(0.25)(2)}{1 + 0.75} \log \left[\frac{208.99 + 9.2}{208.99} \right] = 0.0054 \text{ m} = \textbf{5.4 mm}$$

Hence, total settlement =

$$\Delta s_g = 162.4 + 15.7 + 5.4 = \textbf{183.5 mm}$$

8.17   Elastic Settlement of Pile Groups

Several investigations relating to the settlement of pile groups have been reported in the literature with widely varying results. The simplest relation for the settlement of group piles has been given by Vesic (1969) as

$$\boxed{s_{g(e)} = \sqrt{\frac{B_g}{Ds}}} \tag{8.118}$$

where $s_{g(e)}$ = elastic settlement of group piles
B_g = width of pile group section (see Figure 8.34a)
D = width or diameter of each pile in the group
s = elastic settlement of each pile at comparable working load (see Section 8.9)

For pile groups in sand and gravel, Meyerhof (1976) suggested the following empirical relation for elastic settlement:

$$\boxed{s_{g(e)} \text{ (mm)} = \frac{0.92q\sqrt{B_g}I}{N_{cor}}} \tag{8.119}$$

where $q = Q_g/(L_g B_g)$ (8.120)

L_g and B_g = length and width of the pile group section

N_{cor} = average corrected standard penetration number within seat of settlement ($\approx B_g$ deep below the tip of the piles)

I = influence factor = $1 - L/8B_g \geq 0.5$ (8.121)

L = length of embedment of piles

In a similar manner, the pile group settlement can be related to the cone penetration resistance as

$$s_{g(e)} = \frac{qB_g I}{2q_c}$$ (8.122)

where q_c = average cone penetration resistance within the seat of settlement

In Eq. (8.122), all symbols are in consistent units.

8.18 Uplift Capacity of Group Piles

The efficiency of group piles under compressive load was discussed in Section 8.15. However, under certain circumstances, group piles may be used for construction of foundations subjected to uplifting load (Figure 8.38). As in Eq. (8.108), the group efficiency under uplift can be expressed as

$$\eta_T = \frac{T_{un(g)}}{T_{un}}$$ (8.123)

where η_T = group efficiency under uplift

$T_{un(g)}$ = net ultimate uplift capacity of pile group

T_{un} = net ultimate uplift capacity of single pile (Section 8.10)

Note that

$$T_{un(g)} = T_{ug(g)} - n_1 \cdot n_2 W - W_{cap}$$ (8.124)

where $T_{ug(g)}$ = gross ultimate uplift capacity of group piles

W = effective self-weight of each pile

$n_1 \cdot n_2$ = number of piles in the group

W_{cap} = effective weight of pile cap

At this time, very few field and laboratory experimental results relating to the evaluation of η_T are available in the literature. Das and Azim (1985) conducted a limited number of model tests to determine the group effi-

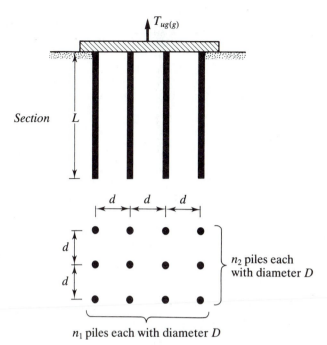

Figure 8.38   Group piles subjected to uplifting load

ciency, η_T, of rigid pile groups embedded in saturated clay. The results of this study are shown in Figure 8.39. From this figure, the following general conclusions can be drawn:

  1. For a given pile group, η_T increases linearly with the d/D ratio until it reaches 100%. The d/D ratio at which η_T reaches a value of 100% is about $\frac{1}{2}(L/D)$.
  2. For given d/D and L/D ratios, the magnitude of η_T decreases with the increase of the number of piles in a group.
  3. For a given d/D ratio and number of piles in a group, the magnitude of η_T decreases with the increase of L/D.

The group efficiency may, however, be a function of the consistency of the clay.

  Figure 8.40 shows the laboratory model test results for group efficiency of rough rigid piles embedded in loose and dense sand (Das, 1984). The group piles in this case had an L/D ratio of 15. As can be seen from Figure 8.40, the magnitude of η_T is a function of L/D, d/D, the number of piles in the group, and the relative density of the sand.

  More studies are required to define quantitatively the parameters controlling the group efficiency, η_T.

Figure 8.39 Efficiency of rigid pile groups embedded in saturated clay and subjected to uplifting force

Figure 8.40 Efficiency of rigid pile groups embedded in sand and subjected to uplifting force [based on laboratory model test results of Das (1984)]

8.19   Negative Skin Friction

Negative skin friction is a downward drag force on the pile by the soil surrounding it. This can happen under several conditions, such as the following:

1. If a fill of clay soil is placed over a granular soil layer into which a pile is driven, the fill will gradually consolidate. This consolidation process will provide a downward drag force on the pile (Figure 8.41a) during the period of consolidation.

2. If a fill of granular soil is placed over a layer of soft clay, as shown in Figure 8.41b, it will induce the process of consolidation in the clay layer and thus provide a downward drag on the pile.

3. Lowering of the ground water table will increase the vertical effective stress on the soil at any depth. This will induce consolidation settlement in clay. If a pile is located in the clay layer, it will be subjected to a downward drag force.

In some cases, the downward drag force could be excessive and cause foundation failure. This section outlines two tentative methods for the calculation of negative skin friction.

Clay Fill over Granular Soil (Figure 8.41a)

Similar to the β method presented in Section 8.6, the negative (downward) skin stress on the pile can be given by the relation

$$f_n = K'\sigma'_v \tan \delta \tag{8.125}$$

where K' = earth pressure coefficient = $K_o = 1 - \sin \phi$
$\quad \sigma'_v$ = vertical effective stress at any depth $z = \gamma'_f z$
$\quad \gamma'_f$ = effective unit weight of fill
$\quad \delta$ = soil-pile friction angle $\approx 0.5–0.7\phi$

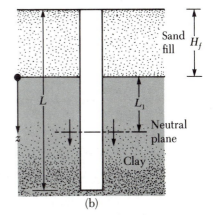

Figure 8.41   Negative skin friction

Hence, the total downward drag force, Q_n, on a pile is

$$Q_n = \int_0^{H_f} (pK'\gamma'_f \tan \delta)z \, dz = \frac{pK'\gamma'_f H_f^2 \tan \delta}{2} \tag{8.126}$$

where H_f = height of the fill

If the fill is above the water table, the effective unit weight, γ'_f, should be replaced by the moist unit weight.

Granular Soil Fill over Clay (Figure 8.41b)

In this case, there is enough evidence to show that the negative skin stress on the pile may exist from $z = 0$ to $z = L_1$, which is referred to as the *neutral depth* (see Vesic, 1977, pp. 25–26 for discussion). The neutral depth may be given as (Bowles, 1982)

$$L_1 = \frac{(L - H_f)}{L_1} \left[\frac{L - H_f}{2} + \frac{\gamma'_f H_f}{\gamma'} \right] - \frac{2\gamma'_f H_f}{\gamma'} \tag{8.127}$$

where γ'_f and γ' = effective unit weights of the fill and the underlying clay layer, respectively

Once the value of L_1 is determined, the downward drag force can be obtained in the following manner: The unit negative skin friction at any depth from $z = 0$ to $z = L_1$ is

$$f_n = K'\sigma'_v \tan \delta \tag{8.128}$$

where $K' = K_o = 1 - \sin \phi$
$\sigma'_v = \gamma'_f H_f + \gamma' z$
$\delta = 0.5 - 0.7\phi$

Hence, the total drag force

$$Q_n = \int_0^{L_1} pf_n \, dz = \int_0^{L_1} pK'(\gamma'_f H_f + \gamma' z) \tan \delta \cdot dz$$
$$= (pK'\gamma'_f H_f \tan \delta)L_1 + \frac{L_1^2 pK'\gamma' \tan \delta}{2} \tag{8.129}$$

If the soil and the fill are above the water table, the effective unit weights should be replaced by moist unit weights. In some cases, the piles can be coated with asphalt in the down-drag zone to avoid this problem.

Example 8.23

Refer to Figure 8.41a. $H_f = 2$ m. The pile is circular in cross section with a diameter of 0.305 m. For the fill that is above the ground water table, $\gamma_f = 16$ kN/m^3 and $\phi = 32°$. Determine the total drag force.

Solution

From Eq. (8.126)

$$Q_n = \frac{pK'\gamma_f^2 \tan \delta}{2}$$

$$p = \pi(0.305) = 0.958 \text{ m}$$

$$K' = 1 - \sin \phi = 1 - \sin 32° = 0.47$$

$$\delta = (0.6)(32) = 19.2°$$

$$Q_n = \frac{(0.958)(0.47)(16)(2)^2 \tan 19.2°}{2} = \textbf{5.02 kN}$$

Example 8.24

Refer to Figure 8.41b. $H_f = 2$ m, pile diameter = 0.305 m, $\gamma_f = 16.5$ kN/m^3, $\phi_{\text{clay}} = 34°$, $\gamma_{\text{sat(clay)}} = 17.2$ kN/m^3, and $L = 20$ m. The water table coincides with the top of the clay layer. Determine the down-drag force.

Solution

The depth of the neutral plane is given by Eq. (8.127) as

$$L_1 = \frac{L - H_f}{L_1}\left(\frac{L - H_f}{2} + \frac{\gamma_f H_f}{\gamma'}\right) - \frac{2\gamma_f H_f}{\gamma'}$$

Note that γ_f' in Eq. (8.127) has been replaced by γ_f, because the fill is above the water table. So

$$L_1 = \frac{(20 - 2)}{L_1}\left[\frac{(20 - 2)}{2} + \frac{(16.5)(2)}{(17.2 - 9.81)}\right] - \frac{(2)(16.5)(2)}{(17.2 - 9.81)}$$

$$L_1 = \frac{242.4}{L_1} - 8.93$$

or

$$L_1 = 11.75 \text{ m}$$

Now, referring to Eq. (8.129)

$$Q_n = (pK'\gamma_f H_f \tan \delta)L_1 + \frac{L_1^2 pK'\gamma' \tan \delta}{2}$$

$$p = \pi(0.305) = 0.958 \text{ m}$$

$$K' = 1 - \sin 34° = 0.44$$

$$Q_n = (0.958)(0.44)(16.5)(2)[\tan (0.6 \times 34)](11.75)$$

$$+ \frac{(11.75)^2(0.958)(0.44)(17.2 - 9.81)[\tan (0.6 \times 34)]}{2}$$

$$= 60.78 + 79.97 = \textbf{40.75 kN}$$

PROBLEMS

8.1 A prestressed concrete pile is 20 m long. The cross section of the pile is 460 mm × 460 mm. The pile is fully embedded in sand. Using Meyerhof's method [Eqs. (8.12), (8.13), and (8.14)], estimate the ultimate point load, Q_p. Given, for the sand: $\gamma = 16.8$ kN/m^3 and $\phi = 30°$.

8.2 Solve Problem 8.1 by Coyle and Castello's method. Use Figure 8.20.

8.3 Solve Problem 8.1 by Vesic's method [Eq. (8.19)]. Use $I_r = I_{rr} = 75$.

8.4 Solve Problem 8.1 by Janbu's method [Eq. (8.29)]. Use $\eta' = 90°$.

8.5 On the basis of the results of Problems 8.1–8.4, estimate an allowable value for the point load. Use $FS = 4$.

8.6 Refer to Problem 8.1. Calculate the total frictional resistance for the pile. Use Eqs. (8.32), (8.33), and (8.36). Given: $K = 1.5$ and $\delta = 0.8\phi$.

8.7 Refer to Problem 8.1. Calculate the total frictional resistance by using Coyle and Castello's method.

8.8 A concrete pile is 50 ft long and 16 in. × 16 in. in cross section. The pile is fully embedded in sand. Given: for sand, $\gamma = 110$ lb/ft^3 and $\phi = 30°$. Calculate the following:
  a. The ultimate total point load, Q_p, by Meyerhof's method [Eqs. (8.12), (8.13), and (8.15)]
  b. The total frictional resistance [Eqs. (8.32), (8.33), and (8.36)]—use $K = 1.3$ and $\delta = 0.08\phi$

8.9 Solve Problem 8.8 by Coyle and Castello's method (Figures 8.20 and 8.21).

8.10 Solve Problem 8.8, part (a), by Vesic's method [Eq. (8.19)]. Use $I_r = I_{rr} = 50$.

8.11 A concrete pile fully embedded in sand has a cross section of 356 mm × 356 mm. The length of embedment is 15.5 m. The sand has an average relative density of 55% and a unit weight of 15.5 kN/m^3. Determine the frictional resistance of the pile. Use Eqs. (8.32), (8.33), (8.34), and (8.36).

8.12 A pipe pile (closed end) that has been driven into a layered sand is shown in Figure P8.12. Calculate the following:

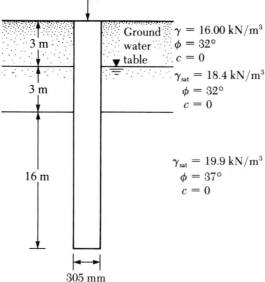

3 m

Ground water table

$\gamma = 16.00$ kN/m^3
$\phi = 32°$
$c = 0$

3 m

$\gamma_{sat} = 18.4$ kN/m^3
$\phi = 32°$
$c = 0$

16 m

$\gamma_{sat} = 19.9$ kN/m^3
$\phi = 37°$
$c = 0$

305 mm

Figure P8.12

a. The ultimate point load by Meyerhof's procedure
b. The ultimate point load by Vesic's procedure ($I_r = I_{rr} = 50$)
c. An approximate ultimate point load on the basis of parts (a) and (b)
d. The ultimate frictional resistance, Q_s—use Eqs. (8.32), (8.33), and (8.36), and $K = 1.4$, $\delta = 0.7\phi$
e The allowable load of the pile—use $FS = 5$

8.13 A concrete pile 60 ft long having a cross section of 15 in. × 15 in. is fully embedded in a saturated clay layer. For the clay, given: $\gamma_{sat} = 122$ lb/ft³, $\phi = 0$, and $c_u = 1450$ lb/ft². Assume that the ground water table is located below the tip of the pile. Determine the allowable load that the pile can carry ($FS = 3$). Use the α method to estimate the skin resistance.

8.14 Redo Problem 8.13 by using the λ method for estimating the skin resistance.

8.15 A concrete pile 405 mm × 405 mm in cross section is shown in Figure P8.15. Calculate the ultimate skin resistance by using

Silty clay
$\gamma_{sat} = 17.8$ kN/m³
$c_u = 28$ kN/m²

Ground water table

Silty clay
$\gamma_{sat} = 19.6$ kN/m³
$c_u = 85$ kN/m²

8 m

14 m

405 mm

Figure P8.15

a. The α method
b. The λ method
c. The β method

Use $\phi_R = 25°$ for all clays. All clays are normally consolidated.

8.16 Redo Problem 8.14 assuming that the ground water table is located at a depth of 30 ft below the ground surface.

8.17 A steel pile (H-section; HP 14 × 102; see Table 8.1b) is driven to a layer of sandstone. The length of the pile is 50 ft. Following are the properties of the sandstone:

$$\text{Unconfirmed compression strength} = q_{u(lab)} = 12,400 \text{ lb/in.}^2$$
$$\text{Angle of friction} = 34°$$

Using a factor of safety of 4, estimate the allowable point load that can be carried by the pile.

8.18 The allowable working load on a prestressed concrete pile 21 m long that has been driven into sand is 502 kN. The pile is octagonal in shape with $D = 356$ mm (see Table 8.3a). 350 kN of the allowable load is carried by the skin resistance and the rest by point bearing. Given:

$$E_p = 21 \times 10^6 \text{ kN/m}^2 \qquad \mu_s = 0.35$$
$$E_s = 25 \times 10^3 \text{ kN/m}^2 \qquad \zeta = 0.55 \text{ [Eq. (8.58)]}$$

Determine the elastic settlement of the pile. [Use 0.8 for I_{wp} in Eq. (8.59).]

8.19 A concrete pile is 50 ft long and has a cross section of 16 in. × 16 in. The pile is embedded in a sand having $\gamma = 117$ lb/ft^3 and $\phi = 37°$. The allowable working load is 170 kips. If 100 kips are contributed by the frictional resistance and 70 kips are from the point load, determine the elastic settlement of the pile. Given:

$$E_p = 3 \times 10^6 \text{ lb/in.}^2 \qquad \mu_s = 0.35$$
$$E_s = 5 \times 10^3 \text{ lb/in.}^2 \qquad \zeta = 0.62$$

8.20 A precast concrete pile with a cross section of 406 mm × 406 mm is embedded in sand. The length of the pile is 10.4 m. Assume that $\gamma_{sand} = 15.8$ kN/m^3, $\phi_{sand} = 30°$, and relative density of sand = 70%. Estimate the allowable pullout capacity of the pile ($FS = 4$).

8.21 Redo Problem 8.20 with the following changes: $\gamma_{sand} = 118$ lb/ft^3, $\phi_{sand} = 37°$, and relative density of compaction of sand = 80%.

8.22 A concrete pile 50 ft long is embedded in a saturated clay with $c_u = 850$ lb/ft^2. The pile is 12 in. × 12 in. in cross section. Using $FS = 4$, determine the allowable pullout capacity of the pile.

8.23 Redo Problem 8.22 with the following changes: The top 5 m of the clay has a $c_u = 25$ kN/m^2, and below that $c_u = 55$ kN/m^2.

8.24 Determine the ultimate pullout capacity of the pile described in Problem 8.15.

8.25 Determine the ultimate pullout capacity of the pile described in Problem 8.13.

8.26 A steel H-pile (Section: HP 14 × 117; see Table 8.1b) is driven by an MKT S-20 hammer (see Table 8.6). Given:

Length of pile = 80 ft
Coefficient of restitution = 0.35
Weight of pile cap = 17.8 kips
Hammer efficiency = 0.84
Number of blows for last 1 in. of penetration = 10

Estimate the ultimate pile capacity by using Eq. (8.97). For the pile, $E_p = 30 \times 10^6$ lb/in.^{2}.

8.27 Redo Problem 8.26 using Eq. (8.98).

8.28 Redo Problem 8.26 using Eq. (8.100).

8.29 The plan of a group pile (friction pile) in sand is shown in Figure P8.29. The piles are circular in cross section and have an outside diameter of 460 mm. The center-to-center spacings of the piles (d) are 920 mm. Using Eq. (8.109), find the efficiency of the pile group.

$\leftarrow\!\!-d\!\!-\!\!\rightarrow$ **Figure P8.29**

8.30 Solve Problem 8.29 using the Converse-Labarre equation.

8.31 Refer to Problem 8.29. If the center-to-center pile spacing is increased to 1200 mm, what will be the group efficiency?

8.32 Solve Problem 8.31 using the Converse-Labarre equation.

8.33 The plan of a group pile is shown in Figure P8.29. Assume that the piles are embedded in a saturated homogeneous clay having a $c_u = 95.8$ kN/m². Given:

> Diameter of piles $(D) = 406$ mm
> Center-to-center spacing of piles = 850 mm
> Length of piles = 18.5 m

Find the allowable load-carrying capacity of the pile group. Use $FS = 3$.

8.34 Redo Problem 8.33 with the following:

> Center-to-center spacing of piles = 30 in.
> Length of piles = 45 ft
> $D = 12$ in.
> $c_u = 860$ lb/ft²

8.35 The section of a group pile (3 × 4) in a layered saturated clay is shown in Figure P8.35. The piles are square in cross section (14 in. × 14 in.). The center-to-center

Figure P8.35

spacing (d) of the piles is 35 in. Determine the allowable load-bearing capacity of the pile group. Use $FS = 4$.

8.36 Figure P8.36 shows a group pile in clay. Determine the consolidation settlement of the group.

Figure P8.36

8.37 Refer to Example 8.22. If the pile group has $L_g = 3.5$ m and $B_g = 3$ m and clay layer 1 is preconsolidated ($p_c = 155$ kN/m², $C_s = \frac{1}{4}C_c$), determine the settlement of the pile group.

8.38 Figure 8.41a shows a pile. Let $L = 50$ ft; diameter of pile $= 18$ in.; height of clay fill, $H_f = 11.5$; unit weight of fill, $\gamma_f = 112$ lb/ft³; and $\phi_{fill} = 28°$. Determine the total down-drag force on the pile. Assume that the fill is located above the ground water table. Assume that $\delta = 0.6\phi_{\text{fill}}$.

8.39 Redo Problem 8.38 assuming that the ground water table coincides with the top of the fill and that $\gamma_{\text{sat(fill)}} = 124.5$ lb/ft^3. Other quantities remaining the same, what would be the down-drag force on the pile? Assume that $\delta = 0.6\phi_{\text{fill}}$.

8.40 Refer to Figure 8.41b. Let $L = 19$ m, $\gamma_{\text{fill}} = 15.2$ kN/m^3, $\gamma_{\text{sat(clay)}} = 19.5$ kN/m^3, $\phi_{\text{clay}} = 30°$, $H_f = 3.2$ m, and pile diameter = 0.460 m. The ground water table coincides with the top of the clay layer. Determine the total down-drag force on the pile. Assume that $\delta = 0.5\phi_{\text{clay}}$.

REFERENCES

American Society of Civil Engineers (1959). "Timber Piles and Construction Timbers," *Manual of Practice*, No. 17.

Bhusan, K. (1982). "Discussion: New Design Correlations for Piles in Sands," *Journal of the Geotechnical Engineering Division*, American Society of Civil Engineers, Vol. 108, No. GT11, pp. 1508–1510.

Bowles, J. E. (1982). *Foundation Design and Analysis*, McGraw-Hill, New York.

Coyle, H. M., and Castello, R. R. (1981). "New Design Correlations for Piles in Sand," *Journal of the Geotechnical Engineering Division*, American Society of Civil Engineers, Vol. 107, No. GT7, pp. 965–986.

Das, B. M. (1984). "Model Uplift Tests on Pile Groups in Sand," *Transportation Research Record No. 998*, National Academy of Sciences, Washington, D.C., pp. 25–28.

Das, B. M. and Azim, M. F. (1985). "Uplift Capacity of Rigid Pile Groups in Clay," *Soils and Foundations*, Vol. 25, No. 4, pp. 56–60.

Das, B. M., and Seeley, G. R. (1975). "Uplift Capacity of Buried Model Piles in Sand," *Journal of the Geotechnical Engineering Division*, American Society of Civil Engineers, Vol. 101, No. GT10, pp. 1091–1094.

Das, B. M., and Seeley, G. R. (1982). "Uplift Capacity of Pipe Piles in Saturated Clay," *Soils and Foundations*, The Japanese Society of Soil Mechanics and Foundation Engineering, Vol. 22, No. 1, pp. 91–94.

Davisson, M. T., and Gill, H. L. (1963). "Laterally Loaded Piles in a Layered Soil System," *Journal of the Soil Mechanics and Foundations Division*, American Society of Civil Engineers, Vol. 89, No. SM3, pp. 63–94.

Goodman, R. E. (1980). *Introduction to Rock Mechanics*, Wiley, New York.

International Conference of Building Officials (1982). "Uniform Building Code," Whittier, Calif.

Janbu, N. (1953). *An Energy Analysis of Pile Driving with the Use of Dimensionless Parameters*, Norwegian Geotechnical Institute, Oslo, Publication No. 3.

Janbu, N. (1976). "Static Bearing Capacity of Friction Piles," *Proceedings*, Sixth European Conference on Soil Mechanics and Foundation Engineering, Vol. 1.2, pp. 479–482.

Matlock, H., and Reese, L. C. (1960). "Generalized Solution for Laterally Loaded Piles," *Journal of the Soil Mechanics and Foundations Division*, American Society of Civil Engineers, Vol. 86, No. SM5, Part I, pp. 63–91.

McClelland, B. (1974). "Design of Deep Penetration Piles for Ocean Structures," *Journal of the Geotechnical Engineering Division*, American Society of Civil Engineers, Vol. 100, No. GT7, pp. 709–747.

Meyerhof, G. G. (1976). "Bearing Capacity and Settlement of Pile Foundations," *Journal of the Geotechnical Engineering Division*, American Society of Civil Engineers, Vol. 102, No. GT3, pp. 197–228.

Meyerhof, G. G., and Adams, J. I. (1968). "The Ultimate Uplift Capacity of Foundations," *Canadian Geotechnical Journal*, Vol. 5, No. 4, pp. 225–244.

Michigan State Highway Commission (1965). *A Performance Investigation of Pile Driving Hammers and Piles*, Lansing, Michigan, 338 pp.

Vesic, A. S. (1961). "Bending of Beams Resting on Isotropic Elastic Solids," *Journal of the Engineering Mechanics Division*, American Society of Civil Engineers, Vol. 87, No. EM2, pp. 35–53.

Vesic, A. S. (1969). "Experiments with Instrumented Pile Groups in Sand," American Society for Testing and Materials; Special Technical Publication, No. 444, pp. 177–222.

Vesic, A. S. (1970). "Tests on Instrumented Piles—Ogeechee River Site," *Journal of the Soil Mechanics and Foundations Division*, American Society of Civil Engineers, Vol. 96, No. SM2, pp. 561–584.

Vesic, A. S. (1977). *Design of Pile Foundations*, National Cooperative Highway Research Program Synthesis of Practice No. 42, Transportation Research Board, Washington, D.C.

Vijayvergiya, V. N., and Focht, J. A., Jr. (1972). *A New Way to Predict Capacity of Piles in Clay*, Offshore Technology Conference Paper 1718, Fourth Offshore Technology Conference, Houston, Texas.

Woodward, R. J., Gardner, W. S., and Greer, D. M. (1972). *Drilled Pier Foundations*, McGraw-Hill, New York.

Chapter *9*

DRILLED-PIER AND
CAISSON FOUNDATIONS

9.1   Introduction

The terms *caisson*, *pier*, *drilled shaft*, and *drilled pier* are often used inter-
changeably in foundation engineering; they all refer to a *cast-in-place pile
generally having a diameter of about 750 mm (≈2.5 ft)* or more, with or
without steel reinforcements and with or without an enlarged bottom.
Sometimes the diameter can be as small as 305 mm (≈1 ft). In order to
avoid confusion, this text will use the following definitions:

The term *drilled pier* is used when a hole is drilled or excavated to the
bottom of a structure's foundation and then filled with concrete. Depending
on the soil conditions, casings or *laggings* (boards or sheet piles) may be used
to prevent the soil around the hole from caving in during construction. The
diameter of the pier shaft is usually large enough for a person to enter for
inspection.

The use of drilled-pier foundations has several advantages. For
example:

1. A single drilled pier can be used to replace a group of piles and the
pile cap.

2. It is easier to construct drilled piers in deposits of dense sand and
gravel than it is to drive piles.

3. Construction of drilled piers can be completed before the completion of grading operations.

4. When piles are driven by a hammer, the ground vibration may
cause damage to the nearby structures. The use of drilled piers does not
present such hazards.

5. Piles driven into clay soils may produce ground heaving and may
also cause previously driven piles to move laterally. Such conditions do
not exist in the construction of drilled piers.

523

6. There is no hammer noise during the construction of drilled piers, as there is during pile driving.

7. Because the base of a drilled pier can be enlarged, it provides great resistance to the uplifting load (Section 9.8).

8. The surface over which the base of the drilled pier is constructed can be visually inspected.

9. Construction of drilled piers generally requires mobile equipment. Under proper soil conditions, they may prove to be more economical than pile foundations.

10. Drilled piers have high resistance to lateral loads.

There are also several drawbacks to the use of drilled-pier construction. The concreting operation always needs very close supervision. This can be delayed to some extent during bad weather. Also, as in the case of braced cuts, deep excavations for drilled piers may induce substantial ground loss and damage to closely located structures.

The term *caisson* refers to a substructure element used at wet construction sites, such as rivers, lakes, and docks. For the construction of caissons, a hollow shaft or a box is sunk into position to rest on firm ground. The lower part of the shaft or the box is provided with a cutting edge to help it penetrate the soft soil layers below the water level and come to rest on a load-bearing stratum. The material inside the shaft or box is dredged through the openings at the top, and then concrete is poured in. Bridge abutments, quay walls, and structures for shore protection can be built over caissons.

DRILLED PIERS

9.2   Types of Drilled Pier

Drilled piers can be classified according to the manner in which they are designed to transfer the structural load to the substratum. Figure 9.1a shows a drilled pier that has a *straight shaft*. It extends through the upper layer(s) of poor soil, and its tip rests on a strong load-bearing soil layer or rock. The shaft can be cased with steel shell or pipe when required (as in the case of cased, cast-in-place concrete piles; Figure 8.4). For these piers, the resistance to the applied load may develop from end bearing and also from side friction at the pier perimeter and soil interface.

A *belled pier* (Figure 9.1b and c) consists of a straight shaft with a bell at the bottom. The bell rests on good bearing soil. The bell can be constructed in the shape of a dome (Figure 9.1b), or it can be angled (Figure 9.1c). For angled bells, the underreaming tools commercially available are such that the sides can make 30° to 45° angles with the vertical. For the majority of drilled piers constructed in the United States, the entire load-carrying capacity is assigned to the end bearing only. However, in certain circumstances, the end-bearing capacity and the side friction are taken into

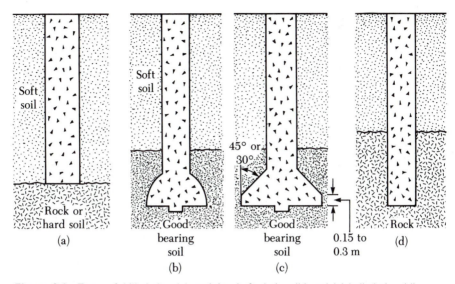

Figure 9.1 Types of drilled pier: (a) straight-shafted pier; (b) and (c) belled pier; (d) straight-shafted pier socketed into rock

account. In Europe, both the side frictional resistance and the end-bearing capacity are always taken into account.

Straight-shafted piers can also be extended into an underlying rock layer (Figure 9.1d). In the calculation of the load-bearing capacity of such piers, the end bearing and the shear stress developed along the pier perimeter and rock interface can be taken into account. This is explained in more detail in Section 9.7.

9.3   Construction Procedures

One of the oldest methods of construction of drilled piers is the *Chicago* method (Figure 9.2a). In this method, circular holes with diameters of 1.1 m (3.5 ft) or more are excavated by hand for depths varying from 0.6–1.8 m (2–6 ft) at a time. The sides of the excavated hole are then lined with vertical boards. These boards are referred to as *laggings*. They are kept tightly in place by two circular steel rings. After the placement of the rings, the excavation is continued for another 0.6–1.8 m (2–6 ft). When the desired depth of excavation is reached, the bell of the pier is excavated. Following the completion of the excavation, the hole is filled with concrete.

In the *Gow* method of pier construction (Figure 9.2b), excavation of the hole is done by hand. Telescopic metal shells are used to maintain the shaft. The shells can be removed one section at a time as the concreting progresses. The minimum diameter of a Gow drilled pier is about 1.22 m (4 ft). Any given section of the shell is about 50 mm (2 in.) less in diameter than the section located immediately above it. Piers as deep as 30 m (100 ft) have been installed by this method.

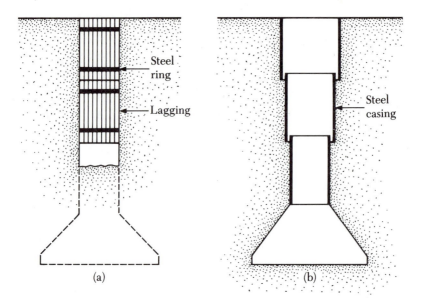

Figure 9.2   (a) Chicago method of pier construction; (b) Gow method of pier construction

Most of the pier shaft excavations nowadays are done mechanically rather than by hand. Open helix augers (flight augers) are common excavation tools. These augers have cutting edges or cutting teeth. The augers with cutting edges are used mostly for drilling in soft, homogeneous soil; those with cutting teeth are for drilling in hard soil and hard pan. The auger is attached to a square shaft referred to as the *Kelly* and pushed into the soil and rotated. When the flights are filled with soil, the auger is raised above the ground surface, and the soil is dumped into a pile by rotating the auger at high speed. These augers are available in various diameters; sometimes they may be as large as 3 m ($\approx$ 10 ft) or more.

When the excavation is extended to the level of the load-bearing stratum, the auger is replaced by underreaming tools to shape the bell, if required. An underreamer essentially consists of a cylinder with two cutting blades that are hinged to the top of the cylinder (Figure 9.3). When the

Figure 9.3   Underreamer

underreamer is lowered into the hole, the cutting blades stay folded inside the cylinder. When the bottom of the hole is reached, the blades are spread outward, and the underreamer is rotated. The loose soil falls inside the cylinder, which is raised periodically and emptied until the bell is completed. Most underreamers can cut bells with diameters as large as three times the diameter of the shaft.

Another common pier-drilling device is the *bucket type drill*. It is essentially a bucket with an opening and cutting edges at the bottom. The bucket is attached to the Kelly and rotated. The loose soil is collected in the bucket, which is periodically raised and emptied. Holes as large as 5–5.5 m (16–18 ft) in diameter can be drilled with this type of equipment.

When rock is encountered during drilling, *core barrels* with *tungsten carbide teeth* attached to the bottom of the barrels are used. *Shot barrels* are also used for drilling into very hard rock. The principle of rock coring by a shot barrel is shown in Figure 9.4. The drill stem is attached to the shot barrel's plate. The barrel has some feeder slots through which chilled steel shots are supplied to the bottom of the bore hole. The steel shots cut the rock when the barrel is rotated. Water is supplied to the drill hole through

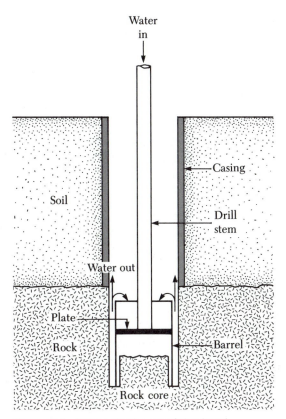

Figure 9.4 Schematic diagram of shot barrel

the drill stem. Fine rock and steel particles (that are produced by the grinding of the steel shots) are washed upward, and they settle on the upper portion of the barrel.

The *Benoto machine* is another type of pier-drilling equipment that is generally used when the drilling conditions are difficult and many boulders are in the soil. It essentially consists of a steel tube that can be oscillated and pushed into the soil. A tool usually referred to as the *hammer grab*, which is fitted with cutting blades and jaws, is used to break up the soil and rock inside the tube and remove them.

Use of Casings and Drilling Mud

When holes for piers are driven in soft clays, the soil tends to squeeze in and close the hole. In such situations, casings may be used to keep the hole open. The casings may have to be driven before excavation begins. Holes made in gravelly and sandy soils also tend to cave in. Excavation of pier holes in these soils can be continued either by casing as the hole progresses or by using *drilling mud*. As Chapter 2 pointed out, drilling mud is also used during field exploration.

Inspection of the Bottom of the Hole

In many instances, the bottom of the hole must be inspected to ensure that the load-bearing stratum is what was anticipated and that the bell is properly done. For these reasons, an inspector must descend to the bottom of the hole. Several safety precautions must be observed during this procedure.

    1. If a casing is not already in the hole, one should be lowered by crane. This will prevent the hole and the bell from collapsing.

    2. The hole should be tested for the presence of poisonous or explosive gasses. This can be done by using a miner's safety lamp.

    3. The inspector should wear a safety harness.

    4. The inspector should also carry a safety lamp and an air tank.

9.4   Other Design Considerations

For the design of ordinary drilled piers without casings, a minimum amount of vertical steel reinforcement is always desirable. Minimum reinforcement equals one percent of the gross cross-sectional area of the shaft. In the state of California, a reinforcing cage having a length of about 3.65 m (≈ 12 ft) is used in the top part of the caisson, and no reinforcement is provided at the bottom. This procedure helps in the construction process in that the cage is placed after most of the concreting is complete.

For piers with nominal reinforcement, most building codes suggest using a value of the design concrete strength (f_c) of the order of $f'_c/4$. Thus, the minimum shaft diameter can be given as

$$f_c = 0.25 f'_c = \frac{Q_w}{A_{gs}} = \frac{Q_w}{\frac{\pi}{4} D_s^2}$$

or

$$D_s = \sqrt{\frac{Q_w}{\left(\frac{\pi}{4}\right)(0.25) f'_c}} = 2.257 \sqrt{\frac{Q_w}{f'_c}} \qquad (9.1)$$

where D_s = diameter of the pier shaft
$\quad f'_c$ = 28-day concrete strength
$\quad Q_w$ = working load of the pier
$\quad A_{gs}$ = gross cross-sectional area of the shaft

Depending on the loading conditions for a given pier, the reinforcement percentage may sometimes be too high. In that case, use of a *single rolled-steel section* at the center of the pier (Figure 9.5b) may be considered. In that case

$$Q_w = (A_{gs} - A_s) f_c + A_s \sigma_{all} \qquad (9.2)$$

where A_s = area of the steel section
$\quad \sigma_{all}$ = allowable strength of steel $\approx 0.5 \sigma_{yield}$

When a permanent steel casing is used for construction instead of a central rolled-steel section (Figure 9.5a), the preceding equation may also be used. However, σ_{all} for steel should be of the order of $0.4 \sigma_{yield}$.

If drilled piers are likely to be subjected to tensile loads, reinforcement should be continued for the entire length of the pier.

Concrete Mix Design

The concrete mix design for piers is not very different from that for any other concrete structure. When a reinforcing cage is used, consideration should be given to the ability of the concrete to flow through the reinforcement. In most cases, a concrete slump of about 150 mm (6 in.) is considered satisfactory. Also, the maximum size of the aggregate should be limited to about 20 mm.

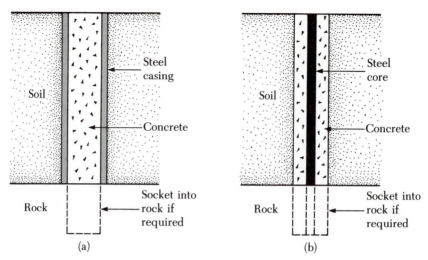

Figure 9.5 Drilled piers with (a) steel casing and (b) a central steel core

9.5   Estimation of Load-Bearing Capacity

The ultimate load-bearing capacity of a drilled pier (Figure 9.6) can be given as

$$Q_u = Q_p + Q_s \tag{9.3}$$

where Q_u = ultimate load
Q_p = ultimate load-carrying capacity at the base
Q_s = frictional (skin) resistance

The ultimate base load can be given by the equation (similar to that for shallow foundations)

$$Q_p = A_p(cN_c^* + q'N_q^* + 0.3\gamma D_b N_\gamma^*) \tag{9.4a}$$

where N_c^*, N_q^*, N_γ^* = the bearing capacity factors
q' = vertical effective stress at the level of the bottom of the pier
D_b = diameter of the base (see Figure 9.6a and b)
A_p = area of the base = $\pi/4 D_b^2$

In most cases, the last term (containing N_γ^*) is neglected except for relatively short piers, so

$$Q_p = A_p(cN_c^* + q'N_q^*) \tag{9.4b}$$

The net load-carrying capacity at the base (that is, the gross load minus the weight of the pier) can be approximated as

$$\boxed{Q_{p(net)} = A_p(cN_c^* + q'N_q^* - q') = A_p[cN_c^* + q'(N_q^* - 1)]} \tag{9.5}$$

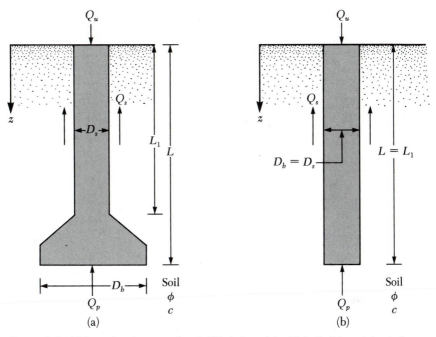

Figure 9.6 Ultimate bearing capacity of drilled piers: (a) with bell; (b) straight shaft

The frictional, or skin, resistance, Q_s, can be expressed in a form similar to that for piles:

$$Q_s = \int_0^{L_1} pf \, dz \qquad\qquad (9.6)$$

where p = shaft perimeter = πD_s

f = unit frictional (or skin) resistance

Piers in Sand

For piers in sand, $c = 0$ and, hence, Eq. (9.5) can be simplified to

$$\boxed{Q_{p(\text{net})} = A_p q'(N_q^* - 1)} \qquad\qquad (9.7)$$

Determination of N_q^* is always a problem for deep foundations, as was shown in the case of piles. Note, however, that all piers are *drilled*, unlike the majority of piles, which are *driven*. For similar initial soil conditions, the actual value of N_q^* may be substantially lower for objects drilled and placed *in situ* compared with objects that are driven. Vesic (1967) has compared the theoretical results obtained by several investigators relating to the variation of N_q^* with soil friction angle. These investigators include DeBeer, Meyer-

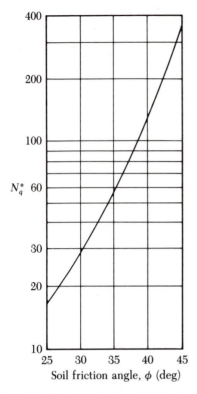

Figure 9.7 Vesic's bearing capacity factor, N_q^*, for deep foundations

hof, Hansen, Vesic, and Terzaghi. The values of N_q^* given by Vesic (1963)
are approximately the lower bound, and, hence, they are recommended for
use in this text (see Figure 9.7). It is recommended that Eq. (8.19) also be
used for calculation of the ultimate point load, Q_p. Thus

$$Q_{p(net)} = A_p(\sigma_o' N_\sigma^* - q')$$

where $\sigma_o' = [(1 + 2K_o)/3]q'$

or

$$Q_{p(net)} = A_p\left[\frac{(1 + 2K_o)}{3} N_\sigma^* - 1\right]q' \tag{9.8}$$

Table 8.8 gave the values of N_σ^* for various values of I_{rr} and soil friction
angles. For ease of calculation, the N_σ^* values given in that table are plotted
in Figure 9.8.

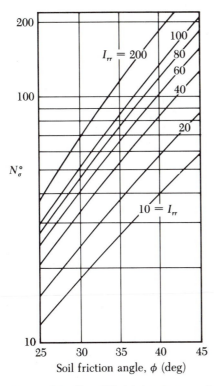

Figure 9.8 Plot of Vesic's bearing capacity factor, N_σ^* (see Table 8.8)

The frictional resistance at ultimate load, Q_s, developed in a drilled pier can be calculated from the relation given in Eq. (9.6). Referring to this equation

$$p = \text{perimeter of shaft} = \pi D_s$$
$$f = K\sigma_v' \tan \delta \tag{9.9}$$

where K = earth pressure coefficient $\approx K_o = 1 - \sin \phi$
σ_v' = effective vertical stress at any depth z

Thus

$$Q_s = \int_0^{L_1} pf\, dz = \pi D_s (1 - \sin \phi) \int_0^{L_1} \sigma_v' \tan \delta \cdot dz \tag{9.10}$$

The value of σ_v' will increase up to a depth of about $15D_s$ and will remain constant thereafter, as shown in Figure 8.16.

An appropriate factor of safety should be applied to the ultimate load to obtain the net allowable load, or

$$Q_{u(net)} = \frac{Q_{p(net)} + Q_s}{FS} \qquad (9.11)$$

It is extremely important that a reliable estimate of the soil friction angle, ϕ, is made in obtaining the net base resistance, $Q_{p(net)}$. Figure 9.9 shows a conservative correlation between the soil friction angle and the corresponding corrected standard penetration resistance numbers in granular soils. However, these friction angles are valid for low confining pressures. At higher confining pressures, which occur in the case of deep foundations, the value of ϕ can decrease substantially for medium to dense sands. This affects the value of N_q^* or N_σ^* (that is, I_{rr}) to be used for estimation of Q_p. For example, Vesic (1977) has shown that for Chattahoochee River sand at a relative density of about 80%, the triaxial angle of friction is about 45° at a confining pressure of 70 kN/m² (10 lb/in.²). However, at a confining pressure of 10.35 MN/m² (1500 lb/in.²), the friction angle is about 32.5°. This will ultimately result in a tenfold decrease of N_q^* or N_σ^*. So, it is the opinion of the author that, for general working conditions of drilled piers, the estimated friction angle determined from Figure 9.9 should be reduced by

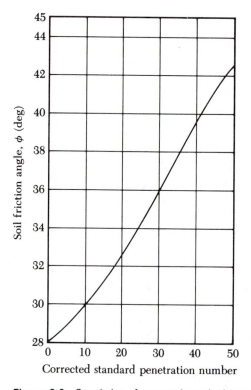

Figure 9.9 Correlation of corrected standard penetration number with the soil friction angle

about 10–15%. In general, the existing experimental values show the follow-ing range of N_q^* for standard drilled piers (or cast-in-place piles):

Sand type	Relative density of sand	Range of N_q^*
Loose	40 or less	10–20
Medium	40–60	25–40
Dense	60–80	30–50
Very dense	>80	75–90

Touma and Reese (1974) have suggested, based on the performance of bored piles in sand with an average diameter of 750 mm (2.5 ft), the follow-ing procedure for calculation of the allowable load-carrying capacity. This can also apply to drilled piers in sand.

For $L > 10D_b$ and a base movement of 25.4 mm (1 in.)

$$Q_{p(net)} = \frac{0.508A_p}{D_b} q_p \qquad (9.12)$$

where $Q_{p(net)}$ is in kN, A_p is in m², D_b is in m, and q_p is the unit point resistance in kN/m².

In English units

$$Q_{p(net)} = \frac{A_p}{0.6D_b} \cdot q_p \qquad (9.13)$$

where $Q_{p(net)}$ is in lb, A_p is in ft², D_b is in ft, and q_p is in lb/ft².

The values of q_p as recommended by Touma and Reese are as follows:

Sand type	q_p (kN/m²)	q_p (lb/ft²)
Loose	0	0
Medium	1530	32,000
Very dense	3830	80,000

For sands of intermediate densities, linear interpolation can be used. The shaft friction resistance can be calculated as

$$Q_s = \int_0^{L_1} (0.7)p\sigma_v' \tan \phi \; dz = 0.7(\pi D_s) \int_0^{L_1} \sigma_v' \tan \phi \; dz$$

$$= 2.2D_s \int_0^{L_1} \sigma_v' \tan \phi \; dz \qquad (9.14)$$

where ϕ = soil friction angle

σ'_v = vertical effective stress at a depth z

With the base resistance and frictional resistance known, a suitable factor of safety (about 3) may be applied to determine the net allowable load.

Piers in Clay

Referring to Eq. (9.5), in saturated clays with $\phi = 0$, the value of N_q^* is equal to one; and, hence, the net base resistance becomes

$$Q_{p(net)} = A_p c_u N_c^* \qquad (9.15)$$

where c_u = undrained cohesion

The bearing capacity factor N_c^* is usually taken to be equal to 9. It can be seen from Figure 7.15 that, when the L/D_b ratio is 4 or more, $N_c^* = 9$. This is the condition in the case of most drilled piers. Experiments by Whitaker and Cooke (1966) have shown that, for belled piers, the full value of $N_c^* = 9$ is realized with a base movement of about 10–15% of D_b. Similarly, for piers with straight shafts (that is, $D_b = D_s$), the full value of $N_c^* = 9$ is obtained with a base movement of about 20% of D_b.

The skin resistance of piers in clay can be expressed by a relation similar to Eq. (8.43), or

$$Q_s = \sum_{L=0}^{L=L_1} \alpha^* c_u p \, \Delta L \qquad (9.16)$$

where p = perimeter of the pier cross section

The value of α^* that can be used in Eq. (9.16) has not yet been fully established. However, the field test results available at this time indicate that α^* may vary from between 0.35 to 0.6. So, conservatively, one can assume that

$$\alpha^* = 0.4 \qquad (9.17)$$

9.6   Settlement of Piers at Working Load

The settlement of drilled piers at working load can be calculated in a manner similar to the one outlined in Section 8.9. In many cases, the load carried by shaft resistance is small compared with the load carried at the base. In such cases, the contribution of s_3 may be ignored. Note that, in Eqs. (8.59) and (8.60), the term D should be replaced by D_b for piers.

Example 9.1

A soil profile is shown in Figure 9.10. A point bearing pier with a bell is proposed for placement in the dense sand and gravel layer. The working load, Q_w, is 2000 kN.

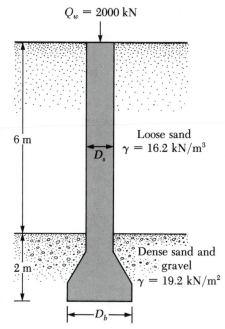

$Q_w = 2000$ kN

6 m

D_s

Loose sand
$\gamma = 16.2$ kN/m^3

2 m

Dense sand and gravel
$\gamma = 19.2$ kN/m^2

D_b

Average corrected standard penetration number $= 40 = N_{cor}$

Figure 9.10

    a. Assuming that $f_c' = 21,000$ kN/m^2, determine the shaft diameter.
    b. Using Eq. (9.7) and a factor of safety of 4, determine the bell diameter, D_b. Ignore the frictional resistance of the shaft.
    c. Using Eq. (9.12), obtain D_b for a settlement of 25.4 mm. Ignore the frictional resistance of the shaft. Use $q_p = 3000$ kN/m^2.
    d. Discuss the differences in the results obtained from Parts b and c.
    e. Choose a bell diameter for a settlement of 25.4 mm.
    f. Estimate Young's modulus for the dense sand layer.
    g. Calculate the possible settlement of the pier by using Eqs. (8.57), (8.58), and (8.59). [Assume s_3 in Eq. (8.57) to be zero.] Use D_s determined in Part a and D_b in Part e; $\mu_s = 0.3$, $E_c = 21 \times 10^6$ kN/m^2.

Solution

Part a: Determination of the Shaft Diameter, D_s

From Eq. (9.1)

$$D_s = 2.257 \sqrt{\frac{Q_w}{f'_c}}$$

Given: $Q_w = 2000$ kN and $f'_c = 21{,}000$ kN/m². So

$$D_s = 2.257 \sqrt{\frac{2000}{21{,}000}} = 0.697 \text{ m}$$

Use $D_s = \mathbf{1}$ **m.**

Part b: Determination of the Bell Diameter Using Eq. (9.7)

$$Q_{p(net)} = A_p q'(N_q^* - 1)$$

Referring to Figure 9.9, for $N_{cor} = 40$, the value of ϕ is approximately equal to 39.5°. To be conservative, we will use a reduction of about 10%. So, assume that $\phi = 35.6$. From Figure 9.6, $N_q^* \approx 60$.

$$q' = 6(16.2) + 2(19.2) = 135.6 \text{ kN/m}^2$$
$$Q_{p(net)} = (Q_w)(FS) = (2000)(4) = 8000 \text{ kN}$$
$$8000 = (A_p)(135.6)(60 - 1)$$
$$A_p = 1.0 \text{ m}^2$$
$$D_b = \sqrt{\frac{1.0}{\frac{\pi}{4}}} = \mathbf{1.13 \text{ m}}$$

Part c: Determination of Bell Diameter Using Eq. (9.12)

$$Q_{p(net)} = \frac{0.508 A_p}{D_b} q_p$$

Because the limit of settlement is 25.4 mm and Eq. (9.12) is for the limiting settlement, a factor of safety over Q_w should not be used to obtain $Q_{p(net)}$. So

$$Q_w = Q_{p(net)}$$

Thus

$$Q_{p(net)} = Q_w = 2000 = \frac{0.508 A_p}{D_b} q_p = \frac{(0.508)\left(\frac{\pi}{4}\right)(D_b^2) q_p}{D_b}$$
$$= 0.399 D_b q_p$$

or

$$D_b = \frac{2000}{(0.399)(3000)} = \mathbf{1.67 \text{ m}}$$

Part d

The value of D_b determined in Part b corresponds to an allowable bearing capacity that is based on the ultimate bearing capacity. The settlement has not been taken into consideration at all. The ultimate bearing capacity of drilled piers may occur at a settlement exceeding 10–15% of the bell diameter. The bell diameter in Part c corresponds to a settlement of 25.4 mm.

Part e

Use a bell diameter of 2 m to be conservative.

Part f

The effective overburden pressure at the surface of the dense sand gravel layer is 135.6 kN/m². Substituting this value of effective overburden pressure and $N_{cor} = 40$ in Eqs. (2.4) and (2.6), the uncorrected value of N is about 47. Using Eq. (3.70), Young's modulus

$$E_s = 766N = 766(47) = \textbf{36,002 kN/m}^2$$

Part g: Calculation of Settlement

From Eq. (8.58), with $\xi = 0$

$$s_1 = \frac{Q_w L}{A_s E_p} = \frac{(2000)(6)}{\dfrac{\pi}{4}(1)^2(21 \times 10^6)} = 0.000728 \text{ m} = 0.728 \text{ mm}$$

From Eq. (8.59)

$$s_2 = \frac{q_{wp} D_b}{E_s}(1 - \mu_s^2)I_{wp}$$

$$I_{wp} = 0.88 \text{ (Figure 3.23)}$$

$$q_{wp} = \frac{2000}{\dfrac{\pi}{4}D_b^2} = \frac{2000}{\dfrac{\pi}{4}(2)^2} = 636.6 \text{ kN/m}^2$$

$$s_2 = \frac{(636.6)(2)}{36,002}(1 - 0.3^2)(0.88) = 0.0283 \text{ m} = 28.3 \text{ mm}$$

So, the total settlement is $28.3 + 0.728 = \textbf{29.03 mm}$

This settlement is higher than 25.4 mm. However, the assumption of E_s is approximate.

≣

Example 9.2

A drilled pier is shown in Figure 9.11. Given: $Q_w = 2800$ kN.

a. Assume that $f_c' = 28,000$ kN/m². Determine if the proposed diameter of the shaft is adequate or not.

b. Determine the net ultimate point load-carrying capacity.

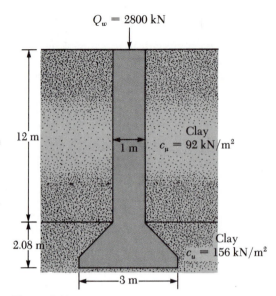

$Q_w = 2800$ kN

12 m

Clay
$c_u = 92$ kN/m²

1 m

2.08 m

Clay
$c_u = 156$ kN/m²

—3 m—

Figure 9.11

    c. Determine the ultimate skin resistance.
    d. Calculate the factor of safety with respect to the working load, Q_w.
    e. Estimate the total elastic settlement of the pier under the working load. Use
Eqs. (8.58), (8.60), (8.63), and (8.64). $E_p = 22 \times 10^6$ kN/m².

Solution

Part a

From Eq. (9.1)

$$D_s = 2.257 \sqrt{\frac{Q_w}{f_c'}} = 2.257 \sqrt{\frac{2{,}800}{28{,}000}} = 0.714 \text{ m}$$

Proposed shaft diameter $D_s = \textbf{1 m} > \textbf{0.714 m—O.K.}$

Part b

$$Q_{p(\text{net})} = c_u N_c^* A_p = (156)(9)\left[\left(\frac{\pi}{4}\right)(3)^2\right] = \textbf{9924.4 kN}$$

Part c

According to Eq. (9.16)

$$Q_s = \pi D_s L_1 c_u \alpha^*$$

Assume that $\alpha^* = 0.4$ [Eq. (9.17)]

$$Q_s = (\pi)(1)(12)(92)(0.4) = \textbf{1387.3 kN}$$

Part d

Factor of safety =

$$\frac{Q_u}{Q_w} = \frac{9924.3 + 1387.3}{2800} = \textbf{4.04}$$

Part e

From Eq. (8.58)

$$s_1 = \frac{(Q_{wp} + \xi Q_{ws})L}{A_p E_p}$$

Assume full mobilization of the skin resistance. So, $Q_{ws} = 1387.3$ kN. Hence $Q_{wp} = 2800 - 1387.3 = 1412.7$ kN. Also, let $\xi = 0.6$.

$$A_p = \left(\frac{\pi}{4}\right)D_s^2 = \frac{\pi}{4}\,(1)^2 = 0.785 \text{ m}^2$$

So

$$s_1 = \frac{[1412.7 + (0.6)(1387.3)](14.08)}{(0.785)(22 \times 10^6)} = 0.00183 \text{ m} = \textbf{1.83 mm}$$

From Eq. (8.60)

$$s_2 = \frac{Q_{wp} C_p}{D_b q_p}$$

The value of C_p for stiff clay is about 0.04 (Table 8.11).

$$q_p = cN_c^* = (156)(9) = 1404 \text{ kN/m}^2$$

So

$$s_2 = \frac{(1412.7)(0.04)}{(3)(1404)} = 0.01342 \text{ m} = \textbf{13.42 mm}$$

Again, from Eq. (8.63)

$$s_3 = \frac{Q_{ws} C_s}{L_1 q_p}$$

$$C_s = \left(0.93 + 0.16\sqrt{\frac{L_1}{D_s}}\right)C_p = \left(0.93 + 0.16\sqrt{\frac{12}{1}}\right)0.04 = 0.0594$$

Hence

$$s_3 = \frac{(1387.3)(0.0594)}{(12)(1404)} = 0.00489 \text{ m} = \textbf{4.89 mm}$$

Total elastic settlement $s = s_1 + s_2 + s_3 = 1.83 + 13.42 + 4.89 = \textbf{20.14 mm}$

9.7   Load-Bearing Capacity of Piers Extending into Rock

Section 9.1 noted that drilled piers can be extended into rock. This section discusses the principles of analysis of the load-bearing capacity of such drilled piers. Figure 9.12 shows a drilled pier whose depth of embedment in the rock is equal to L. Let the total load to be carried by the pier be equal to Q_w. Also, assume that (a) no load transfer takes place along the soil-pier interface and (b) under the pier loading condition, the *bond between the concrete and the rock is broken*. With these assumptions, the compressive stress along the cross section of the pier can be given as

$$\sigma_z = \sigma_{\max} \exp\left[-\frac{2\mu_c \tan \phi_{rc}}{1 - \mu_c + (1 + \mu_r)\dfrac{E_c}{E_r}} \cdot \frac{2z}{D_b} \right] \tag{9.18}$$

where $\sigma_{\max} = \dfrac{Q_{\text{all}}}{\dfrac{\pi}{4} D_b^2}$ (9.19)

μ_c, μ_r = Poisson's ratio in concrete and rock, respectively
E_c, E_r = Young's modulus of concrete and rock, respectively
ϕ_{rc} = angle of friction along the rock-pier interface

The maximum shear stress developed along the rock and concrete pier perimeter can be given conservatively as follows (Goodman, 1980):

$$\tau_{\max} = \frac{q_u}{20} \tag{9.20}$$

where q_u = unconfined compression strength of concrete or rock, whichever is lower

The allowable shear stress can then be determined as

$$\tau_{\text{all}} = \frac{\tau_{\max}}{FS} \tag{9.21}$$

where FS = factor of safety

Once τ_{all} is known or assumed, the following step-by-step procedure should be adopted (Ladanyi, 1977) to determine the length, L, to which the pier should be embedded in rock.

    1. Obtain the size of the shaft $D_s = D_b$ based on the load Q_w and the strength of the concrete. If a rolled-steel section core is used, the diameter D_s will depend on the strength of concrete and steel.

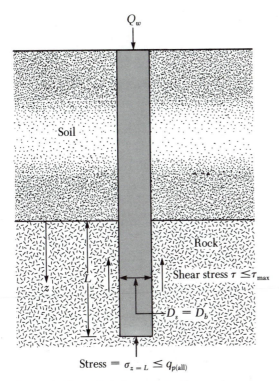

Figure 9.12   Drilled piers socketed into rock

2. Assuming that the load carried at the pier tip is zero, determine the maximum required length of embedment (L_1) of the pier:

$$L_1 = \frac{Q_w}{\pi D_s \tau_{all}} \tag{9.22}$$

3. Now, assume another value of the length of embedment $L_2 < L_1$. Use Eq. (9.18) and calculate σ_z at $z = L_2$. This value of $\sigma_{z=L_2}$ is the unit point resistance at the pier tip.

4. Compare the value of $\sigma_{z=L_2}$ obtained in step 3 with the allowable bearing capacity in rock, $q_{p(all)}$. This can be done by using Eq. (8.55). If $\sigma_{z=L_2} > q_{p(all)}$, go back to step 3 and assume another value of L_2.

5. If $\sigma_{z=L_2} \leq q_{p(all)}$ in step 4, calculate τ developed along the perimeter of the pile as

$$\tau = \left[\underbrace{\left(1 - \frac{\sigma_{z=L_2}}{\sigma_{max}}\right) Q_w}_{\substack{\text{load taken} \\ \text{by shaft}}} \right] \frac{1}{\pi D_b L_2} \tag{9.23}$$

6. Compare the value of τ obtained from Eq. (9.23) with τ_{all} obtained from Eq. (9.21).

7. Repeat steps 3 through 6 to obtain the length L that is most desirable with $\sigma_{z=L_2} \leq q_{p(\text{all})}$ and $\tau \leq \tau_{\text{all}}$.

Example 9.3

Refer to Figure 9.12. The soil located above the rock is soft clay. Neglect the skin resistance developed at the pier perimeter-clay interface. Given: $Q_w = 25{,}000$ kN; $E_r/E_c = 0.7$; $\mu_c = \mu_r = 0.3$; $\phi_{rc} = 38°$; $\tau_{\text{all}} = 490$ kN/m^2; for rock, $q_{p(\text{all})} = 2500$ kN/m^2; for concrete $f_c' = 21{,}000$ kN/m^2.

Determine the required diameter and the length of embedment, L.

Solution

Calculation of $D_s = D_b$

Use $f_c = 0.25f_c' = (0.25)(21{,}000) = 5250$ kN/m^2. So, from Eq. (9.1)

$$D_s = 2.257\sqrt{\frac{Q_w}{f_c'}} = 2.257\sqrt{\frac{25{,}000}{21{,}000}} = 2.46 \text{ m}$$

Adopt $D_s = 2.5$ m.

Calculation of L_1

From Eq. (9.22)

$$L_1 = \frac{Q_w}{\pi D_s \tau_{\text{all}}} = \frac{25{,}000}{(\pi)(2.5)(490)} = 6.496 \approx \textbf{6.5 m}$$

Calculation of L

Assume an embedment length $= L_2 = 5$ m. From Eq. (9.18), with $z = 5$ m

$$\frac{\sigma_{z=L_2}}{\sigma_{\text{max}}} = \exp\left[-\frac{2\mu_c \tan\phi_{rc}}{1 - \mu_c + (1 + \mu_r)\dfrac{E_c}{E_r}}\frac{2z}{D_b}\right]$$

$$= \exp\left\{-\left[\frac{(2)(0.3)(\tan 38°)}{1 - 0.3 + (1 + 0.3)\left(\dfrac{1}{0.7}\right)}\right]\frac{(2)(5)}{2.5}\right\} = 0.48$$

So

$$\sigma_{z=L_2} = (\sigma_{\text{max}})(0.48) = \left(\frac{Q_w}{\dfrac{\pi}{4}D_b^2}\right)(0.48) = \frac{(25{,}000)(0.48)}{(0.785)(2.5^2)}$$

$$\approx 2446 \text{ kN/m}^2 < 2500 \text{ kN/m}^2 = q_{\text{all}} \text{—O.K.}$$

The shear stress developed along the shaft-rock interface can now be checked by using Eq. (9.23) as

$$\tau = \left(1 - \frac{\sigma_{z=L_2}}{\sigma_{\text{max}}}\right)\frac{Q_w}{\pi D_b L_2} = (1 - 0.48)\frac{25{,}000}{(\pi)(2.5)(5)}$$

$$= 331 \text{ kN/m}^2 \leq \tau_{\text{all}} = 490 \text{ kN/m}^2$$

So, $L = L_2 = \textbf{5 m.}$

. Note that if the value of τ had been substantially lower than τ_{all}, the value of L_2 could have been reduced. This would require a new set of calculations of $(\sigma_{z=L_2})/\sigma_{max}$, $\sigma_{z=L_2}$ and τ. However, if τ or $\sigma_{z=L_2}$ becomes less than τ_{all} and $q_{p(all)}$, respectively, the length L_2 must be increased and checked again.

9.8   Uplift Capacity of Piers

Sometimes drilled piers must resist uplifting loads. Field observations of these piers' uplift capacity are relatively scarce. The procedure for determination of the ultimate uplifting load for drilled piers without bells is similar to that for piles described in Chapter 8 (Section 8.10) and will not be described here. When a short drilled pier with a bell is subjected to an uplifting load, the nature of the failure surface in the soil will be like that shown in Figure 9.13. The net ultimate uplift capacity, T_{un}, can be given as

$$T_{un} = T_{ug} - W \qquad (9.24)$$

where T_{ug} = gross ultimate uplift capacity
$\quad\quad W$ = effective weight of the pier

The magnitude of T_{un} for piers in sand can be estimated by the procedure outlined by Meyerhof and Adams (1968) and Das and Seeley (1975):

$$\boxed{T_{un} = B_q A_p \gamma L} \qquad (9.25)$$

where B_q = breakout factor
$\quad\quad A_p = (\pi/4)D_b^2$
$\quad\quad \gamma$ = unit weight of soil above the bell (*Note:* if the soil is submerged effective unit weight should be used.)

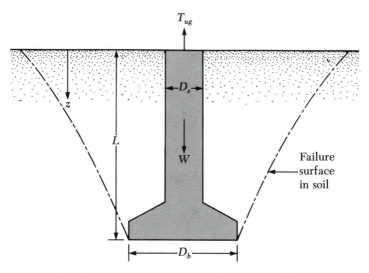

Figure 9.13 Nature of failure surface in soil caused by uplifting force on belled piers

The breakout factor can be given as

$$B_q = 2 \frac{L}{D_b} K'_u \tan \phi \left(m \frac{L}{D_b} + 1 \right) + 1 \qquad (9.26)$$

where K'_u = nominal uplift coefficient
ϕ = soil friction angle
m = shape factor coefficient

The value of K'_u may be taken as 0.9 for all values of ϕ varying from 30–45°. The variation of m has been given by Meyerhof and Adams (1968) as follows:

Soil friction angle, ϕ (deg)	m
30	0.15
35	0.25
40	0.35
45	0.50

It has been shown experimentally that the value of B_q increases with the L/D_b ratio up to a critical value $(L/D_b)_{cr}$ and remains constant thereafter. The critical embedment ratio, $(L/D_b)_{cr}$, increases with the soil friction angle. The approximate ranges are given below:

Soil friction angle, ϕ (deg)	$(L/D_b)_{cr}$
30	4
35	5
40	7
45	9

Hence, piers with $L/D_b \leq (L/D_b)_{cr}$ may be defined as *shallow foundations* with regard to the uplift. Piers with $L/D_b > (L/D_b)_{cr}$ are *deep foundations*. The nature of the failure surface in soil at failure as shown in Figure 9.13 is for the case of a shallow foundation. For deep foundations, local shear failure takes place, and the failure surface in soil *does not extend up to the ground surface*. Based on the preceding considerations, the variation of B_q with L/D_b is shown in Figure 9.14.

Following is a step-by-step procedure for the calculation of the net ultimate uplift capacity of drilled piers with bells in sand.

1. Determine L, D_b, and L/D_b.
2. Estimate $(L/D_b)_{cr}$ and, hence, L_{cr}.
3. If $(L/D_b) \leq (L/D_b)_{cr}$, obtain B_q from Figure 9.14. Now,

$$T_{ug} = B_q A_p \gamma L + W$$

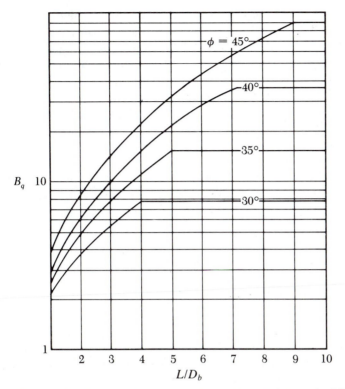

Figure 9.14 Variation of the breakout factor, B_q, with L/D_b and soil friction angle

4. If $(L/D_b) > (L/D_b)_{cr}$

$$T_{ug} = B_q A_p \gamma L + W + \int_0^{L-L_{cr}} (\pi D_s)\sigma_v' K_u \tan \delta \, dz \qquad (9.27)$$

The last term of the preceding equation is for the frictional resistance developed along the soil-shaft interface from $z = 0$ to $z = L - L_{cr}$. This is similar to Eqs. (8.73) and (8.74). The term σ_v' is the effective stress at any depth z, and K_u and δ are to be obtained from Figure 8.28b and c, respectively.

The net ultimate uplift capacity of belled piers in clay can be estimated according to the procedure outlined by Das (1980). Based on this method

$$T_{un} = \{c_u B_c + \gamma L\}A_p \qquad (9.28)$$

where c_u = undrained cohesion
$\qquad B_c$ = breakout factor
$\qquad \gamma$ = unit weight of clay soil above the bell

As in the case of B_q, the value of B_c increases with the embedment ratio up to a critical value of $L/D_b = (L/D_b)_{cr}$ and remains constant thereafter. Beyond the critical depth, the value of B_c is approximately equal to 9. The critical embedment ratio can be related to the undrained cohesion by the relation

$$\left(\frac{L}{D_b}\right)_{cr} = 0.107c_u + 2.5 \leq 7 \tag{9.29a}$$

where c_u is in kN/m^2.

In English units

$$\left(\frac{L}{D_b}\right)_{cr} = 0.738c_u + 2.5 \leq 7 \tag{9.29b}$$

where c_u is in $lb/in.^2$

Following is a step-by-step procedure for determining the net ultimate uplift capacity of belled piers in clay:

1. Determine c_u, L, D_b, and L/D_b.
2. Obtain $(L/D_b)_{cr}$ from Eq. (9.29), and obtain L_{cr}.
3. If $L/D \leq (L/D_b)_{cr}$, refer to Figure 9.15 and obtain the value of B_c.
4. Use Eq. (9.28) to obtain T_{un}.

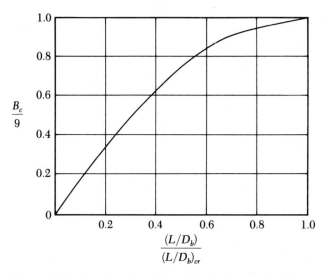

Figure 9.15 Nondimensional plot of the breakout factor, B_c

5. If $L/D > (L/D_b)_{cr}$, $B_c = 9$. The value of T_{un} can be given by the following relation:

$$T_{un} = (9c_u + \gamma L)A_p + \Sigma(\pi D_s)(L - L_{cr})\alpha' c_u \qquad (9.30)$$

The last term of the preceding equation is the skin resistance obtained from the adhesion along the soil-shaft interface. This is similar to Eq. (8.66). The value of α' can be obtained from Eqs. (8.68) and (8.69).

Example 9.4

Refer to Figure 9.13. A drilled pier with a bell has a shaft diameter of 0.76 m and a bell diameter of 1.85 m. Given: $L = 9.5$ m. The bell is supported by a dense sand ($z > 9.5$ m) layer. However, a fine, loose sand layer exists above the bell ($z = 0$–9.5 m). For this sand, $\gamma = 16.4$ kN/m³, $\phi = 32°$, and the approximate relative density is 30%. The entire pier is located above the ground water table. Determine the net allowable uplift capacity of the pier with a factor of safety of 3.

Solution

Given: $L = 9.5$ m and $D_b = 1.85$ m. So, $L/D_b = 9.5/1.85 = 5.14$. For $\phi = 30°$, $(L/D_b)_{cr} = 4$; for $\phi = 35°$, $(L/D_b)_{cr} = 5$. By interpretation, $(L/D_b)_{cr} \approx 4.2$ for $\phi = 32°$. So, $L_{cr} = (4.2)(D_b) = 7.77$. Because $L/D_b = 5.13 > (L/D_b)_{cr} = 4.2$, it is a deep foundation.

According to Eq. (9.26)

$$B_q = 2\left(\frac{L}{D_b}\right)_{cr} K'_u \tan \phi\left[m\left(\frac{L}{D_b}\right)_{cr} + 1\right] + 1$$

Note that $(L/D_b)_{cr}$ has been used in the preceding equation in place of L/D_b because it is a deep foundation. The value of m is approximately equal to 0.17 for $\phi = 32°$. Hence

$$B_q = (2)(4.2)(0.9)(\tan 32°)[(0.17)(4.2) + 1] + 1 = 9.09$$

From Eq. (9.27)

$$T_{un} = T_{ug} - W = B_q A_p \gamma L + \int_0^{L-L_{cr}} (\pi D_s)(\sigma'_v K_u \tan \delta)\, dz$$

$$= B_q A_p \gamma L + \frac{\pi}{2} \gamma D_s K_u \tan \delta (L - L_{cr})^2$$

$$A_p = \left(\frac{\pi}{4}\right)(D_b)^2 = \left(\frac{\pi}{4}\right)(1.85)^2 = 2.687 \text{ m}^2$$

$$L - L_{cr} = 9.5 - 7.77 = 1.73 \text{ m}$$

Also, from Figure 8.28b and c, for $\phi = 32°$ and relative density $= 30\%$, $K'_u = 1.5$ and $\delta/\phi \approx 0.73$. Hence

$$T_{un} = (9.09)(2.687)(16.4)(9.5) + \left(\frac{\pi}{2}\right)(16.4)(0.76)(1.5)$$

$$[\tan (0.73 \times 32)](1.73)^2$$

$$= 3805.4 + 37.96 = 3843.36 \approx 3843$$

So, net allowable capacity $= 3843/FS = 3843/3 = $ **1281 kN**

Example 9.5

Consider the drilled pier described in Example 9.4. If the soil above the bell is clay with an average value of the undrained shear strength equal to 95 kN/m², calculate the net ultimate uplift capacity. Given: $\gamma_{clay} = 17.9$ kN/m³.

Solution

From Eq. (9.29a)

$$\left(\frac{L}{D_b}\right)_{cr} = 0.107c_u + 2.5 = (0.107)(95) + 2.5 = 12.67$$

This value is more than 7, so use $(L/D_b)_{cr} = 7$. Hence, $L_{cr} = (7)(1.85) = 12.95$ m. $L_{cr} = 12.95$ is greater than $L = 9.5$ m, so this is a shallow foundation for uplift consideration.

For shallow foundations [Eq. (9.28)]

$$T_{un} = (c_u B_c + \gamma L)A_p$$

The value of the breakout factor, B_c, can be determined from Figure 9.15:

$$\frac{\left(\dfrac{L}{D_b}\right)}{\left(\dfrac{L}{D_b}\right)_{cr}} = \frac{\left(\dfrac{9.5}{1.85}\right)}{7} = 0.734$$

So, $B_c/9 = 0.92$, or $B_c = 8.28$. $A_p = (\pi/4)D_b^2 = (\pi/4)(1.85)^2 = 2.687$ m². Thus

$$T_{un} = [(95)(8.28) + (17.9)(9.5)]2.687 = \textbf{2570.5 kN}$$

9.9   Lateral Load-Carrying Capacity

The lateral load-carrying capacity of piers can be analyzed in a manner similar to that presented in Section 8.11 for piles. Therefore, it will not be repeated here.

9.10   Types of Caisson

Caissons can be divided into three major types: (1) *open caissons* (Figure 9.16), (2) *box caissons* (or closed caissons; Figure 9.17), and (3) *pneumatic caissons* (Figure 9.18).

   Open caissons are concrete shafts that remain open at the top and bottom during construction. The bottom of the caisson has a cutting edge. The caisson is sunk into place, and soil from the inside of the shaft is removed by grab buckets until the bearing stratum is reached. The shafts can be of any shape—circular, square, rectangular, or oval. Once the bearing stratum is reached, concrete is poured into the shaft (under water) to form a seal at its bottom. When the concrete seal hardens, the water inside the caisson shaft is pumped out. Concrete is then poured into the shaft to fill it. Open caissons can be extended to great depths, and the cost of construction is relatively low. However, one of their major disadvantages is the lack of quality control of the concrete poured into the shaft for the seal. Also, the bottom of the caisson cannot be thoroughly cleaned out. An alternate method of open-caisson construction is to drive some sheet piles to form an

Figure 9.16   Open caisson

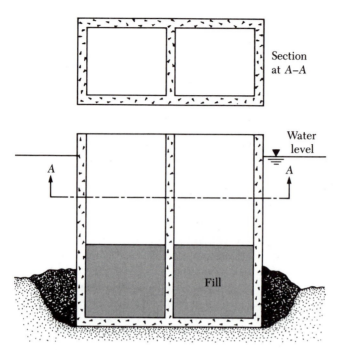

Figure 9.17 Box caisson

enclosed area. The enclosed area is filled with sand, and is generally referred to as a *sand island*. The caisson is then sunk through the sand to the desired bearing stratum. This procedure is somewhat analogous to sinking a caisson when the ground surface is above the water table.

Box caissons are caissons with their bottoms closed. They are constructed on land and then transported to the construction site. They are gradually sunk at the site by filling the inside with sand, ballast, water, or concrete. The cost for this type of construction is low. However, the bearing surface should be level. If the bearing surface is not level, it must be leveled by excavation.

Pneumatic caissons are generally used for depths of about 15–40 m (50–130 ft). This type of caisson is necessary where an excavation cannot be kept open because the soil flows into the excavated area faster than it can be removed. A pneumatic caisson has a work chamber at the bottom that is at least 3 m (≈ 10 ft) high. In this chamber, the workers excavate the soil and place the concrete. The air in the chamber is kept at a specific pressure to prevent water and soil from entering. The workers do not usually encounter any severe discomfort when the chamber pressure is raised to about 100 kN/m^2 (≈ 15 lb/in.^{2}) above the atmospheric pressure. Beyond this pressure, decompression periods are required when the workers leave the chamber. When chamber pressures of about 300 kN/m^2 (≈ 44 lb/in.^{2}) above the atmospheric pressure are required, the workers should not be kept inside for

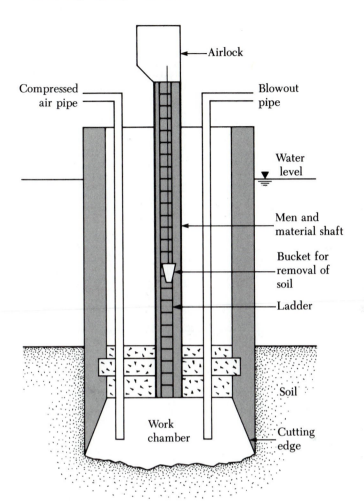

Figure 9.18 Pneumatic caisson

more than $1\frac{1}{2}$–2 hours at a time. The workers enter and exit the work chamber through a steel shaft by means of a ladder. This shaft is also used for the removal of excavated soil and the placement of concrete. For large caisson construction, more than one shaft may be necessary; an airlock is provided for each one. Pneumatic caissons gradually sink with the progress of excavation work. When the bearing stratum is reached, the work chamber is filled with concrete.

9.11   Load-Bearing Capacity of Caissons

Calculation of the load-bearing capacity of caissons is similar to that for drilled piers. Therefore, it will not be further discussed in this section.

9.12   Thickness of Concrete Seal in Open Caissons

Section 9.10 mentioned that, before dewatering the caisson, a concrete seal is placed at the bottom of the shaft and allowed to cure for some time. The concrete seal should be thick enough to withstand an upward hydrostatic force from its bottom after dewatering is complete and before concrete fills the shaft. Based on the theory of elasticity, the thickness, t, (see Figure 9.19) can be given as follows (Teng, 1962):

$$t = 1.18R_i \sqrt{\frac{q}{f_c}} \quad \text{(circular caisson)} \tag{9.31}$$

and

$$t = 0.866B_i \sqrt{\frac{q}{f_c\left[1 + 1.61\left(\dfrac{L_i}{B_i}\right)\right]}} \quad \text{(rectangular caisson)} \tag{9.32}$$

where R_i = inside radius of a circular caisson
q = unit bearing pressure at the base of the caisson
f_c = allowable concrete flexural stress (≈ 0.1–0.2 of f'_c; where f'_c is the 28-day compressive strength of concrete)
B_i, L_i = inside width and length, respectively, of rectangular caisson (Figure 9.19b)

According to Figure 9.19, the value of q in Eqs. (9.31) and (9.32) can be approximated as

$$q \approx H\gamma_w - t\gamma_c \tag{9.33}$$

where γ_c = unit weight of concrete

The thickness of the seal calculated by Eqs. (9.31) and (9.32) will be sufficient to protect it from cracking immediately after dewatering. However, two other conditions should also be checked for safety. They are as follows:

1. *Check for Perimeter Shear at Contact Face of Seal and Shaft:* According to Figure 9.19, the net upward hydrostatic force from the bottom of the seal is equal to $A_i H\gamma_w - A_i t\gamma_c$ (where $A_i = \pi R_i^2$ for circular caissons and $A_i = L_i B_i$ for rectangular caissons). So, the perimeter shear developed can be given by

$$v \approx \frac{A_i H\gamma_w - A_i t\gamma_c}{p_i t} \tag{9.34}$$

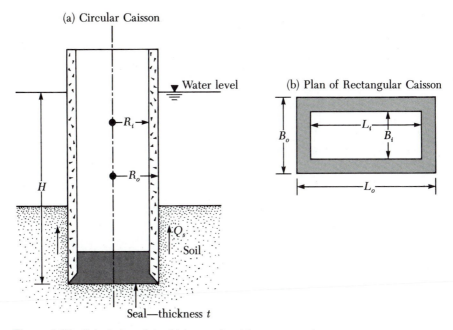

Figure 9.19   Calculation of the thickness of seal for an open caisson

where p_i = inside perimeter of the caisson

Note that

$$p_i = 2\pi R_i \quad \text{(for circular caissons)} \tag{9.35}$$

and

$$p_i = 2(L_i + B_i) \quad \text{(for rectangular caissons)} \tag{9.36}$$

The perimeter shear given by Eq. (9.34) should be less than the permissible shear stress, v_u, as given by Eq. (A.15a) (Appendix A), or

$$v \; (\text{MN/m}^2) \leq v_u \; (\text{MN/m}^2) = 0.17\phi \sqrt{f_c'} \; (\text{MN/m}^2) \tag{9.37a}$$

where $\phi = 0.85$

In English units

$$v \; (\text{lb/in.}^2) \leq v_u \; (\text{lb/in.}^2) = 2\phi \sqrt{f_c'} \; (\text{lb/in.}^2) \tag{9.37b}$$

where $\phi = 0.85$

2. *Check for Buoyancy:* If the shaft is completely dewatered, the buoyant upward force, F_u, can be given as

$$F_u = (\pi R_o^2)H\gamma_w$$

$$\text{(for circular caissons, as shown in Figure 9.19a)} \tag{9.38}$$

and

$$F_u = (B_o L_o) H \gamma_w$$

(for rectangular caissons, as shown in Figure 9.19b)   (9.39)

The downward force (F_d) is caused by the weight of the caisson and the seal and by the skin friction at the caisson-soil interface, or

$$F_d = W_c + W_s + Q_s \qquad (9.40)$$

where W_c = weight of caisson
W_s = weight of seal
Q_s = skin friction

If $F_d > F_u$, the caisson is safe from buoyancy. However, if $F_d < F_u$, it will be unsafe to dewater the shaft completely. For that reason, the thickness of the seal should be increased by Δt [over the thickness calculated by using Eqs. (9.31) or (9.32)], or

$$\Delta t = \frac{F_u - F_d}{A_i \gamma_c} \qquad (9.41)$$

Example 9.6

An open caisson (circular) is shown in Figure 9.20. Determine the thickness of the seal that will enable complete dewatering:

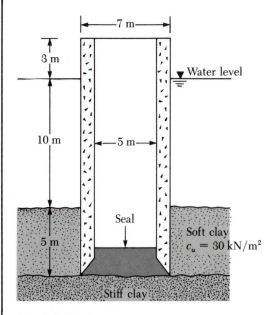

Figure 9.20

Solution

From Eq. (9.31)

$$t = 1.18R_i \sqrt{\frac{q}{f_c}}$$

For this problem, $R_i = 5/2 = 2.5$ m.

$$q \approx (15)(9.81) - t\gamma_c$$

With $\gamma_c = 23.58$ kN/m^3, $q = 147.15 - 23.58t$.

$$f_c = 0.1f'_c = 0.1 \times 21 \times 10^3 \text{ kN/m}^2 = 2.1 \times 10^3 \text{ kN/m}^2$$

So

$$t = (1.18)(2.5) \sqrt{\frac{147.15 - 23.58t}{2.1 \times 10^3}}$$

or

$$t^2 + 0.0978t - 0.61 = 0$$

Solution to the preceding equation gives $t = 0.73$ m. So, let $t \approx 0.75$ m.

Check for Perimeter Shear

According to Eq. (9.34)

$$v = \frac{(\pi)R_i^2 H\gamma_w - (\pi)R_i^2 t\gamma_c}{2\pi R_i t}$$

$$= \frac{(\pi)(2.5)^2[(15)(9.81) - (0.75)(23.58)]}{(2)(\pi)(2.5)(0.75)} = 215.77 \text{ kN/m}^2$$

However, the allowable shear stress

$$v_u = 0.17\phi \sqrt{f'_c} = (0.17)(0.85)\sqrt{21} = 0.662 \text{ MN/m}^2$$
$$v = 215.77 \text{ kN/m}^2 < v_u = \textbf{662 kN/m}^2 \textbf{—O.K.}$$

Check Against Buoyancy

Buoyant upward force $= F_u = \pi R_o^2 H\gamma_w \cdot R_o = 3.5$ m. So

$$F_u = (\pi)(3.5)^2(15)(9.81) = 5663 \text{ kN}$$

Downward force $= F_d = W_c + W_s + Q_s$

$$W_c = \pi(R_o^2 - R_i^2)(\gamma_c)(18) = \pi(3.5^2 - 2.5^2)(23.58)(18) \approx 8000 \text{ kN}$$
$$W_s = (\pi R_i^2)t\gamma_c = (\pi)(2.5)^2(0.75)(23.58) = 347.24 \text{ kN}$$

Assume that $Q_s \approx 0$. So

$$F_d = 8000 + 347.24 = 8347.24 \text{ kN}$$

Because $F_u < F_d$, it is safe. For design, **assume that $t = 1$ m.**

PROBLEMS

9.1 A drilled pier is shown in Figure P9.1. Determine the net allowable point bearing capacity. Given:

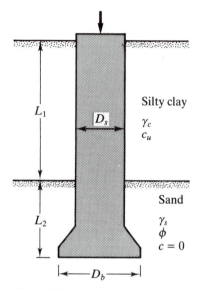

Figure P9.1

$D_b = 6$ ft $\gamma_c = 110$ lb/ft^3

$D_s = 3.5$ ft $r_s = 112$ lb/ft^3

$L_1 = 18$ ft $\phi = 35°$

$L_2 = 10$ ft

$c_u = 720$ lb/ft^2

Factor of safety $= 4$

Reduce the friction angle, ϕ, by 10%. Use Figure 9.7.

9.2 Redo Problem 9.1 by Vesic's method [Eq. (9.8)]. Given: $I_{rr} = 100$.

9.3 For the drilled pier described in Problem 9.1, what skin resistance would develop in the top 18 ft, which are in clay?

9.4 Redo Problem 9.1 with the following:

$D_b = 1.75$ m $\gamma_c = 17.8$ kN/m^3

$D_s = 1$ m $\gamma_s = 18.2$ kN/m^3

$L_1 = 4$ m $\phi = 32°$

$L_2 = 2.5$ m

$c_u = 32$ kN/m^2

9.5 Solve Problem 9.4 by Vesic's method [Eq. (9.8)]. Given: $I_{rr} = 80$.

9.6 For the drilled pier described in Problem 9.4, what skin resistance would develop in the top 4 m (that is, the portion in clay soil)?

9.7 Figure P9.7 shows a drilled pier without a bell. Determine

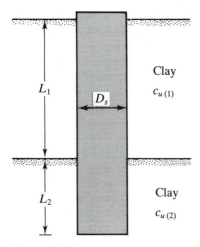

Figure P9.7

a. The net ultimate point bearing capacity
b. The ultimate skin resistance
c. The working load, Q_w (factor of safety $= 3$)

Given: $L_1 = 6$ m $\qquad$ $c_{u(1)} = 45$ kN/m²

$\qquad\quad$ $L_2 = 5$ m $\qquad$ $c_{u(2)} = 74$ kN/m²

$\qquad\quad$ $D_s = 1.5$ m

9.8 For the drilled pier described in Problem 9.7, estimate the total elastic settlement at working load. Use Eqs. (8.58), (8.60), and (8.61). Assume that $E_p = 20 \times 10^6$ kN/m² and $\xi = 0.65$. Make other assumptions as necessary.

9.9 Repeat Problem 9.7 with the following:

$\qquad\quad$ $L_1 = 25$ ft $\qquad$ $c_{u(1)} = 1200$ lb/ft²

$\qquad\quad$ $L_2 = 10$ ft $\qquad$ $c_{u(2)} = 2000$ lb/ft²

$\qquad\quad$ $D_s = 3.5$ ft

9.10 For the pier described in Problem 9.9, estimate the total elastic settlement at working load. Use Eqs. (8.58), (8.60), and (8.61). Assume that $E_p = 3 \times 10^6$ lb/in.² and $\xi = 0.65$. Make other assumptions as necessary.

9.11 Refer to the soil profile shown in Figure P9.11. A drilled pier with a bell is to be constructed. Given: $f_c' = 21,000$ kN/m². The pier has to support a working load, $Q_w = 900$ kN (factor of safety $= 3$).
a. Determine the minimum diameter of the pier shaft required [use Eq. (9.1)]. Make a reasonable assumption for the diameter to be used.
b. With the pier diameter determined in part (a), determine the diameter of the bell needed. Skin friction and point bearing capacity are to be considered.

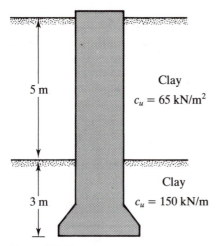

5 m
Clay
$c_u = 65$ kN/m²

3 m
Clay
$c_u = 150$ kN/m

Figure P9.11

9.12 For the pier described in Problem 9.11, estimate the total elastic settlement at working load. Use Eqs. (8.58), (8.60), and (8.61). $E_p = 19.5 \times 10^6$ kN/m² and $\xi = 0.65$. Make other assumptions as necessary.

9.13 A drilled pier in a medium sand is shown in Figure P9.13. Using the method proposed by Touma and Reese, determine the following:

D_s

L_1

L

D_b

Medium sand
γ
ϕ
Relative density $= D_r$

Figure P9.13

a. The net point resistance for a base movement of 25.4 mm
b. The shaft frictional resistance
c. The total load that can be carried by the pier for a total base movement of 25.4 mm

Given: $L = 12$ m $\gamma = 18$ kN/m³
       $L_1 = 11$ m $\phi = 38°$
       $D_s = 1$ m $D_r = 65\%$
       $D_b = 2$ m

9.14 Repeat Problem 9.13 with the following:

       $L = 20$ ft $\gamma = 100$ lb/ft³
       $L_1 = 17$ ft $\phi = 33°$
       $D_s = 4$ ft $D_r = 60\%$
       $D_b = 5.5$ ft

9.15 Assume the pier shown in Figure P9.15 to be a point bearing pier with a working load of 560 kips. Calculate the pier settlement by using Eqs. (8.57) and (8.58). Given $E_p = 3 \times 10^6$ lb/in.², $\mu_s = 0.35$, and $E_s = 5070$ lb/in.²

Silty clay
$\gamma = 110$ lb/ft³
$c_u = 600$ lb/ft²

18 ft

4 ft

Sand
$\gamma = 115$ lb/ft³
$\phi = 39°$
$c = 0$

9 ft

6.5 ft

Figure P9.15

9.16 Repeat Example 9.1 for a working load of $Q_w = 3000$ kN. Make any assumptions necessary.

9.17 Consider the case of the drilled pier socketed into rock, as shown in Figure 9.12. Given: $Q_{all} = 18{,}000$ kN, $E_r/E_c = 0.6$, $\mu_c = \mu_r = 0.35$, $\phi_{rc} = 34°$, $\tau_{all} = 360$ kN/m², allowable point bearing capacity of rock = 2200 kN/m², and f'_c for concrete = 28,000 kN/m². Determine the required length of embedment (L) of the pier in the rock.

9.18 Refer to the drilled pier in Problem 9.17. If the allowable bearing capacity of rock is changed to 3000 kN/m² and other quantities remain the same, what will be the required length of embedment of the pier in the rock?

9.19 Refer to Figure 9.13. Given: $D_s = 1$ m, $D_b = 1.8$ m, and $L = 8$ m. The drilled pier is in a homogeneous sand with $\phi = 38°$ and $\gamma = 18$ kN/m³. Determine the net ultimate uplift capacity (T_{un}) of the pier.

9.20 Repeat Problem 9.19 with $L = 5$ m.

9.21 Refer to Figure 9.13. Given: $D_s = 3$ ft, $D_b = 5$ ft, and $L = 20$ ft. The drilled pier is located in a clay with $c_u = 1480$ lb/ft^2. The unit weight of clay is 115 lb/ft^3. Estimate the net ultimate uplift capacity.

9.22 Repeat Problem 9.21 with $c_u = 700$ lb/ft^2.

9.23 Estimate the ultimate uplift capacity of the drilled pier described in Problem 9.7.

9.24 Estimate the ultimate uplift capacity of the drilled pier described in Problem 9.9.

REFERENCES

Das, B. M. (1980). "A Procedure for Estimation of Ultimate Uplift Capacity of Foundations in Clay," *Soils and Foundations*, The Japanese Society of Soil Mechanics and Foundation Engineering, Vol. 20, No. 1, pp. 77–82.

Das, B. M., and Seeley, G. R. (1975). "Breakout Resistance of Shallow Vertical Anchors," *Journal of the Geotechnical Engineering Division*, American Society of Civil Engineers, Vol. 101, No. GT9, pp. 999–1003.

Goodman, R. E. (1980). *Introduction to Rock Mechanics*, Wiley, New York.

Ladanyi, B. (1977). "Discussion on Friction and Endbearing Tests on Bedrock for High Capacity Socket Design," *Canadian Geotechnical Journal*, Vol. 14, No. 1, pp. 153–156.

Meyerhof, G. G., and Adams, J. I. (1968). "The Ultimate Uplift Capacity of Foundations," *Canadian Geotechnical Journal*, Vol. 5, No. 4, pp. 225–244.

Teng, W. C. (1962). *Foundation Design*, Prentice-Hall, Englewood Cliffs, N.J.

Touma, F. T., and Reese, L. C. (1974). "Behavior of Bored Piles in Sand," *Journal of the Geotechnical Engineering Division*, American Society of Civil Engineers, Vol. 100, No. GT7, pp. 749–761.

Vesic, A. S. (1963). "Bearing Capacity of Deep Foundations in Sand," *Highway Research Record*, No. 39, Highway Research Board, National Academy of Science, Washington, D.C., pp. 112–153.

Vesic, A. S. (1967). "Ultimate Load and Settlement of Deep Foundations in Sand," *Proceedings*, Symposium on Bearing Capacity and Settlement of Foundations, Duke University, Durham, N.C., p. 53.

Vesic, A. S. (1977). "Design of Pile Foundations," *NCHRP No. 42*, Transportation Research Board, National Research Council, Washington, D.C.

Whitaker, T., and Cooke, R. W. (1966). "An Investigation of the Shaft and Base Resistance of Large Bored Piles in London Clay," *Proceedings*, Conference on Large Bored Piles, Institute of Civil Engineers, London, pp. 7–49.

C h a p t e r **10**

FOUNDATIONS
ON DIFFICULT SOILS

10.1   Introduction

In many areas of the United States and other parts of the world, there are soils that make construction of foundations extremely difficult. For example, expansive or collapsible soils may cause high differential movements in structures as the result of excessive heave or settlement. Similar problems can also arise when foundations are constructed over sanitary landfills. Foundation engineers must be able to identify difficult soils when they are encountered in the field. Although it is not possible to solve all the problems caused by all soils, preventive measures can be taken to reduce the possibility of damage to structures built on them.

This chapter is composed of three major parts: *collapsible soils, expansive soils,* and *sanitary landfills.* It outlines the fundamental properties of these soil conditions, along with methods of careful foundation construction.

FOUNDATIONS ON COLLAPSING SOIL

10.2   Definition and Types of Collapsing Soil

Collapsing soils, which are sometimes referred to as *metastable soils,* are unsaturated soils that undergo a large volume change upon saturation. This volume change may or may not be the result of the application of additional load. In order to explain the behavior of collapsing soils under load, one needs to consider the typical nature of void ratio against pressure plot (e against $\log p$) for a collapsing soil, as shown in Figure 10.1. The branch ab is determined from the consolidation test on a specimen at its natural moisture content. At a pressure level of p_w, the equilibrium void ratio is e_1. However, if water is introduced into the specimen for saturation, the soil structure will collapse. After saturation, the equilibrium void ratio at the same pressure

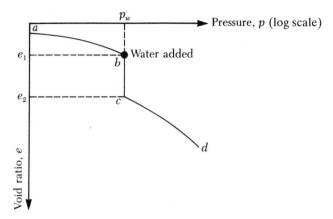

Figure 10.1   Nature of variation of void ratio with pressure for a collapsing soil

level p_w is e_2; cd is the branch of e log p curve under additional load after saturation. Foundations that are constructed on such soils may undergo large and sudden settlement if and when the soil under them becomes saturated with an unanticipated supply of moisture. This moisture may come from several sources, such as (a) broken water pipelines, (b) leaky sewers, (c) drainage from reservoirs and swimming pools, (d) slow increase of ground water table, and so on. This type of settlement generally causes considerable structural damage. Hence, identification of collapsing soils during field exploration is critical.

The majority of naturally occurring collapsing soils are *aeolian*—that is, wind-deposited sand and/or silts, such as loess, aeolic beaches, and volcanic dust deposits. These deposits have high void ratios and low unit weights and are cohesionless or only slightly cohesive. *Loess* is a type of deposit with silt-size particles. The cohesion in loess may be the result of the presence of clay coatings around the silt-size particles, which holds them in a rather stable condition in an unsaturated state. The cohesion may also be caused by the presence of chemical precipitates leached by rain water. When the soil becomes saturated, the clay binders lose their strength and, hence, undergo a structural collapse. Large parts of the midwestern states and arid western United States have such types of deposit. *Loess* deposits are also found over 15–20% of Europe and over large parts of China.

Many collapsing soils may be residual soils that are products of weathering of parent rocks. The weathering process produces soils with a large range of particle-size distribution. Soluble and colloidal materials are leached out by weathering, resulting in large void ratios and thus unstable structures. Many parts of South Africa and Rhodesia have residual soils that are decomposed granites. Sometimes collapsing soil deposits may be left by flash floods and mud flows. These deposits dry out and are very poorly consolidated. An excellent review of collapsing soils is given by Clemence and Finbarr (1981).

10.3   Physical Parameters for Identification

Several investigators have proposed various methods to evaluate the physical parameters of collapsing soils for identification. Some of these are briefly discussed in Table 10.1.

Table 10.1   Reported Criteria for Identification of Collapsing Soil[a]

Investigator	Year	Criteria
Denisov	1951	Coefficient of subsidence: $$K = \dfrac{\text{void ratio at liquid limit}}{\text{natural void ratio}}$$ $K = 0.5\text{–}0.75$: highly collapsible $K = 1.0$: noncollapsible loam $K = 1.5\text{–}2.0$: noncollapsible soils
Clevenger	1958	If dry unit weight is less than 12.56 kN/m^3 ($\approx 80 \text{ lb/ft}^3$), settlement will be large; if dry unit weight is greater than 14.13 kN/m^3 ($\approx 90 \text{ lb/ft}^3$), settlement will be small.
Priklonski	1952	$$K_D = \dfrac{(\text{natural moisture content}) - (\text{plastic limit})}{\text{plasticity index}}$$ $K_D < 0$: highly collapsible soils $K_D > 0.5$: noncollapsible soils $K_D > 1.0$: swelling soils
Gibbs	1961	Collapse ratio, $R = \dfrac{\text{saturation moisture content}}{\text{liquid limit}}$ This was put into graph form.
Soviet Building Code	1962	$L = \dfrac{e_o - e_L}{1 + e_o}$ where e_o = natural void ratio and e_L = void ratio at liquid limit. For natural degree of saturation less than 60%, if $L > -0.1$, it is a collapsing soil.
Feda	1964	$K_L = \dfrac{w_o}{S_r} - \dfrac{PL}{PI}$ where w_o = natural water content, S_r = natural degree of saturation, PL = plastic limit, and PI = plasticity index. For $S_r < 100\%$, if $K_L > 0.85$, it is a subsident soil.
Benites	1968	A dispersion test in which 2 grams of soil are dropped into 12 ml of distilled water and specimen is timed until dispersed; dispersion times of 20 to 30 sec were obtained for collapsing Arizona soils.
Handy	1973	Iowa loess with clay (<0.002 mm) contents: $<16\%$: high probability of collapse $16\text{–}24\%$: probability of collapse $24\text{–}32\%$: less than 50% probability of collapse $>32\%$: usually safe from collapse

[a] Modified after Lutenegger and Saber (1988)

Jennings and Knight (1975) have suggested a procedure to describe the *collapse potential* of a soil. The collapse potential can be determined by taking an undisturbed soil specimen at natural moisture content in a consolidation ring. Step loads are applied to the specimen up to a pressure level of 200 kN/m^2 (≈ 29 lb/in.^{2}). (In Figure 10.1, this is p_w.) At this pressure ($p_w = 200$ kN/m^2), the specimen is flooded for saturation and left for 24 hours. This test will provide the void ratios (e_1 and e_2) before and after flooding. The collapse potential, C_p, can now be calculated as

$$C_p = \Delta\varepsilon = \frac{e_1 - e_2}{1 + e_o} \tag{10.1}$$

where e_o = natural void ratio of the soil
$\Delta\varepsilon$ = vertical strain

The foundation problems associated with a collapsible soil have been correlated with the collapse potential, C_p, by Jennings and Knight (1975). These have been summarized by Clemence and Finbarr (1981) and are given in Table 10.2.

Holtz and Hilf (1961) have suggested that a loessial soil that has a void ratio large enough to allow its moisture content to exceed its liquid limit upon saturation is susceptible to collapse. So, for collapse

$$w_{(saturated)} \geq LL \tag{10.2}$$

However, for saturated soils

$$e_o = wG_s \tag{10.3}$$

where LL = liquid limit
G_s = specific gravity of soil solids

Combining Eqs. (10.2) and (10.3), for collapsing soils

$$e_o \geq (LL)(G_s) \tag{10.4}$$

Table 10.2 Relation of Collapse Potential to the Severity of Foundation Problems[a]

C_p (%)	Severity of problem
0–1	No problem
1–5	Moderate trouble
5–10	Trouble
10–20	Severe trouble
20	Very severe trouble

[a] After Clemence and Finbarr (1981)

The natural dry unit weight, γ_d, of the soil for collapse may be given as

$$\gamma_d \leq \frac{G_s \gamma_w}{1 + e_o} = \frac{G_s \gamma_w}{1 + (LL)(G_s)} \tag{10.5}$$

Assuming an average value of $G_s = 2.65$, the limiting values of γ_d for various liquid limits may now be calculated from Eq. (10.5) as follows:

Liquid limit (%)	Limiting values of γ_d	
	(kN/m³)	(lb/ft³)
10	20.56	130.8
15	18.60	118.3
20	16.99	108.1
25	15.64	99.5
30	14.48	92.1
35	13.49	85.8
40	12.62	80.3
45	11.86	75.4

Figure 10.2 shows a plot of the preceding limiting dry unit weights against the corresponding liquid limits. For any given soil, if the natural dry unit weight falls below the limiting line, the soil is likely to collapse.

Care should be taken to obtain undisturbed samples for determination of the collapse potentials and the dry unit weights just described. It is pref-

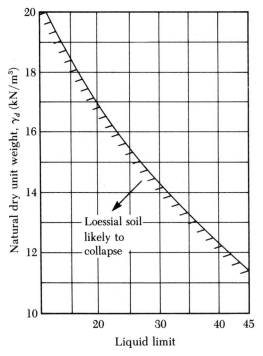

Figure 10.2   Loessial soil likely to collapse

erable to obtain block samples cut by hand. Samples obtained by thin wall tubes may undergo some compression during the sampling process. Note that if this procedure is used, the boreholes should be made *without water*.

10.4   Procedure for Calculation of Collapse Settlement

Jennings and Knight (1975) have proposed the following laboratory procedure to determine the collapse settlement of structures upon saturation of soil.

1. Obtain *two* undisturbed soil specimens for tests in a standard consolidation test apparatus (oedometer).

2. Place the two specimens under 1 kN/m^2 pressure for 24 hours.

3. After 24 hours, saturate one specimen by flooding. The other specimen is kept at natural moisture content.

4. After 24 hours of flooding, resume the consolidation test for both specimens by doubling the load (same procedure as the standard consolidation test) to the desired pressure level.

5. Plot the *e*-log *p* graphs for both of the specimens (Figure 10.3a and b).

6. Calculate the *in situ* effective pressure, p_o. Draw a vertical line corresponding to the pressure p_o.

7. From the *e*-log *p* curve of the soaked sample, determine the preconsolidation pressure, p_c. If $p_c/p_o = 0.8$–1.5, the soil is normally consolidated; however, if $p_c/p_o > 1.5$, it is preconsolidated.

8. Determine e'_o, corresponding to p_o from the *e*-log *p* curve of the soaked sample. (This procedure for normally consolidated and overconsolidated soils is shown in Figures 10.3a and b, respectively.)

9. Through the point p_o, e'_o, draw a curve that is similar to the *e*-log *p* curve obtained from the specimen tested at natural moisture content.

10. Determine the incremental pressure, Δp, on the soil caused by the construction of the foundation. Draw a vertical line corresponding to the pressure of $p_o + \Delta p$ in the *e*-log *p* curve.

11. Now, determine Δe_1 and Δe_2. The settlement of soil without change in the natural moisture content is

$$S_1 = \frac{\Delta e_1}{1 + e'_o} (H) \tag{10.6}$$

Also, the settlement caused by collapse in the soil structure

$$S_2 = \frac{\Delta e_2}{1 + e'_o} (H) \tag{10.7}$$

where H = thickness of soil susceptible to collapse

(a)

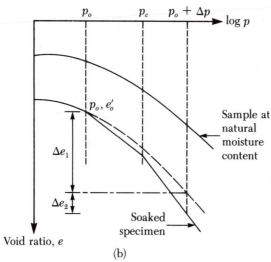

(b)

Figure 10.3 Settlement calculation from double oedometer test: (a) normally consolidated soil; (b) overconsolidated soil

10.5 **Foundation Design in Soils Not Susceptible to Wetting**

For actual foundation design purposes, some standard field load tests may also be conducted. Figure 10.4 shows the results of some field load tests in loess deposits in Nebraska and Iowa. Note that the load-settlement relationships are essentially linear up to a certain critical pressure, p_{cr}, at which there is a breakdown of the soil structure and, hence, a large settlement.

Figure 10.4 Results of standard load test on loess deposits in Iowa and Nebraska (adapted from *Foundation Engineering*, Second Edition, by R. B. Peck, W. E. Hanson, and T. H. Thornburn. Copyright 1974 by John Wiley and Sons. Reprinted by permission.)

Sudden breakdown of soil structure is more common with soils having a high natural moisture content.

If enough precautions are taken in the field to prevent moisture from increasing under structures, spread foundations and raft foundations may be built on potentially collapsible soils. However, the foundations must be proportioned in such a manner that the critical stresses (Figure 10.4) in the field are never exceeded. A factor of safety of about 2.5 to 3 should be used to calculate the allowable soil pressure; or

$$p_{all} = \frac{p_{cr}}{FS} \tag{10.8}$$

where p_{all} = allowable soil pressure
       FS = factor of safety (about 2.5 to 3)

The differential and total settlements of these foundations should be similar to those of foundations designed for sandy soils.

In several instances, continuous foundations may be safer over collapsible soils than isolated foundations, in that they can effectively minimize differential settlement. A typical procedure for construction of continuous foundations is shown in Figure 10.5. This procedure uses footing beams and longitudinal load-bearing beams.

In the construction of heavy structures, such as grain elevators, over collapsible soils, settlements up to about 0.3 m ($\approx$1 ft) are sometimes

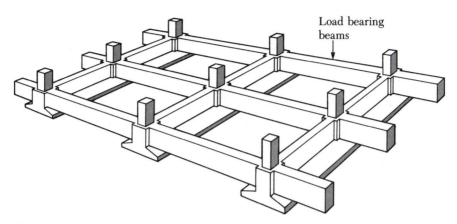

Figure 10.5 Continuous foundation with load-bearing beams (after Clemence and Finbarr, 1981)

allowed (Peck, Hanson, and Thornburn, 1974). In this case, it is assumed that tilting of the foundation is not likely to occur; that is, there is no eccentric loading. The total expected settlement for such structures can be estimated from standard consolidation tests on samples at field moisture contents. If eccentric loading is avoided, the foundations will exhibit uniform settlement over loessial deposits; however, if the soil is of residual or colluvial nature, settlement may not be uniform. This is due to the non-uniformity that is generally encountered in residual soil profiles.

Extreme caution must be used in building heavy structures over collapsible soils. If large settlements are expected, drilled-pier and pile foundations may be considered. These foundations can transfer the load to a stronger load-bearing stratum.

10.6   Foundation Design in Soils Susceptible to Wetting

If it is suspected that the upper layer of soil may get wet and collapse at some time after construction of the foundation, several design techniques may be considered.

1. If the expected depth of wetting is about 1.5–2 m ($\approx$5–6.5 ft) from the ground surface, the soil may be moistened and recompacted by heavy rollers. Spread footings and rafts may be constructed over the compacted soil. An alternative to recompaction by heavy rollers is *heavy tamping*, which is sometimes referred to as *dynamic consolidation*. This consists primarily of dropping a heavy weight repeatedly on the ground. The height of the hammer drop can vary from 7.5–30.5 m ($\approx$25–100 ft). The stress waves generated by the hammer drop help in the densification of the soil.

2. If conditions are favorable, foundation trenches can be flooded with solutions of sodium silicate and calcium chloride. This will chemically stabilize the soil. The soil will behave like a soft sandstone and resist collapse upon saturation. This method is successful only if the solutions can penetrate to the desired depth; thus, it is most applicable to fine sand deposits. Silicates are rather costly and are not generally used. However, in some parts of Denver, silicates have been used very successfully.

3. When the soil layer is susceptible to wetting up to a depth of about 10 m, several techniques may be used to cause collapse of the soil *before* foundation construction. Two of these are *vibroflotation* and *ponding* (also called *flooding*). Vibroflotation is used successfully in free-draining soil (see Chapter 12). The procedure of ponding—by construction of low dikes—is employed at sites that have no impervious layers. Note that, even after saturation and collapse of the soil by ponding, some additional settlement of the soil may occur after the placement of the foundation. Additional settlement may also be caused by incomplete saturation of the soil at the time of foundation construction. Ponding may be used successfully in the construction of earth dams.

4. If precollapsing of soil is not practical, foundations may be constructed beyond the zone of possible wetting. This may require drilled piers and piles. In the design of drilled piers and piles, one must consider the effect of negative skin friction, which is the result of the collapse of the soil structure and the associated settlement of the zone of subsequent wetting.

In some cases, a *rock column type of foundation (vibroreplacement)* may also be considered. Rock columns are built with large boulders that penetrate the potentially collapsible soil layer. They act as piles in transferring the load to a more stable soil layer (also see Section 12.8 on stone columns).

FOUNDATIONS ON EXPANSIVE SOIL

10.7   Expansive Soils—General

There are many plastic clays that swell considerably when water is added to them and then shrink with the loss of water. Foundations constructed on these clays are subjected to large uplifting forces caused by the swelling. These forces will induce heaving, cracking, and breakup of both building foundations and slab-on-grade members. Expansive clays cover large parts of the United States, South America, Africa, Australia, and India. In the United States, these clays are predominant in Texas, Oklahoma, and the upper Missouri Valley. In general, potentially expansive clays have liquid limits and plasticity indices greater than about 40 and 15, respectively.

As noted, an increase in moisture content causes clay to swell. The depth in a soil profile up to which periodic changes of moisture occur is usually referred to as the *active zone*. The depth of the active zone will vary depending on the location of the soil profile. Some typical active-zone depths in American cities are given in Table 10.3. In some clays and clay

Table 10.3 Typical Active-Zone
Depths in Some U.S. Cities[a]

	Depth of active zone	
City	(m)	(ft)
Houston	1.5 to 3	5 to 10
Dallas	2.1 to 4.6	7 to 15
San Antonio	3 to 9	10 to 20
Denver	3 to 4.6	10 to 15

[a] After O'Neill and Poormoayed (1980)

shales in the western United States, the depth of the active zone can be as much as 15 m ($\approx$50 ft). The active-zone depth can be easily determined by plotting the liquidity index against the depth of the soil profile over several seasons. Figure 10.6 shows an example of this from the Beaumont formation in the Houston area.

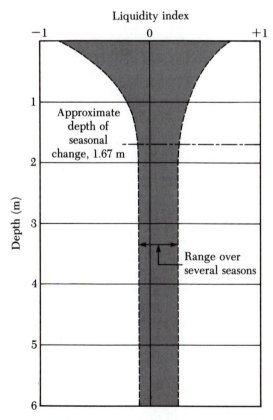

Figure 10.6 Active zone in Houston area—Beaumont formation (after O'Neill and Poormoayed, 1980)

10.8   Laboratory Measurement of Swell

To study the magnitude of possible swell in a clay, simple laboratory oedometer tests can be conducted on undisturbed specimens. Two common tests are the *unrestrained swell test* and *swelling pressure test*.

In the *unrestrained swell test*, the specimen is placed in an oedometer under a small surcharge of about 6.9 kN/m² (1 lb/in.²). Water is then added to the sample, and the expansion of the volume of the specimen (that is, height; the area of cross section is constant) is measured until equilibrium is reached. The percent of free swell may be expressed as a ratio

$$s_{w(free)}(\%) = \frac{\Delta H}{H}(100)$$

(10.9)

where $s_{w(free)}$ = free swell
ΔH = height of swell due to saturation
H = original height of the specimen

Vijayvergiya and Ghazzaly (1973) have analyzed a number of soil test results conducted in this manner and have prepared a correlation chart of the free swell, liquid limit, and natural moisture content (see Figure 10.7). The free surface swell can be calculated by use of this chart (O'Neill and Poormoayed, 1980):

$$\Delta S_F = 0.0033 Z s_{w(free)}$$

(10.10)

where ΔS_F = free surface swell
Z = depth of active zone
$s_{w(free)}$ = free swell, as a percent (Figure 10.7)

More recently, a new test method for obtaining a *modified free swell index* for clays has been suggested by Sivapullaiah, Sitharam, and Rao (1987), which appears to give a better indication for the swelling potential of clayey soils. This test consists of obtaining an oven-dried soil with a mass of about 10 grams. The soil mass is well pulverized and transferred into a 100-ml graduated jar containing distilled water. After 24 hours, the swollen sediment volume is measured. The *modified free swell index* is then calculated as

$$\text{Modified free swell index} = \frac{V - V_s}{V_s}$$

(10.11)

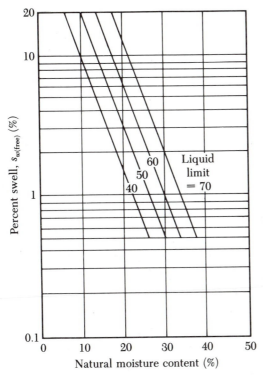

Figure 10.7 Relation between percent free swell, liquid limit, and natural moisture content (after Vijayvergiya and Ghazzaly, 1973)

where V = soil volume after swelling

$$V_s = \text{volume of soil solid} = \frac{W_s}{G_s \gamma_w}$$

W_s = weight of oven-dried soil
G_s = specific gravity of soil solids
γ_w = unit weight of water

Based on the modified free swell index, the swelling potential of a soil can be qualitatively classified as follows:

Modified free swell index	Swelling potential
<2.5	Negligible
2.5 to 10	Moderate
10 to 20	High
>20	Very high

The *swelling pressure test* can be conducted by taking a specimen in a consolidation ring and applying a pressure equal to the effective overburden pressure (p_o) plus the approximate anticipated surcharge caused by the foundation (p_s). Water is then added to the specimen. As the specimen starts

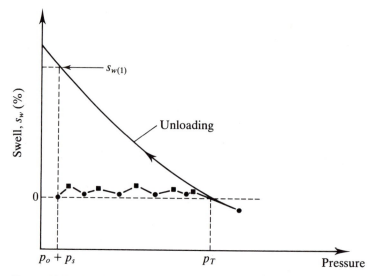

Figure 10.8 Swelling pressure test

to swell, pressure is applied in small increments to prevent swelling. This is continued until full swelling pressure is developed. At that time, the total pressure on the specimen is

$$p_T = p_o + p_s + p_1 \tag{10.12}$$

where p_T = total pressure to prevent swelling, or zero swell pressure
$\quad\quad\; p_1$ = additional pressure added to prevent swelling after addition of water

Figure 10.8 shows the nature of variation of the percentage of swell against pressure during a swelling pressure test. For more information on this type of test, readers are referred to a paper by Sridharan, Rao, and Sivapullaiah (1986).

A value of $p_T \approx$ 20–30 kN/m^2 (0.4–0.65 kips/ft^2) is considered to be low, and a value of 1500–2000 kN/m^2 (30–40 kips/ft^2) is considered to be very high. After zero swell pressure is attained, the soil specimen can be unloaded in steps up to the level of the overburden pressure, p_o. This unloading process will cause the specimen to swell. The equilibrium swell for each pressure level is also recorded. The nature of variation of the swell, in percent ($s_w\%$), and the applied pressure on the specimen will be like that shown in Figure 10.8.

The swelling pressure test can be used to determine the surface heave, ΔS, for a given foundation as (O'Neill and Poormoayed, 1980)

$$\Delta S = \sum_{i=1}^{n} [s_{w(1)}\%](H_i)(0.01) \tag{10.13}$$

where $s_{w(1)}\% =$ swell, in percent, for layer i under a pressure of $p_o + p_s$ (see
   Figure 10.8)
   $\Delta H_i =$ thickness of layer i

The procedure of surface heave calculation using the preceding equation is
shown in Example 10.2.

Example 10.1

A soil profile has an active zone of expansive soil of 2.0 m. The liquid limit and the
average natural moisture content during the construction season are 50% and 20%,
respectively. Determine the free surface swell.

Solution

From Figure 10.7, for $LL = 50\%$ and $w = 20\%$, $s_{w(free)} = 3\%$. From Eq. (10.10)

$$\Delta S_F = 0.0033 Z s_{w(free)}$$

Hence

$$\Delta S_F = 0.0033(2)(3)(1000) = \textbf{19.8 mm}$$

Example 10.2

A soil profile's active-zone depth is 3.5 m. If a foundation is to be placed at a depth
of 0.5 m below the ground surface, what would be the estimated total swell? Given
the following from laboratory tests:

Depth, in	Swell under overburden and estimated foundation surcharge pressure, $s_{w(1)}$ (%)
0.5	2
1	1.5
2	0.75
3	0.25

Solution

The values of $s_{w(1)}$ (%) have been plotted with depth in Figure 10.9a. The area of
this diagram will be the total swell. Using the trapezoidal rule

$$\Delta S = \frac{1}{100} \left[\frac{1}{2} (1)(0 + 0.5) + \frac{1}{2} (1)(0.5 + 1.1) + \frac{1}{2} (1)(1.1 + 2) \right]$$

$$= 0.026 \text{ m} = \textbf{26 mm}$$

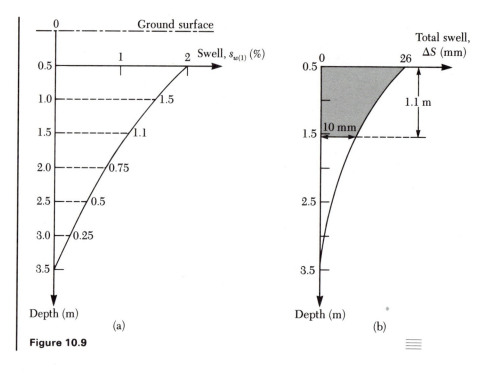

Figure 10.9

Example 10.3

In Example 10.2, if the allowable total swell is 10 mm, what would be the undercut necessary to reduce the total swell?

Solution

Using the procedure outlined in Example 10.2, the total swell at various depths below the foundation can be calculated as follows (from Figure 10.9a):

Depth (m)	Total swell, ΔS (mm)
3.5	0
3	$0 + \left[\frac{1}{2}(0.5)(0.25)\right]\frac{1}{100} = 0.000625$ m $= 0.625$ mm
2.5	$0.000625 + \frac{1}{100}\left[\frac{1}{2}(0.5)(0.25 + 0.5)\right] = 0.0025$ m $= 2.5$ mm
1.5	$0.0025 + \frac{1}{100}\left[\frac{1}{2}(1)(0.5 + 1.1)\right] = 0.0105$ m $= 10.5$ mm
0.5	26 mm

These total settlements have been plotted in Figure 10.9b, and they show that a total swell of 10 mm corresponds to a depth of 1.6 m below the ground surface.

Hence, the undercut below the foundation is **1.6 − 0.5 = 1.1 m.**

This soil should be excavated, replaced by nonswelling soil, and recompacted.

10.9   Classification of Expansive Soil Based on Index Tests

There are several classification systems for expansive soils based on the problems they can create in the construction of foundations. The classification system proposed by the U.S. Army Waterways Experiment Station (Snethen, Johnson, and Patrick, 1977) is the most recent one. It has also been summarized by O'Neill and Poormoayed (1980); see Table 10.4.

Table 10.4   Expansive Soil Classification System[a]

Liquid limit	Plasticity index	Potential swell (%)	Potential swell classification
<50	<25	<0.5	Low
50–60	25–35	0.5–1.5	Marginal
>60	>35	>1.5	High
Potential swell = vertical swell under a pressure equal to overburden pressure			

[a] Compiled from O'Neill and Poormoayed (1980)

10.10   Foundation Considerations on Expansive Soils

If a soil is classified as having a low swell potential, standard construction practices may be followed. However, if the soil possesses a marginal or high swell potential, precautions need to be taken. This may entail the following:

1. Replacing the expansive soil under the foundation

2. Changing the nature of the expansive soil by such measures as compaction control, prewetting, installation of moisture barriers, and chemical stabilization

3. Strengthening the structures to withstand heave, constructing structures that are flexible enough to withstand the differential soil heave without failure, or constructing isolated deep foundations below the depth of the active zone

One particular method may not be the cure-all in every situation. It may become necessary to combine several techniques, and local construction experience should always be taken into consideration. Following are details about some of the commonly used techniques of dealing with expansive soils.

Replacement of Expansive Soil

When moderately expansive soils of low thickness are present at the surface, they can be removed and replaced by less expansive soils and then compacted properly.

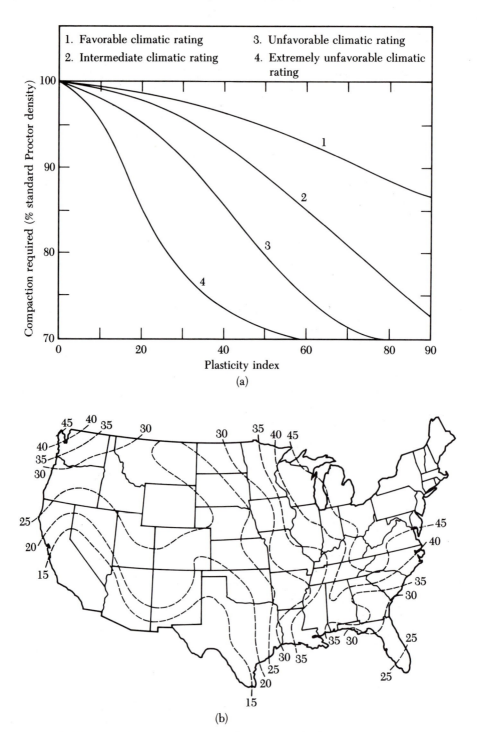

Figure 10.10   (a) Soil compaction requirement based on climatic rating; (b) equivalent climatic rating of the United States (after Gromko, 1974)

Changing the Nature of Expansive Soil

1. *Compaction:* Heave of expansive soils is known to decrease substantially when the soil is compacted to a lower unit weight on the high side of the optimum moisture content (possibly 3–4% above the optimum moisture content). Even under such conditions, a slab-on-ground type of construction should not be encouraged where the total probable heave is expected to be about 35 mm or more. Figure 10.10a shows the recommended limits of soil compaction in the field for reduction of heave. Note that the recommended dry unit weights in Figure 10.10a are based on climatic ratings. According to U.S. Weather Bureau data, a climatic rating of 15 represents an extremely unfavorable climatic condition; a value of 45 represents a favorable climatic rating. The isobars of climatic rating for the continental United States are shown in Figure 10.10b.

2. *Prewetting:* This is a technique for increasing the moisture content of the soil by ponding and, hence, achieving most of the heave before construction. However, this technique may be time consuming because the seepage of water through highly plastic clays is slow. After ponding, 4–5% of hydrated lime may be added to the top layer of the soil to make it less plastic and more workable (Gromko, 1974).

3. *Installation of Moisture Barriers:* The long-term effect of the differential heave can be reduced by controlling the moisture variation in the soil. This can be achieved by providing vertical moisture barriers about 1.5 m ($\approx$5 ft) deep around the perimeter of slabs for the slab-on-grade type of construction. These moisture barriers may be constructed in trenches filled with gravel, lean concrete, or impervious membranes.

4. *Stabilization of Soil:* Chemical stabilization with the aid of lime and cement has often proved useful. A mix of about 5% lime is sufficient in

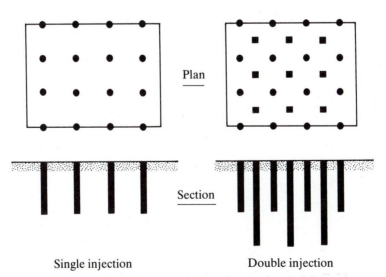

Single injection   Double injection

Figure 10.11 Multiple lime slurry injection planning for a building pad

Figure 10.12 Pressure injection of lime slurry for a building pad (Courtesy of GKN Hayward Baker Inc., Woodbine Division, Ft. Worth, Texas)

most cases. Lime or cement and water are mixed with the top layer of soil and compacted. The addition of lime or cement will decrease the liquid limit, the plasticity index, and the swell characteristics of the soil. This type of stabilization work can be done to a depth of about 1–1.5 m ($\approx$3–5 ft). Hydrated high-calcium lime and dolomite lime are generally used for lime stabilization.

Another method of stabilization of expansive soil is the *pressure injection* of lime slurry or lime–fly ash slurry into the soil, usually up to a depth of

Figure 10.13 Slope stabilization of a canal bank by pressure injection of lime–fly ash slurry (Courtesy of GKN Hayward Baker Inc., Woodbine Division, Ft. Worth, Texas)

4–5 m (12–16 ft) and occasionally deeper to cover the active zone. Further details of the pressure injection technique are described in Section 12.6. Depending on the soil conditions at a given site, single or multiple injections can be planned, as shown in Figure 10.11. Figure 10.12 shows the slurry pressure injection work for a building pad. The stakes marked are the planned injection points. Figure 10.13 shows the lime–fly ash stabilization by pressure injection of the bank of a canal that had experienced sloughs and slides.

10.11 Construction on Expansive Soils

One must be careful in choosing the type of foundation to be used on expansive soils. Table 10.5 shows some recommended construction procedures based on the total predicted heave (ΔS) and the length-to-height ratio of the wall panels.

For example, Table 10.5 proposes the use of waffle slabs as an alternative in designing rigid buildings capable of tolerating movement. A schematic diagram of a waffle slab is shown in Figure 10.14. In this type of construction, the ribs hold the structural load. The waffle voids allow the expansion of soil.

Table 10.5 also suggests the use of foundation piers with a suspended floor slab for the construction of structures independent of movement. Figure 10.15a shows a schematic diagram of such an arrangement. The bottom of the piers should be placed below the active zone of the expansive soil. For the design of the piers, the uplifting force, U, may be estimated (Figure 10.15b) from the equation

$$U = \pi D_s Z p_t \tan \phi_{ps} \tag{10.14}$$

where D_s = diameter of the shaft of the pier
$\quad Z$ = depth of the active zone
$\quad \phi_{ps}$ = effective angle of plinth-soil friction
$\quad p_T$ = pressure for zero horizontal swell (see Figure 10.8;
$\qquad p_T = p_o + p_s + p_1$)

In most cases, the value of ϕ_{ps} varies between 10° and 20°. An average value of the zero horizontal swell pressure must be determined in the laboratory.

Figure 10.14   Waffle slab

Table 10.5 Construction Procedures on Expansive Clay Soils[a]

Total predicted heave (mm)		Recommended construction	Method	Remarks
L/H = 1.25	L/H = 2.5			
0 to 6.35	12.7	No precaution		
6.35 to 12.7	12.7 to 50.8	Rigid building tolerating movement (steel reinforcement as necessary)	*Foundations:* Pads, Strip footings, Raft (waffle) *Floor slabs:* Waffle, Tile *Walls:*	Footings should be small and deep, consistent with the soil-bearing capacity. Rafts should resist bending. Slabs should be designed to resist bending and should be independent of grade beams. Walls on a raft should be as flexible as the raft. No rigid connections vertically. Brick works should be strengthened with tie bars or bands.
12.7 to 50.8	50.8 to 101.6	Building damping movement	*Joints:* Clear, Flexible *Walls:* Flexible, Unit construction, Steel frame *Foundations:* Three point, Cellular, Jacks	Contacts between structural units should be avoided; or flexible, waterproof material may be inserted in the joints. Walls or rectangular building units should heave as a unit. Cellular foundations allow slight soil expansion to reduce swelling pressure. Adjustable jacks can be inconvenient to owners. Three-point loading allows motion without duress.
>50.8	>101.6	Building independent of movement	*Foundation piers:* Straight shaft, Bell bottom *Suspended floor:*	Smallest-diameter and widely spaced piers compatible with load should be placed. Clearance should be allowed under grade beams. Floor should be suspended on grade beams 0.3 to 0.45 m above the soil.

[a] After Gromko, 1974

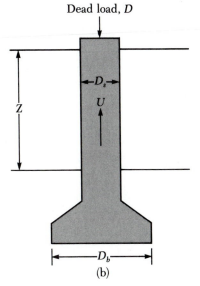

Figure 10.15 (a) Construction of belled piers and grade beam; (b) definition of parameter in Eq. (12.14)

In the absence of laboratory results, the value of $p_t \tan \phi_{ps}$ may be taken to be equal to the undrained shear strength of clay (c_u) in the active zone.

The belled portion of the drilled pier will act as an anchor to resist the uplifting force. Ignoring the weight of the pier

$$Q_{net} = U - D \qquad (10.15)$$

where Q_{net} = net uplift load
D = dead load

Now

$$Q_{net} \approx \frac{c_u N_c}{FS} \cdot \frac{\pi}{4} (D_b^2 - D_s^2) \tag{10.16}$$

where c_u = undrained cohesion of the clay in which the bell of the pier is located

Combining Eqs. (10.15) and (10.16)

$$U - D = \frac{c_u N_c}{FS} \cdot \frac{\pi}{4} (D_b^2 - D_s^2) \tag{10.17}$$

where N_c = bearing capacity factor
FS = factor of safety
D_b = diameter of the bell of the pier

Conservatively, the value of N_c can be taken as follows (Tables 3.2 and 3.4):

$$N_c \approx N_{c(strip)} F_{cs} = N_{c(strip)} \left(1 + \frac{N_q B}{N_c L} \right) \approx 5.14 \left(1 + \frac{1}{5.14} \right) = 6.14$$

An example of a drilled-pier design is given in Example 10.4.

Example 10.4

Figure 10.16 shows a drilled pier. The depth of the active zone is 5 m. The zero swell pressure of the swelling clay (p_T) is 500 kN/m². For the pier:

    Dead load = 600 kN
    Live load = 300 kN
    Dead load + live load = 900 kN

Figure 10.16

    a. Determine the diameter of the bell, D_b.
    b. Determine the reinforcement required for the pier shaft.
    c. Check the bearing capacity of the pier, assuming that the uplift force is zero.

Solution

Part a: Determination of the Bell Diameter, D_b

The uplifting force is [Eq. (10.14)]

$$U = \pi D_s Z p_T \tan \phi_{ps}$$

Assume that $\phi_{ps} \approx 12°$, $Z = 5$ m, and $p_T = 500$ kN/m². So

$$U = \pi(0.75)(5)(500) \tan 12° \approx 1252 \text{ kN}$$

Assume dead load and live load to be zero, and FS in Eq. (10.17) to be 1.25. So, from Eq. (10.17)

$$U = \frac{c_u N_c}{FS} \cdot \frac{\pi}{4}(D_b^2 - D_s^2)$$

$$1252 = \frac{(450)(6.14)}{1.25} \cdot \frac{\pi}{4}(D_b^2 - 0.75^2)$$

$$D_b^2 \approx 1.28$$

or

$$D_b \approx 1.13. \text{ So use } \mathbf{1.25 \text{ m}.}$$

The factor of safety against uplift with the dead load should also be checked. A factor of safety of at least 3 is desirable. So, from Eq. (10.17)

$$FS = \frac{c_u N_c \left(\dfrac{\pi}{4}\right)(D_b^2 - D_s^2)}{U - D}$$

$$U - D = 1252 - 600 = 652 \text{ kN}$$

So

$$FS = \frac{c_u N_c \left(\dfrac{\pi}{4}\right)(D_b^2 - D_s^2)}{U - D} = \frac{(450)(6.14)\left(\dfrac{\pi}{4}\right)(1.25^2 - 0.75^2)}{652}$$

$$= \mathbf{3.33 > 3 - O.K.}$$

Part b: Reinforcement

The reinforcement should be provided for the total uplift load—that is, 1252 kN (assuming that dead load and live load are zero). So, the area of steel

$$A_s = \frac{U}{\left(\dfrac{\text{yield stress of steel}}{\text{factor of safety}}\right)}$$

The preceding factor of safety may be taken as 1.25. The yield stress of steel ≈ 275 MN/m² ($\approx 40,000$ lb/in.²). Hence, the area of steel

$$A_s = \frac{1252}{\left(\dfrac{275 \times 10^3}{1.25}\right)} = 5.69 \times 10^{-3} \text{ m}^2 = 5.69 \times 10^3 \text{ mm}^2 \; (8.82 \text{ in.}^2)$$

Part c: Check for Bearing Capacity

Assume that $U = 0$.

$$\text{Dead load} + \text{live load} = 600 + 300 = 900 \text{ kN}$$

$$\text{Downward load per unit area} = \frac{900}{\left(\dfrac{\pi}{4}\right)(D_b^2)} = \frac{900}{\left(\dfrac{\pi}{4}\right)(1.25)^2} = 733.4 \text{ kN/m}^2$$

Net bearing capacity of the soil under the bell $= q_u =$

$$c_u N_c = 450(6.14) = 2763 \text{ kN/m}^2$$

Hence, factor of safety against bearing capacity failure

$$= \frac{2763}{733.4} = 3.77 > 3 \text{---O.K.}$$

SANITARY LANDFILL

10.12 Sanitary Landfills—General

Sanitary landfills provide a way to dispose of refuse on land without causing danger to public health. Almost all countries use sanitary landfills with varying degrees of success. The refuse disposed in sanitary landfills can contain materials like wood, paper, and fibrous wastes or demolition wastes like bricks and stones. The refuse is dumped and compacted at frequent intervals and then covered with a layer of soil (Figure 10.17). In the com-

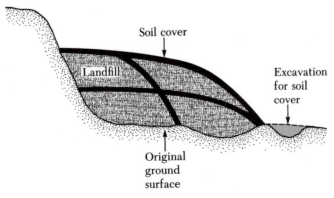

Figure 10.17   Schematic diagram of sanitary landfill in progress

pacted state, the average unit weight of the refuse may vary between 5–10 kN/m³ (32–64 lb/ft³). A typical city in the United States, with a population of one million, can generate about 3.8×10^6 m³ ($\approx 134 \times 10^6$ ft³) of compacted landfill material per year.

As property value continues to increase in densely populated areas, it becomes more and more tempting to construct structures over sanitary landfills. In some instances, a visual site inspection may not be enough to detect an old sanitary landfill. However, construction of foundations over sanitary landfills is generally problematic because of poisonous gases such as methane, excessive settlement, and low inherent bearing capacity. Based on the information available, the next two sections will discuss settlement and bearing capacity problems associated with sanitary landfills.

10.13 Settlement of Sanitary Landfills

Sanitary landfills undergo large continuous settlements over a long period of time. Yen and Scanlon (1975) have documented the settlement of several landfill sites in California. The settlement rate after the completion of the landfill (Figure 10.18) can be expressed as

$$m = \frac{\Delta H_f}{\Delta t} \tag{10.18}$$

where m = settlement rate
$\quad H_f$ = maximum height of the sanitary landfill

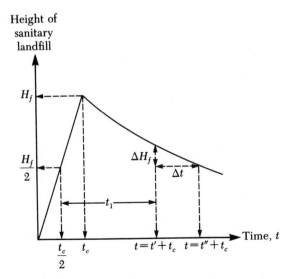

Figure 10.18 Settlement of sanitary landfills

Based on several field observations, the following empirical correlations for the settlement rate have been determined (Yen and Scanlon, 1975).

$$m = 0.0268 - 0.0116 \log t_1 \qquad \text{(for fill heights ranging from 12–24 m)} \qquad (10.19)$$

$$m = 0.038 - 0.0155 \log t_1 \qquad \text{(for fill heights ranging from 24–30 m)} \qquad (10.20)$$

$$m = 0.0433 - 0.0183 \log t_1 \qquad \text{(for fill heights greater than 30 m)} \qquad (10.21)$$

where m is in m/month and t_1 is the median fill age, in months. The medium fill age can be defined as follows (Figure 10.18):

$$t_1 = t - \frac{t_c}{2} \qquad (10.22)$$

where t = time from the beginning of landfill
t_c = time for completion of the landfill

Equations (10.19), (10.20), and (10.21) were based on field data from landfills for which the value of t_c varied from 70 to 82 months. To get an idea of the approximate length of time required for a sanitary landfill to undergo complete settlement, consider Eq. (10.19). For a fill of height of 12 m and t_c = 72 months

$$m = 0.0268 - 0.0116 \log t_1$$

or

$$\log t_1 = \frac{0.0268 - m}{0.0116}$$

If $m = 0$ (that is, for zero settlement rate), the $\log t_1 = 2.31$, or $t_1 \approx 200$ months. Thus, the settlement of the fill will continue for a period of $t_1 - t_c/2 = 200 - 36 = 164$ months (≈ 14 years) after its completion. This is a fairly long time. This calculation shows that one must pay close attention to the settlement of foundations constructed on sanitary landfills.

A comparison of Eqs. (10.19) to (10.21) for rates of settlement shows that the value of m increases with the increase of the height of the fill. However, for fill heights greater than about 30 m, the rate of settlement should not be much different from that from Eq. (10.21). This is because decomposition of the organic matter close to the surface is mainly the result of an anaerobic environment. For deeper fills, the decomposition is slower. Hence, for fill heights greater than about 30 m, the rate of settlement does not accelerate any faster than those that are about 30 m in height.

Sowers (1973) has also proposed a relation for calculation of the settlement of a sanitary landfill:

$$\boxed{\Delta H = \frac{\alpha H_f}{1 + e} \log \left(\frac{t''}{t'} \right)} \qquad (10.23)$$

where H_f = height of the fill

$\quad\quad e$ = void ratio

$\quad\quad \alpha$ = a coefficient for settlement

$\quad t''$, t' = times (see Figure 10.18)

$\quad\quad \Delta H$ = settlement between times t' and t''

The coefficients α are between the following limits:

$$\alpha = 0.09e \quad\quad \text{(for conditions favorable to decomposition)} \quad\quad (10.24)$$

and

$$\alpha = 0.03e \quad\quad \text{(for conditions unfavorable to decomposition)} \quad (10.25)$$

Equation (10.23) is similar to the equation for secondary consolidation settlement.

10.14   Bearing Capacity of Foundations on Sanitary Landfills

Shallow foundations constructed on sanitary landfills with a compacted soil cover may fail in two ways (Sowers, 1968): by punching shear, or by rotational shear, as shown in Figure 10.19. Punching shear failure (Figure

Figure 10.19   Nature of bearing capacity failure and settlement of shallow foundations on sanitary landfills (redrawn after Sowers, 1968)

10.19a) occurs when the width of the foundation, B, is relatively small compared with the thickness of the soil cover, D_C. However, when the thickness of the soil cover is relatively small compared with the foundation width and when the strength of the soil cover is low, rotational shear failure may occur (Figure 10.19b). The allowable bearing capacity of shallow foundations for light residential or office buildings over sanitary landfills should not be greater than about 20–40 kN/m^2 ($\approx$ 400–800 lb/ft^2).

Sometimes the allowable bearing capacity of shallow foundations can be increased by increasing the thickness of the compacted soil cover such that $D_C \geq 1.5$–$2B$. Note, however, that this excess weight of the fill and the compaction process may increase the ultimate settlement of the structure.

When the fill is relatively homogeneous and light structures are constructed over it, the settlement pattern will be somewhat like that in Figure 10.19c. However, when the fill is nonhomogeneous in nature with irregular hard zones, such as boulders, the settlement is nonuniform, as shown in Figure 10.19d. Serious damage to structures is usually caused by this type of settlement.

In several instances, the use of continuous foundations, such as that shown in Figure 10.5, may help reduce the differential settlement problem. If the estimated settlement of a structure is not tolerable, or if heavier structures are to be built, pile or drilled-pier foundations are alternatives. In this case, it will be necessary to use noncorroding materials, because sanitary fills with moisture may corrode metal piles and may also damage concrete. If pile or pier foundations are used, the floor slab should be poured independently of the grade beams to avoid cracking due to differential settlement; otherwise, a structural slab will be necessary.

PROBLEMS

10.1 Refer to Figure 10.2, which is based on Eq. (10.5) and $G_s = 2.65$. Draw a similar curve of γ_d (in lb/ft^3) against liquid limit with $G_s = 2.74$, and show the zone in which the loessial soils are likely to collapse on saturation.

10.2 A natural loessial soil deposit has a dry unit weight of 14.5 kN/m^3. The liquid limit of the soil is 27. Is collapse likely to occur in this soil? [Use Eq. (10.5).] Assume that $G_s = 2.66$.

10.3 A collapsible soil layer in the field has a thickness of 3 m. The average effective overburden pressure on the soil layer is 52 kN/m^2. An undisturbed specimen of this soil was subjected to a double oedometer test (refer to Figure 10.3). The preconsolidation pressure of the specimen as determined from the soaked specimen was 84 kN/m^2. Is the soil in the field normally consolidated or preconsolidated?

10.4 An expansive soil has an active-zone thickness of 8 m. The natural moisture content of the soil is 20%, and its liquid limit is 50. Calculate the free surface swell of the expansive soil upon saturation.

10.5 Repeat Problem 10.4 assuming that the liquid limit of the soil is 60. All other quantities are the same.

10.6 The following are the results of a modified free swell index test:

> Mass of dry soil = 10 grams
> $G_s = 2.71$
> Volume of swollen sediment after 24 hours in water = 26.2 cm^3

a. Determine the modified free swell index.
b. Describe the swell potential.

10.7 An expansive soil profile has an active-zone thickness of 5.2 m. A shallow foundation is to be constructed at a depth of 1.2 m below the ground surface. Based on a swelling pressure test, the following are given:

Depth from ground surface (m)	Swell under overburden and estimated foundation surcharge pressure, $S_{w(1)}$ (%)
1.2	3.0
2.2	2.0
3.2	1.2
4.2	0.55
5.2	0.0

Estimate the total possible swell under the foundation.

10.8 Repeat Problem 10.7 with the following (active zone thickness = 25 ft; depth of shallow foundation = 2.5 ft):

Depth from ground surface (ft)	Swell under overburden and estimated foundation surcharge pressure, $S_{w(1)}$ (%)
2.5	4.7
5.0	3.85
10.0	2.45
15.0	1.3
20.0	0.4
25.0	0

10.9 Refer to Problem 10.7. If the allowable total swell is 20 mm, what would be the necessary undercut?

10.10 Refer to Problem 10.8. If the allowable total swell is 0.8 in., what would be the necessary undercut?

10.11 Refer to Figure 10.15b. For the drilled pier, given:

> Thickness of active zone, $Z = 30$ ft
> Dead load = 300 kips
> Live load = 60 kips
> Diameter of pier shaft, $D_s = 3.5$ ft
> Zero swell pressure for the clay in the active zone = 6 ton/ft^2
> Average angle of plinth-soil friction, $\phi_{ps} = 15°$
> Average undrained cohesion of the clay around the bell = 3020 lb/ft^2

Determine the diameter of the pier bell, D_b. A factor of safety of 2 against uplift is required, with the assumption that dead load plus live load is equal to zero.

10.12 Refer to Problem 10.11. If an additional requirement is placed such that the factor of safety against uplift is at least 3 with the dead load on (live load = 0), what should be the diameter of the bell?

REFERENCES

Benites, L. A. (1968). "Geotechnical Properties of the Soils Affected by Piping Near the Benson Area, Cochise County, Arizona," M.S. Thesis, University of Arizona, Tucson.

Clemence, S. P., and Finbarr, A. O. (1981). "Design Considerations for Collapsible Soils," *Journal of the Geotechnical Engineering Division*, American Society of Civil Engineers, Vol. 107, No. GT3, pp. 305–317.

Clevenger, W. (1958). "Experience with Loess as Foundation Material," *Transactions*, American Society of Civil Engineers, Vol. 123, pp. 151–170.

Denisov, N. Y. (1951). *The Engineering Properties of Loess and Loess Loams*, Gosstroiizdat, Moscow, U.S.S.R.

Feda, J. (1964). "Colloidal Activity, Shrinking and Swelling of Some Clays," *Proceedings*, Soil Mechanics Seminar, Loda, Illinois, pp. 531–546.

Gibbs, H. J. (1961). "Properties Which Divide Loose and Dense Uncemented Soils," *Earth Laboratory Report EM-658*, Bureau of Reclamation, U.S. Department of the Interior.

Gromko, G. J. (1974). "Review of Expansive Soils," *Journal of the Geotechnical Engineering Division*, American Society of Civil Engineers, Vol. 100, No. GT6, pp. 667–687.

Handy, R. L. (1973). "Collapsible Loess in Iowa," *Proceedings*, Soil Science Society of America, Vol. 37, pp. 281–284.

Holtz, W. G., and Hilf, J. W. (1961). "Settlement of Soil Foundations Due to Saturation," *Proceedings*, Fifth International Conference on Soil Mechanics and Foundation Engineering, Paris, Vol. 1, 1961, pp. 673–679.

Jennings, J. E., and Knight, K. (1975). "A Guide to Construction on or with Materials Exhibiting Additional Settlements Due to 'Collapse' of Grain Structure," *Proceedings*, Sixth Regional Conference for Africa on Soil Mechanics and Foundation Engineering, Johannesburg, pp. 99–105.

Lutenegger, A. J., and Saber, R. T. (1988). "Determination of Collapse Potential of Soils," *Geotechnical Testing Journal*, American Society for Testing and Materials, Vol. 11, No. 3, pp. 173–178.

O'Neill, M. W., and Poormoayed, N. (1980). "Methodology for Foundations on Expansive Clays," *Journal of the Geotechnical Engineering Division*, American Society of Civil Engineers, Vol. 106, No. GT12, pp. 1345–1367.

Peck, R. B., Hanson, W. E., and Thornburn, T. B. (1974). *Foundation Engineering*, Wiley, New York.

Priklonski, V. A. (1952). *Gruntovedenia-Vtoraid Chast*, Gosgeolzdat, Moscow, U.S.S.R.

Sivapullaiah, P. V., Sitharam, T. G., and Rao, K. S. S. (1987). "Modified Free Swell Index for Clay," *Geotechnical Testing Journal*, American Society for Testing and Materials, Vol. 11, No. 2, pp. 80–85.

Snethen, D. R., Johnson, L. D., and Patrick, D. M. (1977). "An Evaluation of Expedient Methodology for Identification of Potentially Expansive Soils," *Report No. FHWA-RD-77-94*, U.S. Army Engineers Waterways Experiment Station, Vicksburg, Miss.

Sowers, G. F. (1968). "Foundation Problems in Sanitary Landfills," *Journal of the Sanitary Engineering Division*, American Society of Civil Engineers, Vol. 94, No. SA1, pp. 103–116.

Sowers, G. F. (1973). "Settlement of Waste Disposal Fills," *Proceedings*, Eighth

International Conference on Soil Mechanics and Foundation Engineering, Moscow, pp. 207–210.

Sridharan, A., Rao, A. S., and Sivapullaiah, P. V. (1986). " Swelling Pressure of Clays," *Geotechnical Testing Journal*, American Society for Testing and Materials, Vol. 9, No. 1, pp. 24–33.

Vijayvergiya, V. N., and Ghazzaly, O. I. (1973). "Prediction of Swelling Potential of Natural Clays," *Proceedings*, Third International Research and Engineering Conference on Expansive Clays, pp. 227–234.

Yen, B. C., and Scanlon, B. (1975). " Sanitary Landfill Settlement Rates," *Journal of the Geotechnical Engineering Division*, American Society of Civil Engineers, Vol. 101, No. GT5, pp. 475–487.

Chapter 11

REINFORCED
EARTH STRUCTURES

11.1 Introduction

The use of reinforced earth is a recently developed method of designing foundations and earth-retaining structures. *Reinforced earth* is a construction material consisting of soil that has been strengthened by the inclusion of tensile elements such as metal rods and/or strips, nonbiodegradable fabrics, and the like. The fundamental idea of reinforcing soil is not new; in fact, it goes back to biblical times. However, the present concept of systematic analysis and design was initiated by a French engineer named Vidal (1966). The French Road Research Laboratory has done an extensive amount of research on the applicability and the beneficial effects of the use of reinforced earth as a construction material. This research has been documented in detail by Darbin (1970), Schlosser and Long (1974), and Schlosser and Vidal (1969). These works were conducted by using metallic strips as the reinforcing material.

Retaining walls with reinforced earth have been constructed around the world since Vidal initiated his work. The first reinforced earth retaining wall with metal strips as reinforcement in the United States was constructed in 1972 in Southern California.

This chapter discusses the fundamentals of the theory and design of retaining walls and shallow continuous foundations with reinforced earth.

The beneficial effects of soil reinforcement derive from (a) the soil's increased tensile strength and (b) the shear resistance developed from the friction at the soil-reinforcement interfaces. This is comparable to the reinforcement of concrete structures. At this time, most reinforced earth design is done with *free-draining granular soil only*. Thus one avoids the effect of pore water development in cohesive soils, which will, in turn, reduce the shear strength of the soil.

596

11.2   General Considerations for Soil Reinforcement

Metal Strips

In most instances, galvanized steel strips are used as reinforcement in soil. However, galvanized steel is subject to corrosion. The rate of corrosion will depend on several environmental factors. Binquet and Lee (1975b) have suggested that the average rate of corrosion of galvanized steel strips varies between 0.025 mm/year to 0.050 mm/year. So, in the actual design of reinforcements, allowances must be made to take into account the rate of corrosion. Thus

$$t_c = t_{\text{design}} + r \text{ (life span of structure)}$$

where t_c = actual thickness of reinforcing strips to be used in construction
t_{design} = thickness of strips determined from design calculations
r = rate of corrosion

Further research needs to be done on corrosion-resistant materials such as fiberglass before they can be used as reinforcing strips.

Nonbiodegradable Fabrics

Nonbiodegradable fabrics are generally referred to as *geotextiles*. Since 1970, the use of geotextiles in various construction works has increased tremendously around the world. The fabrics are usually made from petroleum products—that is, polyester, polyethylene, and polypropylene. They may also be made from fiberglass. Geotextiles are not prepared from natural fabrics because these decay too quickly. Geotextiles may be *woven, knitted,* or *nonwoven.*

   Woven geotextiles are made of two sets of parallel filaments or strands of yarn systematically interlaced to form a planar structure. *Knitted geotextiles* are formed by interlocking a series of loops of one or more filaments or strands of yarn to form a planar structure. *Nonwoven geotextiles* are formed from filaments or short fibers arranged in an oriented or random pattern in a planar structure. These filaments or short fibers are, in the beginning, arranged into a loose web. They are then bonded using one or a combination of the following processes:

   1. *Chemical bonding*—by glue, rubber, latex, cellulose derivative, and so forth
   2. *Thermal bonding*—by heat for partial melting of filaments
   3. *Mechanical bonding*—by needle punching

The *needle-punched nonwoven* geotextiles are thick and have high in-plane permeability.
   There are four major uses of geotextiles in foundation engineering:

   1. *Drainage:* The fabrics can rapidly channel water from soil to various

Table 11.1 List and Measurements of Some Common Geotextiles Manufactured in the United States (in SI Units)

Manufacturer or distributor	Fabric trade name	Structure	Fiber	Width (m)	Mass (g/m^2)
Du Pont	Typar 3201, 3341, 3401, 3601	Nonwoven	Polypropylene	3.84 to 5	67.8, 115.3, 135.6, 203.4
Du Pont	Reepav	Nonwoven	Polypropylene	3.76	101.7
Texel	Texel 7607, 7609, 7612, 7618, 7643	Nonwoven	Polyester	Up to 12.19	240 to 2000
Nicolon	40/30	Woven	Polypropylene	3.66 and multiples of 3.66	
Nicolon	70/06, 70/20, 100/08	Woven	Polypropylene	1.83 multiples	
Nicolon	100/05	Woven	Polyvinylidene chloride	1.83 multiples	
Nicolon	Geolon 200	Woven	Polypropylene	3.81 and 5.03	
Nicolon	Geolon 400	Woven	Polypropylene	3.66 multiples	
Nicolon	Geolon 1250	Woven	Polypropylene	5.09	
Enka	Enkamat		Nylon	0.99	265 and 405
Enka	Enkadrain		Nylon and polyester	0.99	735
Enka	Stabilenka	Nonwoven	Polyester	1.07 and 2.14	67.8, 101.7, 135.6
Synflex	Permealiner	Woven and nonwoven	Polypropylene		
Exxon	Industrial Fabrics Style 49862	Woven	Polypropylene	3.81 and 4.72	115.3
Celanese	Mirafi 500X	Woven	Polypropylene	3.81 and 5.33	
Celanese	Mirafi 140N	Nonwoven	Polypropylene	4.57	
Celanese	Mirafi 600X	Woven	Polypropylene	3.81	
Celanese	Mirafi 700X	Woven	Polypropylene	1.83 and 3.66	
Hoechst	Trevira	Nonwoven	Polyester	Up to 4.88	
Crown Zellerbach	Fibretex 150, 200, 300, 400	Nonwoven	Polypropylene	3.96 and 5.33	
Crown Zellerbach	Fibretex Ten-1	Woven	Polypropylene	3.81 and 4.72	115.3
Dominion Textile	Penroad 50, 150, 250, 350, 450, 550	Nonwoven	Polyester	1.11	200, 270, 340, 400, 540, 680

Dominion Textile	Penroad 100, 500	Woven	Polypropylene	1.16	125 and 150
Amoco	ProPex 4545, 4551, 4553	Nonwoven	Polypropylene	4.42	152.6, 203.4, 271.2
Amoco	ProPex 1325, 1369	Nonwoven	Polypropylene	4.42	220.4
Amoco	ProPex 2002	Woven	Polypropylene	4.42	152.6
Collins & Aikman	Diamond 8	Woven	Nylon or polyester	1.52	389.9
Collins & Aikman	Contain 4	Woven	Polyester	1.52	466.1
Collins & Aikman	Retain	Woven	Nylon	1.83	305.1
Collins & Aikman	SB-375	Stitchbound	Polyester	1.42 to 3.66	127.1 and 271.2
Phillips	Petromat	Nonwoven	Polypropylene	1.91 and 3.81	145.8
Phillips	Supac	Nonwoven	Polypropylene	Up to 4.57	135.6, 169.5, 542.4
Carthage Mills	Poly-Filter X	Woven	Polypropylene	1.83 to 2.13 (custom fab.)	244.1
Carthage Mills	Poly-Filter GB	Woven	Polypropylene	1.83 to 21.95 (custom fab.)	223.7
Carthage Mills	Filter-X	Woven	Polyvinylidene chloride		393.2
Rhone-Poulene	Bidim	Nonwoven	Polyester	2.7, 4.2, 5.3	50 to 100
Foss	F-55	Nonwoven	Polyester	0.91, 1.91, 3.81, 4.72 (other widths up to 7.42 available on request)	186.5
Foss	F-65	Nonwoven	Polyester	0.91, 1.91, 3.81, 4.72 (other widths up to 7.42 available on request)	220.4
Foss	F-90	Nonwoven	Polyester	0.91, 1.91, 3.81, 4.72 (other widths up to 7.42 available on request)	305.1
Foss	F-180	Nonwoven	Polyester	0.91, 1.91, 3.81, 4.72 (other widths up to 7.42 available on request)	610.2

Adapted from *Industrial Fabric Products Review*, July, 1981. Copyright by the Industrial Fabric Association International. Reprinted by permission.

Table 11.2 List and Measurements of Some Common Geotextiles Manufactured in the United States (in English Units)

Manufacturer or distributor	Fabric trade name	Structure	Fiber	Width	Weight
Du Pont	Typar 3201, 3341, 3401, 3601	Nonwoven	Polypropylene	151 in. to 197 in.	2 oz, 3.4 oz, 4 oz, 6 oz
Du Pont	Reepav	Nonwoven	Polypropylene	148 in.	3 oz
Texel	Texel 7607, 7609, 7612, 7618, 7643	Nonwoven	Polyester	Up to 40 ft	7.08 to 59 oz/yd^2
Nicolon	40/30	Woven	Polypropylene	12 ft and multiples of 12 ft	
Nicolon	70/06, 70/20, 100/08	Woven	Polypropylene	6-ft multiples	
Nicolon	100/05	Woven	Polyvinylidene chloride	6-ft multiples	
Nicolon	Geolon 200	Woven	Polypropylene	12.5 ft and 16.5 ft	
Nicolon	Geolon 400	Woven	Polypropylene	12-ft multiples	
Nicolon	Geolon 1250	Woven	Polypropylene	16.7 ft	
Enka	Enkamat		Nylon	39 in.	7.82 and 11.95 oz/yd^2
Enka	Enkadrain		Nylon and Polyester	39 in.	21.68 oz/yd^2
Enka	Stabilenka	Nonwoven	Polyester	3.5 ft and 7 ft	2 oz/yd^2 3 oz/yd^2 4 oz/yd^2
Synflex	Permealiner	Woven and nonwoven	Polypropylene		
Exxon	Industrial Fabrics Style 49862	Woven	Polypropylene	12.5 ft and 15.5 ft	3.4 oz/yd^2
Celanese	Mirafi 500X	Woven	Polypropylene	12.5 ft and 17.5 ft	
Celanese	Mirafi 140N	Nonwoven	Polypropylene	15 ft	
Celanese	Mirafi 600X	Woven	Polypropylene	12.5 ft	
Celanese	Mirafi 700X	Woven	Polypropylene	6 ft and 12 ft	
Hoechst	Trevira	Nonwoven	Polyester	Up to 16 ft	
Crown Zellerbach	Fibretex 150, 200, 300, 400	Nonwoven	Polypropylene	13 ft and 17.5 ft	
Crown Zellerbach	Fibretex Ten-1	Woven	Polypropylene	12.5 ft and 15.5 ft	3.4 oz/yd^2
Dominion Textile	Penroad 50, 150, 250, 350, 450, 550	Nonwoven	Polyester	11.97 ft	5.9 to 20.06 oz/yd^2

Manufacturer	Product	Construction	Polymer	Width	Weight
Dominion Textile	Penroad 100, 500	Woven	Polypropylene	12.53 ft	3.69 and 4.43 oz/yd²
Amoco	ProPex 4545, 4551, 4553	Nonwoven	Polypropylene	14.5 ft	4.5 oz/yd² 6 oz/yd² 8 oz/yd²
Amoco	ProPex 1325, 1369	Nonwoven	Polypropylene	14.5 ft	6.5 oz/yd²
Amoco	ProPex 2002	Woven	Polypropylene	14.5 ft	4.5 oz/yd²
Collins & Aikman	Diamond 8	Woven	Nylon or Polyester	60 in.	11.5 oz/yd²
Collins & Aikman	Contain 4	Woven	Polyester	60 in.	13.75 oz/yd²
Collins & Aikman	Retain	Woven	Nylon	72 in.	9 oz/yd²
Collins & Aikman	SB-375	Stitchbound	Polyester	56 in. to 144 in.	3.75 oz/yd² and 8 oz/yd²
Phillips	Petromat	Nonwoven	Polypropylene	75 in. and 150 in.	4.3 oz/yd² 4 oz/yd²
Phillips	Supac	Nonwoven	Polypropylene	Up to 180 in.	5 oz/yd², 16 oz/yd²
Carthage Mills	Poly-Filter X	Woven	Polypropylene	6 ft to 84 ft (custom fab.)	7.2 oz/yd²
Carthage Mills	Poly-Filter GB	Woven	Polypropylene	6 ft to 72 ft (custom fab.)	6.6 oz/yd²
Carthage Mills	Filter-X	Woven	Polyvinylidene chloride		11.6 oz/yd²
Rhone-Poulene	Bidim	Nonwoven	Polyester	2.7 m, 4.2 m, 5.3 m	50 g/m² to 1000 g/m²
Foss	F-55	Nonwoven	Polyester	36 in., 75 in., 150 in., 186 in. (other widths up to 186 in. available on request)	5.5 oz/yd²
Foss	F-65	Nonwoven	Polyester	36 in., 75 in., 150 in., 186 in. (other widths up to 186 in. available on request)	6.5 oz/yd²
Foss	F-90	Nonwoven	Polyester	36 in., 75 in., 150 in., 186 in. (other widths up to 186 in. available on request)	9 oz/yd²
Foss	F-180	Nonwoven	Polyester	36 in., 75 in., 150 in., 186 in. (other widths up to 186 in. available on request)	18 oz/yd²

Adapted from *Industrial Fabric Products Review*, July, 1981. Copyright by the Industrial Fabric Association International. Reprinted by permission.

outlets and thereby provide a higher shear strength of soil and hence stability.

2. *Filtration:* When placed between two soil layers, one coarse grained and the other fine grained, the fabric allows free seepage of water from one layer to the other. However, it protects the fine-grained soil from being washed into the coarse-grained soil.

3. *Separation:* Geotextiles help keep various soil layers separate after construction and during the projected service period of the structure. For example, in the construction of highways, a clayey subgrade can be kept separate from a granular base course.

4. *Reinforcement:* The tensile strength of the geofabrics increases the load-bearing capacity of the soil.

Details of some of the commercially produced geotextiles in the United States are listed in Tables 11.1 and 11.2. In Table 11.1 the units are in the SI system, and in Table 11.2 the units are in the English system.

RETAINING WALLS

11.3 **Retaining Walls with Metallic Strip Reinforcement**

Reinforced earth walls are flexible walls. Their main components are

1. *Backfill,* which is granular soil
2. *Reinforcing strips,* which are thin, wide strips placed at regular intervals
3. *A cover* on the front face, which is referred to as the *skin*

Figure 11.1 is a diagram of a reinforced earth wall. Note that, at any given depth, the reinforcing strips or ties are placed with a horizontal spacing of S_H center-to-center; the vertical spacing of the strips or ties (center-to-center) is S_V. The skin can be constructed with sections of relatively flexible thin material. Lee, Adams, and Vagneron (1973) have shown that, with a conservative design, a 5-mm (≈ 0.2-in.) thick galvanized steel skin would be enough to hold a wall about 14–15 m (45–50 ft) high. In most cases, precast concrete slabs can also be used as skin. The slabs are grooved to fit into each other so that soil does not flow out between the joints. When metal skins are used, they are bolted together, and reinforcing strips are placed between the skins.

Figures 11.2 and 11.3 show a reinforced earth retaining wall under construction that has a precast concrete slab as its skin (facing). Figure 11.4 shows a metallic reinforcement tie attached to the concrete slab.

The simplest and most common method for design of ties is by the *Rankine method.* The following is a detailed discussion of this procedure.

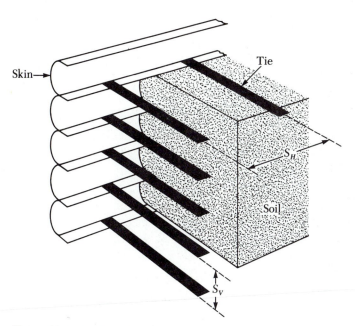

Figure 11.1   Reinforced earth retaining wall

Calculation of Active Horizontal and Vertical Pressure

Figure 11.5a shows a retaining wall with a granular backfill having a unit weight of γ_1 and a friction angle of ϕ_1. Below the base of the retaining wall, the *in situ* soil has been excavated and recompacted, with the granular soil used as the backfill. Below the backfill soil, the *in situ* soil has a unit weight of γ_2, a friction angle of ϕ_2, and cohesion of c_2. A surcharge having an intensity of q per unit area is located at the top of the retaining wall. The wall has reinforcement ties located at depths $z = 0$, S_V, $2S_V$, ..., NS_V. The height of the wall is equal to $NS_V = H$.

According to the Rankine active pressure theory (Section 5.3)

$$\sigma_a = \sigma_v K_a - 2c\sqrt{K_a}$$

where σ_a = Rankine active pressure at any depth z

For dry granular soils with no surcharge at the top, $c = 0$, $\sigma_v = \gamma z$, and $K_a = \tan^2(45 - \phi_1/2)$. Thus

$$\boxed{\sigma_a = \gamma_1 z K_a}$$

(11.1)

Figure 11.2   Reinforced earth retaining wall (with metallic strip) under construction

When a surcharge is located at the top, as shown in Figure 11.5

$$\sigma_v = \sigma_{v(1)} \quad + \quad \sigma_{v(2)} \tag{11.2}$$

$$\uparrow \qquad\qquad \uparrow$$

$= \gamma_1 z$ Due to the
(Due to   surcharge
soil only)

The magnitude of $\sigma_{v(2)}$ can be calculated by using the 2 : 1 method of stress distribution described in Eq. (3.96) (Figure 3.38). This is shown in Figure 11.6a. According to Laba and Kennedy (1986)

$$\sigma_{v(2)} = \frac{qa'}{a' + z} \quad (\text{for } z \leq 2b') \tag{11.3}$$

Figure 11.3   Another view of the retaining wall shown in Figure 11.2

Figure 11.4   Metallic strip attachment to the precast concrete slab used as the skin

and

$$\sigma_{v(2)} = \frac{qa'}{a' + \dfrac{z}{2} + b'} \qquad \text{(for } z > 2b')$$

(11.4)

Also, when a surcharge is located at the top, the lateral pressure at any depth

$$\sigma_a = \sigma_{a(1)} \qquad + \qquad \sigma_{a(2)}$$

(11.5)

$$\qquad\qquad \uparrow \qquad\qquad\qquad \uparrow$$

$$\qquad\qquad = K_a \gamma_1 z \qquad \text{Due to the}$$
$$\qquad\qquad \text{(Due to} \qquad\qquad \text{surcharge}$$
$$\qquad\qquad \text{soil only)}$$

Figure 11.5 Analysis of a reinforced earth retaining wall

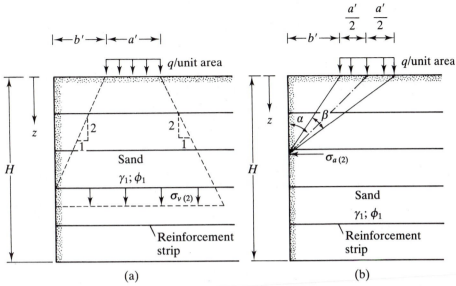

Figure 11.6   (a) Notation of the relationship of $\sigma_{v(2)}$—Eqs. (11.3) and (11.4);
(b) Notation of the relationship of $\sigma_{a(2)}$—Eqs. (11.6) and (11.7)

According to Laba and Kennedy (1986), $\sigma_{a(2)}$ can be expressed as follows
(Figure 11.6b):

$$\sigma_{a(2)} = M\left[\frac{2q}{\pi}\ (\beta - \sin \beta \cdot \cos 2\alpha)\right] \qquad (11.6)$$

$$\uparrow$$

(in radians)

where

$$M = 1.4 - \frac{0.4b'}{0.14H} \geq 1 \qquad (11.7)$$

It is important to note that Eq. (11.6) is the same as Eq. (5.38) along with
the multiplying factor M.

The net active (lateral) pressure distribution on the retaining wall calculated by using Eqs. (11.5), (11.6), and (11.7) is shown in Figure 11.5b.

Tie Force

Referring to Figure 11.5, the tie force per unit length of the wall developed
at any depth z can be given as

T = (active earth pressure at depth z) (area of the wall to be
    supported by the tie)

So

$$T = (\sigma_a)(S_V\, S_H) \tag{11.8}$$

Factor of Safety Against Tie Failure

The reinforcement ties at each level and, thus, the walls could fail by either (a) *tie breaking* or (b) *tie pullout*.

The factor of safety against tie breaking can be determined as

$$\boxed{\begin{aligned} FS_{(B)} &= \frac{\text{yield or breaking strength of each tie}}{\text{maximum tie force in any tie}}\\[6pt] &= \frac{wtf_y}{\sigma_a S_V S_H} \end{aligned}} \tag{11.9}$$

where w = width of each tie

t = thickness of each tie

f_y = yield or breaking strength of the tie material

A factor of safety of about 2.5–3 is generally recommended for ties at all levels.

The reinforcing ties at any given depth, z, will fail by pullout if the frictional resistance developed along their surfaces is less than the force to which the ties are being subjected. The *effective length* of the ties along which the frictional resistance is developed may be conservatively taken as the length that extends *beyond the limits of the Rankine active failure zone*, which is the zone ABC in Figure 11.5. The line BC in Figure 11.5 makes an angle of $45 + \phi_1/2$ with the horizontal. Now, the maximum friction force F_R that can be realized for a given tie at a depth z is equal to

$$F_R = 2l_e w\sigma_v \tan \phi_\mu \tag{11.10}$$

where l_e = effective length

σ_v = effective vertical pressure at a depth z

ϕ_μ = soil-tie friction angle

So, the factor of safety against tie pullout at a given depth z can be given as

$$FS_{(P)} = \frac{F_R}{T} \tag{11.11}$$

where $FS_{(P)}$ = factor of safety against tie pullout.

Substituting Eqs. (11.8) and (11.10) into Eq. (11.11)

$$\boxed{FS_{(P)} = \frac{2l_e w\sigma_v \tan \phi_\mu}{\sigma_a S_V S_H}} \tag{11.12}$$

Total Length of Tie

The total length of ties at any depth is equal to

$$L = l_r + l_e \tag{11.13}$$

where l_r = length within the Rankine failure zone
l_e = effective length

For a given $FS_{(P)}$, from Eq. (11.12)

$$l_e = \frac{FS_{(P)}\sigma_a S_V S_H}{2w\sigma_v \tan \phi_\mu} \tag{11.14}$$

Again, at any depth z

$$l_r = \frac{(H - z)}{\tan\left(45 + \dfrac{\phi_1}{2}\right)} \tag{11.15}$$

So, combining Eqs. (11.13), (11.14), and (11.15)

$$\boxed{L = \frac{(H - z)}{\tan\left(45 + \dfrac{\phi_1}{2}\right)} + \frac{FS_{(P)}\sigma_a S_V S_H}{2w\sigma_v \tan \phi_\mu}} \tag{11.16}$$

General Design Procedure

Following is a step-by-step procedure for the design of reinforced earth retaining walls:

1. Determine the height of the wall, H, and also the properties of the granular backfill material, such as unit weight (γ_1) and angle of friction (ϕ_1).
2. Obtain the soil-tie friction angle, ϕ_μ, and also the required values of $FS_{(B)}$ and $FS_{(P)}$.
3. Assume values for the horizontal and vertical spacings of the ties. Also assume the width of reinforcing strip (w) to be used.
4. Calculate σ_a using Eqs. (11.5), (11.6), and (11.7).
5. Calculate the tie forces at various levels [Eq. (11.8)].
6. With the known values of $FS_{(B)}$, calculate the thickness of ties (t) to resist the tie breakout, or

$$T = \sigma_a S_V S_H = \frac{wtf_y}{FS_{(B)}}$$

or

$$t = \frac{(\sigma_a S_V S_H)[FS_{(B)}]}{w f_y} \qquad (11.17)$$

It is conventional to keep the magnitude of t the same at all levels. So, σ_a in Eq. (11.17) should be equal to $\sigma_{a(max)}$.

7. With known values of ϕ_μ and $FS_{(P)}$, determine the length, L, of the ties at various levels from Eq. (11.16).

8. The values of S_V, S_H, t, w, and L can be changed to obtain the most economical design.

9. Once the reinforcements have been designed, check for the overall stability of the wall; that is, check for overturning, sliding, and bearing capacity failure. This is similar to that done for retaining walls in Chapter 5 (see Figure 11.7).

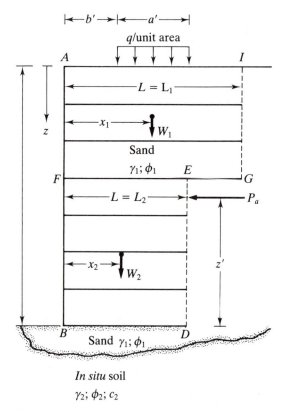

Figure 11.7 Stability check for the retaining wall

Check for Overturning

Taking the moment about B, the overturning moment for the unit length of the wall

$$M_O = P_a z' \tag{11.18}$$

where P_a = active force = $\int_0^H \sigma_a \cdot dz$

The resisting moment per unit length of the wall

$$M_R = W_1 x_1 + W_2 x_2 + \cdots + qa'\left(b' + \frac{a'}{2}\right) \tag{11.19}$$

where W_1 = (area $AFEGI$)(1)(γ_1)
$\quad\quad\; W_2$ = (area $FBDE$)(1)(γ_1)
$\quad\quad\;\; \vdots$

So

$$\boxed{\begin{aligned} FS_{(\text{overturning})} &= \frac{M_R}{M_O} \\[2mm] &= \frac{W_1 x_1 + W_2 x_2 + \cdots + qa'\left(b' + \dfrac{a'}{2}\right)}{\left(\displaystyle\int_0^H \sigma_a \cdot dz\right) z'} \end{aligned}} \tag{11.20}$$

Check for Sliding

As in Eq. (5.64), the factor of safety against sliding can be given as

$$\boxed{FS_{(\text{sliding})} = \frac{(W_1 + W_2 + \cdots + qa')[\tan (k\phi_1)]}{P_a}} \tag{11.21}$$

where $k \approx 2/3$.

Check for Bearing Capacity Failure

Referring to Chapter 3, the ultimate bearing capacity of a shallow foundation can be given as

$$q_{\text{ult}} = cN_c F_{cs} F_{cd} F_{ci} + qN_q F_{qs} F_{qd} F_{qi} + \frac{1}{2}\gamma BN_\gamma F_{\gamma s} F_{\gamma d} F_{\gamma i} \tag{3.16}$$

For the present case, all the shape, depth, and inclination factors are equal to one. Also, $B = L_2$, $c = c_2$, $\gamma = \gamma_2$, and $q = 0$, since the depth of the foundation is zero. So

$$q_{ult} = c_2 N_c + \frac{1}{2} \gamma_2 L_2 N_\gamma \tag{11.22}$$

The bearing capacity factors N_c and N_γ correspond to the soil friction angle ϕ_2 (Table 3.2). The vertical stress at $z = H$ is [from Eq. (11.2)]

$$\sigma_{v(H)} = \gamma_1 H + \sigma_{v(2)} \tag{11.23}$$

So, the factor of safety against bearing capacity failure is

$$\boxed{FS_{(bearing\ capacity)} = \frac{q_{ult}}{\sigma_{v(H)}}} \tag{11.24}$$

Generally, minimum values of $FS_{(overturning)} = 3$, $FS_{(sliding)} = 3$, and $FS_{(bearing\ capacity\ failure)} = 3$ to 5 are recommended.

Example 11.1

A reinforced earth retaining wall is proposed to be 8 m high. The properties of the backfill material are as follows: $\gamma_1 = 16.6$ kN/m³ and $\phi_1 = 30°$. Galvanized steel ties are to be used for the construction of the wall. Design the reinforcements with $FS_{(B)} = 3$, $FS_{(P)} = 3$, $f_y = 2.4 \times 10^5$ kN/m², and $\phi_\mu = 20°$. The properties of the *in situ* soil below the retaining wall are as follows: $\gamma_2 = 18$ kN/m³, $\phi_2 = 28°$, and $c_2 = 52$ kN/m².

Solution
Design of the Thickness of the Ties

Let $S_V = 0.5$, $S_H = 1$ m, and $w = 75$ mm. Given: soil friction angle, ϕ_1, is 30°. So, $K_a = \tan^2 (45 - \phi_1/2) = \tan^2 (45 - 30/2) = 1/3$. Hence, from Eq. (11.8), tie force

$$T = \sigma_a S_V S_H$$

The maximum tie force will be where σ_a is maximum. For this case, $\sigma_{a(max)} = \gamma H K_a$. So

$$T_{max} = \gamma H K_a S_V S_H = (16.6)(8)\left(\frac{1}{3}\right)(0.5)(1) = 22.14 \text{ kN}$$

From Eq. (11.17), the thickness of ties is

$$t = \frac{(T_{max})[FS_{(B)}]}{wf_y}$$

or

$$t = \frac{(22.14)(3)}{\left(\dfrac{75}{1000}\right)(2.4 \times 10^5)} = 0.00369 \text{ m} = 3.69 \text{ mm}$$

If the rate of corrosion is 0.025 mm/year and the life span of the structure is 50 years, then the actual thickness, t, of the ties will be

$$t = 3.69 + (0.025)50 = 4.94 \text{ mm}$$

So, a **tie thickness of 5 mm** would be enough.

Determination of Length of Ties

Referring to Eq. (11.16), for this case $\sigma_a = \gamma_1 z K_a$ and $\sigma_v = \gamma_1 z$. So

$$L = \frac{(H - z)}{\tan\left(45 + \dfrac{\phi_1}{2}\right)} + \frac{FS_{(P)} \gamma_1 z K_a S_V S_H}{2 w \gamma_1 z \tan \phi_\mu}$$

Now the following table can be prepared. (*Note:* $FS_{(P)} = 3$, $H = 8$ m, $w = 0.075$ m, and $\phi_\mu = 20°$.

z	$\dfrac{(H-z)}{\tan\left(45 + \dfrac{\phi_1}{2}\right)}$ (m)	$\dfrac{FS_{(P)} \gamma_1 z K_a S_V S_H}{2 w \gamma_1 z \tan \phi_\mu}$ (m)	L (m)
1	2.34	9.16	12.0
2	2.0	9.16	11.16
3	1.67	9.16	10.83
4	1.34	9.16	10.50
5	1.0	9.16	10.16
6	0.67	9.16	9.83
7	0.33	9.16	9.49

So, use **$L = 12$ m for $z = 0$ to 5 m, and use $L = 10$ m below that** (see Figure 11.8).

Check for Overall Stability

For Overturning: Referring to Eq. (11.20) and Figure (11.8)

$$FS_{(overturning)} = \frac{W_1 x_1 + W_2 x_2}{\left[\displaystyle\int_0^H \sigma_a \cdot dz\right] z'}$$

$$W_1 = (12)(5)(1)(\gamma_1) = (12)(5)(1)(16.6) = 996 \text{ kN}$$
$$W_2 = (10)(3)(1)(\gamma_1) = (10)(3)(1)(16.6) = 498 \text{ kN}$$

$x_1 = 6$ m, $x_2 = 5$ m. Also

$$\int_0^H \sigma_a \cdot dz = P_a = \frac{1}{2}\gamma_1 K_a H^2 = \left(\frac{1}{2}\right)(16.6)\left(\frac{1}{3}\right)(8)^2 = 177 \text{ kN}$$

So

$$FS_{(overturning)} = \frac{(996)(6) + (489)(5)}{(177)\left(\dfrac{8}{3}\right)} = \frac{8466}{472}$$

$$= 17.94 > 3 \text{—O.K.}$$

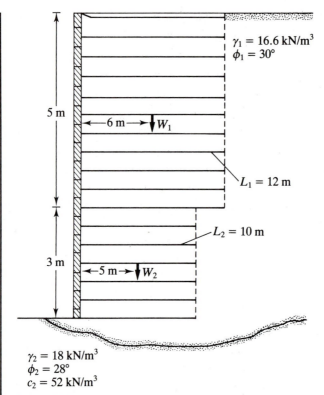

5 m

←6 m→ W_1

$L_1 = 12$ m

$L_2 = 10$ m

3 m

←5 m→ W_2

$\gamma_1 = 16.6$ kN/m^3
$\phi_1 = 30°$

$\gamma_2 = 18$ kN/m^3
$\phi_2 = 28°$
$c_2 = 52$ kN/m^3

Figure 11.8

For Sliding: From Eq. (11.21)

$$FS_{(sliding)} = \frac{(W_1 + W_2) \tan(k\phi)}{P_a}$$

$$= \frac{(996 + 498) \tan\left[\left(\dfrac{2}{3}\right)(30)\right]}{177} = 3.07 > 3\text{—O.K.}$$

For Bearing Capacity Failure: For $\phi_2 = 28°$, $N_c = 25.8$, and $N_\gamma = 16.72$ (Table 3.2). From Eq. (11.22)

$$q_{ult} = c_2 N_c + \frac{1}{2}\gamma_2 L_2 N_\gamma$$

$$= (52)(25.8) + (0.5)(18)(10)(16.72)$$

$$= 1341.6 + 1504.8 = 2846.4 \text{ kN/m}^2$$

From Eq. (11.23)

$$\sigma_{v(H)} = \gamma_1 H = (16.6)(8) = 132.8 \text{ kN/m}^2$$

So

$$FS_{(bearing\ capacity\ failure)} = \frac{q_{ult}}{\sigma_{v(H)}} = \frac{2846.4}{132.8} = 21.43 > 5\text{—O.K.}$$

11.4   Retaining Walls with Geotextile as Reinforcement

Figure 11.9 shows a retaining wall in which layers of geotextile have been used as reinforcement. As in Figure 11.6, the backfill is a granular soil. In this type of retaining wall, the facing of the wall is formed by lapping the sheets as shown. The lap length is equal to l_l. When construction of the wall is finished, the exposed face of the wall must be covered; otherwise, the geotextile will deteriorate due to ultraviolet light exposure. This is done by spraying *bitumen emulsion* or *Gunite* on the wall face. A wire mesh anchored to the geotextile facing may be necessary to keep the coating on the face of the wall.

The design of this type of retaining wall is practically similar to that presented in Section 11.3. Following is a step-by-step procedure for design based on the recommendations of Bell, Stilley, and Vandre (1975) and Koerner (1986).

1. Determine active pressure distribution on the wall as

$$\sigma_a = K_a \sigma_v = K_a \gamma_1 z \qquad (11.25)$$

where K_a = Rankine earth pressure coefficient
$\qquad = \tan^2 (45 - \phi_1/2)$
$\qquad \gamma_1$ = unit weight of the granular backfill
$\qquad \phi_1$ = friction angle of the granular backfill

2. Select a geotextile fabric that has an allowable strength of σ_G (lb/ft or kN/m).

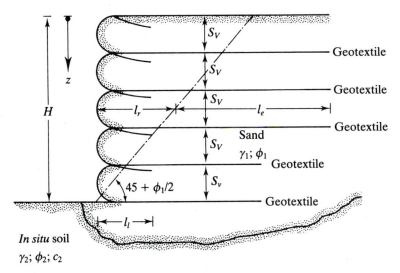

Figure 11.9 Retaining wall with geotextile reinforcement

3. Determine the vertical spacing of the layers at any depth z as

$$S_V = \frac{\sigma_G}{\sigma_a FS_{(B)}} = \frac{\sigma_G}{(\gamma_1 z K_a)[FS_{(B)}]} \tag{11.26}$$

Note that the preceding equation is in a form similar to that of Eq. (11.9). The magnitude of $FS_{(B)}$ is generally taken as 1.3 to 1.5.

4. Determine the length of each layer of geotextile as

$$L = l_r + l_e \tag{11.27}$$

where $l_r = \dfrac{H - z}{\tan\left(45 + \dfrac{\phi_1}{2}\right)}$ \hfill (11.28)

and

$$l_e = \frac{S_V \sigma_a[FS_{(P)}]}{2\sigma_v \tan \phi_F} \tag{11.29}$$

where $\sigma_a = \gamma_1 z K_a$
$\quad\quad\sigma_v = \gamma_1 z$
$\quad FS_{(P)} \approx 1.3$ to 1.5
$\quad\quad\phi_F =$ friction angle at geotextile-soil interface
$\quad\quad\quad\approx \dfrac{2}{3}\phi_1$

Note that Eqs. (11.27), (11.28), and (11.29) are similar to Eqs. (11.13), (11.15), and (11.14), respectively.

5. Determine the lap length, l_l, as follows:

$$l_l = \frac{S_V \sigma_a FS_{(P)}}{4\sigma_v \tan \phi_F} \tag{11.30}$$

The minimum lap length should be 1 m (3 ft).

6. Check the overall stability and determine the factors of safety against overturning, sliding, and bearing capacity failure as described under Item 9 of the design procedure in Section 11.3.

Example 11.2

A geotextile reinforced retaining wall that is 16 ft high is shown in Figure 11.10. For the granular backfill, given: $\gamma_1 = 110$ lb/ft^3 and $\phi_1 = 35°$. For the geotextile: $\sigma_G = 80$ lb/in. For the design of the wall, determine S_V, L, and l_l.

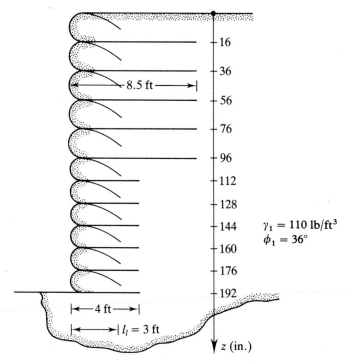

Figure 11.10

Solution

$$K_a = \tan^2\left(45 - \frac{\phi_1}{2}\right) = 0.26$$

Determination of S_V

In order to find S_V, let us make a few trials. From Eq. (11.26)

$$S_V = \frac{\sigma_G}{(\gamma_1 z K_a)[FS_{(B)}]}$$

With $FS_{(B)} = 1.5$ at $z = 8$ ft

$$S_V = \frac{(80 \times 12 \text{ lb/ft})}{(110)(8)(0.26)(1.5)} = 2.8 \text{ ft} \approx 33.6 \text{ in.}$$

At $z = 12$ ft

$$S_V = \frac{(80 \times 12 \text{ lb/ft})}{(110)(12)(0.26)(1.5)} = 1.87 \text{ ft} \approx 22 \text{ in.}$$

At $z = 16$ ft

$$S_V = \frac{(80 \times 12 \text{ lb/ft})}{(110)(16)(0.26)(1.5)} = 1.4 \text{ ft} \approx 16.8 \text{ in.}$$

So, use $S_V = 20$ in. for $z = 0$ to $z = 8$ ft and $S_V = 16$ in. for $z > 8$ ft. This is shown in Figure 11.10.

Determination of L

From Eqs. (11.27), (11.28), and (11.29)

$$L = \frac{(H - z)}{\tan\left(45 + \dfrac{\phi_1}{2}\right)} + \frac{S_V K_a [FS_{(P)}]}{2 \tan \phi_F}$$

Using $FS_{(P)} = 1.5$, $\tan \phi_F = \tan\left[\left(\dfrac{2}{3}\right)(36)\right] = 0.445$

$$L = (0.51)(H - z) + 0.483 S_V$$

Now the following table can be prepared.

z (in.)	(ft)	S_V (ft)	$(0.51)(H - z)$ (ft)	$0.483 S_V$ (ft)	L (ft)
16	1.33	1.67	7.48	0.81	8.29
56	4.67	1.67	5.78	0.81	6.59
76	6.34	1.67	4.93	0.81	5.74
96	8.0	1.67	4.08	0.81	4.89
112	9.34	1.33	3.23	0.64	3.87
144	12.0	1.33	2.04	0.64	2.68
176	14.67	1.33	0.68	0.64	1.32

So, based on the above calculation, use $L = 8.5$ ft for $z \le 8$ ft and $L = 4$ ft for $z > 8$ ft.

Determination of l_l

From Eq. (11.30)

$$l_l = \frac{S_V \sigma_a [FS_{(P)}]}{4\sigma_v \tan \phi_F}$$

With $\sigma_a = \gamma z K_a$, $FS_{(P)} = 1.5$; and with $\sigma_v = \gamma z$, $\phi_F = \frac{2}{3}\phi_1$.

$$l_l = \frac{S_V K_a [FS_{(P)}]}{4 \tan \phi_F} = \frac{S_V(0.26)(1.5)}{(4) \tan\left[\left(\dfrac{2}{3}\right)(36)\right]} = 0.219 S_V$$

At $z = 16$ in.,

$$l_l = 0.219 S_V = (0.219)\left(\frac{20}{12}\right) = 0.365 \text{ ft} \le 3 \text{ ft}$$

So, use $l_l = 3$ ft.

BEARING CAPACITY OF CONTINUOUS FOUNDATION ON GRANULAR SOIL REINFORCED BY METALLIC STRIPS

11.5   Modes of Failure

The bearing capacity of shallow foundations resting on reinforced earth has been studied in detail by Binquet and Lee (1975a, b), who have proposed a rational design method that will be presented in the following sections.

The nature of bearing capacity failure of a shallow strip foundation resting on a compact and homogeneous soil mass was presented in Figure 3.1a. In contrast, if layers of reinforcing strips (sometimes referred to as *ties*) are placed in the soil under a shallow strip foundation, the nature of failure in the soil mass will be like that shown in Figure 11.11a, b, and c.

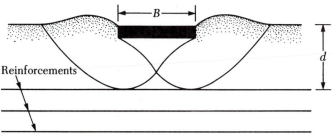

(a) $d/B > 2/3$ shear above reinforcement

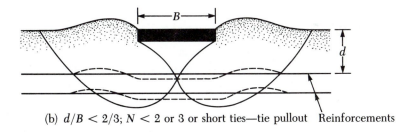

(b) $d/B < 2/3$; $N < 2$ or 3 or short ties—tie pullout   Reinforcements

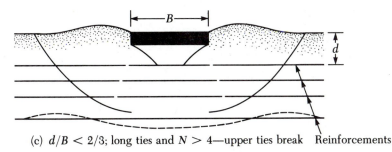

(c) $d/B < 2/3$; long ties and $N > 4$—upper ties break   Reinforcements

Figure 11.11   Three modes of bearing capacity failure in reinforced earth (redrawn after Binquet and Lee, 1975b)

The nature of failure in the soil mass, shown in Figure 11.11a, generally occurs when the first layer of reinforcement is located at a depth, d, greater than about $\frac{2}{3}B$ (B = width of the foundation). If the reinforcements in the first layer are strong and their concentration is sufficiently large, they may act as a rigid base located at a limited depth. The bearing capacity of foundations in such cases can be evaluated by the theory presented by Mandel and Salencon (1972). Experimental laboratory results for the bearing capacity of shallow foundations resting on a sand layer with a rigid rough base at a limited depth have also been provided by Meyerhof (1974), Pfeifle and Das (1979), and Das (1981).

The type of failure shown in Figure 11.11b could occur if d/B is less than about $\frac{2}{3}$ and the number of layers of reinforcement (N) is less than about 2–3. In this type of failure, reinforcement tie pullout occurs.

The most beneficial effect of reinforced earth is obtained when d/B is less than about $\frac{2}{3}$ and the number of reinforcement layers is greater than 4 but no more than 6–7. (The ties must be sufficiently long; see Section 11.7 for length calculation.) In this case, the soil mass fails when the upper ties break (see Figure 11.11c). The following section discusses this type of bearing capacity failure in detail.

11.6   Determination of the Force Induced in Reinforcement Ties

Location of Failure Surface

When designing shallow strip foundations, one must estimate the force that develops in the reinforcing ties as a result of the foundation load. This section presents the analytical procedure for estimation proposed by Binquet and Lee (1975b).

Figure 11.12 shows an idealized condition for the development of the failure surface in soil for the condition shown in Figure 11.11c. It consists of a central zone—Zone I—immediately below the foundation that settles along with the foundation with the application of load. On each side of Zone I, the soil is pushed outward and upward—this is Zone II. The points A', A'', A''', ..., and B', B'', B''', ..., which define the limiting lines between Zones I and II, can be obtained by considering the shear stress distribution, τ_{xz}, in the soil caused by the foundation load. The term τ_{xz} refers to the shear stress developed at a depth z below the foundation at a distance x measured from the center line of the foundation. If integration of Boussinesq's equation is used, τ_{xz} can be given by the relation

$$\tau_{xz} = \frac{4bq_R xz^2}{\pi[(x^2 + z^2 - b^2)^2 + 4b^2z^2]} \tag{11.31}$$

where b = half-width of the foundation = $B/2$
B = width of foundation
q_R = load per unit area on the foundation

The nature of variation of τ_{xz} at any given depth z is shown by the broken lines in Figure 11.12a. The points A', B' refer to the points at which the value of τ_{xz} is maximum at $z = z_1$. Similarly, A'', B'' refer to the points at which τ_{xz} is maximum at $z = z_2$. The distances $x = X_o$ at which the maximum value of τ_{xz} occurs can be given in a nondimensional form; this is shown in Figure 11.12b.

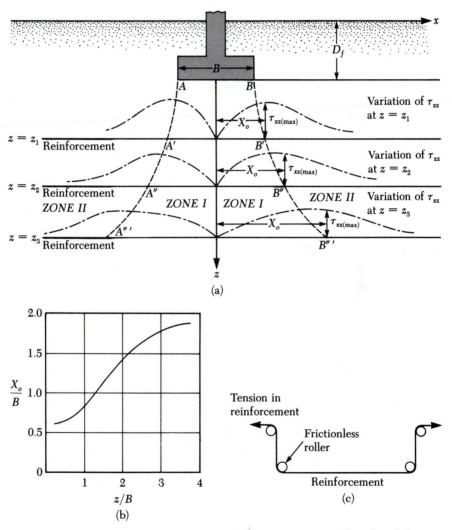

Figure 11.12 Failure mechanism under a foundation supported by reinforced earth [part (b) after Binquet and Lee, 1975b]

Other Assumptions

Other assumptions needed to obtain the tie force at any given depth are as follows:

   1. Under the application of bearing pressure by the foundation, the reinforcing ties at points A', A'', A''', ..., and B', B'', B''', ..., take the shape shown in Figure 11.12c. This means that the ties take two right angle turns on each side of Zone I around two frictionless rollers.

   2. For a given number of reinforcing layers (N), the ratio of the load per unit area on the foundation supported by reinforced earth (q_R) to the load per unit area on the foundation supported by unreinforced earth (q_o) is constant irrespective of the settlement level, s (see Figure 11.13). Binquet and Lee (1975a) have proved this in laboratory experiments.

Derivation of Equation

Figure 11.14a shows a continuous foundation supported by unreinforced soil and subjected to a load of q_o per unit area. Similarly, Figure 11.14b shows a continuous foundation supported by a reinforced soil layer (one layer of reinforcement—that is, $N = 1$) and subjected to a load of q_R per unit area. (Due to symmetry only one-half of the foundation is shown in Figure 11.14.) In both cases—that is, in Figures 11.14a and 11.14b—let the settlement be equal to s. For one-half of each foundation under consideration, the forces per unit length on a soil element of thickness ΔH located at a depth z are as follows:

   Unreinforced Case: F_1 and F_2 are the vertical forces, and S_1 is the shear force. Hence, for equilibrium

$$F_1 - F_2 - S_1 = 0 \tag{11.32}$$

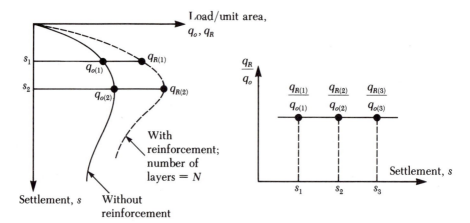

Figure 11.13   Relationship between load per unit area and settlement for foundations resting on reinforced and unreinforced soil

(a) Foundation of unreinforced soil

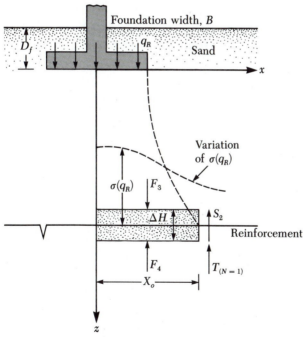

(b) Foundation of reinforced soil
    (one layer reinforcement)

Figure 11.14   Derivation of Eq. (11.50)

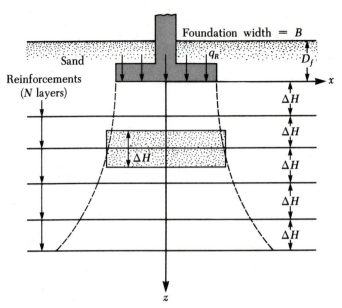

(c) Foundation on reinforced soil (N layers of reinforcement)

Figure 11.14   (Continued)

Reinforced Case: F_3 and F_4 are the vertical forces, S_2 is the shear force, and $T_{(N=1)}$ is the tensile force developed in the reinforcement. The force $T_{(N=1)}$ is vertical due to the assumption made for the deformation of reinforcement as shown in Figure 11.12c. So

$$F_3 - F_4 - S_2 - T_{(N=1)} = 0 \tag{11.33}$$

If the settlement of foundation (s) is the same in both cases

$$F_2 = F_4 \tag{11.34}$$

Subtracting Eq. (11.32) from Eq. (11.33) and using the relationship given in Eq. (11.34), one obtains

$$T_{(N=1)} = F_3 - F_1 - S_2 + S_1 \tag{11.35}$$

Note that the force F_1 is caused by the vertical stress, σ, on the soil element under consideration as a result of the load q_o on the foundation. Similarly, F_3 is caused by the vertical stress imposed on the soil element as a result of the load q_R. Hence

$$F_1 = \int_0^{x_o} \sigma(q_o) \cdot dx \tag{11.36}$$

$$F_3 = \int_0^{x_o} \sigma(q_R) \cdot dx \tag{11.37}$$

$$S_1 = \tau_{xz}(q_o) \cdot \Delta H \tag{11.38}$$

$$S_2 = \tau_{xz}(q_R) \cdot \Delta H \tag{11.39}$$

where $\sigma(q_o)$ and $\sigma(q_R)$ are the vertical stresses at a depth z caused by the loads q_o and q_R on the foundation; and $\tau_{xz}(q_o)$ and $\tau_{xz}(q_R)$ are the shear stresses at a depth z and at a distance X_o from the center line caused by the loads q_o and q_R. Using integration of Boussinesq's solution

$$\sigma(q_o) = \frac{q_o}{\pi}\left[\tan^{-1}\frac{z}{x-b} - \tan^{-1}\frac{z}{x+b} - \frac{2bz(x^2 - z^2 - b^2)}{(x^2 + z^2 - b^2)^2 + 4b^2z^2}\right]$$

(11.40)

$$\sigma(q_R) = \frac{q_R}{\pi}\left[\tan^{-1}\frac{z}{x-b} - \tan^{-1}\frac{z}{x+b} - \frac{2bz(x^2 - z^2 - b^2)}{(x^2 + z^2 - b^2)^2 + 4b^2z^2}\right]$$

(11.41)

$$\tau_{xz}(q_o) = \frac{4bq_o X_o z^2}{\pi[X_o^2 + z^2 - b^2)^2 + 4b^2z^2]}$$

(11.42)

$$\tau_{xz}(q_R) = \frac{4bq_R X_o z^2}{\pi[(X_o^2 + z^2 - b^2)^2 + 4b^2z^2]}$$

(11.43)

where $b = B/2$

The procedure for derivation of Eqs. (11.40) to (11.43) will not be presented here; for this information readers are referred to a soil mechanics text (for example, Das, 1983). Proper substitution of Eqs. (11.40) to (11.43) into Eqs. (11.36) to (11.39) and simplification will yield the following:

$$F_1 = A_1 q_o B \tag{11.44}$$
$$F_3 = A_1 q_R B \tag{11.45}$$
$$S_1 = A_2 q_o \Delta H \tag{11.46}$$
$$S_2 = A_2 q_R \Delta H \tag{11.47}$$

where A_1 and $A_2 = f(z/B)$

The variations of A_1 and A_2 with nondimensional depth z are given in Figure 11.15.

Substitution of Eqs. (11.44) through (11.47) into Eq. (11.35) gives

$$\begin{aligned} T_{(N=1)} &= A_1 q_R B - A_1 q_o B - A_2 q_R \Delta H + A_2 q_o \Delta H \\ &= A_1 B(q_R - q_o) - A_2 \Delta H(q_R - q_o) \\ &= q_o\left(\frac{q_R}{q_o} - 1\right)(A_1 B - A_2 \Delta H) \end{aligned}$$

(11.48)

Note that Eq. (11.48) has been derived with the assumption that there is only one layer of reinforcement under the foundation shown in Figure 11.14b. However, if there are N layers of reinforcement under the foundation with center-to-center spacings of ΔH, as shown in Figure 11.14c, it can be assumed that

$$T_{(N)} = \frac{T_{(N=1)}}{N} \tag{11.49}$$

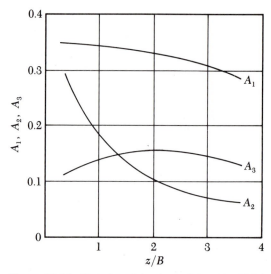

Figure 11.15   Variation of A_1, A_2, and A_3 with z/B (after Binquet and Lee, 1975b)

Combining Eqs. (11.48) and (11.49)

$$T_{(N)} = \frac{1}{N}\left[q_o\left(\frac{q_R}{q_o} - 1\right)(A_1 B - A_2 \Delta H)\right] \tag{11.50}$$

The unit of $T_{(N)}$ in Eq. (11.50) is kN/m (or lb/ft) length of foundation.

11.7   Factor of Safety of Ties Against Breaking and Pullout

Once the tie forces that develop in each layer as the result of the foundation load are determined by Eq. (11.50), one must determine whether or not the ties at any given depth z will fail either by *breaking* or by *pullout*. The factor of safety against tie breaking at any depth z below the foundation can be calculated as

$$FS_{(B)} = \frac{wtnf_y}{T_{(N)}} \tag{11.51}$$

where $FS_{(B)}$ = factor of safety against tie breaking
     w = width of a single tie
     t = thickness of each tie
     n = number of ties per unit length of the foundation
     f_y = yield or breaking strength of the tie material

The term $w \cdot n$ may be defined as the *linear density ratio (LDR)*, so

$$FS_{(B)} = \left[\frac{tf_y}{T_{(N)}}\right](LDR) \tag{11.52}$$

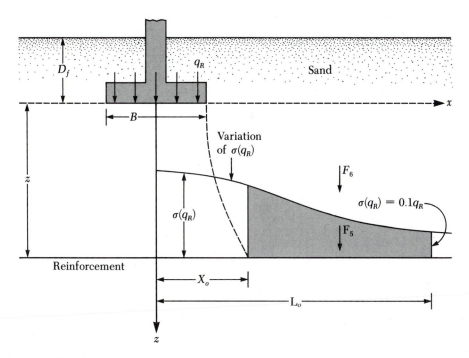

Figure 11.16   Derivation of Eq. (11.54)

The resistance against the tie being pulled out derives from the frictional resistance between the soil and the ties at any given depth. From the fundamental principles of statics, we know that the frictional force per unit length of the foundation resisting tie pullout at any depth z is equal to (Figure 11.16)

$$F_B = 2 \tan \phi_\mu [\text{normal force}]$$

$$= 2 \tan \phi_\mu \left[(LDR) \underbrace{\int_{X_o}^{L_o} \sigma(q_R) \, dx}_{\substack{\text{Due to foundation} \\ \text{load} = F_5}} + \underbrace{(LDR)(\gamma)(L_o - X_o)(z + D_f)}_{\substack{\text{Due to effective} \\ \text{overburden} \\ \text{pressure} = F_6}} \right] \qquad (11.53)$$

$\uparrow$ Two sides of tie, i.e., top and bottom

where γ = unit weight of soil
$\quad D_f$ = depth of foundation
$\quad \phi_\mu$ = tie-soil friction angle

The relation for $\sigma(q_R)$ has been defined in Eq. (11.41). The value of $x = L_o$ is generally assumed to be the distance at which the value of $\sigma(q_R)$ is equal to $0.1q_R$. The value of L_o as a function of depth z is given in Figure

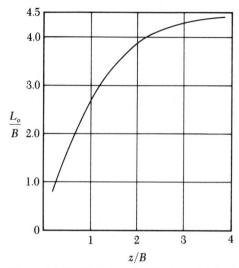

Figure 11.17 Variation of L_o/B with z/B (after Binquet and Lee, 1975b)

11.17. Equation (11.53) may be simplified as

$$F_B = 2 \tan \phi_\mu (LDR) \left[A_3\, Bq_o \left(\frac{q_R}{q_o} \right) + \gamma(L_o - X_o)(z + D_f) \right] \qquad (11.54)$$

where A_3 is a nondimensional quantity and can be expressed as a function of depth (z/B) (see Figure 11.15).

The factor of safety against tie pullout, $FS_{(P)}$, can be given as

$$FS_{(P)} = \frac{F_B}{T_{(N)}} \qquad (11.55)$$

11.8   Design Procedure

Following is a step-by-step procedure for the design of a continuous foundation supported by reinforced earth:

1. Obtain the total load to be supported per unit length of the foundation. Also obtain the following quantities:

    a. Soil friction angle, ϕ
    b. Soil-tie friction angle, ϕ_μ
    c. Factor of safety against bearing capacity failure
    d. Factor of safety against tie breaking, $FS_{(B)}$
    e. Factor of safety against tie pullout, $FS_{(P)}$
    f. Breaking strength of reinforcement ties, f_y
    g. Unit weight of soil, γ
    h. Young's modulus of soil, E_s
    i. Poisson's ratio of soil, μ_s
    j. Allowable settlement of foundation, s
    k. Depth of foundation, D_f

2. Assume a width of foundation, B, and also d and N. The value of d should be less than $\frac{2}{3}B$. Also, the distance from the bottom of the foundation to the lowest layer of the reinforcement should be about $2B$ or less. Calculate ΔH.

3. Assume a value of LDR.

4. For the assumed value of B (Step 2), determine the ultimate bearing capacity (q_u) for the condition of unreinforced soil [Eq. (3.3); *note: c = 0*]. Determine $q_{all(1)}$ as

$$q_{all(1)} = \frac{q_u}{FS \text{ against bearing capacity failure}} \tag{11.56}$$

5. Calculate the allowable load [$q_{all(2)}$] based on tolerable settlement (s), assuming that the soil is not reinforced [Eq. (3.67)].

$$s = \frac{Bq_{all(2)}}{E_s}(1 - \mu_s^2)\alpha_r$$

For $L/B = \infty$, the value of α_r may be taken to be equal to 2, or

$$q_{all(2)} = \frac{E_s s}{B(1 - \mu_s^2)\alpha_r} \tag{11.57}$$

(The allowable load for a given settlement, s, could have been determined from equations in Chapter 3 that relate to standard penetration resistances.)

6. Determine the lower of the two values of q_{all} as obtained from Steps 4 and 5. The lower value of q_{all} is equal to q_o.

7. Calculate the value of q_R for the foundation supported by reinforced earth as

$$q_R = \frac{\text{load on foundation per unit length}}{B} \tag{11.58}$$

8. Calculate the tie force, $T_{(N)}$, in each layer of reinforcement by using Eq. (11.50) (*note:* unit of $T_{(N)}$ as kN/m of foundation).

9. Calculate the frictional resistance of ties for each layer per unit length of foundation (F_B) by using Eq. (11.54). For each layer, determine if $F_B/T_{(N)}$ is equal to or more than the required factor of safety against tie pullout [$FS_{(P)}$]. If $F_B/T_{(N)}$ is less than the required value of $FS_{(P)}$, the length of the reinforcing strips for any given layer may be increased. This will increase the value of F_B and thus $FS_{(P)}$, and so Eq. (11.54) will need to be rewritten as

$$F_B = 2 \tan \phi_\mu (LDR)\left[A_3 Bq_o\left(\frac{q_R}{q_o}\right) + \gamma(L - X_o)(z + D_f)\right] \tag{11.59}$$

where L = the required length to obtain the desired value of F_B

10. Use Eq. (11.52) to obtain the tie thickness for each layer. Some allowance should be made to take into account the corrosion effect of the reinforcements during the life span of the structure.

11. If the design is unsatisfactory, the entire process may be repeated—that is, Steps 2 through 10.

The following example demonstrates the steps outlined above.

Example 11.3

Design a continuous foundation that will carry a load of 1.8 MN/m. Given the following:

Soil: $\gamma = 17.3$ kN/m^3; $\phi = 35°$; $E_s = 3 \times 10^4$ kN/m^2; $\mu_s = 0.35$

Reinforcement ties: $f_y = 2.5 \times 10^5$ kN/m^2; $\phi_\mu \doteq 28°$; $FS_{(B)} = 3$; $FS_{(P)} = 2.5$

Foundation: $D_f = 1$ m; factor of safety against bearing capacity failure $= 3$, tolerable settlement $= s = 25$ mm; desired life span $= 50$ years.

Solution

Let

$$B = 1 \text{ m}$$
$$d = \text{depth from the bottom of the foundation to the first reinforcing}$$
$$\text{layer} = 0.5 \text{ m}$$
$$\Delta H = 0.5 \text{ m}$$
$$N = 5$$
$$LDR = 65\%$$

If the reinforcing strips used are 75 mm wide, then

$$w \cdot n = LDR$$

or

$$n = \frac{LDR}{w} = \frac{0.65}{0.075 \text{ m}} = \textbf{8.67/m}$$

Hence, there will be 8.66 strips in each layer per meter length of the foundation.

Determination of q_o

For an unreinforced foundation

$$q_u = \gamma D_f N_q + \frac{1}{2} \gamma B N_\gamma$$

From Table 3.2, for $\phi = 35°$, $N_q = 33.30$ and $N_\gamma = 48.03$. Thus

$$q_u = (17.3)(1)(33.3) + \frac{1}{2}(17.3)(1)(48.03)$$

$$= 576.09 + 415.46 = 991.55 \approx 992 \text{ kN/m}^2$$

$$q_{all(1)} = \frac{q_u}{FS} = \frac{992}{3} = 330.7 \text{ kN/m}^2$$

From Eq. (11.57)

$$q_{all(2)} = \frac{(E_s)(s)}{B(1 - \mu_s^2)\alpha_r} = \frac{(30{,}000 \text{ kN/m}^2)(0.025 \text{ m})}{(1 \text{ m})(1 - 0.35^2)(2)} = 427.35 \text{ kN/m}^2$$

Since $q_{all(1)} < q_{all(2)}$, $q_o = q_{all(1)} = \textbf{330.7 kN/m}^2$

Determination of q_R

From Eq. (11.58)

$$q_R = \frac{1.8 \text{ MN/m}}{B} = \frac{1.8 \times 10^3}{1} = 1.8 \times 10^3 \text{ kN/m}^2$$

Calculation of Tie Force

From Eq. (11.50)

$$T_{(N)} = \left(\frac{q_o}{N}\right)\left(\frac{q_R}{q_o} - 1\right)(A_1 \cdot B - A_2 \cdot \Delta H)$$

The tie forces for each layer are given in the following table.

Layer no.	$\left(\dfrac{q_o}{N}\right)\left(\dfrac{q_R}{q_o} - 1\right)$	z (m)	$\dfrac{z}{B}$	$A_1 B$	$A_2 \Delta H$	$A_1 B - A_2 \Delta H$	$T_{(N)}$ (kN/m)
1	293.7	0.5	0.5	0.35	0.125	0.225	66.08
2	293.7	1.0	1.0	0.34	0.09	0.25	73.43
3	293.7	1.5	1.5	0.34	0.065	0.275	80.77
4	293.7	2.0	2.0	0.33	0.05	0.28	82.24
5	293.7	2.5	2.5	0.32	0.04	0.28	82.24

Note: A_1 from Figure 11.15; $B = 1$ m; $\Delta H = 0.5$ m; A_2 from Figure 11.15; $q_R/q_o = 1.8 \times 10^3/330.7 \approx 5.45$

Calculation of Tie Resistance Due to Friction, F_B

This can be done by using Eq. (11.54):

$$F_B = 2 \tan \phi_\mu(LDR)\left[A_3 B q_o\left(\frac{q_R}{q_o}\right) + \gamma(L_o - X_o)(z + D_f)\right]$$

The following table shows the value of F_B for each layer.

Quantity	Layer no.				
	1	2	3	4	5
$2 \tan \phi_\mu(LDR)$	0.691	0.691	0.691	0.691	0.691
A_3	0.125	0.14	0.15	0.15	0.15
$A_3 B q_o(q_R/q_o)$	225.0	252.0	270.0	270.0	270.0
z (m)	0.5	1.0	1.5	2.0	2.5
z/B	0.5	1.0	1.5	2.0	2.5
L_o (m)	1.55	2.6	3.4	3.85	4.2
X_o (m)	0.55	0.8	1.1	1.4	1.65
$L_o - X_o$ (m)	1.0	1.8	2.3	2.45	2.55
$z + D_f$ (m)	1.5	2.0	2.5	3.0	3.5
$\gamma(L_o - X_o)(z + D_f)$	25.95	62.28	99.48	127.16	154.4
F_B (kN/m)	173.4	217.2	255.1	274.4	293.3
$FS_{(P)} = F_B/T_{(N)}$	2.62	2.96	3.16	3.34	3.57

Note: A_3 from Figure 11.15; X_o from Figure 11.12; L_o from Figure 11.17; $T_{(N)}$ from the preceding table

The minimum factor of safety is greater than the required value of $FS_{(P)}$, which is 2.5.

Calculation of Tie Thickness to Resist Tie Breaking

From Eq. (11.52)

$$FS_{(B)} = \frac{tf_y}{T_{(N)}} (LDR)$$

or

$$t = \frac{FS_{(B)} T_{(N)}}{(LDR)(f_y)}$$

Given: $f_y = 2.5 \times 10^5 \text{ kN/m}^2$; $LDR = 0.65$; $FS_{(B)} = 3$. So

$$t = \left[\frac{3}{(2.5 \times 10^5)(0.65)} \right] T_{(N)} = (1.846 \times 10^{-5}) T_{(N)}$$

So, for layer 1

$$t = (1.846 \times 10^{-5})(66.08) = 0.00122 \text{ m} = \textbf{1.22 mm}$$

For layer 2

$$t = (1.846 \times 10^{-5})(73.43) = 0.00136 \text{ m} = \textbf{1.36 mm}$$

Similarly, for layer 3

$$t = 0.00149 = \textbf{1.49 mm}$$

For layer 4

$$t = \textbf{1.52 mm}$$

For layer 5

$$t = \textbf{1.52 mm}$$

So, in each layer, ties with a thickness of 1.6 mm will be sufficient. However, if galvanized steel is used, the rate of corrosion is about 0.025 mm/year. So, t should be equal to $1.6 + (0.025)(50) = \textbf{2.85 mm.}$

Calculation of Minimum Length of Ties

The minimum length of ties in each layer should be equal to $2L_o$. Following is the length of ties in each layer.

Layer no.	Minimum length of the tie, $2L_o$ (m)
1	3.1
2	5.2
3	6.8
4	7.7
5	8.4

Figure 11.18 is a diagram of the foundation with the ties. The design could have been changed by varying B, d, N, and ΔH to determine the most economical combination.

Figure 11.18

Example 11.4

Refer to Example 11.3. For the loading given, determine the width of the foundation that is needed for an unreinforced earth condition. Note that the factor of safety against bearing capacity failure is 3, and the tolerable settlement is 25 mm.

Solution
Bearing Capacity Consideration

For a continuous foundation

$$q_u = \gamma D_f N_q + \frac{1}{2} \gamma B N_\gamma$$

For $\phi = 35°$, $N_q = 33.3$ and $N_\gamma = 48.03$. So

$$q_{all} = \frac{q_u}{FS} = \frac{1}{FS} \left[\gamma D_f N_q + \frac{1}{2} \gamma B N_\gamma \right]$$

or

$$q_{all} = \frac{1}{3} \left[(17.3)(1)(33.3) + \frac{1}{2} (17.3)(B)(48.03) \right] \qquad \text{(a)}$$
$$= 192.03 + 138.5B$$

However

$$q_{all} = \frac{1.8 \times 10^3 \text{ kN}}{(B)(1)} \qquad \text{(b)}$$

Equating the right sides of Eqs. (a) and (b)

$$\frac{1800}{(B)(1)} = 192.03 + 138.5B$$

From the solution of the preceding equation, $B \approx 3$ m. So, with $B = 3$ m, q_{all} is equal to **600 kN/m².**

Settlement Consideration

For a friction angle of $\phi = 35°$, the average standard penetration number is about 10–15 (Table 2.4). Let us assume the higher value, $N = 15$. From Eq. (3.107b)

$$q_{all} = 11.98N\left(\frac{3.28B + 1}{3.28}\right)^2\left(1 + \frac{0.33D_f}{B}\right) \quad \text{for a settlement of about 25 mm}$$

Now, we can make a few trials as shown below:

Assumed B (m) (1)	$q_{all} = 11.98N\left(\dfrac{3.28B + 1}{3.28B}\right)^2\left(1 + \dfrac{0.33D_f}{B}\right)$ (kN/m²) (2)	$Q = (B)(q_{all})$ = Col. 1 × Col. 2 (kN/m)
6	209	1254
9	199	1791[a]

Note: $D_f = 1$ m;
[a] Required 1800 kN/m

If $N = 15$, the width of the foundation should be 9 m or more. Based on the consideration of bearing capacity failure and tolerable settlement, the latter settlement criteria will control. So, B is about **9 m.**

Note: At first, the results of this calculation may show the use of reinforced earth for foundation construction to be desirable. However, several factors must be considered before a final decision is made. For example, reinforced earth needs over-excavation and backfilling. It is possible, under many circumstances, that proper material selection and compaction may make the construction of foundations on *unreinforced* soils more economical.

BEARING CAPACITY OF FOUNDATION ON SOIL REINFORCED BY LAYERS OF GEOTEXTILE

11.9   Foundation on Sand with Geotextile Reinforcement

Laboratory model tests for determination of the bearing capacity of a *square* foundation supported by loose sand (relative density = 50%) and reinforced by layers of nonwoven heat-bonded geotextile have been reported by Guido, Biesiadecki, and Sullivan (1985). Some of the results of these tests are shown in Figure 11.19b. For these tests, several parameters were varied. These parameters were as follows (Figure 11.19a): d; ΔH; L_o; number of layers of geotextile, N; and tensile strength of geotextile, σ_G. These test results, in general, show that when the geotextile layers are located within a depth equal to the width of the foundation they produce beneficial effects in increasing the load-bearing capacity of the foundation, but only after a measurable settlement has occurred. This is somewhat to be expected, since the geotextile layers have to deform before their reinforcing benefits are realized. More research needs to be conducted for establishing foundation design procedures under this condition.

(a)

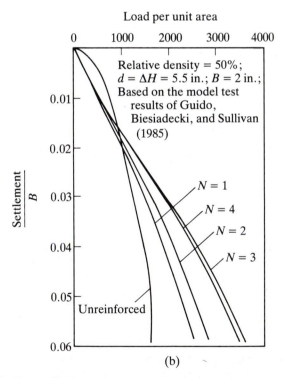

(b)

Figure 11.19   Bearing capacity test on square footing on loose sand with geotextile reinforcement (*note: N* = number of layers of reinforcement)

11.10   Foundation on Saturated Clay ($\phi = 0$ Condition) with Geotextile Reinforcement

Studies relating to determination of the bearing capacity of a shallow foundation supported by a saturated clay layer reinforced by geotextile, similar to that described in Section 11.9, are rather limited at this time. More recently, Sakti and Das (1987) have provided some model test results on the bearing capacity of a *strip* foundation on saturated clay. For their tests, a heat-bonded nonwoven geotextile was used for reinforcement (grab tensile strength = 534 N). Some of the load-settlement curves thus derived are shown in Figure 11.20.

Figure 11.20 Bearing capacity test on strip footing on saturated clay with geotextile reinforcement (*note:* N = number of layers of reinforcement)

From their tests, the following general conclusions can be drawn:

1. Beneficial effects of geotextile reinforcement are realized when reinforcement is placed within a distance equal to the width of the foundation.

2. The first layer of geotextile reinforcement should be placed at a distance, $d = 0.35B$ (B = foundation width) for maximum benefit.

3. The most economical value of L_o/B is about 2. (See Figure 11.19a for a definition of L_o.)

More studies of this type have to be conducted for developing proper foundation design criteria.

PROBLEMS

11.1 Refer to Figure 11.5. Given:

$$Wall: H = 6 \text{ m}$$
$$Soil: \gamma_1 = 16.5 \text{ kN/m}^3, \phi_1 = 35°$$
$$Reinforcement: S_V = 1 \text{ m}, S_H = 1.5 \text{ m}$$
$$Surcharge: q = 50 \text{ kN/m}^2, a' = 1.5 \text{ m}, b' = 2 \text{ m}$$

Calculate the vertical stress, σ_v [Eqs. (11.2), (11.3), and (11.4)] at $z = 1$ m, 2 m, 3 m, 4 m, 5 m, and 6 m.

11.2 For Problem 11.1, calculate the lateral pressure σ_a at $z = 1$ m, 2 m, 3 m, 4 m, 5 m, and 6 m. Use Eqs. (11.5), (11.6), and (11.7).

11.3 A reinforced earth retaining wall (Figure 11.5) is to be 30 ft high. Given the following:

Backfill: unit weight, $\gamma_1 = 119$ lb/ft^3, soil friction angle, $\phi_1 = 34°$
Reinforcement: vertical spacing, $S_V = 3$ ft; horizontal spacing, $S_H = 4$ ft, width of reinforcement $= 4.75$ in., $f_y = 38{,}000$ lb/in.^{2}; $\phi_\mu = 25°$; factor of safety against tie pullout $= 3$; factor of safety against tie breaking $= 3$

Determine the following:
a. The required thickness of ties
b. The required maximum length of ties

11.4 In Problem 11.3, assume that the ties of the same length at all depths [that is, the length determined in part (b) of Problem 11.3]. Given, for *in situ* soil: $\phi_2 = 25°$, $\gamma_2 = 116$ lb/ft^3, and $c_2 = 650$ lb/ft^2. Calculate the factor of safety against (a) overturning, (b) sliding, and (c) bearing capacity failure.

11.5 Redo Problem 11.4 for a retaining wall with a height of 24 ft.

11.6 In Problem 11.5, assume that the ties are of the same length at all depths [that is, the length determined in part (b) of Problem 11.5]. Given, for the *in situ* soil: $\phi_2 = 28°$, $\gamma_2 = 121$ lb/ft^3, and $c_2 = 500$ lb/ft. Calculate the factor of safety against (a) overturning, (b) sliding, and (c) bearing capacity failure.

11.7 Redo Problem 11.3 and change S_V to 1.5 ft.

11.8 A geotextile reinforced retaining wall is 6 m high. For the granular backfill, given: $\gamma_1 = 15.9$ kN/m^3 and $\phi_1 = 30°$. Given, for the geotextile: $\sigma_G = 16$ kN/m. For the design of the wall, determine S_V, L, and l_1. Use $FS_{(B)} = FS_{(P)} = 1.5$.

11.9 Refer to Problem 11.8. With proper assumption of S_V, L, and l_1, check the overall stability (that is, factor of safety against overturning, sliding, and bearing capacity failure). For the *in situ* soil, given: $\gamma_2 = 16.8$ kN/m^3, $\phi_2 = 20°$, and $c_2 = 55$ kN/m^2.

11.10 Check the overall stability (that is, factor of safety against overturning, sliding, and bearing capacity failure) of the geotextile reinforced retaining wall given in Example 11.2. Use $\gamma_2 = 108$ lb/ft^3, $\phi_2 = 20°$, and $c_2 = 1200$ lb/ft^2.

11.11 Figure 11.14c shows a continuous foundation on reinforced soil. Given: $B = 0.9$ m; $D_f = 1$ m; number of layers of reinforcement, $N = 5$; $\Delta H = 0.4$ m. Make the necessary calculations and plot the lines on both sides of the foundation that define the point of maximum shear stress $[\tau_{xz(max)}]$ on the reinforcements.

11.12 The tie forces under a continuous foundation are given by Eq. (11.50). For the foundation described in Problem 11.11, given: $q_o = 200$ kN/m² and $q_R/q_o = 4.5$. Determine the tie forces, $T_{(N)}$, in kN/m for each layer of reinforcement.

11.13 Repeat Problem 11.12 with $q_o = 300$ kN/m² and $q_R/q_o = 6$.

11.14 A continuous foundation (see Figure 11.14c) is to be built on reinforced earth to carry a load of 82.3 kips/ft. Given:

> *Foundation:* $B = 4$ ft, $D_f = 2.6$ ft, factor of safety against bearing capacity failure $= 3$, tolerable settlement $= 0.8$ in.
> *Soil:* $\gamma = 116$ lb/ft³, $\phi = 37°$, $E_s = 5200$ lb/in.², $\mu_s = 0.30$
> *Reinforcement:* $\Delta H = 1.3$ ft, $N = 5$, $LDR = 70\%$, width of reinforcement strips $= 0.23$ ft

Calculate:
a. The number of reinforcement strips per foot length of the foundation
b. The allowable load per unit area of the foundation (q_o) without reinforcement
c. The ratio of q_R/q_o
d. The tie forces for each layer of reinforcement under the foundation (kips/ft)

11.15 Refer to Problem 11.14. For the reinforcements, given: $f_y = 38,000$ lb/in.², $\phi_\mu = 25°$, factor of safety against tie breaking $= 2.5$, factor of safety against tie pullout $= 2.5$. Calculate:
a. The minimum thickness of ties needed to resist tie breaking
b. The minimum length of ties necessary for each layer of reinforcement

REFERENCES

Bell, J. R., Stilley, A. N., and Vandre, B. (1975). "Fabric Retaining Earth Walls," *Proceedings*, Thirteenth Engineering Geology and Soils Engineering Symposium, Moscow, Idaho.

Binquet, J., and Lee, K. L. (1975a). "Bearing Capacity Tests on Reinforced Earth Mass," *Journal of the Geotechnical Engineering Division*, American Society of Civil Engineers, Vol. 101, No. GT12, pp. 1241–1255.

Binquet, J., and Lee, K. L. (1975b). "Bearing Capacity Analysis of Reinforced Earth Slabs," *Journal of the Geotechnical Engineering Division*, American Society of Civil Engineers, Vol. 101, No. GT12, pp. 1257–1276.

Darbin, M. (1970). "Reinforced Earth for Construction of Freeways" (in French), *Revue Générale des Routes et Aerodromes*, No. 457, Sept.

Das, B. M. (1981). "Bearing Capacity of Eccentrically Loaded Surface Footings on Sand," *Soils and Foundations*, Vol. 21, No. 1, pp. 115–119.

Das, B. M. (1983). *Advanced Soil Mechanics*, McGraw-Hill, New York.

Guido, V. A., Biesiadecki, G. L., and Sullivan, M. J. (1985). "Bearing Capacity of a Geotextile Reinforced Foundation," *Proceedings*, Eleventh International Conference on Soil Mechanics and Foundation Engineering, San Francisco, Vol. 3, pp. 1777–1780.

Koerner, R. B. (1986). *Design with Geosynthetics*, Prentice-Hall, Englewood Cliffs, N.J.

Laba, J. T., and Kennedy, J. B. (1986). "Reinforced Earth Retaining Wall Analysis and Design." *Canadian Geotechnical Journal*, Vol. 23, No. 3, pp. 317–326.

Lee, K. L., Adams, B. D., and Vagneron, J. J. (1973). "Reinforced Earth Retaining Walls," *Journal of the Soil Mechanics and Foundations Division*, American Society of Civil Engineers, Vol. 99, No. SM10, pp. 745–763.

Mandel, J., and Salencon, J. (1972). "Force portante d'un sol sur une assise rigide (étude theorizué)," *Geotechnique*, Vol. 22, No. 1, pp. 79–93.

Meyerhof, G. G. (1974). "Ultimate Bearing Capacity of Footings on Sand Layer Overlying Clay," *Canadian Geotechnical Journal*, Vol. 11, No. 2, pp. 223–229.

Pfeifle, T. W., and Das, B. M. (1979). "Bearing Capacity of Surface Footings on Sand Layer Resting on a Rigid Rough Base," *Soils and Foundations*, Vol. 19, No. 1, pp. 1–11.

Sakti, J., and Das, B. M. (1987). "Model Tests for Strip Foundation on Clay Reinforced with Geotextile Layers," *Transportation Research Record No. 1153*, National Academy of Sciences, Washington, D.C., pp. 40–45.

Schlosser, F., and Long, N. (1974). "Recent Results in French Research on Reinforced Earth," *Journal of the Construction Division*, American Society of Civil Engineers, Vol. 100, No. CO3, pp. 113–237.

Schlosser, F., and Vidal, H. (1969). "Reinforced Earth" (in French), *Bulletin de Liaison des Laboratoires Routier*, Ponts et Chaussées, Paris, France, Nov., pp. 101–144.

Vidal, H. (1966). "La terre Armee," *Anales de l'Institut Technique du Bâtiment et des Travaux Publiques*, France, July–August, pp. 888–938.

Chapter *12*

SOIL IMPROVEMENT

12.1   Introduction

The existing soil at a construction site may not always be totally suitable for supporting structures such as buildings, bridges, highways, and dams. For example, in granular soil deposits the *in situ* soil may be very loose and could present a large elastic settlement. In such a case, the soil needs to be densified to increase its unit weight and thus the shear strength.

Sometimes the top layers of soil are undesirable and must be removed and replaced with a better quality soil, on which the structural foundation can be built. The soil that is used as a fill needs to be well compacted to sustain the desired structural load. Compacted fills may also be required in low-lying areas to raise the ground elevation for foundation construction.

Soft saturated clay layers are often encountered at shallow depths below the foundation(s). Depending on the structural load and the depth of the clay layer(s), unusually large consolidation settlement may occur. Special soil-improvement techniques are required to overcome such settlement problems.

Chapter 10 mentioned that the properties of expansive soils could be substantially altered by the addition of stabilizing agents such as lime. Improving *in situ* soils by using additives is usually referred to as *stabilization*.

This chapter briefly discusses various techniques for improving the soil to

1. Reduce the settlement of structures

2. Improve the shear strength of soil and thus increase the bearing capacity of shallow foundations

3. Increase the factor of safety against possible slope failure of embankments and earth dams

4. Reduce the shrinkage and swelling characteristics of soils

The chapter is divided into seven major parts: (1) compaction, (2) vibroflotation, (3) precompression, (4) sand drains, (5) soil stabilization by admixtures, (6) use of stone columns and granular trenches in weak clays for foundation construction, and (7) use of geogrids.

12.2 Compaction—General Principles

If a small amount of water is added to a soil that is then subjected to compaction by a given amount of energy, the soil will be compacted to a certain unit weight. If the moisture content of the same soil is gradually increased and the compaction is done in the same way, the dry unit weight of the soil will gradually increase. This is because water behaves as a lubricant between the soil particles; under compaction it helps rearrange the solid particles into a denser state. The increase in dry unit weight with increase of moisture content for a given soil will reach a limiting value beyond which further addition of water to the soil will result in a *reduction* of dry unit weight. The moisture content at which the *maximum dry unit weight* is obtained is referred to as the *optimum moisture content*.

The standard laboratory tests that are used for evaluation of maximum dry unit weights and optimum moisture contents for various soils are (a) the *standard Proctor test* [ASTM designation D-698; Title: Moisture Density Relations of Soils and Soil-Aggregate Mixtures Using 5.5 lb (2.49 kg) Rammer and 12 in. (305 mm) Drop] and (b) the *modified Proctor test* [ASTM designation D-1557; Title: Moisture Density Relations of Soils and Soil-Aggregate Mixtures Using 10 lb (4.54 kg) Rammer and 18 in. (457 mm) Drop]. Both of these tests are conducted in a mold having a volume of 0.944×10^{-3} m^3 (1/30 ft^3). The soil is compacted in several layers by a hammer. The moisture content of the soil (w) is changed, and the dry unit weight (γ_d) of compaction for each test is determined. The maximum dry unit weight of compaction and the corresponding optimum moisture content are determined by plotting a graph of γ_d against w (%). The standard specifications for the two types of Proctor test are given in Table 12.1.

Table 12.1 Specifications for Standard and Modified Proctor Tests

No.	Item	Specifications	
		Standard Proctor	Modified Proctor
1	Volume of mold	0.944×10^{-3} m^3 (1/30 ft^3)	0.944×10^{-3} m^3 (1/30 ft^3)
2	Mass of hammer	2.495 kg (5.5 lb)	4.536 kg (10 lb)
3	Height of drop of the hammer	304.8 mm (12 in.)	457 mm (18 in.)
4	Number of hammer blows per layer of soil	25	25
5	Number of layers of compaction	3	5
6	Energy of compaction	593 kilojoules/m^3 (12,375 ft-lb/ft^3)	2695 kilojoules/m^3 (56,250 ft-lb/ft^3)

Figure 12.1 Standard and modified Proctor compaction curves for a clayey silt

Figure 12.1 shows the plot of γ_d against w (%) for a clayey silt as obtained for standard and modified Proctor tests. From this figure, the following observations can be made:

1. The maximum dry unit weight and the optimum moisture content are dependent on the level of energy of compaction.

2. The higher the energy of compaction, the higher is the maximum dry unit weight.

3. The higher the energy of compaction, the lower is the optimum moisture content.

4. No portion of the compaction curve can lie to the right side of the zero-air-void line. The zero-air-void dry unit weight (γ_{zav}) at a given moisture content is the theoretical maximum value of γ_d, which means that all of the void spaces of the compacted soil are filled with water. This can be given by the equation

$$\gamma_{zav} = \frac{\gamma_w}{\dfrac{1}{G_s} + w} \tag{12.1}$$

where γ_w = unit weight of water
$\quad\quad G_s$ = specific gravity of the soil solids
$\quad\quad w$ = moisture content

5. The values of the maximum dry unit weight of compaction and the corresponding optimum moisture content will vary from soil to soil.

Using the results of the laboratory compaction (γ_d against w), specifications can be written for the compaction of a given soil in the field. In most cases, the contractor is required to achieve a relative compaction of 90% or more on the basis of a specific laboratory test (that is, standard or modified Proctor compaction test). Relative compaction (RC) is defined as

$$RC = \frac{\gamma_{d(\text{field})}}{\gamma_{d(\text{max})}} \tag{12.2}$$

Chapter 1 introduced the concept of relative density (for compaction of granular soils). Relative density was defined as

$$D_r = \left[\frac{\gamma_d - \gamma_{d(\text{min})}}{\gamma_{d(\text{max})} - \gamma_{d(\text{min})}} \right] \frac{\gamma_{d(\text{max})}}{\gamma_d}$$

where D_r = relative density
$\quad\quad \gamma_d$ = dry unit weight of compaction in the field
$\quad \gamma_{d(\text{max})}$ = maximum dry unit weight of compaction as determined in the laboratory
$\quad \gamma_{d(\text{min})}$ = minimum dry unit weight of compaction as determined in the laboratory

For granular soils in the field, the degree of compaction obtained is often measured in terms of relative density. Comparing the expressions for relative density and relative compaction, one can see that

$$RC = \frac{A}{1 - D_r(1 - A)} \tag{12.3}$$

where $A = \dfrac{\gamma_{d(\text{min})}}{\gamma_{d(\text{max})}}$

Lee and Singh (1971) have reviewed 47 different soils, and, based on that, they have given the following correlation:

$$D_r\,(\%) = \frac{(RC - 80)}{0.2} \tag{12.4}$$

One-Point Method of Obtaining $\gamma_{d(max)}$

The state highway department of Ohio has developed a family of standard curves of their various soil types; these are shown in Figure 12.2. Note that they are plots of *wet unit weight* (γ) against moisture content ($w\%$). These curves can be used to obtain $\gamma_{d(max)}$ in the field. This technique is referred to as the *one-point* method; it serves as a rapid means of field compaction control. In this method one first conducts a standard Proctor test with the soil in use and determines the moist unit weight of compaction and the corresponding moisture content. Then one plots the values of γ and w (%) according to Figure 12.2 and identifies the compaction curve number that corresponds to the test results. Using this curve number, one can refer to

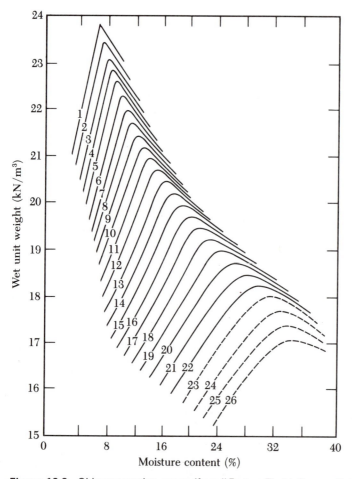

Figure 12.2 Ohio compaction curves (from " Factors That Influence Field Compaction of Soils," by A. W. Johnson and J. R. Sallberg, *Bulletin No. 262*, Highway Research Board, 1960. Reprinted by permission)

Table 12.2 to determine the maximum dry unit weight and the corresponding optimum moisture content.

The one-point method just described appears to be simple and easy to use. However, this may not always be the case. During the past several years researchers have determined that not all soils yield the bell-shaped compaction curves shown in Figure 12.2. Lee and Suedkamp (1972) performed 700 compaction tests on 35 soil samples. Their results show that, depending on the property of the soil, the plot of γ_d against w (%) may exhibit one of four different shapes. These shapes are shown in Figure 12.3 and are marked Types I, II, III, and IV. Type I is a standard bell-shaped curve. Type II is a curve showing one and one-half peaks. Type III is a double-peak curve. Type IV is an oddly shaped curve that shows no distinct optimum moisture content. Lee and Suedkamp (1972) developed the following guidelines to

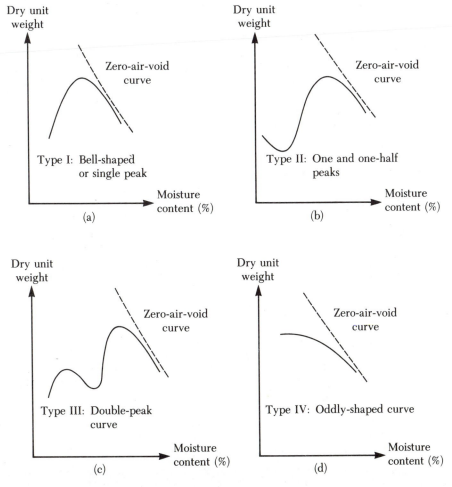

Figure 12.3 Various types of compaction curve

Table 12.2   Maximum Dry Unit Weight and Optimum Moisture Content for the Compaction Curves in Figure 12.2

Curve no.	Maximum dry unit weight (kN/m^3)	Optimum moisture content (%)	Curve no.	Maximum dry unit weight (kN/m^3)	Optimum moisture content (%)
1	22.29	6.6	14	17.23	16.9
2	21.87	7.2	15	16.84	18.1
3	21.43	7.9	16	16.46	19.2
4	21.08	8.5	17	16.10	20.3
5	20.75	9.0	18	15.71	21.5
6	20.33	9.7	19	15.31	22.7
7	19.90	10.5	20	14.87	24.4
8	19.53	11.2	21	14.48	25.8
9	19.13	11.9	22	14.13	27.4
10	18.76	12.7	23	13.76	29.5
11	18.39	13.5	24	13.36	30.5
12	18.02	14.6	25	13.05	31.5
13	17.61	15.8	26	12.75	32.5

From " Factors That Influence Field Compaction of Soils," by A. W. Johnson and J. R. Sallberg, *Bulletin No. 262*, Highway Research Board, 1960. Reprinted by permission.

help predict the nature of compaction curves that may be obtained from various soils.

Liquid limit of soil	Nature of compaction curve to be expected
30 to 70	Type I
Less than 30	Types II and III
Greater than 70	Types III and IV

Example 12.1

A soil was compacted in the field by the standard Proctor test procedure. Following are the results:

> Weight of compacted wet soil in the mold = 17.2 N
> Moisture content = 14%

Using Figure 12.2, determine $\gamma_{d(min)}$ and the optimum moisture content.

Solution

Moist unit weight of compaction =

$$\gamma = \frac{17.2 \text{ N}}{\text{volume of Proctor mold}} = \frac{17.2}{0.944 \times 10^{-3}} = 18.22 \text{ kN/m}^3$$

According to Figure 12.2 (with $\gamma = 18.22$ kN/m^3 and $w = 14\%$), it appears that the soil falls between the curve numbers 15 and 16. Now, according to Table 12.2, $\gamma_{d(max)}$ is between 16.84 kN/m^3 and 16.48 kN/m^3, and the optimum moisture content is

between 18.1–19.2%. So, for this soil

$$\gamma_{d(\max)} \approx 16.6 \text{ kN/m}^3$$

and optimum moisture content $\approx$ **18.5** %.

Field Compaction

Ordinary compaction in the field is done by means of rollers. Several types of roller are used, and the most common types include the following:

1. Smooth wheel rollers (or smooth drum rollers)
2. Pneumatic rubber-tired rollers
3. Sheepsfoot rollers
4. Vibratory rollers

Figure 12.4 shows a *smooth wheel roller* that can also create vertical vibration during compaction. Smooth wheel rollers are suitable for proof-rolling subgrades and for finishing the construction of fills with sandy or clayey soils. They provide 100% coverage under the wheels, and the contact pressure can be as high as 300–400 kN/m² ($\approx$45–60 lb/in²). However, they do not produce uniform unit weight of compaction when used on thick layers.

Figure 12.4   Vibratory smooth wheel rollers (courtesy of Tampo Manufacturing Co., Inc., San Antonio, Texas)

Figure 12.5 Pneumatic rubber-tired roller (courtesy of Tampo Manufacturing Co., Inc., San Antonio, Texas)

Pneumatic rubber-tired rollers (Figure 12.5) are better in many respects than smooth wheel rollers. These rollers, which may weigh up to 2000 kN (450 kips), consist of a heavily loaded wagon with several rows of tires. These tires are closely spaced—about four to six in a given row. The contact pressure under the tires may range up to 600–700 kN/m²($\approx$85–100 lb/in²), and they produce about 70–80% coverage. Pneumatic rollers can be used for sandy and clayey soil compaction. They produce a combination of pressure and kneading action.

Sheepsfoot rollers (Figure 12.6) consist basically of drums with large numbers of projections. The area of each of the projections may be 25–85 cm². These rollers are *most effective in compacting cohesive soils*. The contact pressure under the projections may range from 1500 to 7500 kN/m² ($\approx$215–1090 lb/in²). During compaction in the field, the initial passes compact the lower portion of a lift. Later, compaction is produced in the middle and top of the lift.

Vibratory rollers are efficient in compacting granular soils. Vibrators can be attached to smooth wheel, pneumatic rubber-tired, or sheepsfoot rollers to send vibrations into the soil under compaction. Figures 12.4 and 12.6 show vibratory smooth wheel rollers and a vibratory sheepsfoot roller.

In general, compaction in the field depends on several factors, such as the type of compactor, soil type, moisture content, lift thickness, towing speed of the compacter, and the number of roller passes. Figure 12.7 shows

Figure 12.6   Vibratory sheepsfoot roller (courtesy of Tampo Manufacturing Co., Inc., San Antonio, Texas)

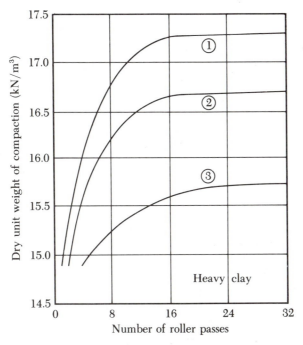

Figure 12.7   Relation between dry unit weight of compaction for the upper 150 mm of soil and the number of passes of pneumatic-tired roller (from "Factors That Influence Field Compaction of Soils," by A. W. Johnson and J. R. Sallberg, *Bulletin No. 262*, Highway Research Board, 1960. Reprinted by permission)

Table 12.3 Details of the Variables for the Three
Curves Shown in Figure 12.7[a]

Curve no.	1	2	3
Moisture content as rolled (%)	19.0	20.0	24.0
Optimum moisture content—standard Proctor test (%)	22.8	22.8	22.8
Roller rating (kN)	416.0	416.0	120.0
Wheel load (kN)	99.6	49.8	13.3
Tire pressure (kN/m²)	966.0	621.0	248.4
Loose lift thickness (mm)	305.0	305.0	229.0

[a] After Johnson and Sallberg (1960)

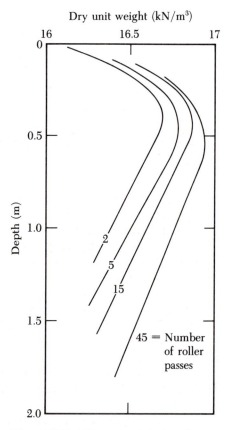

Figure 12.8 Vibratory compaction of a sand—variation of dry unit weight with depth and number of roller passes. Lift thickness = 2.44 m (after D'Appolonia, Whitman, and D'Appolonia, 1969)

the variation of the dry unit weight of a heavy clay with the number of passes of pneumatic-tired rollers. Table 12.3 gives the details of the variables for the three curves shown in Figure 12.7.

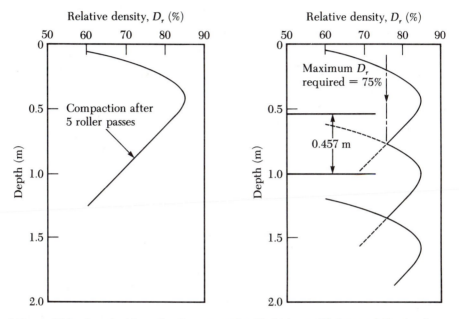

Figure 12.9 A method for estimating compaction lift thickness. Minimum relative density required is 75% after five roller passes (after D'Appolonia, Whitman, and D'Appolonia, 1969)

Figure 12.8 shows the variation of the unit weight of compaction with depth for a poorly graded dune sand that was compacted by a vibratory drum roller. Vibration was produced by mounting an eccentric weight on a single rotating shaft within the drum cylinder. The weight of the roller used for this compaction was 55.7 kN and the drum diameter was 1.19 m. The lifts were kept at 2.44 m. Note that, at any given depth, the dry unit weight of compaction increases with the number of roller passes. However, the rate of increase of unit weight gradually decreases after about 15 passes. Note also (Figure 12.8) the nature of variation of dry unit weight with depth for any given number of roller passes. The dry unit weight and, hence, the relative density (D_r) reaches a maximum value at a depth of about 0.5 m and then gradually decreases as the depth decreases. This is due to the lack of confining pressure toward the surface. Once the depth against relative density (or dry unit weight) relation for a given soil with a given number of roller passes is determined, it is easy to estimate the approximate thickness of each lift. This procedure is shown in Figure 12.9 (D'Appolonia, Whitman, and D'Appolonia, 1969).

12.3   Vibroflotation

Vibroflotation is a technique developed in Germany in the 1930s for *in situ* densification of thick layers of loose granular soil deposits. Vibroflotation was first used in the United States about 10 years later. The process

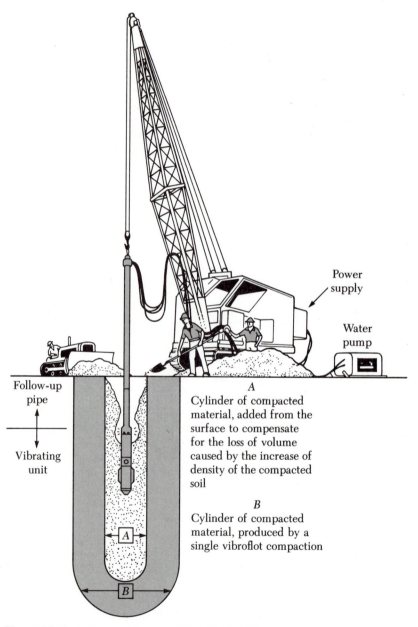

Figure 12.10   Vibroflotation unit (after Brown, 1977)

Figure 12.11   Compaction by the vibroflotation process (after Brown, 1977)

involves the use of a *vibroflot* (called the *vibrating unit*), as shown in Figure 12.10, which is about 2 m (6 ft) in length. This vibrating unit has an eccentric weight inside it and can develop a centrifugal force. The weight enables the vibrating unit to vibrate horizontally. There are openings at the bottom and top of the vibrating unit for water jets. The vibrating unit is attached to a follow-up pipe. Figure 12.10 shows all the vibroflotation equipment necessary for conducting field compaction.

The entire compaction process in the field can be divided into four stages (Figure 12.11):

Stage 1. The jet at the bottom of the vibroflot is turned on, and the vibroflot is lowered into the ground.

Stage 2. The water jet creates a quick condition in the soil, which allows the vibrating unit to sink.

Stage 3. Granular material is poured into the top of the hole. The water from the lower jet is transferred to the jet located at the top of the vibrating unit. This water carries the granular material down the hole.

Stage 4. The vibrating unit is gradually raised in about 0.3-m (1-ft) lifts and held vibrating for about 30 seconds at a time. This process compacts the soil to the desired unit weight.

Table 12.4 gives the details of various types of vibroflot unit used in the United States. Note that the 30-horsepower electric units have been used since the latter part of the 1940s. The 100-horsepower units were introduced in the early 1970s.

The zone of compaction around a single probe will vary according to the type of vibroflot used. The cylindrical zone of compaction will have a

Table 12.4   Types of Vibroflot Units[a]

	100-hp electric and hydraulic motors	30-hp electric motors
(a) Vibrating tip		
Length (m)	2.1	1.86
Diameter (mm)	406.4	381.0
Weight (kN)	17.8	17.8
Maximum movement when full (mm)	12.45	7.62
Centrifugal force (kN)	160.0	89.0
(b) Eccentric		
Weight (kN)	1.16	0.76
Offset (mm)	38.1	31.75
Length (mm)	610.0	387.0
Speed (rpm)	1800.0	1800.0
(c) Pump		
Operating flow rate (m^3/min)	0–1.6	0.6
Pressure (kN/m^2)	690–1035	690–1035
(d) Lower follow-up pipe and extensions		
Diameter (mm)	305.0	305.0
Weight (kN/m)	3.65	3.65

[a] After Brown (1977)

radius of about 2 m for a 30-hp unit. This radius may extend to about 3 m (10 ft) for a 100-hp unit.

Compaction by vibroflotation is done in various probe spacings, depending on the zone of compaction mentioned above (see Figure 12.12).

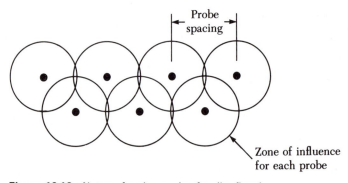

Figure 12.12   Nature of probe spacing for vibroflotation

Figure 12.13 Effective range of grain-size distribution of soil for vibroflotation

Mitchell (1970) and Brown (1977) have listed several successful cases of foundation design using vibroflotation.

The capacity of successful densification of *in situ* soil depends on several factors, the most important of which is the grain-size distribution of the soil and also the nature of backfill used to fill the holes during the withdrawal period of the vibroflot. The range of the grain-size distribution of *in situ* soil marked Zone 1 in Figure 12.13 is most suitable for compaction by vibroflotation. Soils that contain excessive amounts of fine sand and silt-size particles are difficult to compact; in these cases, considerable effort is needed to reach proper relative density of compaction. Zone 2 in Figure 12.13 is the approximate lower limit of grain-size distribution for compaction by vibroflotation. Soil deposits whose grain-size distribution falls in Zone 3 contain appreciable amounts of gravel. For these soils, the rate of probe penetration may be rather slow, and so compaction by vibroflotation might prove to be uneconomical in the long run.

The grain-size distribution of the backfill material is one of the factors that control the rate of densification. Brown (1977) has defined a quantity called *suitability number* (S_N) for rating a backfill material, or

$$S_N = 1.7 \sqrt{\frac{3}{(D_{50})^2} + \frac{1}{(D_{20})^2} + \frac{1}{(D_{10})^2}} \qquad (12.5)$$

where D_{50}, D_{20}, and D_{10} are the diameters (in mm) through which, respectively, 50%, 20%, and 10% of the material is passing.

The smaller the value of S_N, the more desirable is the backfill material. Following is a backfill rating system as proposed by Brown (1977):

Range of S_N	Rating as backfill
0–10	Excellent
10–20	Good
20–30	Fair
30–50	Poor
>50	Unsuitable

12.4   Precompression—General Considerations

When highly compressible, normally consolidated clayey soil layers are located at a limited depth, and large consolidation settlements are expected as the result of the construction of large buildings, highway embankments, or earth dams, precompression of soil may be used to eliminate postconstruction settlement problems. The principles of precompression can be explained by means of Figure 12.14. As shown in Figure 12.14a, let the proposed structural load per unit area be $\Delta p_{(p)}$ and the thickness of the clay layer undergoing consolidation be H_c. Hence, the maximum primary consolidation settlement caused by the structural load, $S_{(p)}$, can be given as

$$S_{(p)} = \frac{C_c H_c}{1 + e_o} \log \frac{p_o + \Delta p_{(p)}}{p_o} \tag{12.6}$$

The nature of the settlement-time relationship under the structural load will be like that shown in Figure 12.14b. However, if a surcharge of $\Delta p_{(p)}$

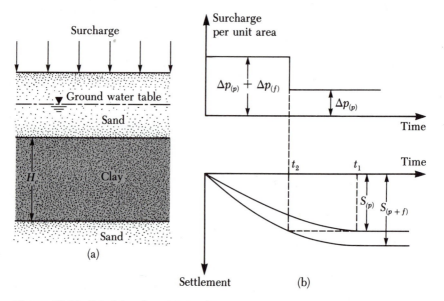

Figure 12.14   Principles of precompression

$+ \Delta p_{(f)}$ is placed on the ground, the primary consolidation settlement $S_{(p+f)}$ will be

$$S_{(p+f)} = \frac{C_c H_c}{1 + e_o} \log \frac{p_o + [\Delta p_{(p)} + \Delta p_{(f)}]}{p_o} \tag{12.7}$$

The settlement-time relationship under a surcharge of $\Delta p_{(p)} + \Delta p_{(f)}$ is also shown in Figure 12.14b. Note that a total settlement of $S_{(p)}$ would occur at a time t_2. Time t_2 is much smaller than t_1. So, if a temporary total surcharge of $\Delta p_{(f)} + \Delta p_{(p)}$ is applied on the ground surface for a time t_2, the settlement will be equal to $S_{(p)}$. At that time, if the surcharge is removed and a structure with a permanent load per unit area of $\Delta p_{(p)}$ is built, no appreciable settlement will occur. The procedure just described is *precompression*. The total surcharge $\Delta p_{(p)} + \Delta p_{(f)}$ can be applied by means of temporary fills.

Derivation of Equations to Obtain $\Delta p_{(f)}$ and t_2

Figure 12.14b shows that, under a surcharge of $\Delta p_{(p)} + \Delta p_{(f)}$, the degree of consolidation at time t_2 after load application can be expressed as

$$U = \frac{S_{(p)}}{S_{(p+f)}} \tag{12.8}$$

Substitution of Eqs. (12.6) and (12.7) into Eq. (12.8) yields

$$U = \frac{\log \left[\dfrac{p_o + \Delta p_{(p)}}{p_o} \right]}{\log \left[\dfrac{p_o + \Delta p_{(p)} + \Delta p_{(f)}}{p_o} \right]} = \frac{\log \left[1 + \dfrac{\Delta p_{(p)}}{p_o} \right]}{\log \left\{ 1 + \dfrac{\Delta p_{(p)}}{p_o} \left[1 + \dfrac{\Delta p_{(f)}}{\Delta p_{(p)}} \right] \right\}} \tag{12.9}$$

Figure 12.15 gives the values of U for various combinations of $\Delta p_{(p)}/p_o$ and $\Delta p_{(f)}/\Delta p_{(p)}$. The degree of consolidation referred to in Eq. (12.9) is actually the average degree of consolidation at time t_2, as shown in Figure 12.14b. However, if the average degree of consolidation is used to determine time t_2, then there might be some construction problems. This is because, after the removal of the surcharge and placement of the structural load, the portion of clay close to the drainage surface will continue to swell, and the soil close to the midplane will continue to settle (Figure 12.16). In some cases, net continuous settlement might result. A conservative approach may solve this problem—that is, assume that U in Eq. (12.9) is the midplane degree of consolidation (Johnson, 1970a). Now, from Eq. (1.77)

$$U = f(T_v)$$

where T_v = time factor = $C_v t_2 / H^2$ $\qquad$ (1.73)
$\qquad C_v$ = coefficient of consolidation
$\qquad t_2$ = time
$\qquad H$ = maximum drainage path ($= H_c/2$ for two-way drainage and equal to H_c for one-way drainage)

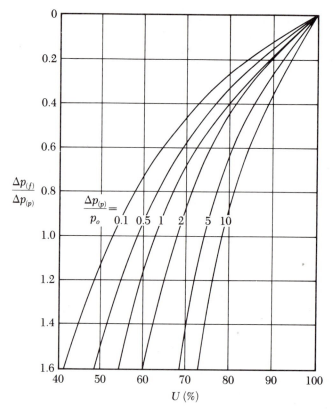

Figure 12.15   Plot of $\Delta p_{(f)}/\Delta p_{(p)}$ against U for various values of $\Delta p_{(p)}/p_o$—Eq. (12.9)

Figure 12.16

Figure 12.17   Plot of midplane degree of consolidation against T_v

The variation of U (midplane degree of consolidation) with T_v is given in Figure 12.17.

Procedure for Obtaining Precompression Parameters

Two problems may be encountered by engineers during precompression work in the field:

1. The value of $\Delta p_{(f)}$ is given; but one must obtain t_2. In such a case, obtain p_o, $\Delta p_{(p)}$, and solve for U using Eq. (12.9) or Figure 12.15. For this value of U, obtain T_v from Figure 12.17. Now

$$t_2 = \frac{T_v H^2}{C_v} \tag{12.10}$$

2. One may need to obtain $\Delta p_{(f)}$ for a specified value of t_2. In such a case, calculate T_v. Then refer to Figure 12.17 to obtain the midplane degree of consolidation, U. With the estimated value of U, go to Figure 12.15 to get the required $\Delta p_{(f)}/\Delta p_{(p)}$, from which the required value of $\Delta p_{(f)}$ can be calculated.

Examples of Precompression and General Comments

Johnson (1970a) has presented an excellent review of the use of pre-compression for improving foundation soils for several projects—for example, the Morganza Floodway Control Structure near Baton Rouge, Louisiana; the Old River Low-Sill Control Structure near Natchez, Mississippi; and the Old River Overbank Control Structure, Port Elizabeth Marine Terminal, New York.

There are several variables involved in proper precompression design and performance. The information obtained from only a handful of borings is used in the calculation of both the surcharge load and the time necessary for removal of the surcharge. For that reason, precise numbers for precompression design may be difficult to obtain. Settlement observations should be continued during the period of surcharge application. These observations may dictate changes in the design.

Example 12.2

Refer to Figure 12.14. During the construction of a highway bridge, it is expected that the average permanent load on the clay layer will increase by about 115 kN/m². The average effective overburden pressure at the middle of the clay layer is 210 kN/m². Given: $H_c = 6$ m, $C_c = 0.28$, $e_o = 0.9$, and $C_v = 0.36$ m²/month. The clay is normally consolidated.

a. Determine the total primary consolidation settlement of the bridge without precompression.

b. Determine the surcharge, $\Delta p_{(f)}$, needed to eliminate by precompression the entire primary consolidation settlement in 9 months.

Solution
Part a

The total primary consolidation settlement can be calculated from Eq. (12.6).

$$S_{(p)} = \frac{C_o H_c}{1 + e_o} \log \left[\frac{p_o + \Delta p_{(p)}}{p_o} \right] = \frac{(0.28)(6)}{1 + 0.9} \log \left[\frac{210 + 115}{210} \right]$$
$$= 0.1677 \text{ m} = \textbf{167.7 mm}$$

Part b

$$T_v = \frac{C_v t_2}{H^2}$$

$C_v = 0.36$ m²/month

$H = 3$ m (two-way drainage)

$t_2 = 9$ months

Hence

$$T_v = \frac{(0.36)(9)}{3^2} = 0.36$$

According to Figure 12.17, for $T_v = 0.36$, the value of U is 47%. Now

$$\Delta p_{(p)} = 115 \text{ kN/m}^2$$
$$p_o = 210 \text{ kN/m}^2$$

So

$$\frac{\Delta p_{(p)}}{p_o} = \frac{115}{210} = 0.548$$

According to Figure 12.15, for $U = 47\%$ and $\Delta p_{(p)}/p_o = 0.548$, $\Delta p_{(f)}/\Delta p_{(p)} \approx 1.8$. So

$$\Delta p_{(f)} = (1.8)(115) = 207 \text{ kN/m}^2$$

12.5 Sand Drains

The use of *sand drains* is another way to accelerate the consolidation settlement of soft, normally consolidated clay layers and achieve precompression before foundation construction. Sand drains are constructed by drilling holes through the clay layer(s) in the field at regular intervals. The holes are then backfilled with highly permeable sand (see Figure 12.18a). After backfilling the drill holes with sand, a surcharge is applied at the ground surface. This surcharge will increase the pore water pressure in the clay. The excess pore water pressure in the clay will be dissipated by drainage—both in the vertical direction and in the radial direction to the sand drains, which accelerates the rate of settlement of the clay layer.

Note that the radius of the sand drains is equal to r_w (Figure 12.18a). Figure 12.18b shows the plan of the layout of the sand drains. As shown in this figure, the effective zone from which the radial drainage will be directed toward a given sand drain can be approximated to be cylindrical with a diameter equal to d_e.

To determine the surcharge that needs to be applied at the ground surface and the length of time that it has to be maintained, refer to Figure 12.14 and use the corresponding equation [(12.9)]:

$$U_{v,r} = \frac{\log\left[1 + \dfrac{\Delta p_{(p)}}{p_o}\right]}{\log\left\{1 + \dfrac{\Delta p_{(p)}}{p_o}\left[1 + \dfrac{\Delta p_{(f)}}{\Delta p_{(p)}}\right]\right\}} \tag{12.11}$$

The notations $\Delta p_{(p)}$, p_o, and $\Delta p_{(f)}$ are the same as in the case of Eq. (12.9). However, unlike Eq. (12.9), the left side of the preceding equation is the *average degree* of consolidation instead of the degree of consolidation at midplane. This average degree of consolidation is due to the contribution of *radial* and *vertical* drainage. If $U_{v,r}$ can be determined for any given time t_2 (see Figure 12.14b), then the total surcharge $\Delta p_{(f)} + \Delta p_{(p)}$ can be easily

Figure 12.18   Sand drains

obtained from Figure 12.15. The procedure for determination of the average degree of consolidation $(U_{v,r})$ is given in the following sections.

The successful use of sand drains has been described in detail by Johnson (1970b). As in the case of precompression technique described earlier, constant field settlement observations may be necessary during the period of surcharge application.

Average Degree of Consolidation Due to Radial Drainage Only

The theory for equal-strain consolidation due to radial drainage only (with no smear) was developed by Barron (1948). It assumes that there is *no drain-*

age in the vertical direction. According to this theory

$$U_r = 1 - \exp\left(\frac{-8T_r}{m}\right) \qquad\qquad (12.12)$$

where U_r = average degree of consolidation due to radial drainage only

$$m = \left(\frac{n^2}{n^2 - 1}\right) \ln(n) - \frac{3n^2 - 1}{4n^2} \qquad\qquad (12.13)$$

$$n = \frac{d_e}{2r_w} \qquad\qquad (12.14)$$

T_r = nondimensional time factor for radial drainage only

$$= \frac{C_{vr}\, t_2}{d_e^2} \qquad\qquad (12.15)$$

C_{vr} = coefficient of consolidation for radial drainage

$$= \frac{k_h}{\left[\dfrac{\Delta e}{\Delta p(1 + e_{\mathrm{av}})}\right]\gamma_w} \qquad\qquad (12.16)$$

Note that the preceding equation is similar to Eq. (1.71). In Eq. (1.71), k was the coefficient of permeability in the vertical direction of the clay layer. In Eq. (12.16), k has to be replaced by k_h, which is the coefficient of permeability for flow in the horizontal direction. In some cases, k_h can be assumed to be equal to k; however, for soils like varved clay, $k_h > k$.

Figure 12.19 shows the plot of U_r against T_r for various values of n.

Average Degree of Consolidation Due to Vertical Drainage Only

The average degree of consolidation due to vertical drainage only can be obtained from Eqs. (1.78) and (1.79) (or Figure 1.21), or

$$T_v = \frac{\pi}{4}\left[\frac{U_v(\%)}{100}\right] \qquad \text{(for } U_v = 0\text{–}60\%) \qquad\qquad (1.78)$$

and

$$T_v = 1.781 - 0.933 \log(100 - U_v\%) \qquad \text{(for } U_v > 60\%) \qquad\qquad (1.79)$$

where U_v = average degree of consolidation due to vertical drainage only

$$T_v = \frac{C_v t_2}{H^2} \qquad\qquad (1.73)$$

C_v = coefficient of consolidation for vertical drainage

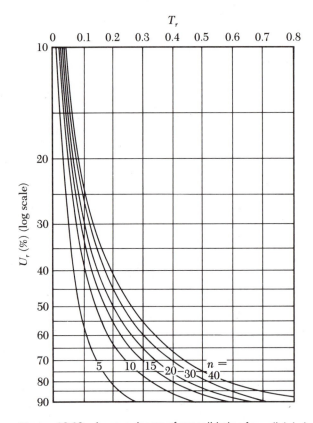

Figure 12.19   Average degree of consolidation for radial drainage only—Eq. (12.12)

Average Degree of Consolidation Due to Vertical and Radial Drainage

For a given surcharge and duration t_2, the average degree of consolidation due to drainage in the vertical and radial directions can be given by the expression

$$U_{v,r} = 1 - (1 - U_r)(1 - U_v)$$ (12.17)

Example 12.3

Redo Example 12.2 with the addition of some sand drains. Assume that $r_w = 0.1$ m, $d_e = 3$ m, and $C_v = C_{vr}$.

Solution
Part a

The total primary consolidation settlement will be 167.7 mm as before.

Part b

From Example 12.2, $T_v = 0.36$. The value of U_v, from Figure 1.20, is about 67%. From Eq. (12.14)

$$n = \frac{d_e}{2r_w} = \frac{3}{2 \times 0.1} = 15$$

Again,

$$T_r = \frac{C_{vr}t_2}{d_e^2} = \frac{(0.36)(9)}{(3)^2} = 0.36$$

From Figure 12.19, with $n = 15$ and $T_r = 0.36$, the value of U_r is about 77%. Hence

$$U_{v,r} = 1 - (1 - U_v)(1 - U_r) = 1 - (1 - 0.67)(1 - 0.77)$$
$$= 0.924 = 92.4\%$$

Now, referring to Figure 12.15, with $\Delta p_{(p)}/p_o = 0.548$ and $U_{v,r} = 92.4\%$, the value of $\Delta p_{(f)}/\Delta p_{(p)} \approx 0.12$. Hence

$$\Delta p_{(f)} = (115)(0.12) = \mathbf{13.8 \ kN/m^2}$$

≡

Wick Drains

The *wick drain* was recently developed as an alternative to the sand drain for inducing vertical drainage in saturated clay deposits. Wick drains appear to be better, faster, and cheaper. They essentially consist of paper or plastic strips that are held in a long tube. The tube is pushed into the soft clay deposit, then withdrawn, leaving behind the strips. These strips will act as vertical drains and induce rapid consolidation. Wick drains can be placed at desired spacings like sand drains. The main advantage of wick drains over sand drains is that they do not require drilling, and, thus, the process of installation is much faster.

12.6   Soil Stabilization by Admixtures

As mentioned in Section 12.1, admixtures are occasionally used to stabilize soils in the field—particularly fine-grained soils. Most common of these admixtures are *lime*, *lime–fly ash*, and *cement*. The main purposes of soil stabilization are to (a) modify the soil, (b) expedite construction, and (c) improve the strength and durability of the soil. The general principles of lime, cement, and lime–fly ash stabilization are discussed in the following sections.

Lime Stabilization

The limes commonly used for stabilization of fine-grained soils are hydrated high-calcium lime [$Ca(OH)_2$], calcitic quick lime (CaO), monohydrated dolomitic lime [$Ca(OH)_2 \cdot MgO$], and dolomitic quick lime. The quantity of lime used for stabilization of most soils usually varies in the range of 5–10%. When lime is added to clayey soils, several chemical reactions are initiated. These reactions are *cation exchange* and *flocculation-agglomeration*, and they are also *pozzolanic*. In the cation exchange and flocculation-agglomeration reactions, the *monovalent* cations generally associated with clays are replaced by the *divalent* calcium ions. The cations, based on their affinity for exchange, can be arranged in a series as shown below:

$$Al^{3+} > Ca^{2+} > Mg^{2+} > NH_4^+ > K^+ > Na^+ > Li^+$$

Based on this series, any given cation can replace the ions to the right of it. For example, calcium ions can replace potassium and sodium ions from a clay. Flocculation and agglomeration produce a change in the texture of clay soils. The clay particles tend to clump together to form larger-sized particles. These reactions tend to (a) decrease the liquid limit, (b) increase the plastic limit, (c) decrease the plasticity index, (d) increase the shrinkage limit, (e) increase the workability, and (f) improve the strength and deformation properties of a soil.

Pozzolanic reaction between soil and lime involves reaction between lime and the silica and alumina of soil to form cementing material. For example

$$Ca(OH)_2 + SiO_2 \rightarrow CSH$$

$$\uparrow$$

Clay silica

where $C = CaO$
$\quad\quad S = SiO_2$
$\quad\quad H = H_2O$

The pozzolanic reaction may continue for a long period of time.

Figure 12.20 shows the variation of the liquid limit, the plasticity index, and the shrinkage limit of a clay with percentage of lime admixture. The first 2–3% lime (on the dry weight basis) has a substantial influence in improving the workability and the property of the soil. Addition of lime to clayey soils has an effect on their compaction characteristics.

Figure 12.21a shows the results of standard Proctor tests for Vicksburg clay without any additives and also with 4% high-calcium hydrated lime additive (uncured). Note that the addition of lime has an influence in reducing the maximum compacted dry unit weight and increasing the optimum moisture content. Figure 12.21b also shows the change of the unconfined compressive strength (q_u) of uncured Vicksburg clay with the percentage of high-calcium hydrated lime. The value of q_u with 6% lime

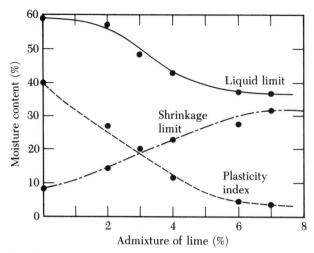

Figure 12.20   Variation of liquid limit, plasticity index, and shrinkage limit of a clay with lime additive

is about six times the value obtained with no additive. Note that the specimens prepared for the determination of q_u were all at a moisture content of 29–29.5%. This is shown as the molding moisture content in Figure 12.21a.

Arman and Munfakh (1972) also evaluated the lime stabilization of organic clays found in Louisiana. Figure 12.22a shows the change of plasticity with $Ca(OH)_2$ content for an organic soil with 22% organic material. The curing time for these soil specimens was 48 hours. The effects of lime are generally similar to those seen in Figure 12.20. The change of the unconfined compression strength of the same soil with lime additives is shown in Figure 12.22b. Based on this study, Arman and Munfakh concluded that (a) the presence of organic matter does not block the pozzolanic reaction that helps in changing the fundamental soil properties and making the soil more workable, and (b) about 2% of lime is sufficient to satisfy the base exchange capacity of organic matters.

Lime stabilization in the field can be done in three ways:

1. The *in situ* material and/or the borrowed material can be mixed with the proper amount of lime at the site and then compacted after the addition of moisture.

2. The soil can be mixed with the proper amount of lime and water at a plant and then hauled back to the site for compaction.

3. Lime slurry can be pressure injected into the soil to a depth of 4–5 m (12–16 ft). Figure 12.23 shows a vehicle used for pressure injection of lime slurry. The slurry injection mechanical unit is mounted to the injection vehicle. A common injection unit is a hydraulic-lift mast with cross beams that contain the injection rods. The injection rods are pushed into the ground by the action of the lift mast beams. The slurry is generally mixed in a batching tank about 3 m (10 ft) in diameter and 12 m (36 ft) long, and

(a)

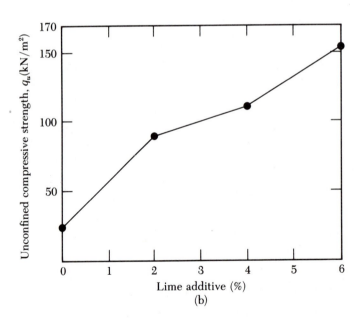

(b)

Figure 12.21 Lime stabilization of Vicksburg clay (% less than 2μ size = 46; liquid limit = 59, plasticity index = 30, A-7-6(20), pH = 6): (a) laboratory dry unit weight against moisture content curve; (b) change of unconfined compression strength with percent of lime. *Note:* The specimens for (b) were molded at a moisture content shown in (a) (from "Stability Properties of Uncured Lime-Treated Fine Grained Soil," by C. H. Neubauer and M. R. Thompson. In *Highway Research Record No. 381*, Highway Research Board, 1972, pp. 20–26. Reprinted by permission)

Figure 12.22   Lime stabilization of organic soil: (a) variation of liquid limit and plasticity index with 48 hours' curing; (b) variation of unconfined compression strength (from "Lime Stabilization of Organic Soil," by A. Arman and G. A. Munfakh. In *Highway Research Record No. 381*, Highway Research Board, 1972, pp. 37–45. Reprinted by permission)

Figure 12.23 Equipment for pressure injection of lime slurry (courtesy of GKN Hayward Baker Inc., Woodbine Division, Ft. Worth, Texas)

it is pumped at high pressure to the injection rods. Figure 12.24 is a photograph of the lime slurry pressure injection process. The ratio typically specified for preparation of lime slurry is 2.5 lb of dry lime to a gallon of water. For more information on this technique, readers are referred to a paper by Blacklock and Pengelly (1988).

Because the addition of hydrated lime to soft clayey soils results in an immediate increase of plastic limit, thus changing the soil from plastic to solid and making it appear to "dry up," limited amounts of it can be thrown on muddy and troublesome construction sites. This improves the trafficability at the construction sites and may save money and time.

Quick limes have also been successfully used in drill holes having diameters of 100 mm to 150 mm (4 in. to 6 in.) for stabilization of subgrades and slopes. For this type of work, holes are drilled in a grid pattern and then filled with quick lime.

Cement Stabilization

Cement is increasingly used as a stabilizing material for soil, particularly for the construction of highways and earth dams. The first controlled soil-cement construction in the United States was carried out near Johnsonville, South Carolina, in 1935. Cement can be used to stabilize sandy and clayey soils. As in the case of lime, cement has an effect in decreasing the liquid limit and increasing the plasticity index and workability of clayey soils. For

Figure 12.24 Pressure injection of lime slurry (courtesy of GKN Hayward Baker Inc., Woodbine Division, Ft. Worth, Texas)

clayey soils, cement stabilization is effective when the liquid limit is less than 45–50 and the plasticity index is less than about 25. The optimum requirements of cement by volume for effective stabilization of various types of soil are given in Table 12.5.

Like lime, cement helps increase the strength of soils; and strength increases with curing time. Table 12.6 presents some typical values of the unconfined compressive strength of various types of untreated soil and soil-cement mixture made with approximately 10% cement by weight.

On the basis of the preceding discussion and the data presented in Tables 12.5 and 12.6, it is obvious that granular soils and clayey soils with low plasticity are more suitable for cement stabilization. Experience has shown that calcium clays are more easily stabilized by addition of cement,

Table 12.5   Cement Requirement by Volume for Effective Stabilization of Various Soils[a]

Soil type		Percent cement by volume
AASHTO classification	Unified classification	
A-2 and A-3	GP, SP, and SW	6–10
A-4 and A-5	CL, ML, and MH	8–12
A-6 and A-7	CL, CH	10–14

[a]After Mitchell and Frietag (1959)

whereas sodium and hydrogen clays, which are expansive in nature, respond better to lime stabilization. For these reasons, proper care should be given in the selection of the stabilizing material.

For field compaction, the proper amount of cement can be mixed with soil, either at the site or in a mixing plant (and then carried to the site). The soil is compacted to the required unit weight with a predetermined amount of water.

Similar to lime injection, cement slurry made of Portland Cement and water (water-cement ratio = 0.5 to 5) can be used for pressure grouting of poor soils under foundations of buildings and other structures. Grouting decreases the permeability of soils and increases the strength and the load-bearing capacity. For design of low-frequency machine foundations subjected to vibrating forces, it is sometimes necessary to stiffen the foundation soil by grouting and thereby increase the resonant frequency.

Table 12.6   Typical Compressive Strengths of Soils and Soil-Cement Mixtures[a]

Material	Unconfined compressive strength range (kN/m^2)[b]
Untreated soil:	
Clay, peat	Less than 350
Well-compacted sandy clay	70–280
Well-compacted gravel, sand, and clay mixtures	280–700
Soil-cement (10% cement by weight):	
Clay, organic soils	Less than 350
Silts, silty clays, very poorly graded sands, slightly organic soils	350–1050
Silty clays, sandy clays, very poorly graded sands, and gravels	700–1730
Silty sands, sandy clays, sands, and gravels	1730–3460
Well-graded sand-clay or gravel-sand-clay mixtures and sands and gravels	3460–10,350

[a]After Mitchell and Freitag (1959)
[b] Rounded off

12.7 Fly Ash Stabilization

Fly ash is a by-product of the pulverized coal combustion process usually associated with electric power generating plants. Fly ash is a fine-grained dust and is primarily composed of silica, alumina, and various oxides and alkalies. It is pozzolanic in nature and can react with hydrated lime to produce cementitious products. For that reason, lime–fly ash mixtures can be used for stabilization of highway bases and subbases. Effective mixes can be prepared with 10–35% of fly ash and 2–10% of lime. Soil–lime–fly ash mixes are compacted under controlled conditions with proper amounts of moisture to obtain stabilized soil layers.

A certain type of fly ash is obtained from the burning of coal primarily from the western United States, and it is referred to as "Type C" fly ash. It contains a fairly large proportion (up to about 25%) of free lime that, with the addition of water, will react with other fly ash compounds to form cementitious products. This may eliminate the need to add manufactured lime.

12.8 Stone Columns

A method now being used to increase the load-bearing capacity of shallow foundations on soft clay layers is the construction of *stone columns*. The construction of a stone column generally consists of water jetting a vibroflot (Section 12.3) into the soft clay layer to make a circular hole that extends through the clay to firmer soil. The hole is then filled with an imported gravel. The gravel in the hole is gradually compacted as the vibrator is withdrawn. The gravel used for the stone column has sizes ranging from 6–40 mm (0.25–1.5 in.). Stone columns usually have diameters ranging from 0.5–0.75 m (1.6–2.5 ft) and are spaced at about 1.5–3 m (5–10 ft) center-to-center.

After the construction of stone columns, it is always desirable that a fill material be placed over the ground surface and compacted before the construction of the foundation. The stone columns tend to reduce the settlement of foundations at allowable loads. Several case histories for construction projects using stone columns can be found in the papers of Hughes and Withers (1974), Hughes, Withers, and Greenwood (1975), Mitchell and Huber (1985), and others.

At this time there is no standard to estimate the settlement of foundations constructed over stone columns. However, based on the recommendation of Greenwood and Thompson (1984) and on observations of the author, a tentative chart for estimation of settlement is given in Figure 12.25. In order to use Figure 12.25, the following procedure needs to be adopted:

1. Determine the cross-sectional area of the stone column, A_S.
2. Determine the average area of the column foundation, A_F.

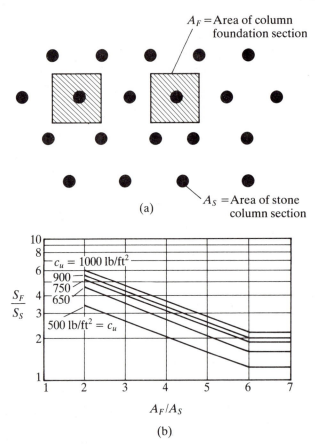

A_F = Area of column foundation section

A_S = Area of stone column section

(a)

(b)

Figure 12.25   Settlement of foundation built on stone columns

3. Calculate the ratio of A_F/A_S.

4. Estimate the undrained shear strength of the clay, c_u, and the probable settlement of a column foundation assuming that it was constructed without the stone columns, S_F.

5. With known values of A_F/A_S and c_u, determine the ratio of S_F/S_S (S_S = probable settlement of the foundation constructed over stone columns) from Figure 12.25b.

6. With known values of S_F and S_F/S_S, calculate S_S.

Example 12.4

Stone columns were constructed in a deposit of soft clay soil 20 ft thick. Given:

Clay: undrained cohesion, $c_u = 750$ lb/ft^2
        modulus of elasticity, $E_s = 1400$ lb/in.^{2}
Stone column: diameter of each column = 1.5 ft
        center-to-center spacing = 8 ft

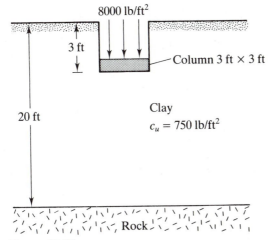

Figure 12.26

Column foundation (Figure 12.26): plan $= 3$ ft $\times$ 3 ft

center-to-center spacing of
columns $= 10$ ft
depth of foundation, $D_f = 3$ ft
load on columns $= 8000$ lb/ft^2

Estimate the settlement of the column foundations.

Solution
Calculation of the Settlement of Columns Without the Stone Columns

Referring to Eq. (3.68)

$$S_F = A_1 A_2 \left(\frac{q_o B}{E_s} \right)$$

Given: $B = 3$ ft, $q_o = 8000$ lb/ft^2, and $E_s = 1400$ lb/in.^{2}. Referring to Figure 3.24, $H/B = 17/3 = 5.67$ and $D_f/B = 1$. So $A_2 = 0.6$, $A_1 = 0.94$

$$S_F = (0.94)(0.6) \left[\frac{(8000)(3)}{(1400)(144)} \right] = 0.067 \text{ ft} \approx \textbf{0.81 in.}$$

Settlement with Stone Columns

$$A_s = \left(\frac{\pi}{4} \right) (1.5)^2 = 1.676 \text{ ft}^2$$

$$A_F = (3)^2 = 9 \text{ ft}^2$$

$$\frac{A_F}{A_s} = \frac{9}{1.767} \approx 5.1$$

Referring to Figure 12.25, for $c_u = 750$ lb/ft^2 and $A_F/A_S = 5.1$, the value of $S_F/S_S \approx 2$. So

$$S_S = \frac{S_F}{2} = \frac{0.81}{2} = \textbf{0.405 in.}$$

12.9   Foundations on Granular Trenches

It was pointed out in Chapter 3 that the allowable bearing capacity of a shallow foundation is controlled by two factors: (a) ultimate bearing capacity and (b) tolerable settlement. When the width of a foundation is relatively small, the ultimate bearing capacity is the controlling factor. These foundations, when constructed over soft clay soil, will still undergo undesirable settlement. The allowable load-bearing capacity of these foundations can be substantially enhanced, accompanied by a reduction of settlement, by constructing them over granular (sand) trenches made in the soft clay, as shown in Figure 12.27. In this figure, a foundation having dimensions of $B \times L$ is supported by a granular trench having dimensions of $B \times L \times H$ (height). The parameters for the compacted sand in the trench are as follows:

> Dry unit weight: γ_s
>
> Angle of friction: ϕ_s

For the soft clay, the unit weight and the undrained shear strength are γ_c and c_u, respectively.

It has been shown that the ultimate bearing capacity of a foundation as shown in Figure 12.27 increases linearly with the H/B ratio up to a maximum at $H/B = (H/B)_{cr}$ and remains constant thereafter (Figure 12.28)

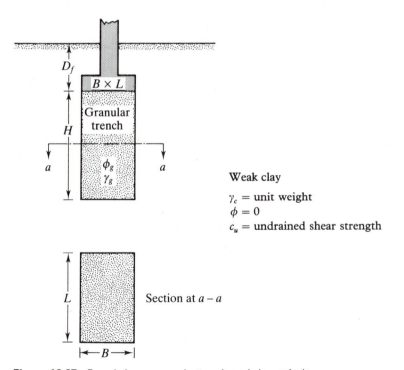

Weak clay
γ_c = unit weight
$\phi = 0$
c_u = undrained shear strength

Section at $a - a$

Figure 12.27   Foundation on granular trench made in weak clay

Figure 12.28   Nature of variation of q_u with H/B for foundations on granular trenches

(Das and Hanna, 1988; Das, 1989). The maximum value of the ultimate bearing capacity can be referred to as $q_{u(max)}$. The magnitude of $q_{u(max)}$ can be expressed as

$$q_{u(max)} = \left(\frac{1 + \sin \phi_s}{1 - \sin \phi_s}\right)(\gamma_c D_f + 2F_s c_u) \qquad (12.18)$$

where F_s = shape factor = $1 + B/L$ $\qquad\qquad\qquad\qquad\qquad$ (12.19)

The magnitude of $(H/B)_{cr}$ is approximately equal to 3. Laboratory model tests have shown that the settlement of a foundation at ultimate load with $H/B = 3$ is about one-third the settlement obtained at ultimate load without the granular trench.

Example 12.5

Refer to Figure 12.27. Given:

Foundation: $B = 2.5$ ft, $L = 5$ ft, $D_f = 3$ ft
Clay: $c_u = 480$ lb/ft^2, $\gamma_c = 120$ lb/ft^3
Sand in granular trench: $\phi_s = 38°$, $\gamma_s = 115$ lb/ft^3

Estimate the ultimate bearing capacity of the foundation.

Solution

From Eqs. (12.18) and (12.19)

$$q_u = \left(\frac{1 + \sin \phi_s}{1 - \sin \phi_s}\right)\left[\gamma D_f + 2\left(1 + \frac{B}{L}\right)c_u\right]$$

$$= \left(\frac{1 + \sin 38}{1 - \sin 38}\right)\left[(120)(3) + (2)\left(1 + \frac{2.5}{5}\right)(480)\right]$$

$$= (4.2)(360 + 1440) = \mathbf{7560\ lb/ft^2}$$

12.10   Geogrids

In Chapter 11, the use of metallic strips and geotextiles as soil reinforcement for the construction of retaining walls and shallow foundations was discussed in some detail. Like geotextiles, geogrids are now used extensively for soil reinforcement. Geogrids are bidimensional structures in the shape of a sheet with apertures. They are made from petroleum products such as

(a)

(b)

Figure 12.29   Geogrids: (a) uniaxial; (b) biaxial (note: 1—longitudinal rib, 2—transverse bar, 3—transverse rib, 4—junction)

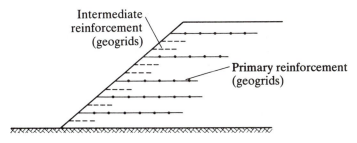

Figure 12.30 Schematic diagram of a slope constructed with geogrid reinforcement (after The Tensar Corporation, 1987)

polypropylene and polyethylene by a process of extrusion and longitudinal stretching. The apertures in geogrids are of sufficient size to allow interlocking with the surrounding soil and/or rock to perform the functions of reinforcement and/or segregation. Generally, geogrids can be of two types: (a) *biaxial geogrids* and (b) *uniaxial geogrids*.

Figures 12.29a and 12.29b are photographs of the two types of geogrids described above. Based on the commercial geogrids presently available for soil reinforcement, the nominal thickness of the ribs can vary from about 0.5 mm to about 1.5 mm (0.02 in. to 0.06 in.), and that of the junctions can vary from about 2.5 mm to 5 mm (0.1 in. to 0.2 in.). The grids used for soil reinforcement usually have apertures that are rectangular or elliptical in shape. The dimensions of the apertures may vary from about 25 mm to about 150 mm (1 in. to 6 in.). Geogrids are manufactured in such a way that the open areas of the grids are greater than 50%, and they develop reinforcing strength at low strain levels, such as 2% (Carroll, 1988). They are used as a soil reinforcement for construction of steep slopes and retaining walls

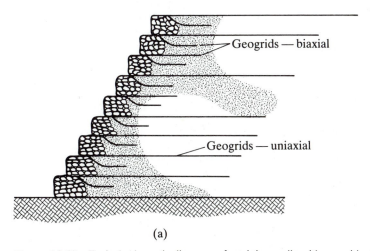

(a)

Figure 12.31 Typical schematic diagrams of retaining walls with geogrid reinforcement: (a) geogrid wraparound wall; (b) wall with gabion facing; (c) concrete panel–faced wall (after The Tensar Corporation, 1986) and stabilization of highway subgrades. Figure 12.30 shows a schematic diagram of a slope constructed with geogrid reinforcement. Figure 12.31 shows typical schematic diagrams of retaining walls with geogrid reinforcement.

(b)

(c)

Figure 12.31 (Continued)

PROBLEMS

12.1 Make the necessary calculations and prepare the zero-air-void unit weight (in kN/m^3) curves as related to a Proctor compaction test for $G_s = 2.6$, 2.65, 2.7, and 2.75.

12.2 The maximum dry unit weight of a soil was determined by a Proctor compaction test to be $18.08 \ kN/m^3$. If the same soil is used in the compaction of an embankment to a unit weight of $15 \ kN/m^3$, what would be the relative compaction?

12.3 A sandy soil has the following:

$$\text{Maximum dry unit weight} = 109.5 \ lb/ft^3$$
$$\text{Dry unit weight of compaction in the field} = 99 \ lb/ft^3$$

Estimate the following:
a. The relative compaction in the field
b. The relative density in the field
c. The minimum dry unit weight of the soil
Use Eqs. (12.2), (12.3), and (12.4).

12.4 A soil was compacted in the field by the standard Proctor test procedure. For the test, given:

$$\text{Moist unit weight of compaction} = 20.5 \ kN/m^3$$
$$\text{Moisture content} = 6\%$$

Estimate $\gamma_{d(max)}$ and the optimum moisture content for the soil according to the Ohio one-point method.

12.5 Repeat Problem 12.4 with the following:

$$\text{Moist unit weight of compaction} = 114 \ lb/ft^3$$
$$\text{Moisture content} = 12\%$$

12.6 According to the Ohio one-point method, a soil will have a maximum dry unit weight of $99.9 \ lb/ft^3$ at an optimum moisture content of 21.5%. Estimate the dry unit weight of the soil when compacted to a moisture content of 18%.

12.7 For a vibroflotation work, the backfill to be used has the following characteristics:

$$D_{50} = 1 \ mm$$
$$D_{20} = 0.5 \ mm$$
$$D_{10} = 0.08 \ mm$$

Determine the suitability number of the backfill. How would you rate the material?

12.8 Repeat Problem 12.7 with the following:

$$D_{50} = 1.8 \ mm$$
$$D_{20} = 0.75 \ mm$$
$$D_{10} = 0.12 \ mm$$

12.9 Refer to Figure 12.14. For the construction of an airport, a large fill operation is required. For the work, the average permanent load $[\Delta p_{(p)}]$ on the clay layer will increase by about $75 \ kN/m^2$. The average effective overburden pressure on the clay layer before the fill operation is $110 \ kN/m^2$. For the clay layer, which is normally consolidated and drained at top and bottom, given:

$$H_c = 8 \text{ m} \qquad e_o = 1.02$$
$$C_c = 0.27 \qquad C_v = 0.52 \text{ m}^2/\text{month}$$

Determine the following:
a. The primary consolidation settlement of the clay layer caused by the additional permanent load, $\Delta p_{(p)}$
b. The time required for 90% of primary consolidation settlement under the additional permanent load only
c. The temporary surcharge, $\Delta p_{(f)}$, that will be required to eliminate the entire primary consolidation settlement in 6 months by the precompression technique

12.10 Redo part (c) of Problem 12.9. Given: the time for elimination of primary consolidation settlement is 12 months.

12.11 Repeat Problem 12.9 with the following: $\Delta p_{(p)} = 1200 \text{ lb/ft}^2$, average effective overburden pressure on clay layer = 1000 lb/ft^2

$$H_c = 15 \text{ ft} \qquad e_o = 1.0$$
$$C_c = 0.3 \qquad C_v = 1.5 \times 10^{-2} \text{ in.}^2/\text{min}$$

12.12 The diagram of a sand drain is shown in Figure 12.18. If $r_w = 0.25$ m, $d_e = 4$ m, $C_v = C_{vr} = 0.28$ m^2/month, and $H_c = 8.4$ m, determine the degree of consolidation caused by a sand drain only after 6 months of surcharge application.

12.13 Estimate the degree of consolidation for the clay layer described in Problem 12.12 that is caused by the combination of vertical drainage (drained on top and bottom) and radial drainage after 6 months of the application of surcharge.

12.14 A 20-ft thick clay layer is drained at top and bottom. Given:

$$C_{vr} = C_v \text{ (for vertical drainage)} = 0.055 \text{ ft}^2/\text{day}$$
$$r_w = 9 \text{ in.}$$
$$d_e = 10 \text{ ft}$$

Estimate the degree of consolidation of the clay layer caused by the combination of vertical and radial drainage at $t = 0.2, 0.4, 0.8,$ and 1 year.

12.15 Repeat Example 12.4 with the following:

Clay: thickness of clay layer = 15 ft
undrained shear strength, $c_u = 650$ lb/ft^2
modulus of elasticity = 1250 lb/in.^{2}
Stone columns: diameter of each column = 20 in.
center-to-center spacing = 60 in.
Column foundations: plan = 4 ft × 4 ft
center-to-center spacing = 15 ft
depth of foundation = 3.5 ft
load on each column = 10,000 lb/ft^2

12.16 Estimate the ultimate bearing capacity of the following foundation constructed on a granular trench.

Foundation: $B = 1.2$ m, $L = 2$ m, $D_f = 1$ m
Clay: $c_u = 21$ kN/m^2, $\gamma = 18$ kN/m^3
Sand in granular trench: $\phi_s = 40°$, $\gamma_s = 17$ kN/m^3
Height of granular trench: 4 m

REFERENCES

Arman, A., and Munfakh, G. A. (1972). "Lime Stabilization of Organic Soils," *Highway Research Record, No. 381*, National Academy of Sciences, pp. 37–45.

Barron, R. A. (1948). "Consolidation of Fine-Grained Soils by Drain Wells," *Transactions*, American Society of Civil Engineers, Vol. 113, pp. 718–754.

Blacklock, J. R., and Pengelly, A. D. (1988). "Soil Treatment for Foundations on Expansive Clay," *Special Topics in Foundations*, GSP No. 16 (ed. B. M. Das), American Society of Civil Engineers, pp. 73–92.

Brown, R. E. (1977). "Vibroflotation Compaction of Cohesionless Soils," *Journal of the Geotechnical Engineering Division*, American Society of Civil Engineers, Vol. 103, No. GT12, pp. 1437–1451.

Carroll, R., Jr. (1988). "Specifying Geogrids," *Geotechnical Fabric Report*, Industrial Fabric Association International, St. Paul, March/April.

D'Appolonia, D. J., Whitman, R. V., and D'Appolonia, E. (1969). "Sand Compaction with Vibratory Rollers," *Journal of the Soil Mechanics and Foundations Division*, American Society of Civil Engineers, Vol. 95, No. SM1, pp. 263–284.

Das, B. M. (1989). "Bearing Capacity of Shallow Foundation on Granular Column in Weak Clay," *Proceedings*, Geotechnical Engineering Congress, American Society of Civil Engineers, Vol. 2, pp. 1252–1263.

Das, B. M., and Hanna, A. M. (1988). "Model Tests for Shallow Strip Foundation on Granular Trench," *Special Topics in Foundations*, GSP No. 16 (ed. B. M. Das), American Society of Civil Engineers, pp. 110–124.

Giroud, J. P. (1980). "Introduction to Geotextiles and Their Application," *Proceedings*, First Canadian Symposium on Geotextiles, Sept., pp. 1–32.

Greenwood, D. A., and Thompson, G. H. (1984). *Ground Stabilization: Deep Compaction and Grouting*, ICE Works Construction Guides, Thomas Telford Ltd., London.

Hughes, J. M. O., and Withers, N. J. (1974). "Reinforcing of Soft Cohesive Soil with Stone Columns," *Ground Engineering*, Vol. 7, pp. 42–49.

Hughes, J. M. O., Withers, N. J., and Greenwood, D. A. (1975). "A Field Trial of Reinforcing Effects of Stone Columns in Soil," *Geotechnique*, Vol. 25, No. 1, pp. 31–34.

Johnson, A. W., and Sallberg, J. R. (1960). "Factors That Influence Field Compaction of Soils," *Bulletin No. 272*, Highway Research Board, National Academy of Sciences, Washington, D.C.

Johnson, S. J. (1970a). "Precompression for Improving Foundation Soils," *Journal of the Soil Mechanics and Foundations Division*, American Society of Civil Engineers, Vol. 96, No. SM1, pp. 114–144.

Johnson, S. J. (1970b). "Foundation Precompression with Vertical Sand Drains," *Journal of the Soil Mechanics and Foundations Division*, American Society of Civil Engineers, Vol. 96, No. SM1, pp. 145–175.

Lee, K. L., and Singh, A. (1971). "Relative Density and Relative Compaction," *Journal of the Soil Mechanics and Foundations Division*, American Society of Civil Engineers, Vol. 97, No. SM7, pp. 1049–1052.

Lee, P. Y., and Suedkamp, R. J. (1972). "Characteristics of Irregularly Shaped Compaction Curves of Soils," *Highway Research Record No. 381*, National Academy of Sciences, Washington, D.C., pp. 1–9.

Mitchell, J. K. (1970). "In-Place Treatment of Foundation Soils," *Journal of the Soil Mechanics and Foundations Division*, American Society of Civil Engineers, Vol. 96, No. SM1, pp. 73–110.

Mitchell, J. K., and Freitag, D. R. (1959). "A Review and Evaluation of Soil-Cement Pavements," *Journal of the Soil Mechanics and Foundations Division*, American Society of Civil Engineers, Vol. 85, No. SM6, pp. 49–73.

Mitchell, J. K., and Huber, T. R. (1985). "Performance of a Stone Column Foundation," *Journal of Geotechnical Engineering*, American Society of Civil Engineers, Vol. 111, No. GT2, pp. 205–223.

Neubauer, C. H., Jr., and Thompson, M. R. (1972). "Stability Properties of Uncured Lime-Treated Fine-Grained Soils," *Highway Research Record No. 381*, National Academy of Sciences, pp. 20–26.

The Tensar Corporation (1986). *Tensar Technical Note No. TTN:RW1*, August.

The Tensar Corporation (1987). *Tensar Technical Note No. TTN:SR1*, August.

REINFORCED CONCRETE DESIGN

A.1   Fundamentals of Reinforced Concrete Design

At the present time, most reinforced concrete designs are based on the recommendations of the building code prepared by the American Concrete Institute (1987)—that is, ACI 318-86. The basis for this code is the *ultimate strength design* or *strength design*. Some of the fundamental recommendations of the code are briefly summarized in the following sections.

Load Factors

According to ACI Code Section 9.2, the ultimate load-carrying capacity of a structural member should be as follows:

$$\boxed{U = 1.4D + 1.7L} \tag{A.1}$$

where U = ultimate load-carrying capacity of a member
$\qquad D$ = dead load
$\qquad L$ = live load

If the wind load is to be considered for design, then

$$\boxed{U = 0.75(1.4D + 1.7L + 1.7W)} \tag{A.2}$$

where W = wind load.

If there is no live load, then

$$\boxed{U = 0.9D + 1.3W} \tag{A.3}$$

However, in no case should the value of U be less than that given in Eq. (A.1).

Strength Reduction Factor

The design strength provided by a structural member is equal to the theoretical strength times a strength reduction factor, ϕ, or

$$\text{Design strength} = \phi(\text{theoretical strength})$$

The reduction factor, ϕ, takes into account the inaccuracies in the design assumptions, changes in property or strength of the construction materials, and so on. Following are some of the recommended values of ϕ (ACI Code Section 9.3):

Condition	Value of ϕ
a. Axial tension; flexure with or without axial tension	0.9
b. Shear or torsion	0.85
c. Axial compression with or without flexure, spiral reinforcement	0.75
d. Axial compression with or without flexure, tied reinforcement	0.7
e. Bearing on concrete	0.7
f. Flexure in plain concrete	0.65

Design Concepts for a Rectangular Section in Bending

Figure A.1a shows a section of a concrete beam having a width b and a depth h. The assumed stress distribution across the section at ultimate load is shown in Figure A.1b. The following notations have been used in this figure:

f'_c = compressive strength of concrete at 28 days
A_s = area of steel tension reinforcement
f_y = yield stress of reinforcement in tension
d = effective depth
l = location of the natural axis measured from the top of the compression face
$a = \beta l$
$\beta = 0.85$ for f'_c of 28 MN/m^2 (4000 lb/in.^{2}) or less and decreases at the rate of 0.05 for every 7 MN/m^2 (1000 lb/in.^{2}) increase of f'_c. However, it cannot be less than 0.65 in any case (ACI Code Section 10.2.7).

From the principles of statics, for the section

$$\Sigma \text{ compressive force, } C = \Sigma \text{ tensile force, } T$$

Thus

$$0.85f'_c\, ab = A_s f_y$$

or

$$a = \frac{A_s f_y}{0.85f'_c\, b} \tag{A.4}$$

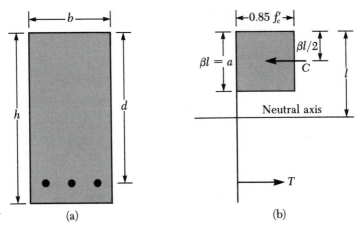

Figure A.1 Rectangular section in bending

Also, for the beam section, the nominal ultimate moment can be given as

$$M_n = A_s f_y \left(d - \frac{a}{2} \right) \tag{A.5}$$

where M_n = theoretical ultimate moment

The design ultimate moment, M_u, can be given as

$$M_u = \phi A_s f_y \left(d - \frac{a}{2} \right) \tag{A.6}$$

Combining Eqs. (A.4) and (A.6)

$$M_u = \phi A_s f_y \left[d - \left(\frac{1}{2} \right) \frac{A_s f_y}{0.85 f'_c b} \right] = \phi A_s f_y \left(d - \frac{0.59 A_s f_y}{f'_c b} \right) \tag{A.7}$$

The steel percentage is defined by the equation

$$s = \frac{A_s}{bd} \tag{A.8a}$$

In a balanced beam, failure would occur by sudden simultaneous yielding of tensile steel and crushing of concrete. The balanced percentage of steel (for modulus of elasticity of steel, $E_s = 200 \text{ MN/m}^2$) can be given as

$$s_b = \frac{0.85 f'_c}{f_y} (\beta) \left(\frac{600}{600 + f_y} \right) \tag{A.8b}$$

where f'_c and f_y are in MN/m^2.

Table A.1 Values of s_{max} [Eq. (A.9)]

f_y		f'_c			
MN/m²	lb/in.²	21 MN/m² (≈ 3000 lb/in.²)	28 MN/m² (≈ 4000 lb/in.²)	35 MN/m² (≈ 5000 lb/in.²)	42 MN/m² (≈ 6000 lb/in.²)
276	40,000	0.0284	0.0378	0.0445	0.0501
345	50,000	0.0209	0.0279	0.0329	0.0370
414	60,000	0.0163	0.0217	0.0255	0.0287
		$\beta = 0.85$	$\beta = 0.85$	$\beta = 0.80$	$\beta = 0.75$

In conventional English units (with $E_s = 29 \times 10^6$ lb/in.²)

$$s_b = \frac{0.85 f'_c}{f_y}(\beta)\left(\frac{87,000}{87,000 + f_y}\right) \tag{A.8c}$$

where f'_c and f_y are in lb/in.²

To avoid sudden failure without warning, ACI Code Section 10.3.3 recommends that the maximum steel percentage (s_{max}) should be limited to $0.75 s_b$, or

$$s_{max} = 0.75 s_b \tag{A.9}$$

Table A.1 gives the values of s_{max} for various values of f'_c and f_y.
The nominal or theoretical shear strength of a section, V_n, can be given as

$$V_n = V_c + V_s \tag{A.10}$$

where V_c = nominal shear strength of concrete
V_s = nominal shear strength of reinforcement

The permissible shear strength, V_u, can be given by

$$V_u = \phi V_n = \phi(V_c + V_s) \tag{A.11}$$

The values of V_c can be given by the following equations (ACI Code Sections 11.3 and 11.11):

$$V_c = 0.17\sqrt{f'_c}\, bd \quad \text{(for member subjected to shear and flexure)} \tag{A.12a}$$

and

$$V_c = 0.34\sqrt{f'_c}\, bd \quad \text{(for member subjected to diagonal tension)} \tag{A.12b}$$

where f'_c is in MN/m², V_c is in MN, and b and d are in m.

In conventional English units, Eqs. (A.12a) and (A.12b) take the following form:

$$V_c = 2\sqrt{f'_c}\, bd \qquad (A.13a)$$
$$V_c = 4\sqrt{f'_c}\, bd \qquad (A.13b)$$

where V_c is in lb, f'_c is in lb/in.^{2} and b and d are in inches.

Note that

$$v_c = \frac{V_c}{bd} \qquad (A.14)$$

where v_c is the shear stress

Now, combining Eqs. (A.11), (A.12a), and (A.14), one obtains

$$\text{Permissible shear stress} = v_u = \frac{V_u}{bd} = 0.17\phi\sqrt{f'_c} \qquad (A.15a)$$

Similarly, from Eqs. (A.11), (A.12b), and (A.14)

$$v_u = 0.34\phi\sqrt{f'_c} \qquad (A.15b)$$

Table A.2 gives the values of v_u/ϕ for various values of f'_c.

In a similar manner, combining Eqs. (A.11), (A.13a), (A.13b), and (A.14), one obtains

$$v_u = 2\phi\sqrt{f'_c} \qquad (A.16a)$$

and

$$v_u = 4\phi\sqrt{f'_c} \qquad (A.16b)$$

Table A.3 gives the values of v_u/ϕ for various values of f'_c.

Table A.2 Values of v_u/ϕ [Eqs. (A.15a) and (A.15b)]

Equation number	f'_c (MN/m^2)			
	21	28	35	42
Eq. (A.15a)	0.78	0.90	1.0	1.1
Eq. (A.15b)	1.56	1.80	2.0	2.2

Table A.3 Values of v_u/ϕ [Eqs. (A.16a) and (A.16b)]

Equation number	f_c' (lb/in.^{2})			
	3000	4000	5000	6000
Eq. (A.16a)	109.5	126.5	141.4	154.9
Eq. (A.16b)	219.0	253.0	282.8	309.8

A.2   Reinforcing Bars

The nominal sizes of reinforcing bars commonly used in the United States are given in Table A.4.

Table A.4 Nominal Sizes of Reinforcing Bars Used in the United States

Bar no.	Diameter		Area of cross section	
	(mm)	(in.)	(mm^2)	(in.^{2})
3	9.52	0.375	71	0.11
4	12.70	0.500	129	0.20
5	15.88	0.625	200	0.31
6	19.05	0.750	284	0.44
7	22.22	0.875	387	0.60
8	25.40	1.000	510	0.79
9	28.65	1.128	645	1.00
10	32.26	1.270	819	1.27
11	35.81	1.410	1006	1.56
14	43.00	1.693	1452	2.25
18	57.33	2.257	2580	4.00

The details regarding standard metric bars used in Canada are as follows:

Bar number	Diameter (mm)	Area (mm^2)
10	11.3	100
15	16.0	200
20	19.5	300
25	25.2	500
30	29.9	700
35	35.7	1000
45	43.7	1500
55	56.4	2500

Reinforcing-bar sizes in the metric system have been recommended by UNESCO (1971):

Bar diameter (mm)	Area (mm^2)
6	28
8	50
10	79
12	113
14	154
16	201
18	254
20	314
22	380
25	491
30	707
32	804
40	1256
50	1963
60	2827

Sections A.6–A.8 use the standard bar diameters recommended by UNESCO.

A.3   Development Length

The development length, L_d, is the length of embedment required to develop the yield stress in the tension reinforcement for a section in flexure. ACI Code Section 12.2.2 lists the basic development lengths for tension reinforcement as

a. 35-mm bar and smaller $0.019 A_b f_y / \sqrt{f_c'}$

   but not less than $0.058 d_b f_y$

b. 43-mm bar $26 f_y / \sqrt{f_c'}$

c. 57-mm bar $34 f_y / \sqrt{f_c'}$

where A_b = area of the individual bar (mm^2)

   d_b = nominal diameter of the bar (mm)

The units f_y and f_c' in the preceding expressions are in MN/m^2, and L_d is in millimeters.

In conventional English units, L_d is expressed in inches. The expressions for L_d are as follows:

a. No. 11 (U.S.) bar and smaller $0.04 A_b f_y / \sqrt{f_c'}$

   but not less than $0.0004 d_b f_y$

b. No. 14 (U.S.) bar $0.085 f_y / \sqrt{f_c'}$

c. No. 18 (U.S.) bar $0.11 f_y / \sqrt{f_c'}$

The units of A_b and d_b are in in.^{2} and in., respectively. The values of f_y and f_c' are expressed in lb/in.^{2}.

The basic development length must be multiplied by appropriate factors given by ACI Code Sections 12.2.3 and 12.2.4:

a. Top reinforcement   1.4

b. Reinforcement with $f_y > 414$ MPa $2 - \dfrac{414}{f_y \ (\mathrm{MN/m^2})}$

   In English units, for $f_y > 60{,}000$ lb/in.^{2} $2 - \dfrac{60{,}000}{f_y \ (\mathrm{lb/in.^2})}$

   c. For lightweight concrete   1.33

   d. Reinforcement spaced at least 152 mm
       on center and at least 76 mm in from
       all sides   0.8

   e. Reinforcement in excess of that required $\left(\dfrac{\text{as required}}{\text{as provided}}\right)$

In any case, the basic development length should not be less than 305 mm (12 in.).

A.4   Summary of ACI 318-83 Code Requirements

For convenience, Table A.5 gives a summary of the strength design principles that apply to foundation design. The table has been divided into two parts: (1) general design principles that apply to strength design and (2) principles that are specifically applicable to foundation design.

Table A.5   Summary of ACI 318-83 Code Requirements (Revised 1986)

Principles	Design items	Code requirements	Code section
General	Load	$U = 1.4D + 1.7L$ $U = 0.75(1.4D + 1.7L + 1.7W)$ $U = 0.9D + 1.3W$ (See Article A.1.1 for explanations)	9.2
	Load factor, ϕ	Flexure: 0.9 Shear and torsion: 0.85 Bearing: 0.7 Flexure in plain concrete: 0.65 (See Article A.1.2)	9.3
	Flexure	$M_u = \phi A_s f_y \left[d - \dfrac{0.59 A_s f_y}{f'_c b} \right]$ [See Eq. (A.7)]	10.2
	Maximum flexure reinforcement	$s_{\max} = 0.75 s_b$ [See Eqs. (A.8b), (A.8c), and (A.9)]	10.3.3
	Minimum flexure reinforcement—steel percentage	$s_{\min} = \dfrac{1.4}{f_y \, (\text{MN/m}^2)}$ $s_{\min} = \dfrac{200}{f_y \, (\text{lb/in.}^2)}$ Uniform thickness: use steel percentage equal to that required for shrinkage and temperature	10.5
	Shrinkage and temperature reinforcement—steel percentage	For $f_y = 275$ MN/m² (40 ksi) or 345 MN/m² (50 ksi); $s_s = (0.002)(\text{gross concrete area})$ For $f_y = 414$ MN/m² (60 ksi): $s_s = (0.0018)(\text{gross concrete area})$	7.12
	β factor (See Fig. A.1)	$\beta = 0.85$ for $f'_c \le 28$ MN/m² (4000 lb/in.²) and reduces by 0.05 for every 7 MN/m² (1000 lb/in.² in excess of 28 MN/m²). *Minimum value* = 0.65	10.2.7

Table A.5—continued

Principles	Design items	Code requirements	Code section
	Shear reinforcement	Refer to ACI Code	11.11
	Development length, L_d	See Section A.3	12.2
	Reinforcement spacing	Clear distance not less than diameter of bar or 25.4 mm (1 in.)	7.6.1
		Walls and slabs: not to be spaced farther apart than 3 times the wall or slab thickness or 457 mm (18 in.)	7.6.5
	Minimum reinforcement cover	76 mm (3 in.) for concrete cast against and permanently exposed to earth	7.7.1
	Modulus of elasticity of concrete, E_c	*SI system:* $E_c \,(\text{MN/m}^2) = (W_c^{1.5})(0.043)\sqrt{f_c' \,(\text{MN/m}^2)}$ W_c = density of concrete (for 1440 kg/m^3 to 2480 kg/m^3) $E_c \,(\text{MN/m}^2) = 4730\sqrt{f_c' \,(\text{MN/m}^2)}$ (for normal weight concrete) *English system:* $E_c \,(\text{lb/in.}^2) = (W_c^{1.5})(33)\sqrt{f_c' \,(\text{lb/in.}^2)}$ for W_c = 90 to 155 lb/ft^3 $E_c \,(\text{lb/in.}^2) = 57{,}000\sqrt{f_c' \,(\text{lb/in.}^2)}$ (for normal weight concrete)	8.5
	Shear strength	$v_n = \dfrac{V_n}{bd}$	
		Wide beam: $v_c \,(\text{MN/m}^2) = 0.17\sqrt{f_c' \,(\text{MN/m}^2)}$ $v_c \,(\text{lb/in.}^2) = 2\sqrt{f_c' \,(\text{lb/in.}^2)}$	11.3
		Diagonal tension: $v_c \,(\text{MN/m}^2) = 0.34\sqrt{f_c' \,(\text{MN/m}^2)}$ $v_c \,(\text{lb/in.}^2) = 4\sqrt{f_c' \,(\text{lb/in.}^2)}$ [See Eqs. (A.12) and (A.13)]	11.11
	Bearing strength	Bearing strength $= 0.85\phi f_c' A_1$ $\phi = 0.7$ *Exceptions:* when supporting surface is wider on all sides than the loaded area, the bearing strength on the loaded area is equal to $0.85\phi f_c' A_1\sqrt{A_2/A_1}$. Limit of $\sqrt{A_2/A_1} \le 2$. A_1 = loaded area; A_2 = area of the portion of the supporting surface that is concentric and geometrically similar to the loaded area	10.15
FOUNDATIONS			
Footings	General considerations	See ACI Code	15
	Maximum moment	See ACI Code	15.4
	Shear	$v_c \,(\text{MN/m}^2) = 0.083\left(2 + \dfrac{4}{\beta_c}\right)\sqrt{f_c' \,(\text{MN/m}^2)}$ $\le 0.34\sqrt{f_c' \,(\text{MN/m}^2)}$ $v_c \,(\text{lb/in.}^2) = \left(2 + \dfrac{4}{\beta_c}\right)\sqrt{f_c' \,(\text{lb/in.}^2)}$ $\le 4\sqrt{f_c' \,(\text{lb/in.}^2)}$	11.11.2.1, 15.5

Table A.5—continued

Principles	Design items	Code requirements	Code section
		β_c = ratio of long side to short side of concentrated load or reaction area	
	Minimum footing depth	Not less than 152 mm (6 in.) above the bottom of reinforcement for footing on soil. Not less than 305 mm (12 in.) for footing on piles	15.7
	Transfer of force at base of column or reinforced pedestal	Area of reinforcement ≥ 0.005 gross area of supported member	15.8.2.1
	Round columns	Treat as square columns with same area for location of critical sections for moment, shear, and development of reinforcement in footings	15.3
WALLS	General considerations	Refer to ACI Code	14
	Minimum thickness	Not less than $\frac{1}{25}$ the supported height or length, whichever is shorter; not less than 102 mm (4 in.)	14.5.3.1
		Exterior basement walls and foundation walls not less than 191 mm (7.5 in.)	14.5.3.2
	Grade beam	See ACI Code	14.7
	Reinforcement	Horizontal: $A_s \geq 0.0025 A_g$ of wall Vertical: $A_s \geq 0.0015 A_g$ of wall A_s = area of reinforcement A_g = gross area of the wall	14.3.3 14.3.2

A.5   Design Example of a Continuous Wall Foundation

This section describes the design of a load-bearing wall with the following characteristics

Dead load = D = 3000 lb/ft
Live load = L = 1200 lb/ft
Gross allowable bearing capacity of soil = 2000 lb/ft^2
Depth of the top of foundation from the ground surface = 4 ft
f_y = 60 kips/in.^{2}
f_c' = 3 kips/in.^{2}
Unit weight of soil = γ = 110 lb/ft^3
Unit weight of concrete = γ_c = 150 lb/ft^3

General Considerations

For this design, assume that the foundation thickness is 12 in., the minimum cover over the steel reinforcement (ACI Code Section 7.7.1) is 3 in., and the steel bars to

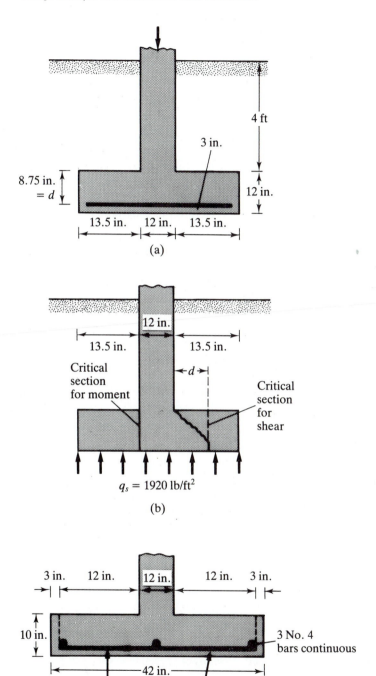

Figure A.2   Continuous wall foundation

be used are U.S. No. 4 bars (diameter = 0.5 in.). From Figure A.2a

$$d = 12 - 3 - \frac{0.5}{2} = 8.75 \text{ in.}$$

Weight of foundation per foot length =

$$\left(\frac{12 \text{ in.}}{12}\right)\gamma_c = \left(\frac{12}{12}\right)(150) = 150 \text{ lb/ft}^2$$

Weight of soil above the foundation =

$$(4)(110) = 440 \text{ lb/ft}^2$$

So, the net allowable soil-bearing capacity

$$q_{\text{net(all)}} = 2000 - 150 - 400 = 1410 \text{ lb/ft}^2$$

Hence, required width of foundation =

$$B = \frac{D + L}{q_{\text{net(all)}}} = \frac{3000 + 1200}{1410} = 2.98 \text{ ft}$$

So, assume that $B = 3.25$ ft. According to ACI Code Section 9.2

$$U = 1.4D + 1.7L = (1.4)(3000) + (1.7)(1200) = 6240 \text{ lb/ft}$$

Converting the net allowable soil pressure to an ultimate (factored value)

$$q_s = \frac{U}{(B)(1 \text{ ft})} = \frac{6240}{3.25} = 1920 \text{ lb/ft}^2$$

Investigation of Shear Strength of the Foundation

The critical section for shear occurs at a distance d from the face of the wall (ACI Code Sections 15.5 and 11.11), as shown in Figure A.2b. So, shear at critical section

$$V_u = \frac{(13.5 \text{ in.} - d)}{12} q_s(1) = \frac{(13.5 - 8.75)}{12} (1920)(1 \text{ ft})$$

$$= 760 \text{ lb/ft}$$

From Eq. (A.13a)

$$V_c = 2\sqrt{f'_c}\, bd = 2\sqrt{3000}(12 \text{ in.})(8.75 \text{ in.}) = 11{,}502 \text{ lb/ft}$$

Also, from Eq. (A.11)

$$\phi V_c = (0.85)(11{,}502) = 9777 \text{ lb/ft} > V_u = 760 \text{ lb/ft}\text{—O.K.}$$

Because $V_u < \phi V_c$, the total thickness of the foundations could be reduced to 10 in. So, the modified

$$d = 10 - 3 - \frac{0.5}{2} = 6.75 \text{ in.} > 6 \text{ in.} = d_{\text{min}} \quad (\text{ACI Code Section 15.7})$$

If $d = 6.75$ in.

$$\phi V_c = (0.85)(2)\sqrt{3000}(12 \text{ in.})(6.75 \text{ in.}) = 7542 \text{ lb/ft} > V_u\text{—O.K.}$$

Flexural Reinforcement

For steel reinforcement, the factored moment at the face of the wall has to be determined (ACI Code Section 15.4). The bending of the foundation will be in one direction only. So, according to Figure A.2b, the design ultimate moment

$$M_u = \frac{q_s l^2}{2}$$

$$l = 13.5 \text{ in.}$$

So

$$M_u = \frac{(1920)\left(\frac{13.5}{12}\right)^2}{2} = 1215 \text{ lb-ft/ft} = 14.58 \times 10^3 \text{ lb-in./ft}$$

From Eqs. (A.4) and (A.5)

$$M_n = A_s f_y\left(d - \frac{a}{2}\right)$$

$$a = \frac{A_s f_y}{0.85 f'_c b} = \frac{A_s (60{,}000)}{(0.85)(3000)(12 \text{ in.})} = 1.961 A_s$$

Thus

$$M_n = (A_s)(60{,}000)\left(6.75 - \frac{1.961}{2} A_s\right)$$

$$= 405 \times 10^3 A_s - 58.83 \times 10^3 A_s^2$$

Again, from Eq. (A.6)

$$M_u = \phi M_u$$

where $\phi = 0.9$

Thus

$$14.58 \times 10^3 = (0.9)(10^3)(405 A_s - 58.83 A_s^2)$$

From the above equation

$$A_{s(1)} = 6.85 \text{ in.}^2; \qquad A_{s(2)} = 0.04 \text{ in.}^2$$

Hence, the steel percentage with $A_{s(1)}$ is

$$s_1 = \frac{A_{s(1)}}{bd} = \frac{6.85 \text{ in.}^2}{(12 \text{ in.})(6.75 \text{ in.})} = 0.0846$$

Similarly

$$s_2 = \frac{A_{s(2)}}{bd} = \frac{0.04}{(12 \text{ in.})(6.75 \text{ in.})}$$

$$= 0.000494 < s_{\min} = 0.0018 \text{ (ACI Code Section 7.12.2)}$$

The maximum steel percentage that can be provided is given in Eqs. (A.8c) and (A.9). Thus

$$s_{\max} = (0.75)(0.85)\frac{f'_c}{f_y} \beta\left(\frac{87{,}000}{87{,}000 + f_y}\right)$$

Note that $\beta = 0.85$. Substituting the proper values of β, f'_c, and f_y in the preceding equation, one obtains

$$s_{max} = 0.016$$

Note that $s_1 = 0.0846 > s_{max} = 0.016$. So use $s = s_{min} = 0.0018$. So

$$A_s = (s_{min})(b)(d) = (0.0018)(12)(6.75) = 0.1458 \text{ in.}^2$$

Use U.S. No. 4 bars (0.5-in. diameter) at 12 in. center-to-center. Hence

$$A_{s(provided)} = \left[\frac{(1 \text{ ft})(12)}{(12 \text{ in.})}\right]\left(\frac{\pi}{4}\right)(0.5)^2 = 0.196 \text{ in}^2$$

Development Length of Reinforcement Bars

According to Section A.3, for No. 4 bars, $L_d = 0.04A_b(f_y/\sqrt{f'_c})$ but not less than $0.0004d_b f_y$ or 12 in. For this case, $A_b = 0.196$ in.^{2} and $d_b = 0.5$ in. So

$$L_d = 0.04A_b\left(\frac{f_y}{\sqrt{f'_c}}\right) = (0.04)(0.196)\left(\frac{60,000}{\sqrt{3000}}\right) = 8.59 \text{ in.}$$

$$L_d = 0.004d_b f_y = (0.0004)(0.5)(60,000) = 12 \text{ in.}$$

Hence

$$L_d = 12 \text{ in.}$$

Now, assuming a 3-in. cover to be on the sides of the foundation, the development length would be 13.5 in. $- 3$ in. $= 10.5$ in. < 12 in. To achieve the proper development length, increase the width of the foundation (B) to 42 in. This will give 15-in. cantilevers, as shown in Figure A.2c. This increase of B will not greatly affect the structural steel provided. It will reduce the value of soil-bearing pressure (q_s) while the shear and moment capacities of the cross section remain constant.

Minimum reinforcement should be furnished in the long direction to offset shrinkage and temperature effects (ACI Code Section 7.12). So

$$A_s = (0.0018)(b)(d) = (0.0018)[(15)(2) + 12](6.75) = 0.51 \text{ in.}^2$$

Provide three No. 4 bars ($A_s = 0.588$ in.^{2}).

The final design sketch is shown in Figure A.2c.

A.6   Design Example of a Square Foundation for a Column

Figure A.3a shows a square column foundation with the following conditions:

Live load $= L = 675$ kN
Dead load $= D = 1125$ kN
Allowable gross soil-bearing capacity $= q_{all} = 145$ kN/m^2
Column size $= 0.5$ m $\times$ 0.5 m
$f'_c = 20.68$ MN/m^2
$f_y = 413.7$ MN/m^2

The design of the foundation can be done in the following manner.

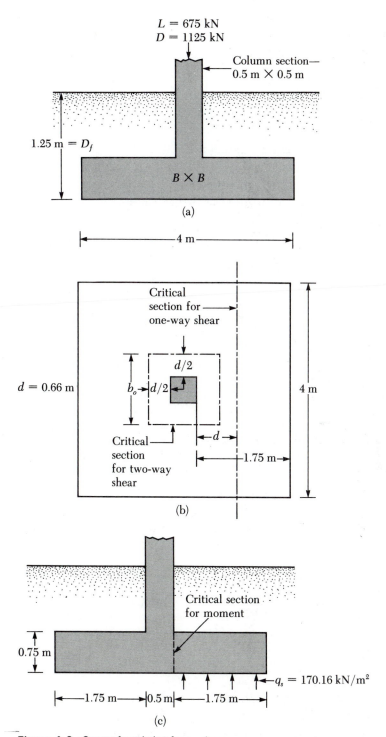

(a)

(b)

(c)

Figure A.3   Square foundation for a column

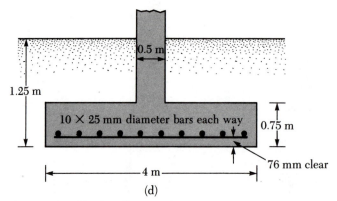

Figure A.3 (Continued)

General Considerations

Let the average unit weight of concrete and soil above the base of the foundation be 21.97 kN/m³. So, the net allowable soil-bearing capacity

$$q_{all(net)} = 145 - (D_f)(21.97) = 145 - (1.25)(21.97) = 117.54 \text{ kN/m}^2$$

Hence, the required foundation area =

$$A = B^2 = \frac{D + L}{q_{all(net)}} = \frac{675 + 1125}{117.54} = 15.31 \text{ m}^2$$

Use a foundation with dimensions (B) of 4 m × 4 m.

The factored load for the foundation

$$U = 1.4D + 1.7L = (1.4)(1125) + (1.7)(675) = 2722.5 \text{ kN}$$

Hence, the factored soil pressure

$$q_s = \frac{U}{B^2} = \frac{2722.5}{16} = 170.16 \text{ kN/m}^2$$

Assume the thickness of the foundation to be equal to 0.75 m. With a clear cover of 76 m over the steel bars and an assumed bar diameter of 25 mm

$$d = 0.75 - 0.076 - \frac{0.025}{2} = 0.6615 \text{ m}$$

Check for Shear

As we have seen in Section A.5, V_u should be equal to or less than ϕV_c. For one-way shear

$$V_u \le \phi(0.17)\sqrt{f_c'} \, bd$$

The critical section for one-way shear is located at a distance d from the edge of the column (ACI Code Section 11.11.1.1) as shown in Figure A.3b. So

$$V_u = q_s \times \text{critical area} = (170.16)(4)(1.75 - 0.6615) = 740.9 \text{ kN}$$

Also

$$\phi V_c = (0.85)(0.17)(\sqrt{20.68})(4)(0.6615)(1000) = 1738.7 \text{ kN}$$

So

$$V_u = 749.9 \text{ kN} \leq \phi V_c = 1738.7 \text{ kN—O.K.}$$

For two-way shear, the critical section is located at a distance of $d/2$ from the edge of the column (ACI Code Section 11.11.1.2). This is shown in Figure A.3b. For this case

$$\phi V_c = \phi(0.34)\sqrt{f_c'} b_o d$$

The term b_o is the perimeter of the critical section for two-way shear. Or, for this design

$$b_o = 4[0.5 + 2(d/2)] = 4[0.5 + 2(0.3308)] = 4.65 \text{ m}$$

Hence

$$\phi V_c = (0.85)(0.34)(\sqrt{20.68})(4.65)(0.6615) = 4.042 \text{ MN} = 4042 \text{ kN}$$

Also

$$V_u = (q_s)(\text{critical area})$$
$$\text{Critical area} = (4 \times 4) - (0.5 + 0.6615)^2 = 14.65 \text{ m}^2$$

So

$$V_u = (170.16)(14.65) = 2492.84 \text{ kN}$$
$$V_u = 2492.84 \text{ kN} < \phi V_c = 4042 \text{ kN—O.K.}$$

So, the assumed depth of foundation is more than adequate.

Flexural Reinforcement

According to Figure A.3c, the moment at critical section (ACI Code Section 15.4.2) is

$$M_u = (q_s B)\left(\frac{1.75}{2}\right)^2 = \frac{[(170.16)(4)](1.75)^2}{2} = 1042.23 \text{ kN-m}$$

From Eq. (A.4)

$$a = \frac{A_s f_y}{0.85 f_c' b} \quad (note: b = B)$$

or

$$A_s = \frac{0.85 f_c' B a}{f_y} = \frac{(0.85)(20.68)(4)a}{413.7} = 0.17a$$

From Eq. (A.6)

$$M_u \leq \phi A_s f_y \left(d - \frac{a}{2}\right)$$

With $\phi = 0.9$ and $A_s = 0.17a$

$$M_u = 1042.23 = (0.9)(0.17a)(413,700)\left(0.6615 - \frac{a}{2}\right)$$

Solution of the preceding equation gives $a = 0.0254$ m. Hence

$$A_s = 0.17a = (0.17)(0.0254) = 0.0043 \text{ m}^2$$

Percentage of steel

$$s = \frac{A_s}{bd} = \frac{A_s}{Bd} = \frac{0.0043}{(4)(0.6615)} = 0.00163 < s_{min}$$
$$= 0.0018 \text{ (ACI Code Section 7.12.2)}$$

So

$$A_{s(min)} = (0.0018)(B)(d) = (0.0018)(4)(0.6615)$$
$$= 0.004762 \text{ m}^2 = 47.62 \text{ cm}^2$$

Provide 10×25-mm ($\approx$ No. 8 U.S.) diameter bars each way [$A_s = (4.91)(10) = 49.1$ cm^2].

Check for Development Length

From Section A.3, the development length

$$L_d = 0.019A_b \frac{f_y}{\sqrt{f'_c}} = (0.019)(491 \text{ mm}^2)\left(\frac{413.7}{\sqrt{20.68}}\right) = 848.68 \text{ mm}$$

Also

$$L_d \geq 0.058d_b f_y = (0.058)(25)(413.7) = 599.87 \text{ mm}$$

So, $L_d = 848.68$-mm controls. Actual L_d provided is $(4 - 0.5/2) - 0.076$ (cover) $= 1.674$ m > 599.87 mm—O.K.

Check for Bearing Strength

Table A.5 indicates that the bearing strength should be at least $0.85\phi f'_c A_1 \sqrt{A_2/A_1}$ with a limit of $\sqrt{A_2/A_1} \leq 2$. For this problem, $\sqrt{A_2/A_1} = \sqrt{(4 \times 4)/(0.5 \times 0.5)} = 8$. So, use $\sqrt{A_2/A_1} = 2$. Also, $\phi = 0.7$. Hence, the design bearing strength $= (0.85)(0.7)(20.68)(0.5 \times 0.5)(2) = 6.15$ MN $= 6150$ kN. However, the factored column load $U = 2722.5$ kN < 6150 kN—O.K.

The final design section is shown in Figure A.3d.

A.7   Design Example of a Rectangular Foundation for a Column

This section describes the design of a rectangular foundation to support a column having dimensions of 0.4 m × 0.4 m in cross section. Other details are as follows:

Dead load $= D = 290$ kN
Live load $= L = 110$ kN

Depth from the ground surface to the top of the foundation = 1.2 m
Allowable gross soil-bearing capacity = 120 kN/m²
Maximum width of foundation = $B = 1.5$ m
$f_y = 413.7$ MN/m²
$f_c' = 20.68$ MN/m²
Unit weight of soil = $\gamma = 17.27$ kN/m³
Unit weight of concrete = $\gamma_c = 22.97$ kN/m³

General Considerations

For this design, let us assume a foundation depth of 0.45 m (Figure A.4a). The weight of foundation/m² = $0.45\gamma_c = (0.45)(22.97) = 10.34$ kN/m², and the weight of soil above the foundation/m² = $(1.2)\gamma = (1.2)(17.27) = 20.72$ kN/m². Hence, the net allowable soil-bearing capacity $[q_{net(all)}] = 120 - 10.34 - 20.72 = 88.94$ kN/m².

The required area of the foundation = $(D + L)/q_{net(all)} = (290 + 110)/88.94 = 4.5$ m². Hence, the length of the foundation is 4.5 m²/$B = 4.5/1.5 = 3$ m.

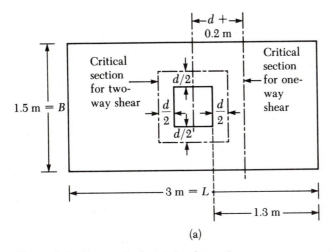

(a)

Figure A.4 Rectangular foundation for a column

(b)

Figure A.4 (Continued)

The factored column load = $1.4D + 1.7L = 1.4(290) + 1.7(110) = 593$ kN.

The factored soil-bearing capacity, q_s = factored load/foundation area = $593/4.5 = 131.78$ kN/m².

Shear Strength of Foundation

Assume that the steel bars to be used have a diameter of 16 mm. So, the effective depth $d = 450 - 76 - 16 = 358$ mm. (Note that the assumed clear cover is 76 mm.)

Figure A.4a shows the critical section for one-way shear (ACI Code Section 11.11.1.1). According to this figure

$$V_u = \left(1.5 - \frac{0.4}{2} - 0.358\right)Bq_s = (0.942)(1.5)(131.78) = 186.21 \text{ kN}$$

The nominal shear capacity of concrete for one-way beam action

$$V_c = 0.17\sqrt{f'_c}\,Bd = 0.17(\sqrt{20.68})(1.5)(0.358) = 0.4152 \text{ MN} = 415.2 \text{ kN}$$

Now

$$V_u = 184.22 \leq \phi V_c = (0.85)(415.2) = 352.92 \text{ kN—O.K.}$$

The critical section for two-way shear is also shown in Figure A.4a. This is based on the recommendations given by ACI Code Section 11.11.1.2. For this section

$$V_u = q_s[(1.5)(3) - 0.758^2] = 517.3 \text{ kN}$$

The nominal shear capacity of the foundation can be given as (ACI Code Section 11.11.2.1; also see Table A.5 of the text)

$$V_c = v_c b_o d = 0.083\left(2 + \frac{4}{\beta_c}\right)\sqrt{f_c'} b_o d$$

where b_o = perimeter of the critical section
β_c = length of foundation/width of foundation

or

$$V_c = 0.083\left[2 + \frac{4}{\dfrac{3}{1.5}}\right]\sqrt{20.68}(4 \times 0.758)(0.358) = 1.638 \text{ MN}$$

So, for two-way shear condition

$$V_u = 517.3 \text{ kN} < \phi V_c = (0.85)(1639) = 1393 \text{ kN}$$

Therefore the section is adequate.

Check for Bearing Capacity of Concrete Column at the Interface with Foundation

According to ACI Code Section 10.15 (also see Table A.5), bearing strength is equal to $0.85\phi f_c' A_1 (\phi = 0.7)$. For this problem, $U = 593$ kN < bearing strength = $(0.85)(0.7)(20.68)(0.4)^2 = 1.969$ MN.

So, a minimum area of dowels should be provided across the interface of the column and the foundation. Based on ACI Code Section 15.8.2.1

$$\text{Minimum area of steel} = (0.005)(\text{area of column})$$
$$= (0.005)(400^2) = 800 \text{ mm}^2$$

So use 4 × 16-mm ($\approx$No. 5 U.S.) diameter bars as dowels.

The minimum required length of development (L_d) of dowels into the foundation is $(0.24 f_y d_b)/\sqrt{f_c'}$, but not less than $0.044 f_y d_b$ (ACI Code Section 12.3.2). So

$$L_d = \frac{0.24 f_y d_b}{\sqrt{f_c'}} = \frac{(0.24)(413.7)(16)}{\sqrt{20.68}} = 349.33 \text{ mm}$$

Also

$$L_d = 0.044 f_y d_b = (0.044)(413.7)(16) = 291.25 \text{ mm}$$

Hence, $L_d = 349.33$-mm controls.

Available depth for the dowels (Figure A.4a) is $450 - 76 - 16 - 16 = 342$ mm. Since hooks cannot be used, the foundation depth must be increased. Let the new depth be equal to 480 mm to accommodate the required $L_d = 349.33$ mm. Hence, the new value of d is equal to $480 - 76 - 16 = 388$ mm.

Flexural Reinforcement in the Long Direction

According to Figure A.4a, the design moment about the column face =

$$M_u = \frac{(q_s B)1.3^2}{2} = \frac{(131.78)(1.5)(1.3)^2}{2} = 167.07 \text{ kN-m}$$

From Eq. (A.4)

$$a = \frac{A_s f_y}{0.85 f'_c b} = \frac{(A_s)(413.7)}{(0.85)(20.68)(1.5)} = 15.69 A_s$$

Again, from Eq. (A.6)

$$M_u = \phi M_n = \phi A_s f_y \left(d - \frac{a}{2} \right)$$

or

$$167.07 = (0.9)(A_s)(413.7 \times 10^3)\left[0.388 - \frac{15.69}{2}(A_s) \right]$$

$$167.07 = 144{,}464 A_s - 2{,}920{,}928 A_s^2$$

The solution of the preceding equation gives

$$A_{s(1)} = 0.0483 \text{ m}^2 \left[\text{that is, steel percentage} = \frac{A_{s(1)}}{Bd} = \frac{0.0483}{(1.5)(0.388)} \right.$$

$$\left. = 0.0829 = s_1 \right]$$

and

$$A_{s(2)} = 0.0012 \text{ m}^2 \left[\text{that is, steel percentage} = \frac{A_{s(2)}}{Bd} = \frac{0.00120}{(1.5)(0.388)} \right.$$

$$\left. = 0.00206 = s_2 \right]$$

Section A.5 (for similar values of f'_c and f_y) indicated that $s_{max} = 0.016$. Also, from ACI Code Section 7.12.2, $s_{min} = 0.0018$. Note that $s_1 > s_{max}$ and $s_2 > s_{min}$. So, $A_s = A_{s(2)}$ may be used. Hence, provide 6 × 16 mm diameter bars (A_s provided is 0.001206 m²).

Flexural Reinforcement in the Short Direction

According to Figure A.4a, the moment at the face of the column =

$$M_u = \frac{(q_s L)(0.55)^2}{2} = \frac{(131.78)(3)(0.55)^2}{2} = 59.8 \text{ kN-m}$$

From Eq. (A.4)

$$a = \frac{A_s f_y}{0.85 f'_c b} = \frac{(A_s)(413.7)}{(0.85)(20.68)(3)} = 7.845 A_s$$

From Eq. (A.6)

$$M_u = \phi A_s f_y \left(d - \frac{a}{2} \right)$$

or

$$59.8 = (0.9)(A_s)(413.7 \times 10^3)\left[0.388 - \frac{7.845}{2}(A_s) \right]$$

The solution of the preceding equation gives

$$A_{s(1)} = 0.0985 \text{ m}^2 \qquad (\text{thus } s_1 > s_{max})$$
$$A_{s(2)} = 0.0004 \text{ m}^2 \qquad (\text{thus } s_2 < s_{min})$$

So, use $s = s_{min}$, or

$$A_s = s_{min} bd = (0.0018)(3)(0.388) \approx 0.0021 \text{ m}^2$$

Use 12×16-mm ($\approx$ No. 5 U.S.) diameter bars.

Final Design Sketch

According to ACI Code Section 15.4.4.2, a portion of the reinforcement in the short direction shall be distributed uniformly over a bandwidth equal to the smallest dimension of the foundation. The remainder of the reinforcement should be distributed uniformly outside the central band of the foundation. The reinforcement in the central band can be given to be equal to $2/(\beta_c + 1)$ (where $\beta_c = L/B$). For this problem, $\beta_c = 2$. Hence, $\frac{2}{3}$ of the reinforcing bars (that is, 8 bars) should be placed in the center band of the foundation. The remaining 4 bars should be placed outside the central band. However, one needs to check the steel percentage in the outside band, or

$$s = \frac{A_s}{bd} = \frac{(2)(201 \text{ mm}^2)}{\left(\dfrac{3000 - 1500}{2}\right)(388)} = 0.0014 < s_{min} = 0.0018$$

So, use $A_s = (s_{min})(b)(d) = (0.0018)(750)(388) = 523.8 \text{ mm}^2$. Hence, 3×16-mm ($\approx$ No. 5 U.S.) diameter bars on each side of the central band will be sufficient. The final design sketch is shown in Figure A.4b.

A.8   Design Example of a Retaining Wall

In Example 5.12 (refer to Figure 5.33), the overall stability of a cantilever retaining wall was checked. This section uses the same retaining wall for the design of stem, toe, and heel. It will be assumed that $f_y = 413.7 \text{ MN/m}^2$ and $f'_c = 20.68 \text{ MN/m}^2$.

Stem Design

The lateral earth pressure distribution behind the stem of the wall given in Figure 5.33 is shown in Figure A.5a. Note that, at any given depth z from the top of the wall

$$\sigma_a = \gamma_1 z K_a$$

The horizontal component of the lateral pressure can be given as

$$\sigma_{a(H)} = \gamma_1 z K_a \cos \alpha \tag{A.17}$$

Hence the moment at any depth z is

$$M = \frac{1}{6} \gamma_1 z^3 K_a \cos \alpha \tag{A.18}$$

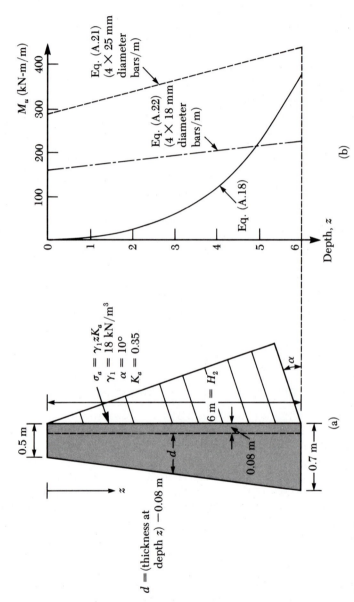

Figure A.5 Stem design of a retaining wall

Table A.6 Stem Design for a Retaining Wall

z (m)	Thickness of stem (m)	d (m)	M_u (kN-m/m)	a (m)	A_s (cm^2)
0	0.5	0.42	0	0	0^a
2	0.57	0.49	14.06	0.0018	0.77^a
4	0.64	0.56	112.5	0.0128	5.44^a
6	0.7	0.62	379.7	0.0400	17.00

Note: d = thickness of stem $-$ 0.08 m; M_u from Eq. (A.19); a from Eq. (A.21); A_s from Eq. (A.20)
^{a} $A_s < A_{s(min)}$; so use $A_{s(min)} = 10.5$ cm^2

The ultimate design moment (ACI Code Section 9.2.4) can be given as

$$M_u = 1.7M = \frac{1.7}{6}\gamma_1 z^3 K_a \cos \alpha \tag{A.19}$$

The variations of M_u at $z = 0, 2, 4,$ and 6 m have been calculated by using Eq. (A.19) and are tabulated in Table A.6. From Eq. (A.4)

$$A_s = \frac{0.85af'_c b}{f_y} = \frac{(0.85)(a)(20.68)(1)}{413.7} = 0.0425a \tag{A.20}$$

Also, from Eq. (A.6)

$$M_u = \phi A_s f_y\left(d - \frac{a}{2}\right)$$

With $\phi = 0.9$ and $A_s = 0.0425a$

$$M_u = (0.9)(0.0425a)(413.7 \times 10^3 \text{ kN/m}^2)\left(d - \frac{a}{2}\right)$$

or

$$M_u = 15{,}824.03ad - 7912.01a^2 \tag{A.21}$$

Assume that the value of d at any given depth z is equal to the thickness of the stem at that depth minus a cover of 80 mm. These values of d at $z = 0, 2, 4,$ and 6 m are shown in Table A.6. Now, with known values of d and M_u at any given depth, the magnitude of a can be obtained from Eq. (A.21) and thus A_s from Eq. (A.20). This is also shown in Table A.6. Note that at $z = 0, 2,$ and 4 m, A_s is less than $A_{s(min)}$. Therefore use (ACI Code Section 14.3.2)

$$A_s = A_{s(min)} = (0.0015)(\text{gross wall area}) = (0.0015)(1)(0.7)$$
$$= 0.00105 \text{ m}^2 = 10.5 \text{ cm}^2$$

For selection of reinforcement bars, try 4×25-mm ($\approx$No. 8 U.S.) diameter bars per meter for the entire height of the stem. Thus, $A_s = (4)(\pi/4)(2.5)^2 = 19.63$ cm^2. For this reinforcement, from Eq. (A.20)

$$a = \frac{A_s}{0.0425} = \frac{19.64 \times 10^{-4} \text{ m}^2}{0.0425} = 0.0462 \text{ m}$$

From Eq. (A.21)

$$M_u = 15{,}824.03ad - 7912.01a^2 = (15{,}824.03)(0.0462)d - (7912.01)(0.0462)^2$$
$$= 731.07d - 16.888 \tag{A.22}$$

In a similar manner, 4×18-mm ($\approx$No. 6 U.S.) diameter bars [$A_s = 10.15$ cm$^2 \approx A_{s(\min)}$] can be tried for the entire height of the wall. Next one obtains

$$M_u = 378.19d - 4.52 \tag{A.23}$$

Figure A.5b shows the variation of actual M_u from Table A.6 and also the variation of M_u from the relations given by Eqs. (A.22) and (A.23). From this figure, it can be seen that, for economy, the 25-mm bars should be cut off at a close distance from the base of the stem. In order to lap the 18-mm diameter bars to the 25-mm diameter bars, refer to ACI Code Section 12.15. Because 100% of the bars are to be lapped, the lap splice class is C. Thus

$$\text{Lap distance} = 1.7L_d \tag{A.24}$$

where L_d = tensile development length (see Section A.3)

$$= 0.019A_b f_y / \sqrt{f'_c}$$

So, lap distance $= (1.7)(0.019)\left(\dfrac{\pi}{4} \cdot 25^2\right)\dfrac{413.7}{\sqrt{20.68}} = 1440$ mm $= 1.44$ m.

Therefore, the 25-mm diameter bars must be extended beyond the cutoff point by 1.44 m. The cutoff point can be determined by combining Eqs. (A.19) and (A.23), or

$$\frac{1.7}{6} \gamma_1 z^3 K_a \cos \alpha = 378.19d - 4.52 \tag{A.25}$$

where $d = 0.42 + 0.0333z$ (A.26)

Substituting $\gamma_1 = 18$ kN/m^3, $K_a = 0.35$, and $\alpha = 10°$ in Eq. (A.25) and combining Eqs. (A.25) and (A.26), one obtains $z = 4.98$ m. So, extend the 25-mm bars to a height of $6 - 4.98 + 1.44 = 2.46$ m above the base of the stem.

The cutting off of bars in the tension zone (ACI Code Section 12.10.5.1) also needs to be checked. According to the ACI Code, in that section V_u should be less than or equal to $\frac{2}{3}$ of the shear capacity of the section. The shear capacity at a distance of $z = 3.54$ m above the base of the stem is $0.17\phi\sqrt{f'_c} bd = 0.17(0.85)(\sqrt{20.68})(1)(0.538) = 0.3535$ MN/m $= 353.5$ kN/m. Also, at that depth

$$V_u = (1.7)\left(\frac{1}{2}\right)\gamma_1 z^2 K_a \cos \alpha = (1.7)\left(\frac{1}{2}\right)(18)(3.54)^2(0.35)(\cos 10°)$$

$$= 66.08 \text{ kN} < \left(\frac{2}{3}\right)353.5 \text{ kN—O.K.}$$

Determination of the Length of Development of the Main Reinforcement Bars into the Foundation: According to ACI Code Section 12.2.2, the development length

$$L_d = 0.019A_b \frac{f_y}{\sqrt{f'_c}}$$

For the 25-mm diameter bars

$$L_d = (0.019)\left(\frac{\pi}{4}25^2\right)\frac{413.7}{\sqrt{20.68}} = 848.7 \text{ mm}$$

The $L_d = 848.7$ mm calculated above is more than the thickness of the base slab, which is 700 mm (Figure 5.33). So, in order to have the proper development length in tension, standard hooks may be used. A standard hook develops a tensile stress that may be given as

$$f_h = \alpha\sqrt{f_c'} \tag{A.27}$$

For a 25-mm diameter bar, $\alpha = 30$. So

$$f_h = 30\sqrt{20.68} = 136.43 \text{ MN/m}^2$$

Hence, the remaining stress to be developed is equal to $f_y - 136.43 = 413.7 - 136.43 = 277.27$ MN/m². The extra embedment length required to develop the stress of 277.27 MN/m² is $(277.27/f_y)L_d = (277.27/413.7)(848.7) = 569$ mm ≈ 0.57 m. Figure A.6a shows a sketch of the hook and the embedded length of a bar. Note that the distance from the top of the base slab to the bottom of the hook is 0.645 m. If a cover of 76 mm is to be provided, the minimum thickness of the base slab should be $0.645 + 0.076 = 0.721$ m. So, for final design purposes, the thickness of the base slab should be increased to 0.75 m (from 0.7 m, as shown in Figure 5.33).

Check for Shear Strength at the Base of the Wall: From Eq. (A.17), the shear at the base of the wall can be given as

$$V_u = (1.7)(V) = (1.7)\left(\frac{1}{2}\gamma_1 H_2^2 K_a \cos \alpha\right)$$

$$= (1.7)\left(\frac{1}{2}\right)(18)(6)^2(0.35)(\cos 10°) = 189.9 \text{ kN/m}$$

A shear key, usually 50 mm × 100 mm, is provided at the base of the stem. The shear stress in the key should not exceed a nominal stress of $0.2f_c'$ or 5.52 MN/m². So

$$\frac{V_u}{A_{key}} = \frac{189.9 \text{ kN}}{(10 \times 10^{-2} \text{ m})(1)} = 1899 \text{ kN/m}^2 < 0.2\phi f_c'$$

$$= (0.2)(0.85)20.68 = 3.52 \text{ MN/m}^2\text{---O.K.}$$

So, provide a 50 mm × 100-mm shear key at the base of the stem.

Temperature and Shrinkage Steel: Horizontal temperature and shrinkage steel must be provided in the wall; according to ACI Code Section 14.3.3

$$A_s = (0.0025)(\text{gross area of the wall})$$
$$= (0.0025)(1)(0.7) = 0.00175 \text{ m}^2/\text{m} = 1750 \text{ mm}^2/\text{m of wall}$$

Hence, provide 18-mm diameter bars at 250 mm center-to-center at each face of the wall ($A_s = 2032$ cm²). Also provide 14-mm diameter bars at 300 mm center-to-center at the outside face of the wall.

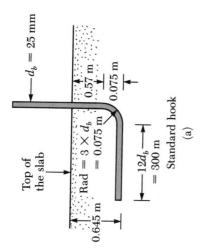

Figure A.6 Final design sketch of the retaining wall

Heel Design

For the design of the heel, refer to Figure A.7. Note that the thickness of the base slab has been increased to 0.75 m, per the findings from the stem design. (In Figure 5.33, it is 0.7 m). The equations for load, shear, and moment for the heel of the retaining wall are as follows:

$$q = q_1 + q_2 - q_3 \tag{A.28}$$

where q_1 = load caused by the soil above the heel = γH_{av}
   = 18(6 + 6.458)/2 = 112.12 kN/m²
q_2 = load caused by the concrete slab = $\gamma_c(0.75)$ = (23.58)(0.75)
   = 17.69 kN/m²
q_3 = soil reaction (refer to Problem 5.12) = $q_{heel} + mx$
   = 45.99 + 35.8x

Figure A.7   Design of heel and toe of a retaining wall

Substitution of proper values of q_1, q_2, and q_3 into Eq. (A.28) gives

$$q = 112.12 + 17.69 - 45.99 - 35.8x = 83.82 - 35.8x \tag{A.29}$$

Shear, $V = \int q \; dx = 83.82x - \dfrac{35.8x^2}{2} + C_1$

where C_1 = a constant. At $x = 0$, $V = P_v$ (see Figure A.7) = 28.03 kN/m.

So

$$V = 83.82x - \frac{35.8x^2}{2} + 28.03 \tag{A.30}$$

Also

$$\text{Moment, } M = \int V \; dx = \frac{83.82x^2}{2} - \frac{35.8x^3}{6} + 28.03x \tag{A.31}$$

The critical section for shear and moment is at $x = 2.6$ m. At the critical section, from Eqs. (A.30) and (A.31)

$$V = 124.96 \text{ kN/m}$$

and

$$M = 251.32 \text{ kN-m/m}$$

Hence

$$V_u = (1.7)V = (1.7)(124.96) = 212.43 \text{ kN/m}$$

and

$$M_u = (1.7)M = (1.7)(251.32) = 427.24 \text{ kN-m/m}$$

Check for Shear: From Eq. (A.12a)

$$V_c = 0.17\sqrt{f'_c}\, bd$$

With $b = 1$ m, $d = 0.75 - 0.076 - 0.025/2 \approx 0.662$ m

$$V_c = (0.17)(\sqrt{20.68})(1)(0.662) = 0.5118 \text{ MN/m} = 511.8 \text{ kN/m}$$

Also

$$\phi V_c = (0.85)(511.8 \text{ kN/m}) = 435.03 \text{ kN/m} > V_u = 212.43 \text{ kN/m}$$

Hence, the section is adequate for shear.

Flexural Reinforcement: From Eq. (A.6)

$$M_u = \phi A_s f_y \left(d - \frac{a}{2} \right)$$

For this design, $M_u = 427.24$ kN-m/m and $a = [(A_s)(413.7)]/[(0.85)(20.68)(1)]$. Or

$$A_s = 0.0425a$$

Thus

$$427.24 = (0.9)(0.0425a)(413.7 \times 10^3)\left(0.662 - \frac{a}{2} \right)$$

Solution of the preceding equation gives $a = 0.0461$ m. Hence, $A_s = (0.0425)a = (0.0425)(0.0421) = 0.00178$ m^2. This gives a steel percentage that is higher than s_{min}. So, provide 25-mm diameter bars at 250 mm center-to-center, which gives $A_s = 1000/250 \times (\pi/4)(2.5)^2 = 19.63$ cm^2.

Toe Design

As in the case of the heel design, refer to Figure A.7. The equation for load, shear, and moment at any point of toe can be written

$$q = -q_2 + q_3 \tag{A.32}$$

where $q_2 = $ load caused by the concrete slab $= \gamma_c(0.75) = (23.58)(0.75)$
$\qquad = 17.69$ kN/m^2
$\qquad q_3 = $ soil reaction $= 189.2 - mx' = 189.2 - 35.8x'$

Note that, in Eq. (A.32), the weight of the soil above the toe has been ignored; thus,

$$q = 189.2 - 35.8x' - 17.69 = 171.51 - 35.8x' \tag{A.33}$$

$$\text{Shear force, } V = \int q\ dx' = 171.51x' - \frac{35.8x'^2}{2} \tag{A.34}$$

$$\text{Moment, } M = \int V\ dx' = \frac{171.51x'^2}{2} - \frac{35.8x'^3}{6} \tag{A.35}$$

For critical section, $x' = 0.7$ m. So, at the critical section,

$$V = 111.29 \text{ kN/m}$$

and

$$V_u = 1.7V = 189.2 \text{ kN/m}$$
$$M = 39.98 \text{ kN-m/m}$$

and

$$M_u = 1.7M = 67.97 \text{ kN-m/m}$$

Check for Shear: Because V_u at the critical section of the toe is less than V_u at the critical section of the heel, the section is adequate for shear.

Flexural Reinforcement: It has been shown in the design of the heel that

$$A_s = 0.0425a$$

and

$$M_u = (0.9)(0.0425a)(413.7 \times 10^3)\left(0.6 - \frac{a}{2}\right)$$

The solution of the preceding equation gives

$$a = 0.0072 \text{ m}$$

So

$$A_s = 0.0425a = (0.0425)(0.0072) = 0.000306 \text{ m}^2$$

Hence, the steel percentage

$$s = \frac{A_s}{bd} = \frac{0.000306}{(1)(0.662)} = 0.000462 < s_{min}$$

Use $A_{s(min)} = s_{min}\,bd = (0.0018)(1)(0.75) \approx 0.00135 \text{ m}^2 = 13.5 \text{ cm}^2$

Hence, provide 25-mm diameter bars at 300 mm center-to-center, which gives $A_s = 16.35 \text{ cm}^2$.

Shrinkage and Temperature Reinforcement for Heel and Toe: For shrinkage and temperature, minimum steel should be used. So

$$A_s = s_{min}\,bd = (0.0018)(1)(0.75) = 0.00135 \text{ m}^2$$

So, provide 18-mm diameter bars at 300 mm center-to-center.

The final design sketch of a retaining wall is shown in Fig. A.6.

REFERENCES

American Concrete Institute (1987). *ACI Standard—Building Code Requirements for Reinforced Concrete, ACI 318-83* (Revised 1986), Detroit.

UNESCO (1971). *Reinforced Concrete: An International Manual*, Butterworth, London.

CONVERSION FACTORS

B.1 Conversion Factors from English to SI Units

Length:
$$1 \text{ ft} = 0.3048 \text{ m}$$
$$1 \text{ ft} = 30.48 \text{ cm}$$
$$1 \text{ ft} = 304.8 \text{ mm}$$
$$1 \text{ in.} = 0.0254 \text{ m}$$
$$1 \text{ in.} = 2.54 \text{ cm}$$
$$1 \text{ in.} = 25.4 \text{ mm}$$

Area:
$$1 \text{ ft}^2 = 929.03 \times 10^{-4} \text{ m}^2$$
$$1 \text{ ft}^2 = 929.03 \text{ cm}^2$$
$$1 \text{ ft}^2 = 929.03 \times 10^2 \text{ mm}^2$$
$$1 \text{ in.}^2 = 6.452 \times 10^{-4} \text{ m}^2$$
$$1 \text{ in.}^2 = 6.452 \text{ cm}^2$$
$$1 \text{ in.}^2 = 645.16 \text{ mm}^2$$

Volume:
$$1 \text{ ft}^3 = 28.317 \times 10^{-3} \text{ m}^3$$
$$1 \text{ ft}^3 = 28.317 \text{ cm}^3$$
$$1 \text{ in.}^3 = 16.387 \times 10^{-6} \text{ m}^3$$
$$1 \text{ in.}^3 = 16.387 \text{ cm}^3$$

Section Modulus:
$$1 \text{ in.}^3 = 0.16387 \times 10^5 \text{ mm}^3$$
$$1 \text{ in.}^3 = 0.16387 \times 10^{-4} \text{ m}^3$$

Coefficient of Permeability:
$$1 \text{ ft/min} = 0.3048 \text{ m/min}$$
$$1 \text{ ft/min} = 30.48 \text{ cm/min}$$
$$1 \text{ ft/min} = 304.8 \text{ mm/min}$$
$$1 \text{ ft/sec} = 0.3048 \text{ m/sec}$$
$$1 \text{ ft/sec} = 304.8 \text{ mm/sec}$$
$$1 \text{ in./min} = 0.0254 \text{ m/min}$$
$$1 \text{ in./sec} = 2.54 \text{ cm/sec}$$
$$1 \text{ in./sec} = 25.4 \text{ mm/sec}$$

Coefficient of Consolidation:
$$1 \text{ in.}^2/\text{sec} = 6.452 \text{ cm}^2/\text{sec}$$
$$1 \text{ in.}^2/\text{sec} = 20.346 \times 10^3 \text{ m}^2/\text{yr}$$
$$1 \text{ ft}^2/\text{sec} = 929.03 \text{ cm}^2/\text{sec}$$

717

Force:

1 lb	= 4.448 N
1 lb	= 4.448×10^{-3} kN
1 lb	= 0.4536 kgf
1 kip	= 4.448 kN
1 U.S. ton	= 8.896 kN
1 lb	= 0.4536×10^{-3} metric ton
1 lb/ft	= 14.593 N/m

Stress:

1 lb/ft^2	= 47.88 N/m^2
1 lb/ft^2	= 0.04788 kN/m^2
1 U.S. ton/ft^2	= 95.76 kN/m^2
1 kip/ft^2	= 47.88 kN/m^2 ✓
1 lb/in.^{2}	= 6.895 kN/m^2

Unit Weight:

1 lb/ft^3	= 0.1572 kN/m^3
1 lb/in.^{3}	= 271.43 kN/m^3

Moment:

1 lb-ft = 1.3558 N · m
1 lb-in. = 0.11298 N · m

Energy

1 ft-lb = 1.3558 J

Moment of Inertia:

1 in.^{4} = 0.4162×10^6 mm^4
1 in.^{4} = 0.4162×10^{-6} m^4

B.2   Conversion Factors from SI to English Units

Length:

1 m	= 3.281 ft
1 cm	= 3.281×10^{-2} ft
1 mm	= 3.281×10^{-3} ft
1 m	= 39.37 in.
1 cm	= 0.3937 in.
1 mm	= 0.03937 in.

Area:

1 m^2	= 10.764 ft^2
1 cm^2	= 10.764×10^{-4} ft^2
1 mm^2	= 10.764×10^{-6} ft^2
1 m^2	= 1550 in.^{2}
1 cm^2	= 0.155 in.^{2}
1 mm^2	= 0.155×10^{-2} in.^{2}

Volume:

1 m^3	= 35.32 ft^3
1 cm^3	= 35.32×10^{-4} ft^3
1 m^3	= 61,023.4 in.^{3}
1 cm^3	= 0.061023 in.^{3}

Section Modulus:	$1 \text{ mm}^3 = 6.102 \times 10^{-5} \text{ in.}^3$ $1 \text{ m}^3 \ \ = 6.102 \times 10^4 \text{ in.}^3$
Coefficient of Permeability:	$1 \text{ m/min} \ \ = 3.281 \text{ ft/min}$ $1 \text{ cm/min} = 0.03281 \text{ ft/min}$ $1 \text{ mm/min} = 0.003281 \text{ ft/min}$ $1 \text{ m/sec} \ \ \ = 3.281 \text{ ft/sec}$ $1 \text{ mm/sec} = 0.03281 \text{ ft/sec}$ $1 \text{ m/min} \ \ = 39.37 \text{ in./min}$ $1 \text{ cm/sec} \ = 0.3937 \text{ in./sec}$ $1 \text{ mm/sec} = 0.03937 \text{ in./sec}$
Coefficient of Consolidation:	$1 \text{ cm}^2/\text{sec} \ = 0.155 \text{ in.}^2/\text{sec}$ $1 \text{ m}^2/\text{yr} \ \ \ = 4.915 \times 10^{-5} \text{ in.}^2/\text{sec}$ $1 \text{ cm}^2/\text{sec} \ = 1.0764 \times 10^{-3} \text{ ft}^2/\text{sec}$
Force:	$1 \text{ N} \qquad \ \ = 0.2248 \text{ lb}$ $1 \text{ kN} \qquad = 224.8 \text{ lb}$ $1 \text{ kgf} \qquad = 2.2046 \text{ lb}$ $1 \text{ kN} \qquad = 0.2248 \text{ kip}$ $1 \text{ kN} \qquad = 0.1124 \text{ U.S. ton}$ $1 \text{ metric ton} = 2204.6 \text{ lb}$ $1 \text{ N/m} \qquad = 0.0685 \text{ lb/ft}$
Stress:	$1 \text{ N/m}^2 \ \ = 20.885 \times 10^{-3} \text{ lb/ft}^2$ $1 \text{ kN/m}^2 = 20.885 \text{ lb/ft}^2$ $1 \text{ kN/m}^2 = 0.01044 \text{ U.S. ton/ft}^2$ $1 \text{ kN/m}^2 = 20.885 \times 10^{-3} \text{ kip/ft}^2 \ \checkmark$ $1 \text{ kN/m}^2 = 0.145 \text{ lb/in.}^2$
Unit Weight:	$1 \text{ kN/m}^3 = 6.361 \text{ lb/ft}^3$ $1 \text{ kN/m}^3 = 0.003682 \text{ lb/in.}^3$
Moment:	$1 \text{ N} \cdot \text{m} = 0.7375 \text{ lb-ft}$ $1 \text{ N} \cdot \text{m} = 8.851 \text{ lb-in.}$
Energy:	$1 \text{ J} = 0.7375 \text{ ft-lb}$
Moment of Inertia:	$1 \text{ mm}^4 = 2.402 \times 10^{-6} \text{ in.}^4$ $1 \text{ m}^4 \ \ = 2.402 \times 10^6 \text{ in.}^4$

ANSWERS TO
EVEN-NUMBERED PROBLEMS

Chapter 1

1.2
  a. 0.448
  b. 69.5%
  c. 17.58 kN/m³
  d. 14.53 kN/m³

1.4
  a. 128.83 lb/ft³
  b. 9.33 lb/ft³
  c. 124.4 lb/ft³

1.6
  a. 0.76
  b. 0.432
  c. 14.93 kN/m³
  d. 17.17 kN/m³
  e. 52.9%

1.8

Soil	Classification
A	A-7-6(9)
B	A-5(5)
C	A-3(0)
D	A-4(5)
E	A-2-6(1)
F	A-7-6(19)

1.10

Equation	k (cm/sec)
(1.25)	0.102
(1.26)	0.177
(1.27)	0.08

1.12 1.5×10^{-5} cm/sec

1.14

Point	σ (kN/m²)	u (kN/m²)	σ' (kN/m²)
A	0	0	0
B	67.36	0	67.36
C	108.98	19.62	89.36
D	220.28	78.48	141.8

1.16 253.7 kN/m²
1.18 1.19 in.
1.20 879.46 days
1.22 6.23 days
1.24 $c = 0$; $\phi = 38°$
1.26 $c = 30$ kN/m²; $\phi = 28°$
1.28 $\phi_u = 23.58°$; $\phi = 31.07°$
1.30 14.7 kN/m²

Chapter 2

2.2 7.54%

2.4

Depth (m)	N_{corr}
1.5	10
3	9
4.5	10
6	8
7.5	11
9	9

2.6 $D_r \approx 30\%$; $\phi \approx 34°$

2.8 8.07 m

2.10 a. 31.4 kN/m^2

   b. 3.31

2.12 a. 5.07 lb/in^2

   b. 4.31 lb/in^2

2.14 56%

2.16 2151.4 kN/m^2

2.18 $v_1 = 678.4$ ft/sec;

   $v_2 = 1748.25$ ft/sec;

   $Z_1 = 32.87$ ft

4.12

Point	Pressure (kips/ft^2)
A	0.4874
B	0.4712
C	0.4549
D	0.5230
E	0.5890
F	0.6052
G	0.6214
H	0.5544

4.16 16.6 lb/in^2

Chapter 3

3.2 480 kN/m^2

3.4 2.6 m

3.6 498.2 kN/m^2

3.8 333.8 kN/m^2

3.10 4.1 ft × 4.1 ft

3.12 506 kips

3.14 442 kN

3.16 B = 2.75 ft

3.18 1446.85 kN

3.20 434.9 kN

3.22 4279 lb/ft^2

3.24 396.5 kN

3.26 0.51 in.

3.28 0.34 in.

3.30 $S_e = 1.554$ mm; t = 0.68×10^{-3} deg

3.32 22.22 kN/m^2

3.34 a. 72 kN/m^2

   b. 54.85 kN/m^2

3.36 37.7 kN/m^2

3.38 56 mm

3.40 3.88 kips/ft^2

3.42 $B = 1.12$ m

Chapter 4

4.2 14.68 kips/ft^2

4.4 168 kN/m^2

4.6 6.38 m

4.8 1.33 m

4.10 0.161 m

Chapter 5

5.2 $P_o = 77.31$ kN/m; $\bar{z} = 1.39$ m

5.4 $P_o = 143.16$ kN/m; $\bar{z} = 1.3$ m

5.6 $P_o = 255.52$ kN/m; $\bar{z} = 1.83$ m

5.8 a. $\sigma_p = 1200$ lb/ft^2 at $z = 0$;

   $\sigma_p = 3160$ lb/ft^2 at $z = 18$ ft

   b. $P_p = 37,440$ lb/ft; $\bar{z} = 7.44$ ft

5.10 1212.13 kN/m

5.12 13.22 kips/ft

5.14 98.38 kips/ft

5.16 65.32 kips/ft

5.18 59.36 kips/ft

5.20 5.53 ft

5.22 a. 800.8 lb/ft

   b. 6.75 ft

5.24 $FS_{(overturning)} = 3.47$;

   $FS_{(sliding)} = 1.59$;

   $FS_{(bearing\ capacity)} = 4.18$

5.26 1.35

5.28 $FS_{(overturning)} = 4.66$;

   $FS_{(sliding)} = 1.86$

Chapter 6

6.2 $M_{max} = 2096.664$ kip-in/ft;

   S = 69.9 in^3/ft ($\sigma_{all} = 30$ kips/in^2)

6.4 $M_{max} = 1480.88$ kN-m;

   S = 8.58×10^{-3} m^3/m

6.6 2762 kN-m/m

6.8 $D = 3.175$ m; $M_{max} = 59.8$ kN-m/m

6.10 $D = 2.2$ m; $M_{max} = 38.72$ kN-m/m

6.12 $M_{max} = 367.04$ kN-m/m;
$S = 2.13 \times 10^{-3}$ m/m^3
$(\sigma_{all} = 172.5 \times 10^3$ kN/m$^2)$

6.14 $M_{max} = 43{,}796.7$ lb-ft/ft;
$S = 17.52$ in^3/ft ($\sigma_{all} = 30$ kips/in^2)

6.16 $D = 0.93$ m; $M_{max} = 18.13$ kN-m/m

6.18 a. 938.2 kN-m/m
    b. PZ-38

6.20 PZ-27

6.22 a. 1.15 ft
    b. 3319.2 lb/ft

6.24 $D = 5.14$ m; $F = 178.5$ kN/m
$M_{max} = 314.59$ kN-m/m

6.26

Part	P_u (kN/m)
a	10.92
b	16.27
c	21.62

6.28 25.5 kips

Chapter 7

7.2 a. $S = 0.129 \times 10^{-6}$ m^3/m
    b. $S = 0.495 \times 10^{-6}$ m^3/m

7.4 $S = 0.109 \times 10^{-6}$ m^3/m

7.6 a. $c_{av} = 1719$ lb/ft^2
    b. $\gamma_{av} = 108.6$ lb/ft^3
    c. $p_a = 814.5$ lb/ft^2 (Figure 7.6)

7.8 $S = 8.82$ in^3/ft; section PMA-22

7.10 1.975

7.12 2.89

Chapter 8

8.2 1810.5 kN
8.4 1495.6 kN
8.6 2356.9 kN
8.8 a. 56.43 kips
    b. 271.6 kips
8.10 234.77 kips
8.12 a. 413 kN
    b. 852 kN
    c. 632.5 kN
    d. 1113.5 kN
    e. 349.2 kN
8.14 354 kips
8.16 325.5 kips
8.18 18.64 mm
8.20 98.25 kN
8.22 27.4 kips

8.24 1034.21 kN
8.26 3024 kips
8.28 960 kips
8.30 60.64%
8.32 68.9%
8.34 232.52 kips
8.36 220 mm
8.38 5584 lb
8.40 188 kN

Chapter 9

9.2 834.7 kips
9.4 1758 kN
9.6 160.85 kN
9.8 11.74 mm
9.10 0.288 in.
9.12 13.71 mm
9.14 a. 70.22 kips
    b. 82.6 kips
    c. 152.82 kips
9.16 a. $D_s = 1$ m
    b. $D_b = 1.4$ m
    c. 2.5 m
    e. 2.5 m
    g. 40.09 mm
9.18 5.5 m
9.20 2519 kN
9.22 150.3 kips
9.24 266.1 kips

Chapter 10

10.2 Collapse is likely to occur
10.4 79.2 mm
10.6 a. 6.1
    b. moderate
10.8 0.41 ft
10.10 11.5 ft
10.12 13 ft

Chapter 11

11.2

z (m)	σ_a (kN/m^2)
1	80.37
2	65.02
3	50.47
4	39.76
5	32.25
6	27.01

11.4 a. 22.05
b. 4.29
c. 11.47

11.6 a. 29.97
b. 4.96
c. 18.8

11.8 $S_v = 0.336$ m; $L = 2.25$ m; $1_1 = 0.115$ (use 1 m)

11.10

Criteria	FS
Overturning	1.99
Sliding	1.3
Bearing capacity	10.84

11.12

z (m)	$T_{(N)}$ (kN/m)
0.4	29.1
0.8	32.9
1.2	35.6
1.6	36.4
2.0	38.2

11.14 a. 3.04/ft
b. 6857.14 lb/ft^2
c. 3

d.

z (ft)	$T_{(N)}$ (kips/ft)
1.3	2.825
2.6	3.06
3.9	3.09
5.2	3.23
6.5	3.32

Chapter 12

12.2 82.96%

12.4 $\gamma_{d(max)} = 20.75$ kN/m^3; $w = 9\%$

12.6 15.25 kN/m^3

12.8 $S_N = 25.48$; rating → fair

12.10 112.5 lb/ft^2

12.12 46%

12.14

t (year)	$U_{v,r}$
0.2	0.404
0.4	0.601
0.8	0.812
1.0	0.870

12.16 391.83 kN/m^2

Index